How to access the supplemental web resource

We are pleased to provide a one-year subscription to a web resource that supplements your textbook, *Clinical Mechanics and Kinesiology*. This resource includes hundreds of 3-D images of the human body, a regional review of human anatomy, and pretest and posttest evaluations to test retention.

Accessing the web resource is easy!
Follow these steps if you purchased a new book:

1. Visit **www.HumanKinetics.com/MusculoskeletalAnatomyReview**.

2. Click the <u>first edition</u> link next to the first edition cover.

3. Click the Sign In link on the left or top of the page. If you do not have an account with Human Kinetics, you will be prompted to create one.

4. If the online product you purchased does not appear in the Ancillary Items box on the left of the page, click the Enter Key Code option in that box. Enter the key code that is printed at the right, including all hyphens. Click the Submit button to unlock your online product.

5. After you have entered your key code the first time, you will never have to enter it again to access this product. Once unlocked, a link to your product will permanently appear in the menu on the left. For future visits, all you need to do is sign in to the product's website and follow the link that appears in the left menu!

6. Your access to *Musculoskeletal Anatomy Review* will expire one year after you first access it. Your key code can only be used once.

→ Click the Need Help? button on the product's website if you need assistance along the way.

How to access the web resource if you purchased a used book:

You may purchase access to the web resource by visiting the product's website, **www.HumanKinetics.com/MusculoskeletalAnatomyReview**, or by calling the following:

800-747-4457 . . . . U.S. customers
800-465-7301 . . . . .Canadian customers
+44 (0) 113 255 5665 . . . . European customers
08 8372 0999 . . . . Australian customers
0800 222 062 . . . . .New Zealand customers
217-351-5076 . . . . .International customers

For technical support, send an e-mail to:
support@hkusa.com . . . . U.S. and international customers
info@hkcanada.com . . . . Canadian customers
academic@hkeurope.com . . . . European customers
keycodesupport@hkaustralia.com . . . . . Australian and New Zealand customers

HUMAN KINETICS
The Information Leader in Physical Activity & Health

02-2013

Product: Clinical Mechanics and Kinesiology web resource

Key code: MAR-IZTI26-OSG

Clinical Mechanics and Kinesiology

Janice K. Loudon, PhD, PT, SCS, ATC

Duke University Medical Center

Robert C. Manske, DPT, SCS, ATC

Wichita State University

Michael P. Reiman, DPT, OCS, ATC

Duke University Medical Center

Human Kinetics

Library of Congress Cataloging-in-Publication Data

Loudon, Janice K. (Janice Kaye), 1959-
  Clinical mechanics and kinesiology / Janice Loudon, Robert Manske, Michael Reiman.
      p. ; cm.
  Includes bibliographical references and index.
  I. Manske, Robert C. II. Reiman, Michael P., 1965- III. Title.
    [DNLM: 1. Biomechanics. 2. Kinetics. 3. Musculoskeletal Physiological Phenomena. 4. Musculoskeletal System--anatomy & histology. WE 103]
    612'.01441--dc23

                                2012025724

ISBN-10: 0-7360-8643-9 (print)
ISBN-13: 978-0-7360-8643-1 (print)

The web addresses cited in this text were current as of October 2, 2012, unless otherwise noted.

Acquisitions Editors: Loarn D. Robertson, PhD, and Amy N. Tocco; **Developmental Editor:** Amanda S. Ewing; **Assistant Editors:** Kali Cox and Casey A. Gentis; **Copyeditor:** Patricia L. MacDonald; **Indexer:** Nancy Ball; **Permissions Manager:** Dalene Reeder; **Graphic Designer:** Nancy Rasmus; **Graphic Artist:** Dawn Sills; **Cover Designer:** Keith Blomberg; **Photograph (cover):** © Human Kinetics; **Artist (cover):** Jen Gibas; **Photographs (interior):** © Human Kinetics, unless otherwise noted; **Photo Production Manager:** Jason Allen; **Art Manager:** Kelly Hendren; **Associate Art Manager:** Alan L. Wilborn; **Art Style Development:** Joanne Brummett; **Illustrations:** © Human Kinetics; **Printer:** Courier Companies, Inc.

Printed in the United States of America   10 9 8 7 6 5 4 3 2 1

The paper in this book was manufactured using responsible forestry methods.

Human Kinetics
Website: www.HumanKinetics.com

United States: Human Kinetics
P.O. Box 5076
Champaign, IL 61825-5076
800-747-4457
e-mail: humank@hkusa.com

Canada: Human Kinetics
475 Devonshire Road Unit 100
Windsor, ON N8Y 2L5
800-465-7301 (in Canada only)
e-mail: info@hkcanada.com

Europe: Human Kinetics
107 Bradford Road
Stanningley
Leeds LS28 6AT, United Kingdom
+44 (0) 113 255 5665
e-mail: hk@hkeurope.com

Australia: Human Kinetics
57A Price Avenue
Lower Mitcham, South Australia 5062
08 8372 0999
e-mail: info@hkaustralia.com

New Zealand: Human Kinetics
P.O. Box 80
Torrens Park, South Australia 5062
0800 222 062
e-mail: info@hknewzealand.com

This book is first and foremost dedicated to our families, friends, and professional colleagues who have supported us in our endeavors to become not only better therapists but also—and more importantly—better human beings. Second, we would like to dedicate this book to our past, present, and future students at the University of Kansas, Wichita State University, and Duke University who have allowed us to become better educators, which in turn has made us each better clinicians. Last, to our patients who over the years have put trust and faith in our ability to treat their various musculoskeletal ailments and conditions. We are indebted to each of you, as every new patient brings a new learning experience. Although this book is written for rehabilitation professionals, ultimately it is for you, the patient—those we serve!

CONTENTS

PREFACE

Clinical Mechanics and Kinesiology provides an introduction to biomechanics and functional anatomy for student clinicians (physical therapy, occupational therapy, athletic training) who will deal with the rehabilitation of different pathologies and populations. The text was written by three faculty instructors in physical therapy and athletic training with more than 40 years of combined teaching experience. Additionally, each of the authors remains active in the clinical practice of orthopedics and sports medicine. This knowledgeable team provides a text that not only is informative on kinesiology basics but also offers clinically applicable examples to supplement the material.

The study of biomechanics and kinesiology is the foundation for many of the examination and treatment techniques used in rehabilitation. *Clinical Mechanics and Kinesiology* provides that foundation in an easy-to-read and consistent format. This text will remain a reference text for students as they move into the advanced orthopedic courses.

Organization

The material contained within *Clinical Mechanics and Kinesiology* is sufficient to enable a thorough understanding of biomechanics and kinesiology in a course spanning two quarters or one full semester. Reviewing the material in chronological order will provide a clear pathway, starting with basic biomechanical principles followed by specific regional areas of the body. The contents of this book are divided into four parts.

Part I contains two chapters that deal with the foundation of kinematics and kinetics. Chapter 1 is dedicated to kinematics, the description of a body segment's displacement, velocity, and acceleration. Introductory material such as planes of motion, arthrokinematics, and osteokinematics is discussed. Chapter 2 focuses on kinetics and includes descriptions of human movement and internal and external forces as they relate to general clinical experiences. These two chapters contain practice activities that students can complete to help them master these biomechanical concepts. The book's appendix, which supplements these two chapters, summarizes geometry and trigonometry principles and is provided as a reference.

Part II contains three chapters. Chapter 3 explores muscle and nerve physiology. Within this chapter, muscle and nerve structure and function are detailed. Relevance to rehabilitation is emphasized with several clinical examples. Chapter 4 reviews the topic of muscle performance: morphology, force generation capacities, contractile properties, and various methods used to increase or decrease strength and endurance. The muscle's roles as a prime mover, in reverse action, and in stabilization as well as the effects of muscle injury, immobilization, and aging are also included in this chapter. Chapter 5 focuses on joint structure by describing the various forms of joints in the human body. Specific detail is dedicated to the composition of a joint and includes discussion on bones, ligaments, capsules, and cartilage.

Part III is devoted to regional anatomy and clinical kinesiology. This part includes 10 chapters (chapters 6 to 15), beginning with four chapters on the axial skeleton: cervical spine, craniomandibular joint, thoracic spine, and lumbar spine and pelvic girdle. The next three chapters contain the upper extremity: shoulder, elbow and forearm, and wrist and hand. The last three chapters in this part focus on the lower extremity: hip, knee, and foot and ankle. Each chapter in part III follows a consistent format. The osteology section starts every chapter with a detailed review of the bones that make up the joints of the anatomical region. This is followed by joint articulations. The joint articulation subheadings cover the joint types more explicitly. Joint anatomy is next, and this section presents the joint capsule, ligaments, and other soft-tissue structures that are imperative to the stability and function of the joint. Joint function is discussed in the next section and describes the axes of joint movement, arthrokinematics, range of motion, closed and loose packed positions, end feel, and capsular pattern of each joint. The final portion of these regional chapters is a discussion on the muscles that act on the described joints. Besides the traditional presentation of origin, insertion, action, and innervation, this muscle section discusses how the muscles are used functionally.

Part IV provides insight into elementary movement patterns of particular concern to clinicians and includes posture, walking gait, running gait, and cutting and jumping. Students should emerge with a strong understanding of mechanical principles governing human motion, with special understanding of both normal and abnormal functional motions from diverse rehabilitation fields. Chapter 16 comprises the specifics of posture and posture examination, which is key information for the student learning to become a rehabilitation specialist. In detail, this chapter reviews normal and abnormal standing posture in the sagittal and frontal views. Chapter 17 focuses on walking gait. Walking is one of the most common functional activities, and therefore gait assessment should be a component of every musculoskeletal examination. Material is broken down in a step-by-step manner for easy understanding and is presented in both qualitative and quantitative formats. Chapters 18 and 19 discuss more advanced functional activities: running, cutting, and jumping. An emphasis contained within these two chapters is the discussion of injury mechanisms associated with these tasks. The goal is that students will understand the con-

eBook

available at your campus bookstore or HumanKinetics.com

nection between faulty biomechanics and injury.

Features

Several special features throughout the book will help students understand and retain the information presented in each chapter.

- **Objectives:** Each chapter begins with a list of objectives that help set the stage for the discussion and give a glimpse of the topics that will be covered. The chapter objectives quickly help the reader review the main points of the chapter.

- **Clinical correlations:** Clinical correlations are contained within all chapters to bridge theory to practice. The clinical correlations include common clinical examples that support the didactic material within the text.

- **Key points:** Another feature that should help with material retention is the key points that are placed throughout the chapters. The key points summarize important aspects.

- **Artwork:** The book is heavily illustrated with superb full-color art. The medical art shows proper locations of bones, muscles, and ligaments. Arthrokinematic motions are clearly shown, with the addition of the appropriate skeletal location when needed for clarity, making it easier for students to see how a motion relates to the rest of the body.

- **Review questions:** Review questions covering the chapter material are included at the end of every chapter. These questions should cue the students on important concepts and help with classroom test preparation.

- **Glossary:** Glossary terms are highlighted throughout the text, giving students a quick reference to terms as they are introduced. Full definitions are provided in the glossary at the end of the book.

- **References:** References are separated by chapter at the end of the book. These resources will help the reader with additional sources for information.

- **Musculoskeletal Anatomy Review:** In addition to these text features, the book is also accompanied by the *Musculoskeletal Anatomy Review* web resource. More information on this resource can be found on page xv. Students can access *Musculoskeletal Anatomy Review* by visiting www.HumanKinetics.com/ MusculoskeletalAnatomyReview.

Instructor Resources

Instructors have access to a full array of ancillary materials that support the text. These materials, provided to help instructors teach and test the book's content, include the instructor guide, image bank, and test package.

• **Instructor guide.** The instructor guide includes an overview of how all the ancillaries can be used by instructors. It provides a syllabus as well, showing how this text can be covered in a semester-long course. The instructor guide also provides chapter-specific files that include the following features: chapter summary, chapter outline, class assignments, lab activities, tips on presenting hard-to-grasp concepts, and answers to the end-of-chapter review questions.

• **Image bank.** The image bank includes all the figures, tables, and photos from the text. Instructors can use these images to supplement lecture slides, create handouts, or develop other teaching materials for their classes.

• **Test package.** The test package includes more than 420 multiple-choice, true-or-false, short-answer, and fill-in-the-blank questions. Instructors can use these questions to create or supplement tests or quizzes.

Instructors can access these ancillary resources by visiting www.HumanKinetics. com/ClinicalMechanicsAndKinesiology.

ACKNOWLEDGMENTS

The motivation behind this book comes from retired acquisitions editor Loarn Robertson. It was his goal that Human Kinetics develop academic physical therapy products, and *Clinical Mechanics and Kinesiology* is one of the first. We would like to thank Loarn for giving us the opportunity to be a part of this venture. Special thanks to Amy Tocco and Amanda Ewing at Human Kinetics for their editorial assistance. We would also like to recognize Joanne Brummett and her team of artists for their excellent addition of the detailed artwork.

Gratefully, we would like to thank Sue Klein and Neena Sharma for their expertise and willingness to write the chapters on the craniomandibular complex and on the wrist and hand.

And last, but not least, we would like to thank our families and friends for allowing us to take time away from them in order to write this text.

HOW TO USE MUSCULOSKELETAL ANATOMY REVIEW

Musculoskeletal Anatomy Review includes hundreds of 3-D images of the human body to aid students in their study of anatomy. This engaging supplement to the text offers a regional review of structural anatomy with exceptionally detailed, high-quality graphic images—the majority provided by Primal Pictures. Students can mouse over muscles and click for muscle identification. This online feature offers students a self-paced yet self-directed review of the musculoskeletal anatomy, providing an intensely visual interface through which students may gain a clear understanding.

Each chapter of *Musculoskeletal Anatomy Review* features a pretest and posttest evaluation to help students pinpoint knowledge gaps and test their retention. The pretest can be taken multiple times and is generated randomly so it will never be the same, but the posttest may be taken only once. Test results can be printed and turned in so instructors have the option to use the tests as a grading tool.

As students proceed through this review of musculoskeletal anatomy, they will encounter interactive learning exercises that will quiz them on key concepts and help them apply what they've learned about manual muscle testing or range of motion assessment in helping a virtual client.

There may be concepts presented in *Musculoskeletal Anatomy Review* that students have not learned in the past. Whenever possible, a learning aid will be provided to assist students in retention of the material. The learning and review aids may be mnemonics, simple organization of a group of muscles, or just a way to understand the terminology and locations of structures. Please take time to learn using the aids provided; if you do, your retention of the material is apt to surprise you.

Students can access *Musculoskeletal Anatomy Review* by going to www.HumanKinetics.com/MusculoskeletalAnatomyReview.

Basic Biomechanics

Through study of these two chapters of *Clinical Mechanics and Kinesiology*, readers will gain knowledge of the basic biomechanical principles used in the examination and treatment of clients undergoing rehabilitation. The first chapter is dedicated to kinematics, the description of a body segment's displacement, velocity, and acceleration. These components are discussed both as linear and angular motion. The chapter begins with an introduction to the planes of motion, the sagittal, frontal, and transverse planes. Joint osteokinematic motions that occur in each of these planes are presented. Further, joint surface motion is discussed and involves an overview of the arthrokinematic motions of roll, spin, and glide and how they are required for normal joint motion. This chapter concludes with a thorough discussion on position–time data and how it is used to calculate kinematic data for research and clinical practice.

Chapter 2, which introduces the topic of kinetics, covers a wide range of subject matter including Newton's laws of motion, forces, levers, work, power, and energy. This chapter introduces forces in both linear and angular (moment) terms and provides application for the reader. A goal of this chapter is to present the concept of free body diagrams and the composition and resolution of vectors. By working through the vector problems provided, the student should gain a better understanding of muscle and joint reaction forces and how these forces may lead to normal motion or injury.

These two chapters contain ample practice activities that students can complete to help them master these biomechanical concepts. An appendix that summarizes geometry and trigonometry principles is provided at the end of the book.

Kinematics

Assessing human motion

based on biomechanical principles is an integral aspect of the physical therapist's examination. Two subdivisions of **biomechanics** are kinematics and kinetics. Utilizing kinematic and kinetic data helps to describe movement. This chapter focuses on the kinematic aspects of human motion. Kinematic variables describe movement in terms of linear and angular displacements, velocities, and accelerations. The chapter begins by describing the kinematic principles, then relates these principles to joint motion, and finishes with practical application of these kinematic principles. Please refer to the appendix for basic conversions and trigonometric principles.

Movement Mechanics

Kinematics is the branch of **mechanics** that describes the motion of objects. Specifically, kinematics is the description of movement related to displacement, velocity, and acceleration. This movement can be described in two dimensions (planar motion) or three dimensions (spatial motion). Three-dimensional kinematics is preferred when working in the research laboratory. Table 1.1 lists the units

TABLE 1.1 Units of Kinematic Measures

Unit system	Displace-ment	Velocity	Accelera-tion
SI (metric)	m	m/s	m/s²
English	ft	ft/s	ft/s²

for the kinematic measures: displacement, velocity, and acceleration.

Displacement is the change in **position** of an object. This can occur in a linear or rotational manner. An example of angular displacement is the amount of 135° that the knee moves from full flexion to full extension. **Velocity** is the rate at which an object moves in a given direction. Mathematically, velocity is the first derivative with respect to time of displacement and is represented by the following equation: velocity = Δ position / Δ time. (The derivative is used in calculus to determine how much one quantity is changing in response to changes in some other quantity.) Velocity is not equivalent to **speed**, in that speed designates only rate and not direction. **Acceleration** is the rate of change of velocity with respect to time. Mathematically, acceleration is the second time derivative of displacement and the first time derivative of velocity and is represented by the following equation: velocity = Δ velocity / Δ time. Acceleration can be positive, negative, or zero in value. A negative acceleration is commonly termed **deceleration**. Acceleration (positive or negative) is critical as a component in the mechanism of many types of injuries. For example, one of the most devastating injuries to an athlete is a tear to the anterior cruciate ligament of the knee. One mechanism for this injury is a noncontact deceleration force that occurs when the athlete suddenly stops to change direction or score a basket.

Kinematic motion can be either linear or rotational (angular). Therefore, displacement, velocity, and acceleration can be described as linear or angular. Linear kinematics occurs when all points move equidistant in the same direction, at the same time. This can also be termed translation. This motion can occur in a straight line (rectilinear) or with curvature (curvilinear).

Angular kinematics occurs when an object moves along a circular path about one stationary point. This can also be termed rotation and is usually described as moving clockwise or counterclockwise and measured in degrees or radians. Figure 1.1 depicts a ball that can rotate about an axis (causing angular motion), or it can slide along the contact surface (creating translational motion).

Symbols and formulas for linear and angular kinematics are displayed in table 1.2. Most, if not all, joints in the human body rotate and translate at the articulating surfaces in order to complete a range of joint motion.

There is a direct relationship between linear and angular measures, as shown in figure 1.2. Similar equations are used to define linear and angular velocities as well as linear and angular acceleration. In referring to human movement, most motion involves a combination of linear and angular measures. An example of this relationship is described in the Practice It! sidebar.

▶ **KEY POINT**

There is a mathematical relationship between displacement, velocity, and acceleration. Velocity is the first derivative with respect to time of displacement (x) and the first integral with respect to time of acceleration. Acceleration is the second time derivative of displacement and the first time derivative of velocity.

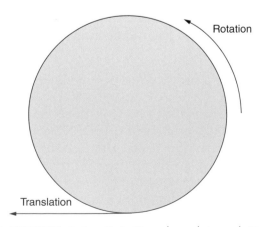

▶ **FIGURE 1.1** Rotational and translational motion.

4

TABLE 1.2 Linear and Angular Kinematics

	Position	Velocity		Acceleration	
		Instantaneous	Average	Instantaneous	Average
Linear	P	$v = dP / dt$	$v = P_2 - P_1 / t_2 - t_1$	$a = dv / dt$	$v = v_2 - v_1 / t_2 - t_1$
Angular	θ	$\Omega = d\theta / dt$	$\Omega = \theta_2 - \theta_1 / t_2 - t_1$	$\alpha = d\Omega / dt$	$\alpha = \Omega_2 - \Omega_1 / t_2 - t_1$

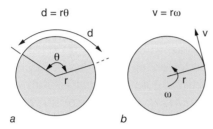

$$d = r\theta \qquad v = r\omega$$

a   *b*

▶ **FIGURE 1.2** Relationship between linear and angular kinematics. *(a)* Linear distance *[d]* moved along the circumference of a circle (radius = *r*) equals *r* × θ. *(b)* Linear velocity *[v]* of a point on the circumference of a circle equals *r* × ω. (Note: Angular measurements of τ, v, and α in these equations must be expressed in units of radians, rad/s, and rad/s², respectively).

It is possible to calculate a person's step length during gait based on the linear–angular relationship of the hip and knee. By knowing the hip and knee range of motion (ROM) during the swing phase of gait and the length of the femur and shank (tibia and fibula), you can use the linear–angular relationship equation d = rθ to estimate the step **distance**. (Other conditions can influence the step length including joint stiffness, muscle power, and inertia.)

Given

Hip ROM is 30°; knee ROM is 45°; femur length is 19 in.; shank length is 17 in.

To solve

1. Convert degrees to radians (1 radian = 57.3°). Hip = 30 / 57.3 = 0.52 radians; knee = 45 / 57.3 = 0.78 radians.

2. Calculate equation: Hip = 0.52 × 19 = 9.88 in.; knee = 0.78 × 17 = 13.26 in.

3. Add the two segments together to derive step distance: 9.88 + 13.26 =

23.14 in. Therefore, 23.14 in. is the estimated step distance based on the range of motion and limb length.

4. To convert inches to centimeters: 1 in. = 2.54 cm, so in this example, 9.88 in. × 2.54 cm/in. = 25.10 cm; 13.26 in. = 33.68 cm; and 23.14 in. = 58.78 cm.

Joint Motion

Joint motion can occur within one of the three cardinal planes. These planes are orthogonal to each other and include the sagittal, frontal (coronal), and transverse (horizontal) planes (figure 1.3). The position of reference is based

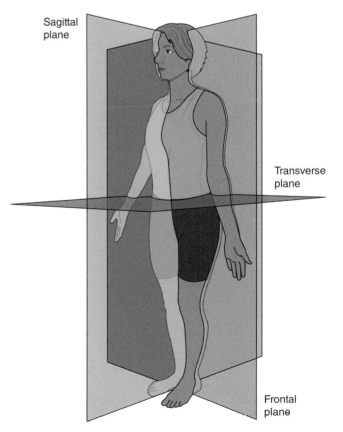

▶ **FIGURE 1.3** Planes of movement.

on the anatomical position, an erect posture with the arms at the side, palms forward, with the fingers and thumbs straight. Based on this anatomical position, the sagittal plane occurs about a coronal axis (medial to lateral). Flexion and extension occur in the sagittal plane. The frontal, or coronal, plane occurs about an anteroposterior axis (front to back). Abduction and adduction occur in the frontal plane. The transverse plane occurs about a vertical, or longitudinal, axis. Internal rotation, external rotation, and horizontal abduction occur in the transverse plane.

▶ **KEY POINT**

Foot abduction and adduction occur in the transverse plane. This is in contrast to the shoulder and hip, in which abduction occurs in the frontal plane. This discrepancy is due to developmental changes that occur at the ankle, creating a right angle between the foot and the tibia and fibula.

The motions defined in table 1.3 describe gross bone movement and are termed osteo-kinematic motions. Clinically, the amount of osteokinematic motion is termed range of motion (ROM) and is commonly documented as an impairment when the ROM is deficient. Osteokinematic ROM is measured in degrees of rotation with a measuring device called a goniometer. Throughout the text we will identify the number of independent osteokinematic movements that occur at a joint. This is termed the degrees of freedom. For example, the knee (tibiofemoral joint) has three rotational degrees of freedom occurring in the cardinal planes: sagittal plane (flexion, extension), transverse plane (internal rotation, external rotation), and frontal plane (abduction, adduction). There are also three translational degrees of freedom at the knee. The three translational degrees of freedom are about the anteroposterior axis (anterior, posterior glide), coronal axis (medial, lateral glide), and longitudinal axis (superior, inferior glide).

The term *arthrokinematics* describes the accessory movement at the joint and includes roll, spin, and glide (Kaltenborn 1980). Figure 1.4 depicts each of these arthrokinematic movements. Roll is when one portion of the

TABLE 1.3   Osteokinematic Motion

Osteokinematic motion	Plane of motion	Definition
Flexion	Sagittal	Ventral surfaces are approximated
Extension	Sagittal	Ventral surfaces move apart
Abduction	Frontal	Distal segment moves away from the midline
Adduction	Frontal	Distal segment moves toward the midline
Lateral flexion	Frontal	Movement of the head away from the midline
Medial rotation	Transverse	Rotation toward the midline of the body
Lateral rotation	Transverse	Rotation away from the midline of the body

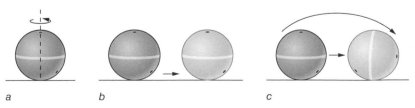

▶ **FIGURE 1.4** Arthrokinematic motions. (a) spinning, (b) gliding, and (c) rolling.

joint surface rolls on the other, like a ball rolling on a tabletop. An example is the femoral condyles rolling over the tibia (via the menisci) with knee flexion and extension. As a bone segment rolls, new equidistant points on the rolling surface come into contact with new equidistant points on the opposing surface. Spin occurs as an object rotates about a fixed axis, much like a top spinning on a tabletop. An anatomical example of joint spin occurs at the superior radioulnar joint with pronation and supination. Glide, or slide, is when one articulating surface translates relative to another. An example of gliding is the relative concave tibia gliding along the femoral condyles with knee flexion and extension. During glide, the same point on the gliding bone comes into contact with new points on the opposing surface. The gliding accessory component is the basis of most joint mobilization techniques.

The articulating surfaces of most joints have a convex surface on one segment and a concave segment on the reciprocal side. Therefore, bone movement usually involves a combination of rolling, gliding, and spinning. For example, at the tibiofemoral joint when the femur is flexing on a fixed tibia (as in squatting), the femur rolls posteriorly and glides anteriorly.

▶ **KEY POINT**

The convex–concave rule was developed by physiotherapist Freddy Kaltenborn. The rule states that when a concave surface moves on a convex surface, roll and glide occur in the same direction. When a convex surface moves on a concave surface, roll and glide occur in the opposite direction. It is important to understand this relationship when performing joint mobilization techniques.

The point about which rotation of an object takes place is termed the **instantaneous center of rotation** (ICR). In the human body, the joints are not simple hinge joints; therefore the instantaneous center of rotation changes throughout the ROM and is referred to as the

pathway of the instantaneous center of rotation (PICR). The change in the ICR through a range of motion is representative of the joint spinning, gliding, and rolling. The closer the center of rotation is to the joint surface, the more gliding takes place. This pathway can be estimated geometrically. An example of the PICR for the tibiofemoral joint is shown in figure 1.5.

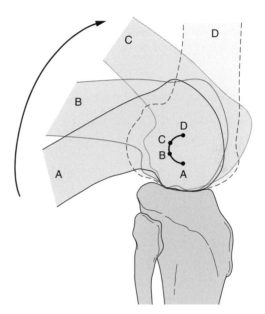

▶ **FIGURE 1.5** Instantaneous center of rotation. Movement of the instantaneous joint center is shown for the knee as the joint extends from a flexed position (A) to full extension (D).

Kinematic Analyses

We will now turn our focus on the quantitative evaluation of joint motion. This can be accomplished by a variety of methods looking at change in displacement, velocity, or acceleration. One example is to plot position by time, resulting in a position–time graph. Another example of a practical application of kinematic analysis is to investigate changes in movement due to injury or pathology (see Clinical Correlation 1.1 later in this chapter). In the following sections, we look specifically at position–time data and a way to graph these data.

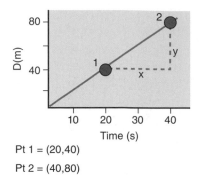

Pt 1 = (20,40)

Pt 2 = (40,80)

▶ **FIGURE 1.6** Graph showing the movement of an object from position 1 to position 2. This graph represents position–time data.

Position–Time Data

Let's say you are interested in looking at an object's position relative to time. By collecting the position location and tracking the position over time, you can calculate velocity. Position versus time can be graphed as displayed in figure 1.6. In figure 1.6, an object moves from position 1 (located at an x position of 20 and a y position of 40) to position 2 (located at an x position of 40 and a y position of 80). The slope of the line between positions 1 and 2 represents the velocity. From the graph you should be able to calculate the average velocity of the object as it moves from position 1 to position 2:

Pt 1 = (20,40)
Pt 2 = (40,80)
x = (40 − 20) = 20
y = (80 − 40) = 40
Slope = y / x
= 40 / 20
= 2 m/s

PRACTICE IT!

This figure illustrates the location of an object at five positions. Practice calculating the velocity between each of the position changes in this figure. We've shown you the steps for calculating the position change from 1 to 2, and you should follow these same steps to see how we got the answers for the other position changes.

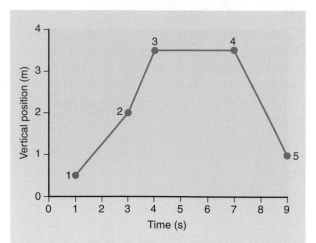

Position change 1 to 2: To calculate the velocity, subtract the position of 1 from 2. The object at position 2 is (3,2), and the object at position 1 is (1,0.5). Thus,

x = 3 − 1 = 2
y = 2 − 0.5 = 1.5
y / x = 1.5 / 2 = 0.75 m/s

Position change 2 to 3: 1.50 m/s (the greater the slope the greater the velocity)

Position change 3 to 4: 0 m/s (no motion taking place here)

Position change 4 to 5: −1.25 m/s (change in direction doesn't mean velocity is slowing)

Now look at the data in table 1.4. Given the position and time, you can calculate average velocity using the following equation: Δ position / Δ time (where Δ = change). You can also calculate acceleration by using the following equation: Δ velocity / Δ time. You should notice that as the average velocity is decreasing in value, the average acceleration is a negative number, which represents deceleration.

Graphic Relationship of Kinematic Data

The points from position–time data can be plotted, resulting in a position–time graph. The position–time graph makes for a visual inspection of the displacement of an object.

This makes interpretation much easier. Figure 1.7 is a graphical representation of the data from table 1.4. You should notice the following relationships between position, velocity, and acceleration:

- The slope of the position–time graph represents velocity.
- The slope of the velocity–time graph represents acceleration.
- As position increases (moves farther away from the start), velocity is positive; as position returns to 0 m, velocity becomes negative, representing a change in direction.

- A similar relationship exists between velocity and acceleration.

In figure 1.8, the position–time data (thigh angle), velocity–time data, and acceleration–time data are depicted. From these three graphs you should be able to describe the kinematic motion of the movement, a runner's stride cycle. We can look at the peaks and valleys (minima, maxima) to describe the kinematics. These peaks and valleys represent a change in direction; the slope is zero at this instant.

In figure 1.8, the thigh angle has one valley (minima) and one peak (maxima). Notice at these instances the velocity is 0 rad/s. The

TABLE 1.4   Position–Time Data

Position (m)	Time (s)	Velocity (Δ position / Δ time) (m/s)	Acceleration (Δ velocity / Δ time) (m/s²)
0.00	0.000		
	0.050	(0.59 – 0.00) / (0.100 – 0.000) = 5.9	
0.59	0.100		(3.6 – 5.9) / (0.150 – 0.050) = –23.0
	0.150	(0.95 – 0.59) / (0.200 – 0.100) = 3.6	
0.95	0.200		(1.0 – 3.6) / (0.255 – 0.150) = –24.7
	0.225	(1.00 – 0.95) / (0.250 – 0.200) = 1.0	
1.00	0.250		(–1.0 – 1.0) / (0.275 – 0.225) = –40.0
	0.275	(0.95 – 1.00) / (0.300 – 0.250) = –1.0	
0.95	0.300		(–3.6 – [–1.0]) / (0.350 – 0.275) = –34.7
	0.350	(0.59 – 0.95) / (0.400 – 0.300) = –3.6	
0.59	0.400		(–5.9 – [–3.6]) / (0.450 – 0.350) = –23.0
	0.450	(0.00 – 0.59) / (0.500 – 0.400) = –5.9	
0.00	0.500		(–5.9 – [–5.9]) / (0.550 – 0.450) = 0.0
	0.550	(–5.9 – 0.00) / (0.600 – 0.500) = –5.9	
–0.59	0.600		(–3.6 – [–5.9]) / (0.650 – 0.550) = 23.0
	0.650	(–0.95 – [–0.59]) / (0.700 – 0.600) = –3.6	
–0.95	0.700		(–1.0 – [–3.6]) / (0.725 – 0.650) = 34.7
	0.725	(–1.00 – [–0.95]) / (0.750 – 0.700) = –1.0	
–1.00	0.750		(1.0 – [–1.0]) / (0.775 – 0.725) = 40.0
	0.775	(–0.95 – [–1.00]) / (0.800 – 0.750) = 1.0	
–0.95	0.800		(3.6 – 1.0) / (0.850 – 0.775) = 34.7
	0.850	(–0.59 – [–0.95]) / (0.900 – 0.800) = 3.6	
–0.59	0.900		(5.9 – 3.6) / (0.950 – 0.850) = 23.0
	0.950	(0.00 – [–0.59]) / (1.000 – 0.900) = 5.9	
0.00	1.000		

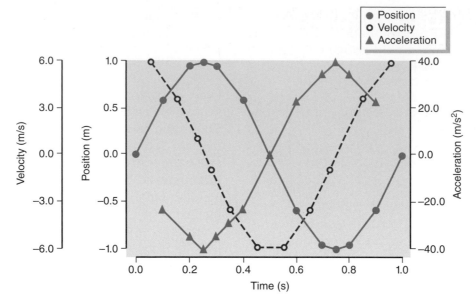

FIGURE 1.7   Graph of the kinematic data in table 1.4.

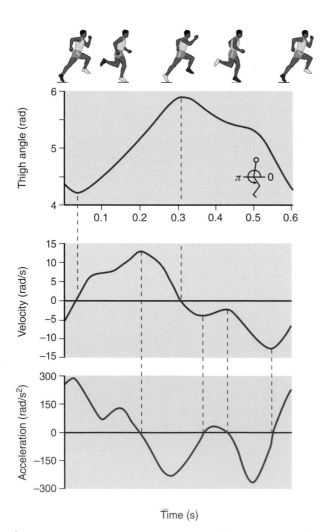

FIGURE 1.8   Kinematic graph of a runner's thigh.

minima and maxima for the thigh angle represent change in direction. The thigh is changing from moving forward to backward and vice versa.

Next, look at the slopes between the maxima and minima. In this figure the first slope is negative—the thigh is moving backward. The corresponding velocity is below the horizontal and represents a negative velocity. As the thigh moves forward, the velocity becomes positive. For this figure, you should notice two intervals of backward thigh rotation separated in time by an interval of forward thigh rotation. The variation in magnitude of the velocity over time indicates the variation in speed of thigh rotation, whereas the sign (positive or negative) indicates the direction (forward or backward) of rotation.

The derivation of the acceleration–time relationship from the velocity–time graph is accomplished by the same two-stage procedure: (1) identifying the maxima and minima and (2) determining the slope intervals. In figure 1.8, the velocity–time curve contains four minima and maxima and thus four instances at which the acceleration–time graph crosses zero (0 rad/s^2). In looking at the slopes of the acceleration–time graph, a positive acceleration indicates an acceleration of the thigh in a forward direction.

CLINICAL CORRELATION 1.1

Kinematic data are commonly used in research studies to investigate changes in movement due to injury or pathology. Kinematic data can be collected using a three-dimensional motion capture system. This instrumentation setup involves placing reflective markers on the subject at precise anatomical landmarks. The kinematic data are captured with multiple cameras as the subject performs some type of movement. Figure 1.9 is an example of an experimental setup in which the investigator is determining the frontal and transverse plane kinematics in a person with patellofemoral pain syndrome (PFPS) performing a single-leg squat. The dynamic Q angle is the angle subtended by the midpatella–ASIS line and the midpatella–tibial tuberosity line. The dynamic Q angle is higher in people with PFPS (Willson & Davis 2008). A treatment goal would be to train the person to land with less of a dynamic Q angle by working on hip internal rotation and adduction strength and control.

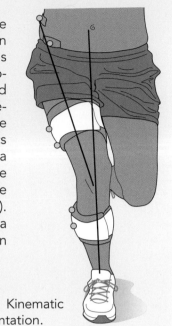

▶ **FIGURE 1.9** Kinematic lab instrumentation.

Conclusion

Human movement can be analyzed using kinematic data. This strategy involves the description of a body segment's displacement, velocity, and acceleration. Using only kinematic data can be limiting. Full analysis requires information on the forces acting on the body. The next chapter describes the forces that occur with movement: kinetics.

REVIEW QUESTIONS

1. What is the equation for the linear–angular relationship?

2. Define the term *instantaneous center of rotation*, and explain what it means in terms of joint movement.

3. Describe the plane and axis of movement for shoulder flexion.

4. Give an example of joint roll and glide.

5. When examining a position–time graph for velocity, what do the maxima and minima points represent?

PROBLEM SETS

1. Mr. Jones is walking with his new prosthesis. He walks 25 m in 30 seconds. What is his average velocity during his 25 m walk?

2. Sam starts at home plate and runs to first base. He reaches a maximal velocity of 4 m/s in 3.5 seconds. What is his average acceleration from standing to maximal velocity?

3. If Angela moves her elbow from a position of full extension to 90° of elbow flexion, and her forearm and hand measure 36 cm from her elbow joint, what is the distance her hand travels?

Kinetics

The study of forces is fundamental to the understanding of body form and motion. The term **kinetics** is used to describe motion in terms of forces. The three laws of motion proposed by Newton form the basis for kinetics. This chapter introduces forces in both linear and angular terms. Development of a free body diagram and resolving force vectors related to the free body diagram are explored. In addition, concepts related to forces are presented including work, power, energy, and pressure.

Basic Kinetic Terms

To begin this chapter, let's introduce a few terms: mass, inertia, and momentum. **Mass** (m) is the amount of matter that makes up an object. The greater the mass, the harder the object is to move. Mass represents only quantity and is measured in pounds or kilograms (kg). Mass is different from the weight of an object. Weight (W) is the force acting on the object due to gravity and is equal to the mass of an object multiplied by the gravitational acceleration constant (W = mg, where g = 9.81 m/s²). **Center of mass** (CoM) is the point about which the body's mass is evenly

OBJECTIVES

After reading this chapter, you should be able to do the following:

> Define the terms *mass*, *inertia*, and *momentum*.

> Define the term *force* and discuss the characteristics of a force.

> Distinguish the differences between Newton's three laws of motion.

> Explain the concept of force applied over a distance.

> Introduce the impulse–momentum relationship and the work–energy relationship.

> Define *torque* and discuss the characteristics of torque.

> Develop a free body diagram, including force vectors, for a variety of clinical applications.

distributed. This can be applied to the whole body or one segment of the body. The CoM changes with different body positions and limb positions.

▶ **KEY POINT**

Center of mass and *center of gravity* are commonly used as interchangeable terms. Technically, center of gravity occurs only in the vertical direction. On the other hand, center of mass is independent of vertical orientation.

Two other terms that need introductory discussion are *inertia* and *momentum*. **Inertia** is the amount of energy required to alter a body's velocity. Inertia is directly related to the mass of the object. This concept can also be stated in angular terms and then is referred to as the mass moment of inertia. The equation for mass moment of inertia is $I = mr^2$, where r is the distance from the axis of rotation to the point mass (radius of gyration). Clinical Correlation 2.1 further expands on this concept with two clinical examples. **Momentum** (p) is the quantity of motion of an object and is equal to the mass and the velocity (v) of an object: $p = mv$. Increasing either the mass or the velocity of an object will increase momentum. Angular momentum (L) is defined as $L = I\omega$, where ω is angular velocity.

Forces

Taking the terms just discussed (mass, weight, inertia, and momentum), we will now proceed

by defining *forces*. A **force** can be described as the action of one body on another. Force is represented as a **vector** with magnitude, direction, point of application, and line of application. Figure 2.1 represents a force vector at the elbow joint. The units for force are newtons or pounds. A newton (N) is the force required to accelerate 1 kg mass at 1 m/s² (1 N = 1 kg × m × s⁻²). Notice this is a pound-force and not a pound-weight.

▶ **KEY POINT**

Injuries are caused from abnormal forces acting on a body. The type and extent of injury can be based on the following force factors: magnitude (the load), location (where the force is applied), direction, duration (how long), frequency (how often), and rate (how quickly applied).

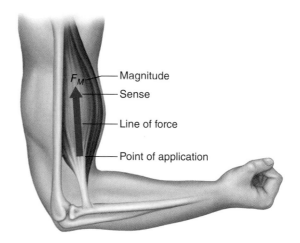

▶ **FIGURE 2.1**   Force is represented as a vector with magnitude, direction, point of application, and line of application.

CLINICAL CORRELATION 2.1

The mass moment of inertia increases with an increase in mass. For example, the risk of ankle sprain potentially increases with an increase in body weight because of increasing mass moment of inertia acting about the ankle. In a study on high school football players, Tyler and colleagues (2006) showed the incidence of ankle sprain was significantly increased in patients with body mass index (BMI) categorized as above normal or overweight when compared with those players with a normal BMI. Waterman and colleagues (2010) reported similar findings in military cadets with ankle sprains who had higher mean weight and BMI than their uninjured counterparts.

Types of Forces

Forces exist in all forms. Forces can be internally generated, such as those that result from tissue tension or muscle contraction. Additionally, forces outside the body exist and are termed external forces. An example of an external force is gravity. The force of gravity is always present. The line of application of gravitational force is given as the line between the center of mass of the object and earth. Therefore, the vector that represents gravity is always acting vertically down. *Weight* is the term used to represent the force of gravity between the earth and an object. The equation W = mg shows weight is equivalent to mass multiplied by gravitational acceleration. The constant for gravitational acceleration is 9.81 m/s^2 or 32.2 ft/s^2.

Specific types of forces the body encounters include tensile, compressive, and shear forces (figure 2.2).

- Tensile forces are **collinear** forces acting in opposite directions to pull an object apart. An example of an injury caused by tensile forces is a ligament tear.
- Compressive forces are collinear forces acting in similar directions to push tissues together. An example of an injury caused by compressive forces is a compression fracture.
- Shear forces are **coplanar** and opposite in direction but not collinear; these forces cause one surface of a body to slide past an adjacent surface. An example of an injury caused by shear forces is a blister. Shearing takes place between the skin and underlying tissue.

These types of forces can be both beneficial and destructive to human tissue. For example, every time we move from a position of non–weight bearing to weight bearing, we impart a compressive force to our joints of the lower extremity. If the compressive force becomes more than cartilage tissue can handle, then the tissue will begin to break down. On the

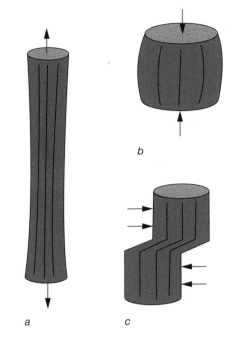

▶ **FIGURE 2.2** Various types of forces (loads): (a) tension, (b) compression, and (c) shear.

other hand, cartilage needs some compression to help nourish it; therefore, some compressive force on the joint is optimal (Hamill & Knutzen 2009).

Force Systems

A **force system** is any group of two or more forces acting in relation to an object. Here are some examples, which are also illustrated in figure 2.3.

- Linear: Two or more forces acting on the same line.
- Parallel: Forces acting on an object in the same plane that are parallel but not collinear.
- Concurrent: Forces with a common origin.
- General: A system of forces that includes a combination of linear, parallel, or concurrent.
- Force couple: Parallel forces (usually two) that are opposite and equal

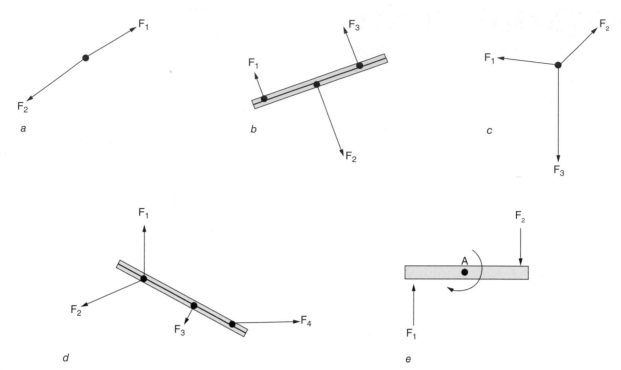

▶ **FIGURE 2.3**   Force systems. *(a)* Linear force system. *(b)* Parallel force system. *(c)* Concurrent force system. *(d)* General force system, the designation given to a force system that does not fall under one of the classifications *a* through *c*. *(e)* Force couple; parallel and oppositely directed forces F_1 and F_2 cause rotation about axis *A*.

in magnitude, with the axis midway between. The force couple creates rotation of an object.

In the human body, examples of muscles that suggest the action of a force couple include the upper trapezius and serratus anterior rotating the scapula.

Friction

Friction is defined as a parallel force that opposes or impedes motion between two contacting bodies. The resistance from friction is a result of molecular bonding between materials. If no movement is occurring between bodies, then the parallel force is termed *static friction*. If movement does take place between bodies, then the frictional force is termed *kinetic friction*, and a sliding frictional force is created. The amount of static friction can be calculated using the formula f = μN, where *f* = friction, *μ* = coefficient of friction, and

N = normal force (perpendicular force). The coefficient of friction is determined for various surfaces, and most tend to range from 0 to 1. The closer this number is to one, the greater the frictional forces. A coefficient of friction that equals 0 indicates a frictionless surface. As an example, the coefficient of friction between ice on ice is 0.05. The coefficient of a basketball shoe on a wood floor is approximately 1.0 (Nigg et al. 2000).

Friction has an important influence on human movement, and both friction and lack of friction may contribute to injury. A lack of friction results in a slippery surface that may contribute to falls. In contrast, high levels of friction can lead to quick deceleration that may be injurious to tissue. In terms of synovial joints, the coefficient of friction between healthy articulating surfaces is very low (0.01 to 0.04) due to the synovial fluid and articular cartilage properties. See Clinical Correlation 2.2 for further details.

CLINICAL CORRELATION 2.2

Healthy synovial joint surfaces exhibit very low friction because of their contact surface design, articular cartilage structure, and self-lubrication. This minimal friction plays an important part in the durability of normal articular cartilage over a long life. Damage to the cartilage can disrupt the normal lubrication process in the joint. This insufficiency may play a role in the development of osteoarthritis (Mow et al. 1992). Artificial joints that are used to replace osteoarthritic joints try to replicate the frictionless mechanics found in the healthy tissue but still do not match the natural mechanics.

Newton's Laws of Motion

The basic laws of motion were formulated by Sir Isaac Newton, an English mathematician (1642-1727). These three laws are referred to as the laws of inertia, acceleration, and action–reaction. Newton's laws describe the relationship between the forces acting on a body and the resultant motion due to those forces. For the clinician, an understanding of these laws is important for explaining several clinical concepts such as injury mechanics and exercise prescription. The three laws are listed here in more detail.

Law of Inertia

The first law of motion—the law of inertia—states that a body at rest will remain at rest (or a body in motion will move in a straight line with constant velocity) unless acted upon by a resultant force. This body is said to be in equilibrium when these conditions exist. Here is an example: A car that is stationary at a stop light will move when hit from behind with enough force by a second moving car. The resting car has inertia, and the second car must overcome this inertia in order to move the resting car.

Law of Acceleration

The second law of motion—the law of acceleration—states that a particle (or body) subjected to a resultant force will accelerate in the direction of that force, and the magnitude of the acceleration will be proportional to the magnitude of the force. The quantity of force (F) can be measured by the product of mass (m) multiplied by acceleration (a), or $F = ma$. We can take the car example just given and add the second law. If the stopped car is hit from behind, it will move forward at a velocity dependent on the speed of the moving car when it hit the stopped car. If the moving car was traveling at 40 miles per hour (mph) (65 km/h), it will have more effect on the stationary car than if it was moving at 5 mph (8 km/h).

Another way to express this second law is to look at force over a period of time as represented by the impulse–momentum relationship. **Impulse** (I) is the time integral of force (F × dt) and equals the difference of momentum. The unit of measure for impulse is newton-seconds (Ns). The equation for impulse–momentum is $I = \Delta mv$. The area under the force–time curve is representative of impulse, which is further discussed at the end of this chapter.

▶ **KEY POINT**

The impulse–momentum equation can be derived from the equation $F = ma$. Because $a = dv / dt$, then $F = m \times dv / dt$; further $F = d(m \times v) / dt$; multiply both sides by dt and the resulting equation is $F \times dt = d(mv)$. The left-hand side is equivalent to impulse, and the right-hand side is change in momentum.

Law of Action–Reaction

The third law of motion—the law of action–reaction—states that to every action there is an equal and opposite reaction. Forces of action and reaction between interacting bodies are equal in magnitude, are opposite in direction, and have the same line of action. For example when you take a step and contact the ground, you are exerting a force onto the ground. At the same time, the ground is exerting a force onto you with equal magnitude; this is called the **ground reaction force** (GRF).

Work

Force applied over a distance describes **work**. Mathematically, the equation is: $W = F \times d$. Work is measured in joules, which is equal to a newton-meter. Work is done only when the object is moving because of an applied force. For example, when holding a 10 lb (4.5 kg) weight in your hand, the biceps muscle is exerting an isometric force, but by definition no work is being performed because the weight is not moving.

In evaluating someone's strength, you should be interested in not only the absolute force or torque the person can exert but also the work performed. For example, in figure 2.4, two force–distance curves are depicted. These two curves represent two people's quadriceps force curves over a distance of 120° to 0° knee flexion (range of motion). You should notice that both people achieve the same peak force, but the second person's total work is less, seen in the sharp decline in the curve as the knee approaches extension. Why does this happen? We can only speculate that the strength of the quadriceps is subpar toward

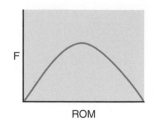 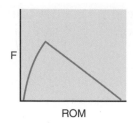

▶ **FIGURE 2.4** Force–distance curves representing work.

extension, possibly due to pain or muscle weakness.

Power

Power is another descriptor you can use to evaluate performance. **Power** is the rate of work; mathematically, power equals work divided by time ($P = W / t$). Power is expressed in units of watts (1 watt = 1 J/s^1). A given amount of work performed in a shorter amount of time will have a greater power output. This concept is exemplified in Clinical Correlation 2.3. Power can also be expressed as $F \times v$, a derivative of the W / t equation.

Energy

Energy is the capacity for doing work. Energy cannot be destroyed, but it can be transformed from one form to another. Examples of energy include kinetic energy, potential energy, strain energy, and heat energy. The unit for energy is the same as for work (the joule). Potential energy is stored energy that under appropriate

CLINICAL CORRELATION 2.3

In athletics, power is an important achievement for most athletes. How do you gain power? Power is achieved by working against a force at a quick speed. So speed is the key. Let's take the example of two football linemen (one on offense and one on defense) who lift the same weight on the bench press, squat, and deadlift. The offensive lineman can run 40 yd in 4.99 seconds, while the defensive lineman can run it in 4.92 seconds. When it comes to the football game, which of the two linemen will probably win the battle of the line of scrimmage? The one able to move his force quicker over a distance is the more powerful one and will probably beat the opposing player.

conditions may be used to do work. Potential energy can take two forms: gravitational energy and strain energy. Kinetic energy is the energy resulting from motion. The mathematical equations for potential and kinetic energy are as follows:

Equation of motion

$$F = ma$$
$$M = I\alpha$$

Work energy

$$W = \tfrac{1}{2}m(v_2^2 - v_1^2)$$

$$W = \tfrac{1}{2}I(\omega_2^2 - \omega_1^2)$$

Kinetic energy

$$E_k = \tfrac{1}{2}mv^2$$

$$E_k = \tfrac{1}{2}I\omega^2$$

Power

$$P = Fv$$
$$P = M\omega$$

> ▶ **KEY POINT**
> The work–energy relationship states that the work done is equal to the change in energy: W = ΔE.

Pressure

The amount of force applied within an area is termed **pressure**. A general principle of injury mechanics suggests that as the area of force application is increased, the likelihood of injury decreases. Usually a sharp object hurts worse than a blunt object because of the pressure involved. The equation for pressure is $p = F / A$, where A is area of contact. The standard unit of pressure is the pascal (Pa), which is equal to 1 newton of force applied to an area 1 meter square (1 Pa = 1 N/m²).

In figure 2.5, the **center of pressure** has been calculated using the force platform and the known dimension of the foot on the force platform. The center of pressure is the point

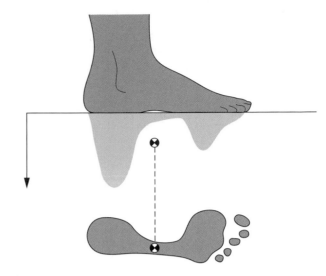

▶ **FIGURE 2.5** Center of pressure.

at which the weight of a body is transferred to the support surface. This may or may not be directly under the body's center of gravity; as seen in the figure, the point falls in between the rearfoot and forefoot.

Levers

A lever is defined as a simple machine used to increase or decrease mechanical advantage, usually of a rigid bar. The components of a lever are the fulcrum, or axis of rotation (A); an effort or force (F); distance of force arm (df); a resistance (R); and distance of resistance arm (dr). Figure 2.6 demonstrates a simple lever setup. The mechanical advantage (MA) of the lever can be found by dividing the force arm distance by the resistance arm distance:

$$MA = FA / RA = df / dr$$

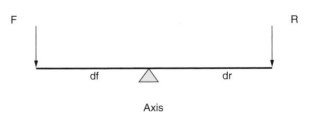

▶ **FIGURE 2.6** Simple lever setup.

A lever may be used to increase force, change the effective direction of the effort force, or gain distance. The type of lever depends on the location of the two forces with respect to the fulcrum. Most human levers create a distance–speed advantage.

Classes of Levers

There are three classes of **levers**.

- *First-class lever*. The first-class lever has the fulcrum located between the force and resistance and can be described by the acronym FAR. A teeter-totter is a good example. A first-class lever can be used to gain force advantage or distance–speed advantage. Its mechanical advantage can be either greater or less than one. An anatomical example is the splenius muscles acting to extend the head across the atlantooccipital joint.

- *Second-class lever*. In the second-class lever, the resistance is located between the force and the fulcrum and can be described by the acronym ARF. Since the resistance arm is always less than the force arm, its mechanical advantage is always greater than one. A second-class lever is used to gain a force advantage such as with a wheelbarrow. An anatomical example is the gastroc–soleus complex when rising up on the toes.

- *Third-class lever*. In the third-class lever, the effort is located between the fulcrum and the resistance and can be described by the acronym AFR. Since the effort arm is always less than the resistance, the mechanical advantage is always less than one. A third-class lever is used to gain a distance–speed advantage. An anatomical example is the biceps muscle.

▶ **KEY POINT**

The three lever types are based on the relationship between the force (F), axis (A), and resistance (R). A first-class lever is represented by FAR. A second-class lever is represented by ARF. A third-class lever is represented by AFR.

Equilibrium of Levers

A lever system is said to be in equilibrium when F × df = R × dr. This primarily occurs with a first-class lever system. Here is an example: Find the value of d in the given example, figure 2.7, in which the force system is in equilibrium.

For this system to stay balanced and not rotate, the forces on the left of the axis must be equivalent to the forces on the right, keeping in account the distance that the force is acting on the system. To solve this problem, we will use the equation F × df = R × dr.

$$50 \times d = 5 \times 100$$
$$50 \times d = 500$$
$$d = 500/50$$
$$d = 10 \text{ m}$$

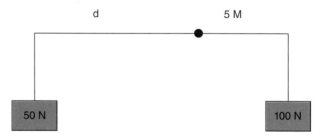

▶ **FIGURE 2.7**   Equilibrium of a force system.

Pulleys

Pulleys may be used in a force system to improve the line of pull or improve the mechanical advantage of the system. They may be fixed or movable. A fixed pulley changes the action line of the force without changing the magnitude; it is used to place the operator in a more favorable position from which to exert the force. In figure 2.8, a fixed pulley is shown in two different setups. Notice that the force output is the same, regardless of the position of the pulley. An example of a fixed pulley in the human body is the lateral malleoli and the peroneal muscles. When the foot is in a supinated or rigid position, there is increased tension on the peroneal muscle to allow this muscle to stabilize the first

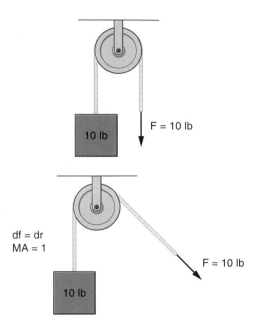

FIGURE 2.8 Fixed pulleys. Note how the force does not change in a fixed pulley system; only the direction of the force changes.

metatarsal. If the foot flattens or pronates, the positional advantage of the peroneus longus is lost, and the first metatarsal becomes unstable.

▶ **KEY POINT**
A fixed pulley is characterized by an axis that is anchored and a pulley wheel that only rotates. A fixed pulley provides a change in direction of force application, resulting in a mechanical advantage that equals one.

In a movable pulley system, one end of the rope is fixed, but the pulley is able to move on the suspended rope, and the supporting effort is exerted on the other end of the rope. In other words, the attachment point for force is not fixed. The pulley wheel rotates and translates to provide a change in direction of force application and to improve the mechanical advantage of the system. Each movable pulley improves the mechanical advantage by two. Therefore a system with two movable pulleys will require one-quarter the force to move the resistance. A system with three movable pulleys requires one-sixth the force.

▶ **KEY POINT**
For each movable pulley, a mechanical advantage of two is gained. In other words, if you are trying to move a 1,000 N load with a movable pulley, it would require only 500 N of force.

Figure 2.9 is an example of a pulley setup with one fixed pulley and one movable pulley. Because there is one movable pulley, the force required to move the 10 lb weight is cut in half and would equal 5 lb of force.

Figure 2.10 shows a common pulley exercise setup for the quadriceps muscle using a fixed pulley. Notice as the leg changes knee

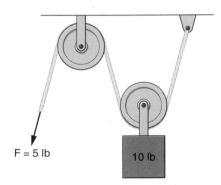

FIGURE 2.9 Fixed and movable pulleys.

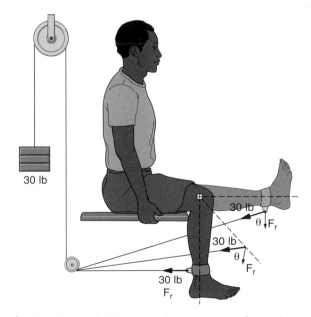

FIGURE 2.10 A pulley system for a knee extension exercise.

position (from 0° to 45° to 90°) that the 30 lb does not change but that the rotary force (F_r) required of the limb does because the angle (θ) between the 30 lb force and F_r changes.

Cams

Cams are nonuniform ellipses commonly used in exercise equipment (e.g., Nautilus) to improve the system's mechanical advantage. The cam shape allows for variable resistance throughout the range of motion, with the intention of matching the length–tension curve of the muscle (see chapter 4). When the muscle is at a mechanical disadvantage, the cam will allow for greater mechanical advantage and vice versa. The mechanical advantage of the system is determined by the ratio of the long axis of the ellipse (diameter, or force arm) with respect to the resistance arm. Figure 2.11 depicts the arrangement of the MA with rotation of the cam. The length of dr and df will change depending on the cam position.

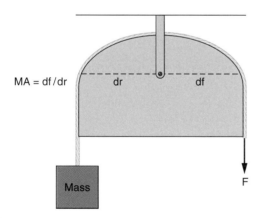

$$MA = df / dr$$

▶ **FIGURE 2.11**   A cam setup.

Moments

The application of force at a distance from the point of pivot provides the concept of moments, or **torque**. A moment of force is the tendency of a force to produce rotation about a point or axis. Mathematically a moment is equivalent to a force multiplied by the perpendicular distance from the line of application of the force to the axis of rotation about which that force acts. The equation is this: M = Fd

The torque arm (d) is the lever arm only when the lever arm is working at 90° to the axis line, as shown in figure 2.12. A moment is a vector quantity, so it has magnitude and direction. The unit of measure of a moment is the foot-pound or newton-meter. In a Cartesian coordinate system, a counterclockwise moment is positive. When a muscle exerts a force that causes rotation of a joint, an internal moment occurs. Forces outside the body that can cause a joint to rotate are termed external moments. An example of an external moment is the gravitational force that places an extensor moment on the elbow joint. To flex your elbow, you need to exert an internal flexor moment via the elbow flexors. For static equilibrium to occur (no movement), external moments have to be countered by internal moments. Clinical Correlation 2.4 provides an example of how moments are quantified to find probable causes of injury.

▶ **KEY POINT**

The following rules apply to a moment, or torque: quantified with a magnitude and direction; magnitude increases as moment arm increases; can be calculated with respect to any point; several forces and moments can be replaced by a net force and net moment; there is no moment if the line of force passes through the point about which the moment is being calculated.

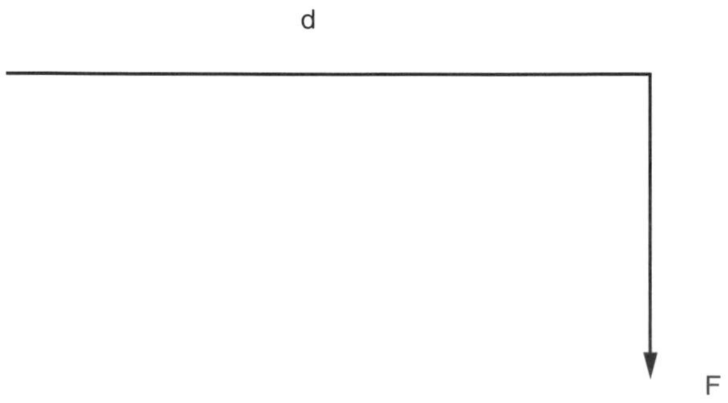

d

F

▶ **FIGURE 2.12**   Graphical representation of a moment.

Here are some examples to help you practice solving for a moment. Remember, the equation used to solve for a moment is $M = Fd$. In figure 2.13, $F = 100$ N and $d = 10$ m; thus the moment = 1,000 Nm ($100 \times 10 = 1,000$). Notice that F is perpendicular to d.

Now, look at figure 2.14 and solve for the moment. Notice here that F is not perpendicular to d, so you need to find the perpendicular vector in order to find the moment. Here is how you do it: The perpendicular vector will be called Fr. Since there is a 30° angle between F and d, and because the angle between Fr and d is 90°, then the angle between F and Fr is 60°. You can then solve for Fr by using trigonometric functions:

$\cos 60° = $ adjacent / hypotenuse
$\cos 60° = Fr / F$
$\cos 60° = Fr / 100$
$100 \cos 60° = Fr$
$Fr = 50$ N

The moment then is $Fr \times d = 50$ N $\times$ 10 m = 500 Nm.

In solving free body diagrams, you need to use the second condition of equilibrium, which states that the sum of the moments about a point equals zero: $\Sigma M = 0$. This is the same equation we used earlier for equilibrium of levers.

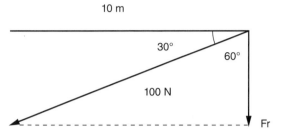

FIGURE 2.14 Solving for a moment, example 2.

Vector Analysis

The composition and resolution of vectors—termed vector analysis—is commonly used in biomechanics to solve force problems. This section reviews the principles of vector analysis. For the majority of this text we will be using a two-dimensional Cartesian coordinate system to define our free body diagrams. Figure 2.15 depicts a two-dimensional coordinate system. Notice the two axes—x and y—intersect to form four quadrants; x is the abscissa, and y is the ordinate. The point of intersection of the two axes is the origin. Measurements along the x-axis to the right of the y-axis are positive. Those to the left of the y-axis are negative. Measurements along the y-axis above the x-axis are positive, and those below the x-axis are negative. Any point in the plane can be identified by an x and y value.

In figure 2.16, point 1 is located two units to the right of the origin and three units above the origin. Both values are positive since they are located in quadrant I. Point 2 has coordinate values of (–2,4). X is always identified first. Notice that the x value is negative since it is located to the left of the origin in quadrant II. What are point 3's coordinates? The answer is (–3,–3), located in quadrant III.

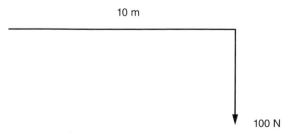

FIGURE 2.13 Solving for a moment, example 1.

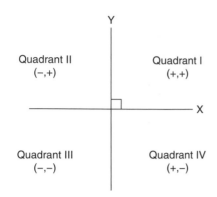

▶ **FIGURE 2.15**   Two-dimensional Cartesian coordinate.

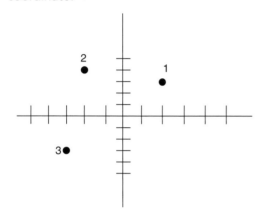

▶ **FIGURE 2.16**   Determining coordinate points.

Now that you have a better understanding of coordinate systems, let's talk about how you solve for vector forces within these coordinate systems. A **scalar** has magnitude alone without direction, and a vector is a quantity that has magnitude and direction. A vector is represented by a straight line with an arrowhead (determines direction). Vectors used within the coordinate system should have units that include both a magnitude and direction, such as 100 N acting northwest.

▶ **KEY POINT**

Examples of scalars include distance traveled (interval); speed (instantaneous); average speed (interval). Scalars can also be represented in angular terms such as angle turned (interval); angular speed (instantaneous); average angular speed (interval). Examples of vector quantities are location (instantaneous); displacement (interval); velocity (instantaneous); acceleration (instantaneous). These vectors can also be angular terms.

Resolving Vectors

Vectors are used to represent forces. To define a vector in terms of its x and y components, you will need to resolve it. In the following example, we take a single vector force, F, and resolve it into its x and y components. Follow these steps when you are resolving vectors.

1. Draw a two-dimensional Cartesian coordinate system. Label the axes (x,y). Define the vectors, and label magnitude and direction.
2. Define the x and y components.
3. Use trigonometric functions and the Pythagorean theorem.

PRACTICE IT!

Given the vector setup in this figure, resolve the F vector into F_x and F_y. F has a magnitude of 100 N acting 30° up from the horizontal.

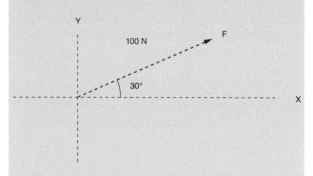

In this figure, a Cartesian coordinate system is displayed with axes (x,y) and force (F) acting 30° up from the x-axis. The F vector is equivalent to 100 N. The x and y components for F are defined and represented as F_x and F_y (see the next figure). These components are parallel to their respective axes.

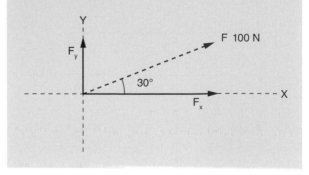

Use trigonometric functions and the Pythagorean theorem to solve for magnitude.

For F_x

You can form a right triangle by using F as the hypotenuse and F_x as the adjacent side. The opposite side would be equivalent to F_y in magnitude. Since you know the value of F and the angle between F and F_x, you can use the cosine function to solve for F_x.

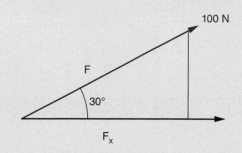

$$\cos 30° = \text{adjacent side / hypotenuse}$$
$$\cos 30° = F_x / 100$$
$$100 \cos 30° = F_x$$
$$86.60 \text{ N} = F_x$$

For F_y

You can utilize the same right triangle to solve for F_y. Here you would use the sine function since F_y is the opposite side.

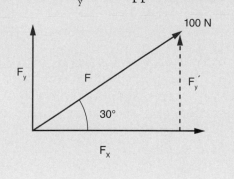

$$\sin 30° = \text{opposite side / hypotenuse}$$
$$\sin 30° = F_y / 100$$
$$100 \sin 30° = F_y$$
$$50.00 \text{ N} = F_y$$

Answer

$$\sin \alpha = y / 100 \rightarrow 100 \sin 30° = F_y = 50.00 \text{ N}$$

$$\cos \alpha = x / 100 \rightarrow 100 \cos 30° = F_x = 86.60 \text{ N}$$

Check

You can double-check your answers by using the Pythagorean theorem. F_x and F_y make up the two other sides of the right triangle, and F is the hypotenuse. To double-check the angle, use the tangent function.

$$F = \sqrt{(86.6)^2 + (50)^2} = 100 \text{ N}$$

$$\text{inv} \tan \alpha = 50 / 86.6 = 30°$$

Composing Vectors

Composition of vectors is used when two or more vectors are acting along the same line (collinear), in the same plane (coplanar), or on the same point (concurrent), and you wish to show their combined effect as a single vector, the resultant.

If the vectors are collinear, simply add up their values, keeping in mind their direction. For example in figure 2.17, three vectors (L,M,N) have magnitudes of 2, 2, and 3 and are acting to the right (and therefore are positive). Two vectors (F,G) have magnitudes of 4 and 2 and are acting to the left (negative). By simple addition, the resultant is:

$$2 + 2 + 3 - 4 - 2 = +1$$

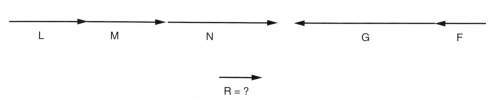

▶ **FIGURE 2.17**   Adding linear vectors.

The resultant has a magnitude of 1 and is acting to the right.

Many times, vectors are concurrent but not collinear; in this case, the following steps should be taken to solve for one resultant vector:

1. Draw a picture; label the axes; and define the vectors, labeling magnitude and direction.
2. Define the x and y components.
3. Determine the x and y values.
4. Add the x components; add the y components.
5. Use the Pythagorean theorem.
6. Define the direction using the inverse tangent.

PRACTICE IT!

Compose one vector from the three vectors given in this figure.

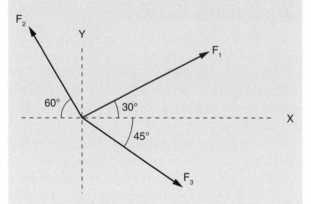

1. Draw a picture; label the axes; and define the vectors, labeling magnitude and direction.
2. Define the x and y components for each of the vectors.
3. Determine the x and y values for each of the vectors, as you did in the section on resolving vectors.
4. Add all the x components together. Add all the y components together.

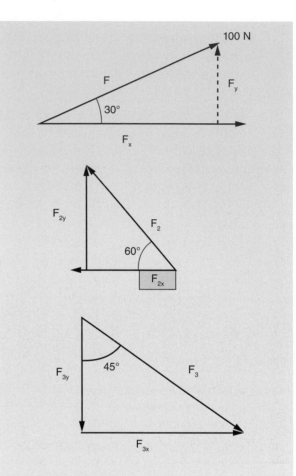

$\cos \alpha = x / 100$
$\quad F_1 = F_{1x} = 100 \cos 30° = 86.6$
$\sin \alpha = y / 100$
$\quad F_{1y} = 100 \sin 30° = 50.0$
$\cos \alpha = x / -120$
$\quad F_2 = F_{2x} = -120 \cos 60° = -60$
$\sin \alpha = y / 120$
$\quad F_{2y} = 120 \sin 60° = 103.92$
$\cos \alpha = x / -50$
$\quad F_3 = F_{3x} = 50 \cos 45° = 35.36$
$\sin \alpha = y / 50$
$\quad F_{3y} = -50 \sin 45° = -35.36$
$\quad x = 61.96$
$\quad y = 118.57$

5. Use the Pythagorean theorem to determine the resultant (hypotenuse).
$R = \sqrt{(61.96)^2 + (118.57)^2} = 133.78 \text{ N}$

6. Define the direction using inverse tangent. The resultant is the hypotenuse and we know the x and y components (adjacent and opposite side), so we can

use the inverse functions to determine the angle between the hypotenuse and the adjacent side. This angle will be the resultant's location relative to the horizontal.

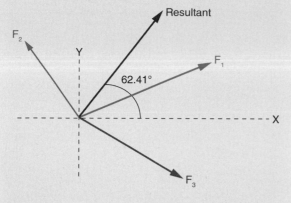

$\tan \theta = y / x = 118.57 / 61.96 = 62.41°$ NE

Or we could use

$\sin \theta = y / h = 118.57 / 133.78 = 62.41°$ NE

Or even

$\cos \theta = x / h = 61.96 / 133.78 = 62.41°$ NE

FIGURE 2.18   Link segment model.

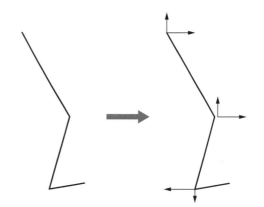

FIGURE 2.19   Free body diagram.

Force Diagrams

To estimate forces acting on rigid bodies, it is helpful to draw a force diagram. The first step is to simplify the object into links and design a link segment model. In figure 2.18 the lower limb has been simplified into three links, one link for the upper leg, one link for the lower leg, and one link for the foot. We take the link segment model diagram a step further in order to calculate forces acting on each of the joints by designing what is called a free body diagram (FBD). In figure 2.19, the free body diagram consists of all the forces drawn in correct proportion acting on the body.

To construct an FBD, follow these rules:

- Identify the forces (GRF, muscle, friction, weight).
- Include the parameters for each force (magnitude, direction, dimension).

- Remember that parallel forces can be added or subtracted to find the resultant.
- When forces are concurrent, the resultant is determined mathematically.

One other issue we need to discuss before solving an FBD is static equilibrium. A body is in equilibrium when, as stated in Newton's first law, it remains at rest or is in motion with constant velocity. If it is at rest, with the velocity equaling zero, it is said to be in static equilibrium. The first condition of static equilibrium is that the sum of the forces equal zero:

$\Sigma F_x = 0$

$\Sigma F_y = 0$

Now, let's construct a free body diagram. Let's use the example of a person holding a 5 kg (10 lb) weight. We want to know how much muscle force is required to hold this weight as well as the force acting on the joint (the joint reaction force). Figure 2.20 is the anatomical figure. We will keep our FBD simple, in a two-dimensional plane and in static equilibrium.

The first force we will define is the gravitational force acting on the forearm. This vector (L = limb) acts vertically down (as do all gravity vectors). It is placed at a point with a known distance from the elbow joint axis, and this point is the center of mass for the forearm.

The next force added is the force vector that represents the 5 kg weight. This vector (W = added weight) also acts straight down because of the gravitational force acting on the weight. The force vector is drawn from the middle of the weight.

The last force vector represents the muscle force. This vector (M = muscle) originates at some known point (point of attachment) and projects in a direction that simulates the direction of muscle pull.

In figure 2.20 you can see how the anatomical model has become the free body diagram.

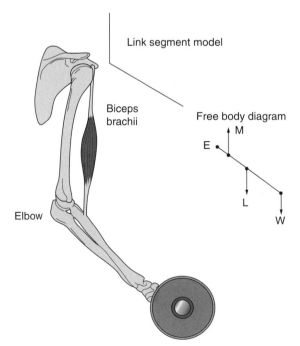

▶ **FIGURE 2.20** Constructing a free body diagram. The text defines the letters used in the free body diagram. E is the axis of the elbow joint.

PRACTICE IT!

Now, let's discuss how to solve the problem mathematically. Take the anatomical figure shown in figure 2.20. Solve the following free body diagram by identifying the muscle force required to hold the weight and the joint reaction force on the elbow. The steps we take will be consistent with how we solve all the FBDs in this text. You are given the following information:

W = 5 kg

L = 6 kg

dW = 30 cm (distance between W and the axis)

dL = 15 cm

joint angle = 30° from the horizontal

muscle angle = 50° (angle of insertion)

dM = 5 cm (distance from muscle insertion to axis)

Step 1

Convert the anatomical figure to a free body diagram, and label all forces and the axis. In the figure that follows, notice the x-axis is along the moving segment, the y-axis is perpendicular to the x-axis, and they intersect at the joint axis. You should also notice that the x and y components for each of the three vectors are identified.

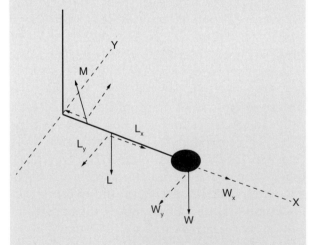

Step 2

Since L and W are given to you as a mass (in kg), you must convert these values to a weight by using the equation W = mg, where the calculated weight is equal to the given mass multiplied by the gravitational constant. Thus,

$$W = 5 \text{ kg} \times 9.81 \text{ m/s}^2 = 49.05 \text{ N}$$
$$L = 6 \text{ kg} \times 9.81 \text{ m/s}^2 = 58.86 \text{ N}$$

Step 3

To solve for the muscle force (which is vector M), use the moment equilibrium equation, $\Sigma M = 0$. By definition, a moment is a perpendicular force and its distance. The perpendicular forces for each of the three vectors will be their y component vectors. So you must solve for these first.

$$L_y = L \cos 30° = 58.86 \times \cos 30° = -50.97 \text{ N}$$
$$W_y = W \cos 30° = 49.05 \times \cos 30° = -42.48$$
$$M_y = M \sin 50° \text{ (you can't solve this yet)}$$

The triangle forming the 30° that you use for L_y and W_y is found in the following figure. You know the angle between W and W_y and L and L_y is 30° based on geometry (see the appendix for further clarification). Since the potentially moving arm is located 30° below the horizontal, you can make a series of right triangles to solve for the angle between the y component vector and the original vector. We will use these geometric principles for all our FBDs.

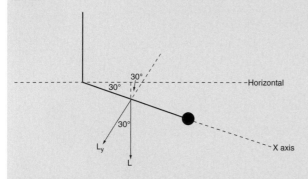

The triangle that forms the muscle vector will not use the 30° because it is based on the insertion angle and not the angle from the horizontal. This triangle is shown in the following figure. Both W_y and L_y are negative because these vectors are acting down relative to the x-axis and y-axis.

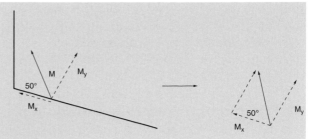

Step 4

Solve the moment equation, which is the following:

$$\Sigma M = 0 = (M_y \times DM) + (L_y \times DL) + (W_y \times DW)$$

Basically you are solving for M_y; all other values are known.

$$\Sigma M = 0 = (M_y \times DM) + (L_y \times DL) + (W_y \times DW)$$
$$\Sigma M = 0 = (M_y \times 5) + (-50.97 \times 15) + (-42.48 \times 30)$$
$$\Sigma M = 0 = (M_y \times 5) - 764.55 - 1{,}274.4$$
$$2{,}038.95 = M_y \times 5$$
$$407.79 \text{ N} = M_y$$

Now to answer the original question. To determine the muscle force required to hold the weight, you must find M. You can use trigonometry functions to do this:

$$M_y = M \sin 50°$$
$$M = 532.33 \text{ N}$$

The joint reaction force (JRF) is a fourth vector that represents the resultant of all forces acting on the joint. For our FBD, that would be the muscle force, gravity, and the lifted weight. To find the joint reaction force, perform the following steps:

1. Use the force equilibrium equation to solve the JRF vector:

$$\Sigma F_x = 0$$
$$\Sigma F_y = 0$$

The full equation is

$$\Sigma F_x = 0 = R_x + M_x + W_x + L_x$$
$$\Sigma F_y = 0 = R_y + M_y + W_y + L_y$$

The R vectors represent the JRF and are the only unknown values for each equation.

2. Solve the x component vectors:

$$W_x = W \sin 30° = 29.43 \text{ N}$$
$$L_x = L \sin 30° = 24.53 \text{ N}$$
$$M_x = M \cos 50° = -342.18 \text{ N}$$
(negative because it is acting to the left)

3. Solve for R_x and R_y:

$$\Sigma R_x = 0 = R_x + M_x + W_x + L_x$$
$$= R_x - 342.18 + 29.43 + 24.53$$
$$= R_x - 288.22$$
$$288.22 \text{ N} = R_x$$
$$\Sigma R_y = 0 = R_y + M_y + L_y + W_y$$
$$= R_y + 407.79 - 50.97 - 42.48$$
$$= R_y + 314.34$$
$$-314.34 \text{ N} = R_y$$

4. Find the R resultant using the Pythagorean theorem:

$$R = \sqrt{R_x^2 + R_y^2}$$
$$R = \sqrt{(288.22)^2 + (-314.34)^2}$$
$$R = 426.47 \text{ N}$$

426.47 N is the joint reaction force.

5. You also need to find the location of this vector relative to the joint. You can use the tangent function to find the angle between R_x and R_y. This will be the location of R relative to the x-axis.

$$\tan \varphi = -314.34 / 288.22 = 47.48° \text{ from the x-axis}$$

The direction of the vector will be southeast (since the x vector is positive and the y vector is negative: the location is the fourth quadrant, or southeast). The final answer is depicted in this figure.

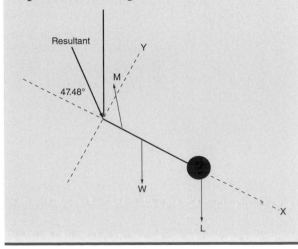

Clinical Application of Kinetics

A common tool used in the research lab is a force plate, which is a rectangular plate usually installed into the floor. As a person walks across the force plate, it measures the vertical and horizontal forces. These forces are called ground reaction forces (GRFs) (Newton's third law). The force plate measures one vertical force and two horizontal forces, one front to back and one left to right. Figure 2.21 shows the forces derived. Figure 2.22 depicts the vertical ground reaction forces during running. The graph has a typical double-peak curve. Now let's look at figure 2.23. This graph shows two vertical ground reaction forces, one of a runner who strikes with the heel, and the second is a runner who strikes with the forefoot. What differences do you notice? The area under the force–time curve is called impulse.

Notice that the person who strikes with the forefoot has slightly less impulse overall than the heel striker. This is one reason so many running coaches advocate a forefoot strike pattern.

To calculate joint-specific kinetics, an inverse dynamics approach can be used. To use this approach, the body is simplified as a rigid-linked FBD. Segments are considered rigid and articulate through frictionless joints. Then ground reaction forces (typically mea-

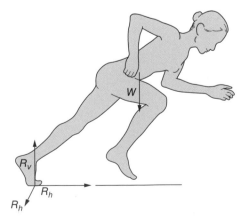

▶ **FIGURE 2.21** Ground reaction forces (GRF) (R_v = vertical GRF; R_h horizontal GRF).

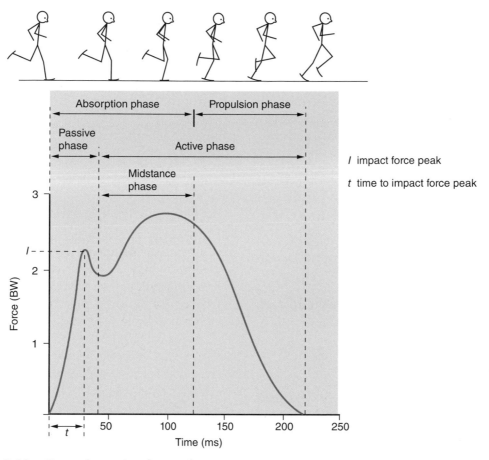

FIGURE 2.22   Ground reaction forces during running.

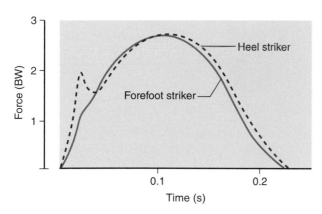

FIGURE 2.23   Ground reaction forces curves.

sured by force platforms) in combination with limb kinematics are used to derive the resultant forces acting at the ankle, the knee, the hip, and so forth. Such a method gives important information on the force produced at each joint during a ballistic task. You can subsequently infer muscle activation patterns that would result in the generation of those

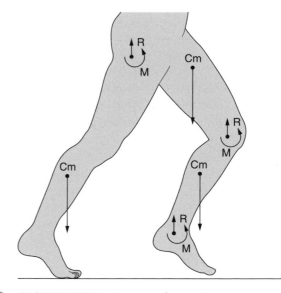

FIGURE 2.24   Inverse dynamics.

forces. Figure 2.24 is a free body diagram of a sagittal view of a person walking, showing the reactive forces and moments of forces at the joints.

Conclusion

This chapter describes the elementary mechanical concepts and principles that underlie linear and angular kinetics. Forces are described and Newton's laws of motion are discussed. Other concepts such as work, power, energy, and pressure are briefly covered. Levers and moments are explained, and the reader has ample opportunity to take this information further by solving free body diagram problems. Many of these concepts will be used in subsequent chapters to describe injury mechanisms.

REVIEW QUESTIONS

1. Differentiate between center of gravity and center of mass.

2. Differentiate between vector and scalar quantities.

3. Define friction, and give an example of how friction can cause injury.

4. Differentiate between work and power.

5. What is the mathematical formula for a moment of force?

PROBLEM SETS

1. A person is performing a straight-leg raise. What is the muscle force needed to hold a 10 kg weight in the air with the leg at an angle of 40° from the floor? The weight is located 70 cm from the hip joint. The leg weighs 70 N, and its center of gravity is located 17 cm from the hip joint. The muscle insertion angle is 15° and is located 5 cm from the hip joint. What is the joint reaction force on the hip?

2. A patient is set up on a pulley exercise machine. What is the magnitude of force necessary for the quadriceps muscle to maintain the exercise load at 0° of flexion? What is the joint reaction force on the knee? The pulley weight is set on 100 N. The location of the pulley strap on the lower leg is 60 cm from the knee axis. The weight of the patient's foot and leg is 30 N, and their combined center of mass is located 20 cm distal to the knee axis. The angle of the patellar tendon inserts at a constant 25° angle 10 cm from the knee axis. The angle of the pulley rope forms an angle of 19° with the leg.

3. Refer to this figure. What is the force required to move this 10 kg mass in the pulley setup that includes one fixed pulley and two movable pulleys?

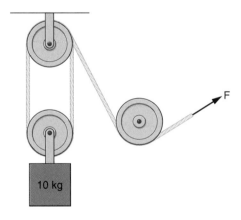

4. Refer to this figure. What is the force required to begin to lift the 30 kg mass?

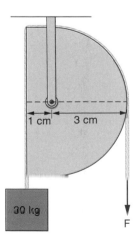

Basic Muscle and Joint Physiology and Function

Part II of *Clinical Mechanics and Kinesiology* contains three chapters that focus on the basic physiology of the muscle and nerve and the function of the muscle and joint systems. These chapters lay the foundation for the next part of the text, which focuses on detailed joint anatomy and biomechanics.

Chapter 3 begins by exploring the structure of muscle. Specific details are included with regard to the functional unit of a muscle—the sarcomere—and its components that are responsible for muscle contraction. Next, the anatomy of the motor unit is introduced, followed by a section on the different muscle fibers. The importance of muscle fiber types and their influence on muscle contraction is emphasized. The majority of this chapter is dedicated to the mechanics of muscle contraction. An understanding of this phenomenon is important for the student who will be working with people who have musculoskeletal injuries. The neurophysiology of a skeletal muscle contraction is quite complex and involves interaction among chemical, electrical, and mechanical stimuli.

Chapter 4 examines the topic of muscle performance. Specific areas included in this chapter are muscle morphology, force generation capacities, contractile properties, and various rehabilitative methods used to increase or decrease strength and endurance. Muscle mechanics, including the muscle's action as a prime mover and stabilizer, are communicated. Another feature that is very relevant to the rehabilitation specialist is the discussion of how the muscle is affected by injury, immobilization, and aging.

The final chapter of this section, chapter 5, discusses the topic of joint structure and function. The different types of joints found in the human body are first presented. This narrative is followed by a detailed look at bones, ligaments, joint capsules, and cartilage. Relevance to rehabilitation is emphasized with several clinical examples.

Muscle and Nerve Physiology

Skeletal muscles serve many purposes throughout the body, primary among them is movement and force production, as well as endurance. Although the muscular system provides the force to move the levers of the skeletal system, the nervous system interprets this information from various senses throughout the body and regulates the movement of the muscular system.

The major functional unit of the nervous system is the motor unit, which is made up of a motor neuron and the muscle fibers it innervates. This chapter details the composition of the nervous system and the innervations of the muscular system. Additionally, motor unit function is dependent on various factors that are discussed in this chapter.

Muscle Structure

Grossly, the contractile unit of a muscle that produces movement and force is the muscle belly and the tendon that attaches the muscle belly to the bone (figure 3.1). Contractile muscle tissue has the ability to develop tension in response to chemical, electrical, or mechanical stimuli. Passive tension is

OBJECTIVES

After reading this chapter, you should be able to do the following:

> List the components of a skeletal muscle.

> List the components of a motor unit.

> List the different muscle fiber types.

> List the two structural divisions of the nervous system and their respective components.

> Describe how the muscle spindle and Golgi tendon organ affect muscle contraction.

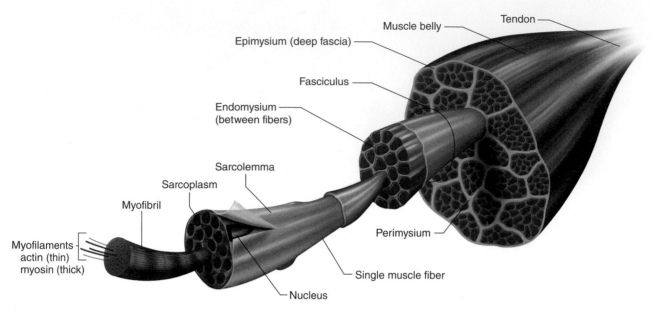

▶ **FIGURE 3.1** Composition of muscle fiber.

provided by noncontractile connective tissue that supports the muscle fiber (Williams et al. 1995).

▶ **KEY POINT**

   Chemical, electrical, and mechanical sources provide the stimuli to produce muscle contraction. Passive tension is provided by noncontractile connective tissue that supports the muscle fiber.

Each skeletal muscle fiber is a long, cylindrical, multinucleated cell composed of smaller units called filaments. The largest of these parallel aligned filaments is called a myofibril. The myofibril is composed of subunits called **sarcomeres** that are arranged end to end along the length of the myofibril. Each myofibril exhibits a characteristic pattern of alternate light and dark transverse bands due to the way the components of the myofibril react to light under an electron microscope. The light and dark bands are referred to as I (isotropic) and A (anisotropic) bands, respectively.

Each sarcomere also contains filaments, known as **myofilaments**. There are two types of myofilaments in each sarcomere, a thicker one composed of **myosin** protein molecules and a thinner myofilament composed of mol-

ecules of the protein **actin**. The basic mechanism of muscle contraction is the sliding of the actin myofilament on the myosin chain.

The **sarcolemma** is a cell membrane that covers each fiber in a single muscle. The muscle fiber is composed of cytoplasm or, as it is called in muscle, **sarcoplasm**. This sarcoplasm contains myofibrils and nonmyofibrillar structures such as ribosomes, glycogen, and mitochondria, which are needed for cell metabolism.

The sarcomere (figure 3.2), containing the contractile proteins myosin and actin, is therefore the basic functional unit of a muscle. As previously mentioned, actin myofilaments are thinner and typically more abundant than the myosin myofilaments. Actin is anchored at each end of the sarcomere at the Z-line and surrounds a thicker myosin myofilament. The region between the ends of the two groups of actin filaments in the middle of a sarcomere is referred to as the H-zone. This arrangement of these two myofilaments is repeated throughout the sarcomere. The amount of these contractile proteins within the cells is strongly related to a muscle's contractile force (Baldwin 1996; Baldwin et al. 1982; Cress et al. 1996).

The multiple fascicles are surrounded by connective tissue (see figure 3.1). The outer-

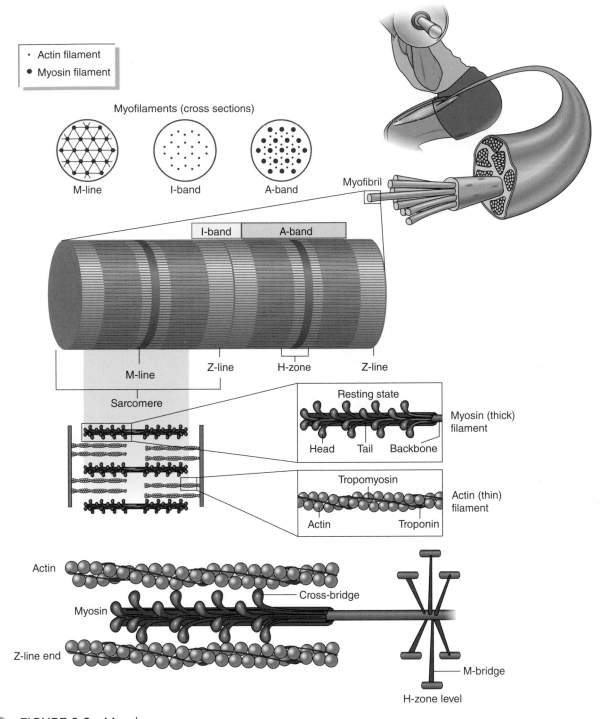

- Actin filament
- Myosin filament

Myofilaments (cross sections)

M-line   I-band   A-band

Myofibril

I-band   A-band

M-line   Z-line   H-zone   Z-line

Sarcomere

Resting state

Myosin (thick) filament

Head   Tail   Backbone

Tropomyosin

Actin (thin) filament

Actin   Troponin

Actin

Cross-bridge

Myosin

Z-line end

M-bridge

H-zone level

▶ **FIGURE 3.2** Muscle sarcomere.

most layer of connective tissue that surrounds the entire muscle belly is known as the epimysium. Fascicles make up the muscle belly and are surrounded by the perimysium. Individual muscle fibers are surrounded by additional connective tissue, the endomysium. The amount of connective tissue in each muscle and the size of the connecting tendons vary from muscle to muscle.

Motor Unit

Although the functional unit of force generation is the sarcomere, the functional unit of

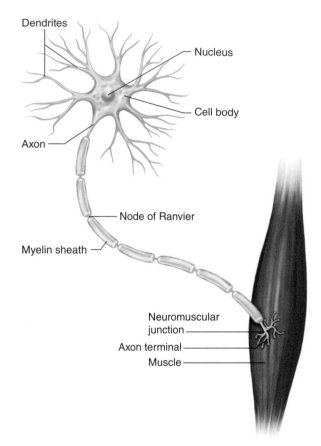

FIGURE 3.3 The motor unit.

movement is the **motor unit** (figure 3.3). A motor unit consists of a motor neuron with an α axon (or alpha motor neuron) and all of the terminal branches of the axon and the muscle fiber they innervate (Gamble 1988). Motor neurons have their cell bodies in the ventral root of the spinal cord. This cell body is responsible for neuronal integrity. The **axon** is the long projection that extends from the cell body. As an axon approaches the muscle, it branches many times.

A single muscle can contain many motor units. The number of muscle fibers belonging to a motor unit (i.e., the innervation ratio) and the number of motor units within a whole muscle vary.

Muscle Fiber Types

Muscle fiber types can be differentiated from each other histochemically, metabolically, morphologically, and mechanically. The three major types of muscle fibers in skeletal muscle include Type I (slow twitch), Type IIa (intermediate), and Type IIb (fast twitch) (Gamble 1988). The muscle fiber types are identified via use of myofibrillar ATPase activity under varying acidic and alkaline conditions. **Type I fibers** have contraction times of 100 to 120 milliseconds, while **Type II fibers** have contraction times of 40 to 45 milliseconds (Gamble 1988; Gregor 1993). Type I fibers are sometimes referred to as slow contracting and fatigue resistant and Type IIa as fast contracting and fatigue resistant, while Type IIb are fast contracting and fatigable. For additional characteristics of these fiber types, refer to table 3.1.

Every skeletal muscle is composed of a combination of the three types of fibers. It is generally believed that this variation in fiber type among people is genetically determined. As one ages, there is generally a decrease in the number of Type II fibers.

▶ **KEY POINT**

Although skeletal muscles are composed of the three types of fibers, the number of Type II fibers decreases as one ages to a greater extent than Type I fibers do.

Stability (or postural) muscles generally have a higher percentage of Type I slow-twitch muscle fibers. Generally, these muscles have relatively small, slow motor units (small cell bodies, small-diameter axons, and a small number of muscle fibers per motor unit) and are almost continually active during postural activity. Dynamic (or mobility) muscles, on the other hand, generally respond much faster to a stimulus but also fatigue more rapidly than Type I fibers do (see Clinical Correlation 3.1).

▶ **KEY POINT**

Stability (postural) muscles generally have a higher percentage of Type I slow-twitch muscle fibers, while dynamic (mobility) muscles generally respond much faster to a stimulus but also fatigue more rapidly than Type I fibers do.

The proportions of muscle fiber types vary between muscles and people. The average

TABLE 3.1   Muscle Fiber Type Characteristics

	Slow twitch (Type I)	Fast-twitch fatigue resistant (Type IIa)	Fast-twitch fatigable (Type IIb)
Histochemical fiber profile	Slow oxidative	Fast oxidative glycolytic	Fast glycolytic
Motor units	Small with low innervation ratio	Medium	Large with high innervation ratio
Twitch response	Slow twitch	Medium	Fast twitch
Order of recruitment	First	Second	Last (usually only when intense effort required)
Fatigue resistance	High	Reasonably resistant	Low
Recovery after exercise	Rapid	Fairly rapid	Slow
Power output	Low	Moderate to high	High
Aerobic capacity	High	Moderate	Low
Anaerobic capacity	Low	High	High

CLINICAL CORRELATION 3.1

Type I slow-twitch muscle fibers generally have low activation thresholds, high aerobic capacity, and low fatigability. Type IIb fast-twitch muscle fibers on the other hand have high activation thresholds, low aerobic capacity, and high fatigability. The training emphasis for Type I muscle fibers should therefore be on high volume (repetition and sets) with low intensity, while training for Type IIb muscle fibers should emphasize low volume and high intensity. Postural muscles are predominantly Type I muscle fibers because of their low energy requirements and need for sustained muscle contractions. Therefore, program design for these muscle types should emphasize high repetition and sets with minimal resistance. Depending on the muscle and its strength, gravity can often serve as sufficient resistance in a properly designed program for these muscles.

person has approximately 50% Type I, 25% Type IIa, and 25% Type IIb fibers in the gastrocnemius muscle (Gamble 1988). Elite distance runners can have a much higher proportion of Type I fibers, while elite sprinters typically have a much higher proportion of Type IIb fibers (Gamble 1988).

Muscle Contraction

Muscle fiber contraction occurs via excitation and activation of the specific muscle fiber's motor neurons. The nervous system recruits a motor unit by altering the voltage potential across the membrane of the cell body of the alpha motor neuron. This process involves a net summation of competing inhibitory and excitatory inputs.

▶ **KEY POINT**

Muscle fiber contraction occurs via excitation and activation of the specific muscle fiber's motor neurons.

Alpha motor neuron activation can come from many sources, first via recruitment and then by a process of rate coding. Recruitment is the initial activation of specific motor neurons, which activate their associated muscle fibers. Motor units are recruited by the nervous system by altering the voltage potential across the membrane of the cell body of the alpha motor neuron. Ions flow across the cell membrane and produce an electrical signal, or action potential. The action potential is disseminated down the axon of the alpha motor neuron to the motor end plate at the neuromuscular junction. A muscle contraction (and corresponding muscle force production) then occurs once the muscle fiber is activated. Additional muscle fiber activation produces increased muscle force throughout the muscle.

Rate coding is the process of modulating the force produced by the associated muscle fibers via the rate of production of sequential action potentials. Since a twitch duration is often longer than the interval between action potential discharges, it is possible for a number of subsequent action potentials to begin during the initial twitch. If a muscle fiber is allowed to completely relax before a subsequent action potential, the second fiber twitch will generate a force equal to that of the first twitch. If the next action potential arrives before the relaxation of the preceding twitch, however, the muscle twitches summate and generate an even greater peak force.

Motor units activated at higher rates are capable of generating greater overall force than the same number of motor units activated at lower rates because of tetanization, or the fusing of successive summated mechanical twitches occurring in a very small time frame. Tetanization represents the greatest force level possible for a single muscle fiber.

The complex interaction of the chemical, electrical, or mechanical stimuli to produce a muscle contraction has been termed the sliding filament theory of muscle contraction (Huxley & Hanson 1954). This theory results from the formation of cross-bridges between the myosin and actin myofilaments, with the actin "sliding" on the myosin chain of myofilaments. The number of cross-bridges determines the strength of the muscle contraction.

▶ **KEY POINT**

Cross-bridging of myosin and actin myofilaments is the primary component of the sliding filament theory of muscle contraction. The strength of the muscle contraction is determined by the number of cross-bridges formed.

An electrical stimulus initiates the muscle contraction from the associated motor neurons, causing depolarization of the muscle fiber. Calcium is then released into the cell and binds with a regulating protein, troponin. This combination of calcium and troponin causes actin to bind with myosin, initiating the muscle contraction. The muscle will relax once a cessation of the nerve's stimulus causes a reduction in the level of calcium in the muscle (Lieber 1992). New cross-bridges are formed once the stimulation of the muscle fiber occurs at a sufficient level. There is a coupling and decoupling of cross-bridges between myosin and actin at different times so that tension can be maintained as the muscle shortens. A successive and ultimately sustained contraction is produced.

During all types of muscular contraction, the myosin and actin filaments remain unchanged in length. During isotonic contractions, the interdigitation between the two sets of filaments changes as the length of the muscle fiber changes. The width of the A-bands stays unchanged, while the width of the I-bands and H-zone varies. As the muscle fibers shorten and the region of the interdigitation increases, the width of the I-bands and H-zones decreases. In contrast, as the length of the muscle increases and the interdigitation decreases, the width of the I-bands and H-zones increases (see figure 3.2). The specific sequence of events in muscular contraction is as follows:

1. Action potential is initiated and disseminated down the motor axon.

2. Acetylcholine is released from the axon terminal at the neuromuscular junction.

3. Acetylcholine binds to receptor sites on the motor end plate.

4. Potassium and sodium ions depolarize the muscle membrane.

5. Muscle action potential is disseminated over membrane surface.

6. Depolarization of T-tubules releases calcium from the lateral sacs of the sarcoplasmic reticulum.

7. Calcium binds to troponin-tropomyosin complex in actin filaments, releasing the

inhibition of actin and myosin binding. The cross-bridge between actin and myosin heads is created.

8. Actin combines with myosin adenosine triphosphate (ATP).

9. The energy created produces movement of the cross-bridge of myosin and actin.

10. Myosin and actin slide relative to each other.

11. The myosin and actin cross-bridge activation continues until the concentration of calcium remains high enough to inhibit the actin of the troponin–tropomyosin system.

12. The cross-bridge is broken when stimulation ceases; calcium moves back into the lateral sacs of the sarcoplasmic reticulum.

Therefore, the sliding filament theory is dependent on the various stimuli. The proper performance of this mechanism of muscle contraction is highly dependent on a properly functioning nervous system.

Nervous System

The two structural divisions of the nervous system are organized into the **central nervous system** (CNS) and the **peripheral nervous system** (PNS). The CNS consists of the brain and spinal cord, while the PNS is made up of 43 pairs of nerves arising from the CNS. Of these 43 pairs, the 12 cranial nerves arise from the base of the brain. The other 31 pairs of spinal nerves originate from the spinal cord.

▶ **KEY POINT**

The two structural divisions of the nervous system are organized into the central nervous system and the peripheral nervous system.

Central Nervous System

The spinal cord is normally 16.5 to 17.5 in (42 to 45 cm) long in adults and has the brain stem and medulla at its upper end. The conus medullaris is the distal end of the spinal cord; in adults, the conus ends at the L1 or L2 level of the vertebral column.

Three membranes envelop the structures of the CNS: **dura mater, arachnoid mater**, and **pia mater**. The dura mater is the outermost and strongest of the membranes. The dura forms the dural sac around the spinal cord. It is separated from the bones and ligaments of the vertebral canal by an epidural space, which can become partly calcified or even ossified with age (Waxman 1996). The pia mater is the deepest of the three layers. It is firmly attached to the outer surface of the spinal cord and nerve roots. The pia mater conveys the blood vessels that supply the spinal cord and houses the denticulate ligaments that anchor the spinal cord to the dura mater (Pratt 1996).

Peripheral Nervous System

The peripheral nervous system is composed of the 12 cranial nerves and the 31 spinal nerves. The cranial nerves consist of the following: olfactory, optic, oculomotor, trochlear, trigeminal, abducens, facial, vestibulocochlear, glossopharyngeal, vagus, accessory, and hypoglossal nerves. Each of these nerves is further discussed in table 3.2.

The spinal nerves are divided topographically into 8 cervical pairs (C1-C8), 12 thoracic pairs (T1-T12), 5 lumbar pairs (L1-L5), 5 sacral pairs (S1-S5), and a coccygeal pair. The posterior (dorsal) and anterior (ventral) roots of the spinal nerves are located within the vertebral canal. The portion of the spinal nerve that occupies the intervertebral foramen, and thus is no longer in the vertebral canal, is referred to as the peripheral nerve.

Nerve fibers are generally categorized according to function: sensory, motor, or mixed (motor and sensory). Sensory nerves carry afferents (nerves conveying impulses from the periphery to the CNS) from a portion of the skin. They also carry efferents (nerves conveying impulses from the CNS to the periphery). The area of sensory nerve distribution is called a dermatome and generally follows the segmental distribution of the underlying nerve distribution (Waxman 1996).

TABLE 3.2 Cranial Nerves and Their Examination

Nerve	Afferent (sensory)	Efferent (motor)	Test
I. Olfactory	Smell	—	Identify familiar odors (e.g., coffee)
II. Optic	Sight	—	Test visual fields
III. Oculomotor	—	Voluntary motor: levator of eyelid; superior, medial, and inferior recti Autonomic: smooth muscle of eyeball	Upward, downward, and medial gaze; reaction to light
IV. Trochlear	—	Voluntary motor: superior oblique muscle of eyeball	Downward and lateral gaze
V. Trigeminal (jaw reflex)	Touch, pain: skin of face, mouth, anterior tongue	Voluntary motor: muscles of mastication	Corneal reflex; face sensation; clench teeth and try to open mouth
VI. Abducens	—	Voluntary motor: lateral rectus muscle of eyeball	Lateral gaze
VII. Facial	Taste: anterior tongue	Voluntary motor: facial muscles Autonomic: lacrimal, submandibular, and sublingual glands	Close eyes tight; smile and show teeth; puff cheeks; identify familiar tastes (e.g., sweet, sour)
VIII. Vestibulocochlear (acoustic nerve)	Hearing: ear Balance: ear	— —	Hear watch ticking; hearing tests; balance and coordination tests
IX. Glossopharyngeal	Touch, pain: posterior tongue, pharynx Taste: posterior tongue	Voluntary motor: unimportant muscle of pharynx Autonomic: parotid gland	Gag reflex; ability to swallow
X. Vagus	Touch, pain; pharynx, larynx Taste: tongue, epiglottis	Voluntary motor: muscles of palate, pharynx, and larynx Autonomic: thoracic and abdominal viscera	Gag reflex; ability to swallow; Say, "Ahhh"
XI. Accessory	—	Voluntary motor: sternocleidomastoid and trapezius muscles	Resisted shoulder shrug
XII. Hypoglossal	—	Voluntary motor: muscles of tongue	Tongue protrusion (if injured deviates toward injured side)

▶ **KEY POINT**

Nerve fibers are generally categorized according to function: sensory, motor, or mixed (motor and sensory).

Motor nerves carry efferents to muscles and return sensation from muscles, ligaments, and associated tissues. Nerves that innervate muscles also mediate the sensation from the joint on which those muscles act. The law of parsimony, stating that the nervous system activates the fewest muscles or muscle fibers possible to control a joint, exists with muscle recruitment for motor function.

A mixed nerve is the combination of skin, sensory, and motor fibers to one trunk. These nerves have the characteristics of each of these types of nerves. Therefore, they would provide the functions of each of these nerve types.

Irrespective of the nerve fiber type, three layers of tissue enclose the peripheral nerves. From outermost to innermost layer they include the epineurium, perineurium, and endoneurium (Fawcett 1984). The attachment of the epineurium with the surrounding connective tissue is loose so that the nerve trunks are relatively mobile, except where it is tethered by entering vessels or exiting nerve branches (Thomas & Olsson 1984).

The four functional parts of the neuron include the following (see figure 3.3):

- **Dendrites**—receive information from other nerve cells or the environment.
- Axon—conducts information to other nerve cells. Axons are often covered by myelin. Along the axon are segments that lack myelin, called nodes of Ranvier. Myelin is a lipid-rich membrane that has a high electrical resistance and increases the nerve conduction velocity of neural transmissions via the process of salutatory conduction. The axon extends from the cell body to the muscle, where it divides into either a few or multiple smaller branches.
- **Cell body**—contains the nucleus of the cell and performs integrative functions. It is located in the anterior horn of the spinal cord.
- **Axon terminal**—the transmission site for action potentials.

Communication among nerve cells occurs at the synapses. It is here that a chemical is released in the form of a neurotransmitter.

Each muscle contains many motor units, each containing a single motor neuron and its composite muscle fibers. Motor unit size is determined by the number of muscle fibers it contains and the size of the motor nerve axon. Fiber number varies from two or three to a few thousand. Muscles performing fine motor control have small-sized motor units. These motor units typically have small cell bodies and small-diameter axons. Muscles that are used to produce large forces and large movements typically have a predominance of large-sized motor units, large cell bodies, and large-diameter axons.

Motor Unit Recruitment

The size principle of motor unit recruitment states that motor units with small cell bodies and few motor fibers are recruited first by the nervous system, and then, as force is increased, larger motor units are recruited (Henneman et al. 1965; Linnamo et al. 2003). Small motor units produce less tension than large motor units and require less energy expenditure, thereby conserving energy. Recruitment strategy is based not only on energy efficiency but also on previous experience, the anticipated magnitude of the required force, and type of muscle action (Linnamo et al. 2003; Howell et al. 1995).

Autonomic Nervous System

The **autonomic nervous system** is the portion of the PNS responsible for innervation of smooth muscle, cardiac muscle, and glands of the body. It primarily functions without conscious control of the person. The two primary components of the autonomic system include the sympathetic and parasympathetic divisions. In general, these systems function in opposition to each other. The sympathetic division typically functions in actions requiring quick responses and is often referred to as "fight or flight." The sympathetic nervous system is generally located in the thoracolumbar region of the spine and has norepinephrine as its principal neurotransmitter (except in sweat glands). The parasympathetic division functions in actions that do not require an immediate response. Some of the primary functions of this division include salivation, lacrimation, urination, food digestion, and defecation. This system is located primarily in the craniosacral region of the spine and has acetylcholine as its principal neurotransmitter.

Neuromuscular Control

Control of coordinated movements and stability throughout the body requires a properly functioning nervous system. The CNS functions in an integrated fashion; in other words, it activates muscles in an integrated fashion to collectively work together to produce the desired motion. The amount of practice required to perform movements in a properly coordinated fashion should be enough to make changes in the cortical activity of the CNS (Vliet & Heneghan 2006). The repetition number required to achieve this is quite variable (Vliet & Heneghan 2006).

Kinesthetic Sense and Proprioception

Proprioception is the sensory abilities required for proper neuromuscular control. Proprioception involves the integration of sensory input concerning static joint position sense; joint movement (**kinesthesia**); velocity of movement; and force of muscular contraction from the skin, muscles, and joints (McCloskey 1978; Borsa et al. 1994).

All synovial joints in the body have mechanoreceptors and nociceptors in articular, muscular, and cutaneous structures. **Mechanoreceptors** are stimulated by mechanical forces (e.g., stretching, relaxation, compression) and mediate proprioception. These structures include Pacinian corpuscles, Ruffini endings, muscle spindles, and Golgi tendon organ–like endings (Chusid 1985; Freeman & Wyke 1967; Wyke 1981). Table 3.3 provides more discussion on these mechanoreceptors. Clinical Correlation 3.2 provides factors that contribute to the destruction of mechanoreceptors.

Mechanoreceptors appear to protect the joint from injury in three primary ways:

1. They prevent movement of the joint in the pathological range. Extremes of joint motion activate the mechanoreceptors of the ligaments, initiating a spinal reflex to contract the antagonistic muscles through a ligamentomuscular reflex (Gardner 1950; Palmer 1938).

TABLE 3.3 Types of Mechanoreceptors

Type	Location	Function
Type I: small Ruffini endings	Joint capsule and ligaments	Contribute to reflex regulation of postural tone, muscle coordination, and perceptional awareness of joint position An increase in joint capsule tension, via range of motion, posture, mobilization, or manipulation, increases their frequency of firing (Wyke 1981, 1985)
Type II: Pacinian corpuscles	Adipose tissue, cruciate and other ligaments, annulus fibrosus, and fibrous capsule	Function primarily in sensing joint motion Regulate motor unit activity of prime movers of the joint Entirely inactive in immobile joints Discharge during active or passive motion of a joint or with applied traction
Type III: large Ruffini endings	Intrinsic and extrinsic joint ligaments, superficial layers of the capsule, but not the anterior or posterior longitudinal ligaments	Detect large amounts of tension Become active only in the extremes of motion or when strong manual techniques are applied to a joint
Type IV: Nociceptors	Free, noncapsulated nerve endings that form a network of unmyelinated nerve fibers (Milne et al. 1981; Vierck et al. 1990)	Inactive in normal circumstances but become active with significant mechanical deformation or tension May also become active in response to direct mechanical or chemical irritation

CLINICAL CORRELATION 3.2

Factors often associated with mechanoreceptor death include the following:

- Immobilization
- Trauma
- Age
- Spondylosis
- Infection

It is believed that a degree of mechanoreceptor regeneration is possible. High-volume repetitions, rhythmic stabilization, and self-mobilization will activate mechanoreceptors without putting undue stress on the damaged tissue. Therefore, the clinician can prescribe or perform these types of activities with minimal stress on the patient to assist with mechanoreceptor regeneration. An example of such a case might be the acute postoperative knee arthroscopy patient. This type of patient is likely to have a large amount of swelling in and around the knee joint, limiting range of motion. As long as it is not contraindicated, the clinician could have the patient lying supine, with both legs on an exercise ball, and perform (with the noninvolved lower extremity) low-level repetitive knee flexion and extension in the tolerable range. This activity not only stimulates and potentially regenerates mechanoreceptors but also can move fluid in the knee joint. Movement of this fluid can help with discomfort, range of motion, and healing.

2. They help in balancing the activity between synergistic and antagonistic muscle forces.
3. They appear to generate an image of body position and movement within the central nervous system.

Muscle Spindle

Muscle spindles are fusiform in shape and widely scattered in the fleshy bellies of skeletal muscles. Each spindle consists of 2 to 10 slender striated muscle fibers, enclosed within a thin connective tissue capsule and attached at both ends to the epimysium or to ordinary striated muscle. These slender muscle fibers, innervated by gamma fibers (3 to 7 μm) are known as intrafusal fibers, and they are tiny compared with the extrafusal fibers that produce contractile tension within a muscle (Carpenter 1977). Muscle spindles are aligned parallel to the extrafusal fibers.

Smaller intrafusal fibers are known as nuclear chain fibers, while the larger fibers are designated nuclear bag fibers (Barker &

Cope 1962). The ends of the nuclear chain fibers are attached to the polar parts of the longer nuclear bag fibers (figure 3.4).

The neuromuscular spindle is arranged in parallel to the extrafusal, or contractile, fibers of the muscle, unlike Golgi tendon organs, which are oriented in series. Therefore, when the tension on the spindle is relaxed, the afferent input from the annulospiral endings ceases, and the muscle relaxes.

The muscle spindle has both sensory and motor components. The purpose of the muscle spindle is to compare the length of the spindle with the length of the muscle that surrounds the spindle.

When a muscle is stretched, the primary sensory fibers of the muscle spindle (type Ia afferent neurons) respond to changes in both muscle length and velocity by transmitting this activity to the spinal cord in the form of changes in the rate of action potentials. Additionally, secondary sensory fibers (type II afferent neurons) respond to muscle length changes and transmit this signal to the spinal cord. The Ia afferent signals are transmitted

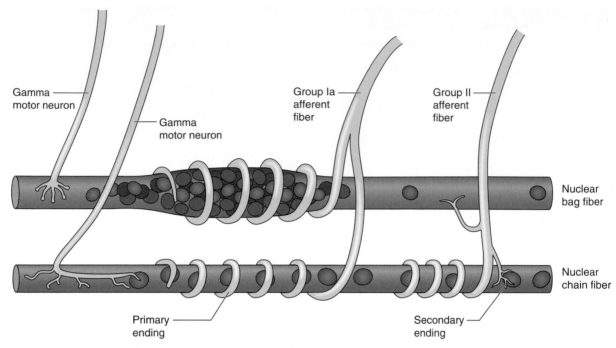

Gamma motor neuron

Gamma motor neuron

Group Ia afferent fiber

Group II afferent fiber

Nuclear bag fiber

Nuclear chain fiber

Primary ending

Secondary ending

▶ **FIGURE 3.4**   Muscle spindle of skeletal muscle.

monosynaptically to many alpha motor neurons of the muscle. The activity of the alpha motor neurons is then transmitted via the efferent axons to the extrafusal muscle fiber, which generates force and thereby resists the stretch. The Ia afferent signal is also transmitted polysynaptically through interneurons that inhibit alpha motor neurons of antagonistic muscles, causing them to relax.

Golgi Tendon Organ

Golgi tendon organs (GTOs) are made up of strands of collagen that are connected at one end to the muscle fiber and at the other end to the tendon proper. Each GTO is innervated by a single afferent type Ib sensory fiber that branches and terminates as spiral endings around the collagen strands.

When the muscle generates force, the sensory terminals are compressed. This stretching deforms the terminals of the Ib afferent axon. As a result, the Ib axon is depolarized and fires nerve impulses that are disseminated to the spinal cord. The action potential frequency signals the force being developed by the 10 to 20 motor units within the muscle. This is representative of whole-muscle force (Prochazka & Gorassini 1998).

The Ib sensory feedback generates spinal reflexes and supraspinal responses that control muscle contraction. Ib afferents synapse with interneurons within the spinal cord that also project to the cerebellum and cerebral cortex. One of the main spinal reflexes associated with Ib afferent activity is the autogenic inhibition reflex, which helps regulate the force of muscle contractions.

Conclusion

The neurophysiology of skeletal muscle contraction is quite complex and involves interaction of chemical, electrical, and mechanical stimuli. Other various factors, such as muscle fiber type, also affect the extent of contraction in skeletal muscle. Nervous system control

of this complex interaction involves both the central and peripheral nervous system. Injury or inefficiency in any of these variables can have significant effect on skeletal muscle contraction.

REVIEW QUESTIONS

1. Which of the following is *not* a commonly accepted type of muscle fiber?
   a. Type I
   b. Type IIa
   c. Type IIb
   d. Type IIc

2. Which of the following detects sense of joint movement?
   a. mechanoreception
   b. kinesthesia
   c. proprioception
   d. touch

3. Which of the following receptor organs inhibits alpha motor neurons, causing the contracting muscle to relax?
   a. Golgi tendon organ
   b. muscle spindle
   c. alpha motor neuron
   d. mitochondria

4. Which contractile protein is the thickest?
   a. actin
   b. myosin
   c. tripson
   d. byosin

5. Which of the following membranes is the deepest, providing a blood supply to the spinal cord?
   a. arachnoid mater
   b. dura mater
   c. pia mater
   d. dental mater

6. Which of the following muscle fiber types is the most fatigable?
   a. Type I
   b. Type IIa
   c. Type IIb
   d. Type IIc

Muscle Performance and Function

This chapter examines the contractile structures of muscle. While chapter 3 discusses muscle architecture basics, sliding filament theory, neurophysiological mechanisms of muscle contractions, and so on, this chapter more specifically details muscle performance and function. Topics included in this chapter are morphology, force generation capacities, contractile properties, and various methods to increase or decrease strength and endurance. The muscle's roles as a prime mover, in reverse action, and in stabilization are also thoroughly discussed. The effects of injury, immobilization, and aging on muscles are also included in this chapter.

Skeletal Muscle Properties

The major properties of skeletal muscle are extensibility, elasticity, irritability, and the ability to develop tension. This section discusses how all these properties contribute to skeletal muscle function.

- **Extensibility** is the ability to be stretched or to increase in length. It might also be described as a muscle's ability to lengthen

OBJECTIVES

After reading this chapter, you should be able to do the following:

> Differentiate the various properties of skeletal muscle.

> Differentiate the characteristics of concentric versus eccentric isotonic muscle contractions.

> List the different types of muscle contractions.

> List some of the major advantages and disadvantages of each type of muscle contraction.

> Describe the major functions of skeletal muscle.

> Describe the major characteristics of each component of muscle performance.

> Describe the major advantages and disadvantages of the various clinical measures of muscle performance.

> List the major effects of injury, immobilization, and aging on skeletal muscle.

or stretch beyond resting length. Extensibility is determined by the connective tissue found in the perimysium, epimysium, and fascia surrounding muscles.

• **Elasticity** is the ability of a tissue to return to its normal resting length after removal of a stretch. Elasticity is determined by the connective tissue about and within the muscle rather than the muscle fibers themselves. Protective mechanisms in the muscle maintain its integrity and basic length. Therefore, a muscle will always return to its original length unless stretched too far, in which case it will tear. Figure 4.1 depicts the stress–strain curve of normal skeletal muscle. The elastic zone of a muscle is the region that demonstrates elasticity (returns to normal length when the stretch is removed). The plastic zone is where the muscle is taken beyond its ability to elastically recover, and permanent plastic deformation of the tissue occurs. Just beyond this zone is the point of tissue failure, where the muscle tears (rupture point).

• **Irritability** is the ability to respond to stimulation from the nervous system. This stimulation is provided by a chemical neurotransmitter. Only nerve tissue is more sensitive to stimulation than muscle tissue.

• The tension that develops in a muscle can be the result of passive (stretch) or active (contraction) tension forces. When an activated muscle develops tension, the amount of tension in the muscle is constant throughout the length of the muscle, in the tendons, and at sites of the musculotendinous attachments to bone (Hall 1999). Tensile force pulls on the attached bones and creates torque at the joints crossed by the muscle.

▶ **KEY POINT**
The major properties of skeletal muscle are extensibility, elasticity, irritability, and the ability to develop tension.

Muscle Contractions

The major types of muscle contractions are isometric, isotonic, and isokinetic. This section describes the different types of muscle contractions and their characteristics, as well as how skeletal muscle contractions differ. Advantages and disadvantages of the different types of contractions are identified so that readers can appreciate the specific characteristics of each contraction type.

▶ **KEY POINT**
The major types of muscle contractions are isometric, isotonic, and isokinetic.

Isometric

Isometric exercises provide a static contraction with a variable and accommodating resistance without producing any appreciable change in muscle length (Luttgens & Hamilton 1997). An example of an isometric contraction is pushing into an immovable wall with your right arm, elbow bent to 90°, wrist in neutral, making a fist, and your fist against the wall. The muscles in your right arm (e.g., anterior deltoid) contract to attempt to produce shoulder flexion, but because of the immovable force (the wall), no appreciable movement is accomplished. Isometric exercises are advantageous when joint movement is restricted, whether it be from pain, casting from a fracture, postsurgical bracing, or another reason. Holding a contraction for 6 seconds at 75% of

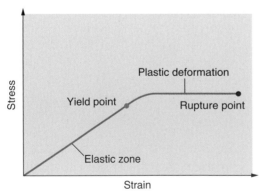

▶ **FIGURE 4.1** Stress–strain curve of skeletal muscle.

maximal resistance has proven sufficient to increase strength when performed repetitively (Hettinger 1964).

Isotonic

Isotonic exercise provides a contraction in which movement is performed and therefore change in muscle length occurs. There are two types of isotonic contractions: concentric and eccentric.

• Concentric—this term describes a shortening muscle contraction. This shortening contraction occurs when the tension generated by the agonist muscle is sufficient to overcome an external resistance and to move the body segment of one attachment toward the segment of its other attachment (Luttgens & Hamilton 1997). The lifting component of a biceps curl (where the elbow bends) is an example of a concentric contraction.

• Eccentric—this term describes a lengthening muscle contraction. Eccentric contractions are capable of producing greater forces than either isometric or concentric contractions (Astrand & Rodahl 1986; Komi 1992; McCardle et al. 1991). These contractions are also involved in activities that require a deceleration to occur. The lowering of the arm and weight during a biceps curl (elbow straightening) is an example of an eccentric contraction.

▶ **KEY POINT**
There are two types of isotonic contractions: concentric and eccentric.

Eccentric muscle contractions differ from concentric contractions in several important ways. Per contractile unit, more tension can be generated eccentrically than concentrically and at a lower metabolic cost (Abbott et al. 1952). Eccentric contractions are the most efficient form, can develop the greatest tension of the various types of muscle actions, and are an important component of functional movement patterns of the body.

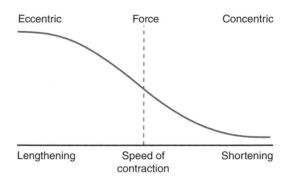

▶ **FIGURE 4.2**   Force–velocity curve for eccentric and concentric isotonic contractions.

During an eccentric contraction, it is generally believed that as the velocity of active muscle lengthening increases, force production in the muscle increases but then quickly levels off (figure 4.2) (Levangie & Norkin 1978; Smith et al. 1996). It is hypothesized that this may be an important aspect for shock absorption or rapid deceleration of a limb during quick changes of direction (Levangie & Norkin 1978). Essentially, eccentric muscle actions produce greater force, with the advantage of requiring less energy per unit of muscle force (Bonde-Peterson et al. 1972; Eloranta & Komi 1980; Komi et al. 1987).

▶ **KEY POINT**
Concentric contractions involve muscle shortening, and eccentric contractions involve muscle lengthening.

Isokinetic

Isokinetic is a term referring to a concentric or eccentric muscle contraction in which a constant velocity is maintained throughout the muscle action. An isokinetic contraction occurs when a muscle is maximally contracting at the same speed throughout the whole range of its related lever (Luttgens & Hamilton 1997). This type of exercise requires specialized equipment that provides an accommodating resistance. The most frequently exercised muscle groups on isokinetic machines are knee flexors and extensors and the shoulder musculature. In most people, repetition is required to adjust to this accommodating resistance since it differs from other forms of resistance. Isokinetic machines are most frequently used for extremity joint strengthening and assessment. The shoulder and knee joints in particular are examples of such extremity joints. Isokinetic knee flexion and extension is frequently used postoperatively as an assessment of knee strength, as is postoperative rotator cuff strength assessment for the shoulder.

Muscle Functions

Muscles serve several important functions. These specific functions are dependent on various factors, including their location, muscle fiber type, and how they interact with other muscles. The primary muscle functions are producing movement, maintaining posture and position, stabilizing the joints, supporting and protecting visceral organs, producing heat to maintain body temperature, and controlling entrance and exits to the body.

Producing Movement

One of the primary functions of muscle is to produce movement of the body and its component parts. Muscle actions generate tension that is then transferred to bones to create movement. As described in chapter 3, a muscle contracts via the sliding filament theory. As the muscle changes its length (either shortening for a concentric contraction or lengthening for an eccentric contraction), tension is developed in the muscle. Since a tendon connects a muscle to bone, the tension that is developed from a muscle contraction is then transmitted via the tendon to the bone. Since the muscle is then directly acting on the bone, it produces movement of that bone and corresponding joint (if applicable).

Maintaining Posture and Positions

Postural muscles are used to maintain upright postures and extremity positions. Type I, slow

twitch oxidative muscle fibers are predisposed to sustained, long-duration postural-type contractions that are required to maintain posture and specific positions. These relatively strong muscles are designed to counter gravitational forces and provide a stable base for other muscles to work from. These muscles are typically poorly recruited in postural dysfunctions (see chapters 6 and 9).

Stabilizing Joints

Muscle tension applied across joints via tendons provides stability where the tendons cross a joint. Muscle cocontraction of agonists and antagonists provides a stabilizing force across the joint. As just described, muscle contraction creates tension and movement on a bone and at a joint, if applicable. When agonistic and antagonistic muscles are contracted simultaneously, these muscles produce tension and therefore contractions that are most likely antagonistic to each other. Therefore, these contractions, working antagonistic to each other, would minimize joint movement. The tension that is then developed, with minimal to no movement, provides a stabilizing force acting across that joint.

Other Functions

Other important functions of muscles include supporting and protecting visceral organs, producing heat to maintain body temperature, and controlling entrance and exits to the body. Areas of the body where muscles control entrances and exits include the eyes, mouth, and anus. Multiple muscles on the anterior, posterior, and lateral thorax and trunk provide vital protection and support of the internal organs in these areas. Shivering is a mechanism that skeletal muscles use to increase and maintain the body's temperature in cold environments.

Role of Muscles

Multiple muscle actions are required to produce movement of the body. The muscle most directly involved in performing the necessary

Advantages and Disadvantages of Different Types of Muscular Contractions

ISOMETRIC

Advantages

- Able to be used early in rehabilitation since there is no joint movement
- Strengthening is joint-angle specific
- No special equipment needed
- Helps decrease swelling
- Short period of training time
- A 20° strengthening overflow throughout ROM

Disadvantages

- Strengthening limited to specific joint angles
- No eccentric work
- Blood pressure concerns with Valsalva maneuver
- Less proprioceptive and kinesthetic training
- No muscle endurance training

ISOTONIC

Advantages

- Concentric and eccentric muscle action
- Can improve muscular endurance
- Multiplanar and multifunctional training with free weights
- Can use body weight for resistance
- Can exercise through full ROM
- Inexpensive and readily available with most types of resistive devices
- Can use manual resistance from rehabilitation specialist
- Various components of the training program can be manipulated to maintain workload (reps, sets, weight)

Disadvantages

- Maximally loads muscle at its weakest point in ROM, especially with elastic tubing
- Muscle is only maximally challenged at one point in ROM with free weights and some machines
- Not safe when patient has pain during movement
- Potential increased risk of injury at increased speeds of movement
- Difficult to perform at fast functional velocities
- Does not provide reciprocal concentric exercise
- Does not allow for rapid force development
- Unable to spread workload evenly over the entire ROM

> *continued*

> Advantages and Disadvantages of Different Types of Muscular Contractions   *(continued)*

ISOKINETIC

Advantages

- Concentric and eccentric strengthening of same muscle group reciprocally or repeatedly
- Reliable measures with equipment
- Wide range of exercise velocities
- Computer-based visual and auditory cues for feedback
- Provides maximum resistance throughout ROM
- Safe to perform high- and low-velocity training
- Accommodation for painful arc of motion
- Decreased joint compressive forces at high speed
- Physiological overflow
- Isolated muscle strengthening
- External stabilization

Disadvantages

- Cannot produce angular velocities of many physical activities
- Large and expensive equipment
- Requires assistance and time for setup
- Cannot use as a home program
- Most units provide only open kinetic chain movement
- Availability of equipment
- Cannot duplicate reciprocal speeds of movement used during most daily and functional activities
- Inconvenience of adjustment of equipment for various joints and ease of various setups
- Some artificial parameters until the limb actually moves at the velocity of the dynamometer on the machine

Based on Fleck and Kraemer 2004; Davies 1992.

movement is termed the prime mover, or agonist. The muscle that can stop this movement, or at least slow it down, is called the antagonist. The agonist and antagonist muscles typically perform opposite motions at a specified joint. The antagonist muscle of a specific movement assists in stabilizing the joint and slowing down the movement performed by the agonist. This is typically a protective rather than a deleterious function.

Synergist muscles assist indirectly in a movement. Synergists are typically required to control body motion when the agonist is a muscle that crosses two joints. These muscles typically provide a stabilizing force that prevents unwanted movements when the person is performing a specific activity. Clinical Correlation 4.1 looks more closely at the interplay of agonist, antagonist, and synergist muscle coordination.

CLINICAL CORRELATION 4.1

Agonist, antagonist, and synergist muscle coordination is necessary in order to perform efficient movement. When a person is rising up from a squat in a low position, hip and knee extension are necessary movement patterns for success. The rectus femoris muscle is a two-joint muscle that crosses anterior to the hip (thereby flexing the hip) and anterior to the knee (thereby extending the knee). Activation of this muscle will elicit both wanted (knee extension) and unwanted (hip flexion) movements. Therefore, muscles that are antagonists to the hip flexion movement of the rectus femoris (gluteus maximus) must act synergistically to counteract the hip flexion movement. The gluteus maximus is therefore preventing unwanted movement (hip flexion) as well as providing the necessary hip extension. This complex interaction of agonist and antagonist muscles accomplishes the task of rising from a squat in a low position.

Muscle Flexibility and Range of Motion

Flexibility can be defined as the ability to move joints fluidly through complete ranges of motion without injury (Heyward 2010). Although it is probably not regarded as a major parameter for assessing human performance, muscle flexibility is integral to dynamic human movement and pain-free function. Prolonged low-load stretching in animals has been shown to increase muscle length and hypertrophy (Goldspink et al. 1995; Lederman 1997) as well as permanently lengthen connective tissue (Sapega et al. 1981). Similar results have been achieved in human subjects with osteoarthritic hips (Leivseth et al. 1989) and joint contractures (Wessling et al. 1987). Evidence of similar effects in healthy shortened muscles has yet to be determined (Weldon & Hill 2003).

Optimal body alignment and proper agonistic–antagonistic muscle relationships have been advocated by many experts (Janda 1994; Sahrmann 2002; Kendall et al. 2005). Functionally, shortened muscles are capable of reciprocally inhibiting their antagonistic counterparts (Janda 1994; Sahrmann 2002; Kendall et al. 2005). This can lead to increased potential for future injury and less than optimal function. The terms *upper crossed syndrome* and *lower crossed syndrome* (Janda 1994) have been used to describe such imbalances (see Clinical Correlation 4.2).

Range of motion (ROM) of a joint can involve muscle flexibility, arthrokinematic and osteokinematic motion of the joint, and neurodynamic mobility of the nervous system. A restriction in joint ROM requires careful assessment of each of these components to determine the limiting factor (see Clinical Correlation 4.3). Range of motion therefore is not the same as flexibility. Flexibility is only a measure of soft-tissue (primarily muscle) excursion of movement. The definition by Heyward (2010) given earlier seems to imply not only muscle and soft-tissue extensibility but also the potential for involvement of joint arthrokinematic and osteokinematic movement. Osteokinematic movement of one bone on a relatively fixed articulating surface and arthrokinematic movement (e.g., rolling, gliding, spinning) of a joint surface are obviously not the same as soft-tissue extensibility. All potential limitations of ROM of a movement should involve assessment of flexibility, arthrokinematics, osteokinematics, and neural tension to most accurately determine appropriate treatment.

Muscle Performance

A definition of *muscle performance* might include the terms *strength*, *power*, and *endurance*. Performance by a muscle, depending on its called-upon action or requirements, may include one or all three of these components.

CLINICAL CORRELATION 4.2

Specific muscles have a predisposition for tightness (e.g., pectoralis major and minor, upper trapezius, suboccipital muscles, hip adductors and flexors, erector spinae) and thereby inhibit their antagonistic muscle groups (e.g., lower trapezius, gluteal muscles, deep neck flexors, rectus abdominis). Clients with this condition will display typical postures (Janda 1994) (figure 4.3) as a result of this muscle imbalance between antagonistic muscle groups. The clinician should assess for these muscle imbalances by testing not only muscle strength but also muscle length. Treatment involves stretching the shortened muscle groups followed by strengthening their antagonists. The antagonist muscles are typically weak, especially in their normal to shortened ranges. Strengthening the muscles specifically in these shortened ranges will assist in maintaining the increased length (of the tightened muscles) gained by stretching.

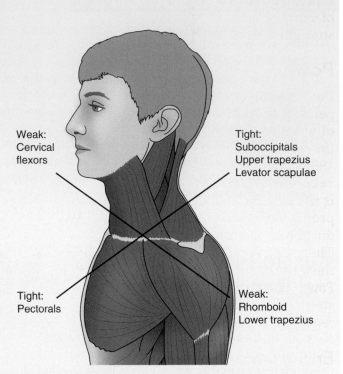

Weak: Cervical flexors

Tight: Suboccipitals Upper trapezius Levator scapulae

Tight: Pectorals

Weak: Rhomboid Lower trapezius

▶ **FIGURE 4.3**   Upper crossed syndrome. Lower crossed syndrome is illustrated in figure 16.11.

CLINICAL CORRELATION 4.3

Determining restricted movement of hip flexion in a client requires moving the hip through flexion ROM with the knee straight as well as bent. Limitation in hip flexion ROM with the knee straight should increase significantly when the knee is bent because of a reduction in both hamstring tension and neural tension in the lower extremity. Neural tension testing with variations of straight-leg-raise testing can help rule in or out the potential of adverse neural dynamic movement. Hamstring flexibility testing with the hip and knee at 90° of flexion in the supine position would help rule in or out hamstring flexibility as a potential limiting factor. Restricted hip joint arthrokinematics can be assessed by measuring hip osteokinematic movement followed by joint-play assessment of this motion.

▶ **KEY POINT**

Muscle performance can include components of strength, power, and endurance.

Strength

Strength is the ability of the muscle to exert a maximal force or torque at a specified or determined velocity. Strength can be measured in terms of force, torque, or work (Hall &

Thein-Brody 1999). Functional strength relates to the ability of the neuromuscular system to produce, reduce, or control forces that are either contemplated or imposed during functional activities in a smooth, coordinated fashion (Fleck & Kraemer 2004).

The effect of strength on function depends on both absolute and relative strength. *Absolute strength* is the most force a muscle can generate, or the maximum amount of weight a person can lift once (1 RM) irrespective of body weight. *Relative strength* is absolute strength divided by the person's body weight.

Power

Power is the rate of work, or amount of work per unit time. Muscular power is the amount of work produced by a muscle per unit time (force × distance / time) (Hall & Thein-Brody 1999; Fleck & Kraemer 2004). Therefore, since velocity is equal to distance over time, power is equal to force × velocity. The rate of force production is an essential element in the production of power (see Clinical Correlation 4.4). The greatest power is produced by exerting the most force in the shortest amount of time. Power can be improved by either increasing strength or by reducing the amount of time required to produce force.

Endurance

Endurance refers to the ability to perform low-intensity, repetitive, or sustained activities over a prolonged period of time without fatigue (McCardle et al. 1991; Powers & Howley 2001). Endurance can be further broken down into local muscle endurance and general endurance. General endurance is often referred to as cardiorespiratory endurance. Local muscle endurance, however, refers to the ability of a muscle to contract repeatedly against a load (resistance), generate and sustain tension, and resist fatigue over an extended period of time (Powers & Howley 2001). Activities requiring cardiorespiratory endurance also require muscular endurance, although tasks requiring muscular endurance do not always require cardiorespiratory endurance.

Factors Affecting Muscle Performance

Muscle functions and contraction types, as previously described, are vital to muscle performance. Muscle performance (whether it be strength, power, endurance, or a combination) is affected by specific factors that need to be manipulated when prescribing exercise to a patient. These factors include neural control and adaptation, muscle fiber arrangement, muscle length, joint angle, muscle fiber type, muscle fiber diameter, the force–velocity relationship, the length–tension relationship of the muscle, and training specificity.

Neural Control and Adaptation

Muscle strength generally increases when more motor units are involved in the contraction, the motor units are greater in size, or the rate of firing is faster. Much of the improvement in strength evidenced in the first few weeks of resistance training is attributable to neural adaptations (Morris et al. 1961). Thus, the strength gains a person encounters when initiating a weight-training program are largely attributable to improved efficiency of movement, muscle coactivation, and other neural adaptations versus muscle hypertrophy.

CLINICAL CORRELATION 4.4

It is necessary to train with heavy loads (85 to 100%) and light loads (30%) at high speeds to develop optimal levels of speed strength (starting, explosive, and reactive). Even though lifting heavy loads (85 to 100% of 1 RM) looks slow, it is done as fast as possible in order to recruit and synchronize as many motor units as possible.

Training with relatively light loads of approximately 30% of maximum performed at high speeds is superior to plyometric training and traditional weight training (80 to 90% of 1 RM) in developing dynamic athletic performance (Wilson et al. 1993). Therefore, once a strength foundation is established, the primary emphasis for improving power should be increased speed of movement.

Muscle Fiber Arrangement

The angle of pennation is the angle created between the fiber direction and the line of pull. Fusiform and longitudinal muscle fibers, where the muscle fibers lie parallel to the long axis of the muscle, have no angle of pennation, while other arrangements of muscle fibers demonstrate some degree of pennation. Although the maximum tension can be improved with pennation because it allows for a higher volume of muscle fibers per area, the range of shortening of the muscle is reduced. Muscle fibers can contract to about 60% of their resting length, and the force is in the same direction as the muscle fiber. If muscle fibers attach parallel to the tendon (e.g., in fusiform muscles), the angle of pennation is 0°, and all the force produced by the muscle would be transmitted to the tendon. This force would then be transmitted across the involved joint. Pennation angles greater than 0° would therefore not transmit the entire muscular force across the tendon and corresponding joint.

Most human muscles have pennation angles that range from 0 to 30° (Lieber & Friden 2001). Only a portion of the force of pennate muscles goes toward producing motion of a bony lever. The more oblique an angle of pennation, the less force the muscle fibers are able to exert. Pennation of the muscle fibers appears to enhance force during high-speed concentric muscle action, particularly at ROM extremes, but may reduce force capability for eccentric, isometric, or low-speed muscle actions (Scott & Winter 1991). Therefore, all the muscles with pennation (as shown in figure 4.4) will have different advantages and disadvantages than fibers without pennation.

This potential decrease in force is offset because pennate muscles usually have larger physiological cross-sectional area (PCSA). Physiological cross-sectional area is the sum total cross-sectional area of all muscle fibers within the muscle. Assuming full activation, the maximal force potential of a muscle is proportional to the sum of the cross-sectional area of all its fibers. For PCSA to be measured accurately, the cross section must be made perpendicular to each of the muscle fibers. Therefore, pennate muscles are able to exert greater force due to their higher volume of muscle fibers.

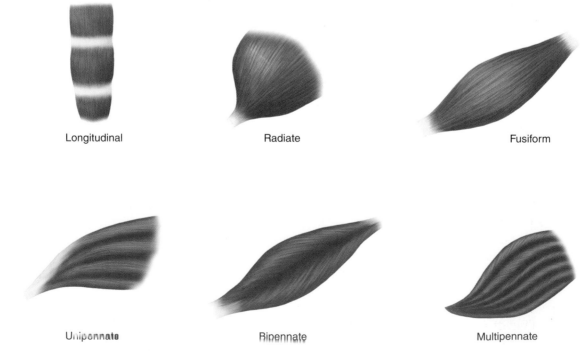

Longitudinal   Radiate   Fusiform

Unipennate   Bipennate   Multipennate

▶ **FIGURE 4.4** Types of fiber pennation.

Muscle Length

A muscle can generate the most force at about resting length and less force when in an elongated or a shortened state (refer to the discussion on active and passive insufficiency later under Length–Tension Relationship). The shape of the total muscle length–tension curve can vary considerably, however, between muscles of different structure and function (Baratta et al. 1993).

Exercise prescription should consider muscle length. Having a muscle contract at its midrange ideally will produce the greatest muscle tension. A muscle contracting at its shortened or lengthened position will produce less force. The clinician must consider this information when prescribing exercise or assessing a patient's strength or function. The clinician also must consider how the other variables described here (joint angle, fiber type, length–tension relationship, and so forth) interact with muscle length when prescribing exercise or assessing muscle strength.

Joint Angle

Changes in strength throughout the joint ROM affect force capability. The amount of torque exerted about a given joint varies throughout the joint's ROM because of the relationship between force and muscle length, as well as the geometric arrangement of muscles, tendons, and joint structures. The joint angle is likely related to the muscle length, length of the lever arm acting on a muscle, and so forth. All these variables require careful consideration on the part of the clinician when planning a treatment or assessment.

Fiber Type

The different muscle fiber types (Type I, Type IIa, and Type IIb), along with their specific primary characteristics, are detailed in chapter 3. Generally, Type I fibers are predisposed for endurance or duration activities, while Type IIb are predisposed for explosive, anaerobic activities. Therefore, exercise prescription should account for the intensity and volume of the exercise prescribed by the clinician. For example, an exercise program for a muscle group that is predominantly Type I fibers should have a much lower intensity and a higher volume than an exercise prescription for a muscle group that is predominantly Type IIb.

Fiber Diameter

The amount of force developed in a maximal static action is independent of the fiber type but is related to the fiber's cross-sectional diameter. Since Type I fibers tend to have smaller diameters than Type II fibers, a high percentage of Type I fibers is believed to be associated with a smaller muscle diameter and therefore lower force development capabilities (Billeter & Hoppeler 1992). All else being equal, the force a muscle can exert is related to its cross-sectional area rather than its volume (Ikai & Fukunaga 1968).

Force–Velocity Relationship

The rate of muscle change, whether it be lengthening or shortening, substantially affects the force a muscle can develop when it contracts. This relationship can be depicted on the force–velocity curve (see figure 4.2). As the speed of a muscle's shortening increases, the force it is capable of producing decreases (Astrand & Rodahl 1986; McCardle et al. 1991). A slower rate of contraction is theorized to produce greater forces by increasing the number of cross-bridges formed. During a lengthening (or eccentric) muscle contraction, the force production differs from a shortening (or concentric) contraction. Rapid eccentric contractions generate more force than do slower eccentric contractions. During slow eccentric muscle contractions, the work produced approximates that of isometric contractions (Astrand & Rodahl 1986; McCardle et al. 1991).

Length–Tension Relationship

A muscle's force production capability is related to its length. The relationship between the length of a muscle and its force capability is referred to as the length–tension curve (figure 4.5). Maximum force is produced near a muscle's normal resting length. The distance through which a muscle can lengthen and shorten is determined by the number of sarcomeres in series. Sarcomere number is not fixed, and in adult muscle, this number can increase or decrease (Tabary et al. 1972).

In a shortened muscle, the overlap of actin and myosin reduces the number of sites available for cross-bridge formation. Active insufficiency of a muscle occurs when a muscle is incapable of shortening to the extent required

to produce full range of motion at all joints crossed simultaneously (Duedsinger 1984; Brownstein et al. 1988).

In a lengthened muscle, the actin filaments are pulled away from the myosin heads so they cannot create as many cross-bridges. Passive insufficiency of a muscle occurs when a two-joint muscle cannot stretch to the extent required for full range of motion in the opposite direction at all joints crossed (Duedsinger 1984; Brownstein et al. 1988).

Training Specificity

Specifically training a muscle or group of muscles to achieve the wanted goals (e.g., increased strength, power, endurance, stabilization) is of paramount importance. It is beyond the scope of this chapter to cover this issue in detail, and the reader is advised to consult other sources (Fleck & Kraemer 2004; Siff & Verkhoshansky 1999; Baechle & Earle 2008; Reiman 2006). General parameters are discussed here.

Training the different parameters of muscle performance (strength, power, and endurance) requires different program designs (table 4.1).

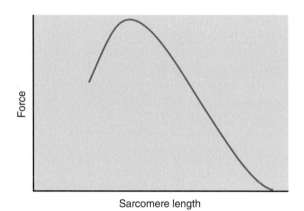

▶ **FIGURE 4.5** Length–tension curve.

TABLE 4.1 Comparison of Training Characteristics for Developing Muscle Performance

	Strength	Power	Strength and endurance	Endurance
Load (% of 1RM)	80-100%	Strength/force (70-100%) Velocity (30-45%) or up to 10% body weight	50-70%	Circuit training (40-60%)
Repetitions	Very low to low 1-6	1-5 (strength) 5-10 (power)	12-25	Moderate to high (15-30+)
Sets	3-5	4-6	2-3	2-5
Rest period	3-6 min	2 to 4-6 min	30-60 s	45-90 s (1:1 work-rest ratio)
Speed of performance	Slow to medium (speed of effort is as fast as possible)	Fast/explosive	Slow to medium	Medium
Primary energy source	Phosphagen Anaerobic glycolysis	Phosphagen	Anaerobic glycolysis/aerobic	Aerobic

Strength training typically involves a load of 80 to 100% of the maximum amount of weight a person can lift in one repetition (1RM), with approximately one to six repetitions. Power training, as previously mentioned, requires a foundation of strength. The primary component of power training after strength is established is termed velocity of movement. Therefore, since velocity is inversely proportional to the amount of load lifted, the load will have to be relatively lighter to accomplish the necessary velocity. Endurance training can involve many methods (e.g., circuit training) but the common theme is high repetitions with lighter loads. The relative work-to-rest ratio for endurance training is the lowest among the three parameters of muscle performance.

Clinical Measures of Muscle Performance

Measuring muscle performance can include both objective measurements (quantitative) and measurements of function or combinations of subjective and objective measures (some qualitative). Measurement of strength specifically can be accomplished by various means.

Quantitative Assessment of Muscle Performance

Quantitative measures of muscle performance include anthropometry and different methods of assessing the strength, power, and endurance capability of a muscle or muscle group. These include manual muscle testing, handheld dynamometry, and isotonic and isokinetic testing.

Anthropometry

Anthropometry is the science of measuring the size, weight, and proportions of the human body. Limb circumference has been used to approximate muscle size. Limb circumference is generally measured with a tape measure and is often assumed to correlate with muscle power and strength, with greater circumference and muscle bulk indicating greater strength. Although limb circumference and muscle strength decrease with muscle atrophy caused by disuse, several studies have found that the assumption that limb girth correlates with strength is misleading. Cooper and colleagues (1981) evaluated the relationship between thigh circumference and muscle strength and power as measured by isokinetic dynamometry and found no correlation between the torques produced at the knee by the knee extensors and flexors and thigh circumference measures at three levels. In addition, Hortobagyi and colleagues (1990) also found that individual differences in muscle strength correlate poorly with segmental girth measurements.

Manual Muscle Testing

Manual muscle testing (MMT) is used to test the strength of individual muscles or muscle groups. MMT is performed by applying manual resistance to a limb or body part. This resistance is typically applied at a point in the limb's ROM where the muscle being tested is most efficient. Muscle strength is graded on a numeric scale of 0 to 5. All the grades above 1 may be scored as the number alone or with a score of the number with a + or −. A break test should be used when testing the strength of muscles with a grade higher than fair (3/5) (Kendall et al. 2005). This is performed by applying pressure to the tested segment, in addition to gravity, to determine the maximal effort the subject can exert. The clinician gradually applies more pressure until the effort by the patient is overcome (Kendall et al. 2005). When a muscle can move a body part only against gravity, its strength is graded as 3/5. When testing weak muscles with less than 3/5 strength, a position that minimizes gravity and that generally involves moving the body part in the horizontal plane is used.

The MMT scale is easy to apply, but it is important to realize that these scores are relative. A score of 4 does not indicate that a muscle is twice as strong as one with a score of 2. Furthermore, validity of high scores may be limited by the strength of the clinician performing the test. MMT is the fastest and most efficient means of assessing muscle strength in the clinical setting and provides information not obtained by other procedures (Kendall et al. 2005).

Handheld Dynamometry

Handheld dynamometers may also be used to test the strength of individual muscles or muscle groups. This small device fits in the examiner's hand and is placed at precise locations on a subject's limb in an effort to assess the force generated by various muscles or groups of muscles. Handheld dynamometers are inexpensive, convenient, and lightweight; require minimal setup time and training; and can be used in a wide variety of settings (Wessel et al. 1999).

Because handheld dynamometers overcome some of the limitations of MMT, particularly the subjectivity and nonlinearity of grading muscle force production, and because they have good to very good reliability, they are popular and well accepted in clinical practice (Agre et al. 1987; Bohannon 1987a, 1988, 1993; Bohannon & Andrews 1989; Byl et al. 1988; Bohannon 1987b). However, as with MMT, consistent locations and patient positions must be used for accurate and reliable results.

Isotonic Testing

Isotonic testing involves lifting a fixed mass against gravity. This testing can be performed using weight machines or free weights. There are many isotonic strength testing protocols including the 1RM and the 10RM. A 10RM is the maximum amount of weight the subject can lift and lower 10 times. The 10RM test was first developed by DeLorme & Wilkins (1951). Isotonic testing of all types has been criticized because it is limited by the "sticking point," or the weakest point, in the ROM and therefore measures only the maximum strength at this point (Sale 1991; DeBries & Housh 1994). The more recent literature calls this type of strength testing dynamic testing because it involves both a concentric (muscle shortening) and eccentric (muscle lengthening) contraction (Escamilla & Wickham 2003).

Isokinetic Testing

Isokinetic strength testing measures force production during fixed-velocity movement with an accommodating resistance (Davies 1992; Davies et al. 1997). The tests are performed using an electrically powered device that maintains a chosen velocity of movement while maximizing the resistance throughout the ROM. Many isokinetic strength testing devices are available. All have components for testing movement of different joints, and some are able to assess open and closed kinetic chain movements. Isokinetic devices can be programmed for velocities from 1 to 500° per second. Refer to the box element earlier in the chapter discussing isokinetic strengthening for advantages and disadvantages of isokinetic training, as well as Clinical Correlation 4.5.

CLINICAL CORRELATION 4.5

Isokinetic testing can provide information about subtle changes in strength that may not be detectable by MMT (Wilk et al. 1992). For example, such testing revealed that some postsurgical knee arthroscopy patients had strength deficits of as much as 23 to 31% despite testing 5/5 for strength with MMT (Wilk et al. 1992). A limitation of isokinetic testing is that it does not isolate specific muscles but rather measures combined strength for moving in a single plane such as knee flexion or extension. Therefore, the clinician is measuring a single movement versus a single muscle. The movement will be limited by the weakest muscle producing that movement. If the clinician wants to assess only the biceps femoris (lateral hamstring), he would not be able to discern its strength simply through isokinetic knee flexion. With MMT, the clinician would lie the patient prone, flex the knee to 90°, and externally rotate the tibia to more accurately isolate the biceps femoris. But again, the limitations of MMT must be appreciated as well.

Qualitative Assessment of Muscle Performance

Qualitative measures of strength, power, and endurance primarily include assessment of the patient's quality of movement. Assessing movement quality in all three planes of movement requires monitoring for compensations in each plane. Monitoring for excessive lumbar lordosis with squatting (figure 4.6a) and medial collapse with single-leg squatting (figure 4.6b) are examples of qualitative measures that can be assessed. These examples can also be measured quantitatively with complex biomechanical analysis not common in the traditional physical therapy clinic. Other qualitative measures of muscle performance could include various subjective questionnaires that clients fill out regarding their functional capabilities.

Effects of Injury, Immobilization, and Aging on Muscle Performance

Muscle strain injuries usually involve muscle failure at the junction between the muscle and the tendon (Garrett et al. 1984; Garrett 1996). After this there is typically localized bleeding, swelling, redness, and pain as a result of the inflammatory process. Various grading scales are used to describe muscle strains. Grade I is often described as a minor strain with minimal muscle tearing, grade II is a moderate strain, and grade III is a complete tear of the muscle.

Immobilization can have various deleterious effects depending on the position of immobilization (shortened or lengthened), the proportion of muscle fiber types in the muscle (slow-twitch versus fast-twitch fibers),

▶ **FIGURE 4.6** *(a)* Demonstration of excessive lumbar lordosis with bilateral squatting. *(b)* Demonstration of medial lower extremity collapse with single-leg squat.

Photos courtesy of Michael Reiman.

and the length of immobilization. Immobilization in a shortened position results in several structural changes, including a thickening of the perimysium and endomysium (Williams & Goldspink 1984); an increase in the amount of collagen; a decrease in the number of sarcomeres (Witzman 1988); and loss of muscle weight and atrophy (Williams & Goldspink 1978; Witzman 1988; Booth 1986).

Immobilization in a lengthened position does not affect the muscle as negatively as immobilization in a shortened position. The primary changes to the muscle immobilized in a lengthened position include an increase in the number of sarcomeres, a decrease in their length, and muscle hypertrophy that may be followed by atrophy (Williams & Goldspink 1984; Williams 1990).

Aging decreases muscle strength as a result of changes in fiber type and motor unit distribution. There is a gradual decrease in the number and size of Type II muscle fibers particularly, leaving the muscle with a relative increase in Type I fibers (Lexell et al. 1988). There is also a decrease in the number of motor units with aging (Lexell et al. 1988).

Conclusion

Skeletal muscle can perform many types of muscle contractions, and the type of contraction is dependent on many factors. The different types of muscle contractions have distinct advantages and disadvantages. Understanding how these factors affect a muscle contraction, as well as the advantages and disadvantages of the different muscle contraction types, is paramount to exercise prescription for patients.

REVIEW QUESTIONS

1. Aging has a greater deleterious effect on what particular type of skeletal muscle fiber?
   a. Type I
   b. Type II
   c. Type III
   d. Type IV

2. Most of the strength gains in the first few weeks of implementing a training program are most likely attributed to
   a. increased capillary density
   b. neural adaptations
   c. collagen synthesis
   d. muscle fiber hyperplasia

3. Skeletal muscle's maximum force capability is produced
   a. near a muscle's normal resting length
   b. near a muscle's maximum lengthened position
   c. near a muscle's maximum shortened position
   d. at an angle of contraction that is directly perpendicular to gravity

4. Which type of muscle contraction is the most efficient, developing the greatest tension of the various types of skeletal muscle actions?
   a. concentric
   b. eccentric
   c. isometric
   d. isotonic

5. Which type of muscle fiber is best suited to perform explosive, dynamic muscle contractions?
   a. Type I
   b. Type IIa
   c. Type IIb
   d. Type III

Human Joint Structure and Function

Extremely important parts

of the human body, joints are defined by their location, coverings, type, and associated structures. Joint structure and function determine the amount of overall body movement and the leverage achieved during active movements that position the extremities and spine. This chapter describes the various forms of joints in the human body. Additionally, internal and external joint structures such as bones, ligaments, capsules, and cartilage are discussed.

Joints

Joints in the human musculoskeletal system are simply the areas where the ends of two or more bones attach. The bones are held together by soft-tissue structures that vary depending on the function of the joint. The type of tissue also helps place the joint into a classification system. Joints are broadly placed into two categories: those without a joint cavity and those with a joint cavity.

OBJECTIVES

After reading this chapter, you should be able to do the following:

> Explain the key differences between joints with a joint cavity and those without a joint cavity.

> List the components that make up a synovial joint cavity.

> Recognize the importance of bone in human structure and function.

> Define the types of articular cartilage in the human body.

> Explain the difference between the open and closed packed positions of a joint.

KEY POINT

Joint types are placed into two broad categories: those without a joint cavity and those with a joint cavity.

Synarthrosis

A joint without a cavity is known as a **synarthrosis** and is named according to the type of connective tissue used to maintain the joint. Synarthrosis joints contain either fibrous or cartilaginous tissue. A fibrous joint maintains opposition of ends that are held together by fibrous connective tissue. The amount of movement that can occur in these joints is very minimal. Fibrous joints commonly hold the sutures together in the skull, the distal forearm via the interosseous membrane (figure 5.1), and the distal tibiofemoral syndesmosis. Since these joints provide minimal overall movement, injuries to an interosseous or syndesmosis joint can create significant problems (see Clinical Correlation 5.1). A **gomphosis** joint is one in which fibrous tissue holds a peg-like portion in a hole. The only gomphosis joints in humans are those that hold the teeth in their sockets.

KEY POINT

Synarthrosis joints are held together by fibrous or cartilaginous tissue and include joints between the forearm and sutures of the skull.

A joint that is held together by cartilaginous tissue is known as an **amphiarthrosis** or a synchondrosis joint. These joints are located in the pubic symphysis region of the pelvis

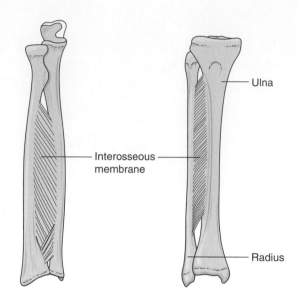

▶ **FIGURE 5.1** The interosseous membrane between the radius and the ulna is a fibrous synarthrotic joint.

(figure 5.2), at each level of the spine, at the junction of the ribs and the sternum, and at the sternomanubrial joint. Like fibrous joints, the cartilaginous joints allow slight movement between opposing ends but also permit bone growth. Once bone growth is finished, these joints may ossify and convert to a bony union called a synostosis.

Diarthrosis Synovial Joints

Joints with a cavity are further broken into several categories including synarthrotic, amphiarthrotic, and diarthrotic. These are all fluid-filled cavities that incorporate articular ends of bones, synovial fluid, a joint capsule,

CLINICAL CORRELATION 5.1

An injury to the interosseous membrane can cause significant damage and delayed recovery. The distal tibia and fibula are held together by several ligaments and an interosseous membrane known as a syndesmosis. A severe injury to these ligaments that also disrupts the syndesmosis is termed a high ankle sprain. This injury allows significant widening of the ankle mortise and requires a significantly longer healing time. Clinically, a patient with a high ankle sprain may not progress as rapidly as another patient with a less severe injury. With standard ankle rehabilitation exercise progression, the patient with a high ankle sprain may continue to have swelling and pain with weight bearing for weeks and in some instances months.

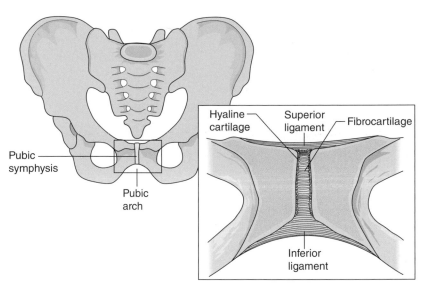

▶ **FIGURE 5.2** The pubic symphysis is a cartilaginous symphysis joint.

and in some instances a meniscal-type structure. All ends of bones in a synovial joint are covered with articular cartilage, and the entire circumference of the joint is covered with loose connective tissues called the joint capsule. The capsule has an outer and inner layer. The outer layer is known as the **stratum fibrosum**. The inner layer, called the **stratum synovium**, is covered with a synovial membrane. This membrane is loose vascularized connective tissue that secretes **synovial fluid**, which is a pale yellow substance that lubricates the joint. Synovial fluid is composed of proteinases, collagenases, hyaluronic acid, and prostaglandins. This fluid nourishes the articular cartilage. Synovial fluid changes its viscosity pending demands. With slow movement of the joint, the fluid is highly viscous, but in response to faster joint movements, the

fluid becomes less viscous, which decreases friction created at the joint surface. Excessive synovial fluid trapped inside a joint will create pain and dysfunction, described in Clinical Correlation 5.2.

▶ **KEY POINT**
Joints with a cavity are synarthrotic, amphiarthrotic, or diarthrotic.

Joint capsules vary in size and thickness and offer various degrees of support. The **ligaments** that connect bones to bones also help support and protect joints from excessive or unwanted movements. Some ligaments are discrete extra-articular structures, while others come as a capsular thickening. Capsular ligaments are thickenings of the joint capsule that are embedded into the actual capsule

CLINICAL CORRELATION 5.2

Synovial fluid is produced to help lubricate the synovial joint surfaces. With injury or irritation to the joint, synovial fluid production can be increased in an attempt to heal the injury. This can result in accumulation of synovial fluid or a synovial effusion. Each joint cavity has a limited amount of space. If the synovial swelling reaches the joint's limited capacity, pain will occur as pressure builds in the joint, irritating the sensitive tissue of the internal joint capsule. Clinically this will cause pain and loss of motion at the given joint. A decrease in activity, often termed relative rest of the affected joint, is helpful for decreasing further synovial irritation. Additionally the use of modalities such as moist heat, cold therapy, compression, elevation, and electrical stimulation may be helpful in lessening joint congestion.

surrounding the joint. Anatomically these are much harder to find and see in the human body. The intracapsular collagen fibers may be oriented toward areas of increased capsular tension. Additionally these ligaments are not usually as stout. The glenohumeral ligaments of the shoulder are examples of capsular ligaments in the human body.

Inside synovial joints it is not uncommon to find an articular disc, or meniscus. Typically these are located in joints that do not have articular surfaces that complement each other. For example, the tibiofemoral joint has on its lateral side a convex femoral condyle, while the tibial plateau is either flat or convex. In this location a large meniscus exists that makes the joint much more congruent by forming a concavity for the condyle to rest on. Structures contained in and around a typical synovial joint are seen in figure 5.3.

Numerous forms of synovial joint receptors are found in joint capsules and synovium. Various peripheral nerves and those from surrounding muscles penetrate the joint capsule to provide sensory information to the central nervous system. **Ruffini endings** are sensitive to stretch at the extremes of joint motion. **Pacinian corpuscles** detect joint compression, pressure, and movement. Golgi tendon organs are found in the capsule and detect

pressure and forceful motion at extremes. Unmyelinated free nerve endings detect noxious and nonnoxious movements and stress in the joint.

▶ **KEY POINT**

Joint receptors are found in ligaments, capsules, and tendons. These organs give sensitive feedback about pressure, position sense, and tension in regard to joint positions and movements.

Joint Classification

When a joint is composed of only two bones, it is considered to be a **simple joint**. When three or more bones make up the joint structure, it is called a **complex joint**. Most joints in the human body are some form of **diarthrosis** (i.e., surrounded by a capsule and filled with synovial fluid). There are seven forms of joints, named and classified according to their available movement (figure 5.4).

1. **Ball-and-socket joint (spheroid)**: A ball-shaped end fits into a cuplike socket, allowing a large range of movement. These forms of joints allow motion in all three planes of flexion–extension, internal–external rotation, and adduction–abduction. The glenohumeral and femoroacetabular joints are ball-and-socket joints.

▶ **KEY POINT**

The glenohumeral joint of the shoulder is a ball-and-socket joint. This form of joint allows extremes of motion in multiple planes. This wide range of motion comes at the expense of joint stability.

2. **Condyloid joint**: An oval-shaped condyle fits into a concavity, allowing angular motion but minimal to no rotation. These joints allow biplanar motion such as flexion–extension and abduction–adduction as well as flexion–extension and slight internal–external rotation. Condyloid joint examples include the tibiofemoral joint and the metacarpophalangeal joints of the fingers.

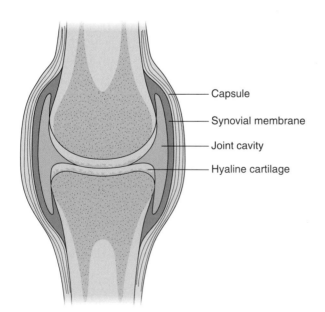

─ Capsule

─ Synovial membrane

─ Joint cavity

─ Hyaline cartilage

▶ **FIGURE 5.3** Elements of a typical human synovial joint.

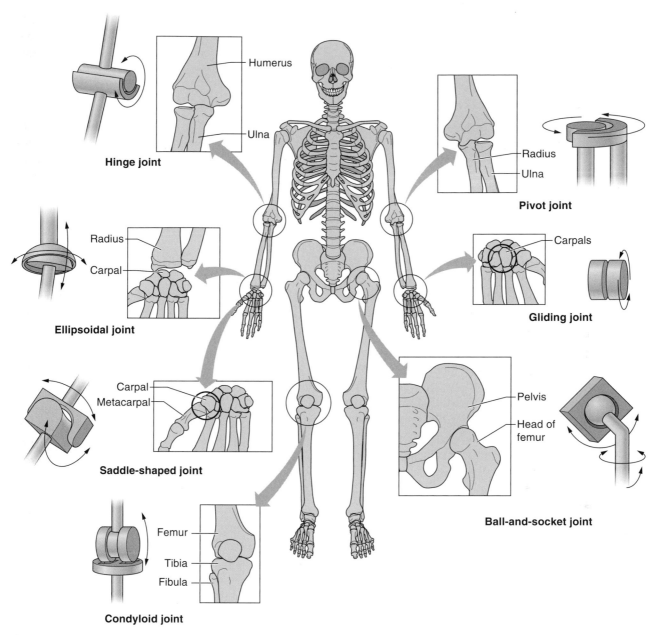

Hinge joint

Humerus

Ulna

Pivot joint

Radius

Ulna

Ellipsoidal joint

Radius

Carpal

Gliding joint

Carpals

Saddle-shaped joint

Carpal

Metacarpal

Ball-and-socket joint

Pelvis

Head of femur

Condyloid joint

Femur

Tibia

Fibula

▶ **FIGURE 5.4**   Seven joint classifications and their movements.

3. **Ellipsoidal joint**: A flattened convex ellipsoid fits into a concave surface. This type of joint allows biplanar motion of flexion–extension and abduction–adduction. The proximal radiocarpal joint and the metacarpophalangeal joint of the finger are examples of ellipsoidal joints.

4. **Gliding joint (plane)**: In this form, also known as a plane joint, two flat or fairly flat surfaces glide against each other, allowing gliding or rotation movement.

Movements in the gliding joint include translation and some degree of rotation. Examples include the carpals in the hand and tarsals in the foot.

5. **Hinge joint (ginglymus)**: A convex portion of one bone fits into a concave depression of another. These basic joints allow only flexion and extension of the segment. Examples of the hinge joint are the humeroulnar joint and interphalangeal joints.

► **KEY POINT**

The elbow joint (humeroulnar) is a hinge joint, one of the most basic forms of joint in the body. Hinge joints are very stable and allow motion only in a single plane. In the elbow, the motion allowed is flexion and extension.

6. **Pivot joint (trochoid):** A pivot joint has a rounded or conical surface on one end of the joint, while the other is concave or a ring. This joint allows spinning of one surface on the other. This spinning occurs around a single axis of motion. The proximal radioulnar joint is an example of a pivot joint, as is the atlantoaxial articulation of the cervical spine.

7. **Saddle-shaped joint (sellar):** Each side of the joint has both a convex and concave portion. These reciprocally shaped surfaces are oriented at right angles to each other, much like a horse rider sitting in a saddle. These joints allow significant biplanar motions; occasionally a small amount of rotation may occur. The carpometacarpal joint of the thumb is an example of a saddle-shaped joint.

Periarticular Tissues

Other tissues in and around joints, such as bone, cartilage, and ligaments, are called periarticular tissues. These tissues help support and stabilize joints, guide joint movements, and approximate joint surfaces. Each of the periarticular tissues is highly organized and has specific structural properties to perform its given functions.

Bone

Although **bone** is often thought of by the layperson as a static, rigid, dead structure as seen in skeletal models, it is the complete opposite—a dynamic, movable, beautifully living structure that can adapt to its changing needs. Bones are essential to support our many muscles, tendons, and ligaments, all of which synergistically use bones as levers to allow even the most basic human movement. Their hard surface allows protection of vital organs such as the brain, lungs, heart, and reproductive organs.

Bone Structure

Bone material is a highly specialized tissue whose components allow stability to maintain its supportive and protective roles. Bone is composed of organic and inorganic materials as well as small amounts of water. The organic component of bone is largely (98%) type I collagen and extracellular matrix. Type I collagen is found in a large abundance throughout the human body, not only in bones but also in tendons, ligaments, joint capsules, and even the cornea of the eye. These collagen fibers are thought to give bone its resilience and flexibility. The other 2% of bone's organic component comes from noncollagenous proteins and other cells.

Multiple other cells exist in human bone. **Osteoblasts** are responsible for synthesis, deposition, and mineralization of bone. These cells are derived from bone marrow mesenchymal cells and are typically located on all active bone surfaces, where they secrete procollagen (the precursor of type I collagen). Once finished forming bone, osteoblasts may become osteocytes. **Osteocytes** are the largest component of mature bone. They secrete enzymes that remove a thin layer of osteoid covering, allowing osteoclasts to bind to bone and begin resorption if needed (Buckwalter et al. 1995). **Osteoclasts** bind to bone surfaces and create an acidic environment that causes bone resorption. They are very similar to osteoblasts, except osteoclasts produce a much higher quantity of lysosomal enzymes in the cell nucleus.

Inorganic bone components include the crystal-like hydroxyapatite, which is composed of calcium phosphate. This is the material that gives bone its rigidity, stiffness, and hardness (Harris 1980).

The functional unit of a portion of bone is known as an osteon (figure 5.5). Each osteon has a center channel, or haversian canal, that contains blood vessels and nerve fibers. Within each osteon is a series of layers of mineralized

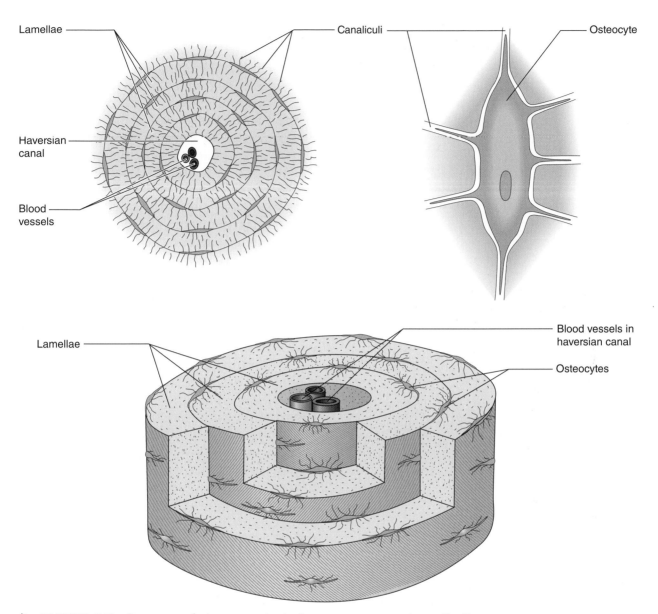

FIGURE 5.5 Structure of an osteon, including osteocytes and canaliculi.

matrix, known as lamellae, that surround the central canal. Around the periphery of each lamella are small cavities known as lacunae, which each contain an osteocyte. Canaliculi are small channels that run from lacunae to adjacent lamellae. Lacunae ultimately reach the haversion canal, allowing nutrients from blood vessels to reach the osteocytes.

All bone comes in one of two forms (figure 5.6): cortical, also known as compact, and cancellous, also known as trabecular or spongy. The outer portion of bone is cortical (compact) bone, which varies in thickness depending on its location in the body. **Cortical bone** is very

dense and hard and is formed for protection and structure (Khan et al. 2001). Calcium makes up approximately 80 to 90% of cortical bone volume (Khan et al. 2001). The spongy **cancellous bone** is formed very differently. Cancellous bone contains trabeculae, thin plates formed into a very loose network of tissue that houses red blood marrow. These calcified plates undergo a self-regulating remodeling process that allows the ability to alter and change force distribution optimally. Because of these functions, only 15 to 25% of cancellous bone is made from calcium (Khan et al. 2001). Alterations of bone density may

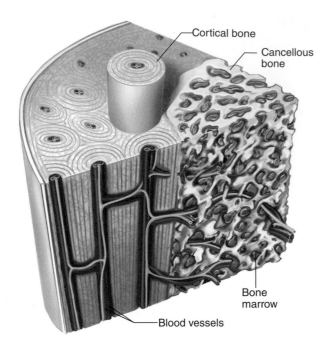

FIGURE 5.6 Compact cortical bone and spongy cancellous (trabecular) bone.

occur through either increased or decreased bone formation depending on stresses applied to the bone.

▶ **KEY POINT**
Bone comes in two forms: cortical (compact) and cancellous (trabecular).

Breaking bone down microscopically, it can be described as either woven or lamellar. Bone found in newborns, newly callused fracture sites, and regions of growth, such as the metaphyseal sites, are woven. Woven bone is immature bone formed by random collagen arrangements; it is present throughout the body at birth. Lamellar bone replaces the softer, more compliant woven bone. This mature bone begins formation at about 1 month after birth (Frankel & Nordin 2001) and makes up most of the skeleton by age 4 (Bennell & Kannus 2003).

A dense fibrous membrane called the **periosteum** covers all bones except the ends of long bones, which are covered with articular cartilage. The periosteum also contains an inner layer of tissue known as the endosteum, which houses both osteoblasts and osteoclasts

and may be important in the processes of remodeling and resorption.

It is clinically important to understand the process of bone remodeling because problems can occur from either a lack of remodeling or from an excessive amount of remodeling. Bone remodeling occurs from the time one is born until death. It is a dynamic process that leads to increases or decreases in mass, geometry, and length. Such changes can occur because of internal (tension via muscles and tendons) or external (weight-bearing loads due to gravity) forces placed on the body. This activity in which cancellous or cortical bone is either lost or gained depending on the level of stress applied is summarized as Wolff's law, which simply states that remodeling of bone is influenced by the mechanical stress placed on it (Wolff 1982).

For those in rehabilitation, remodeling is more of a problem when it does not occur fast enough or at all. Prolonged unloading can result in significant bone loss and resorption. During healing of fractures, occasionally delayed union, slow union, or nonunion of the repairing bone occurs. During all these processes, normal cellular repair is either delayed (fails to appear in the normal time period—a pathological problem), slowed (may not be pathological—simply due to age and other chronic problems), or absent (normal bone repair has ceased permanently). These problems with remodeling will certainly delay or slow functional progress in a patient being seen after injury to bony structures.

▶ **KEY POINT**
Wolff's law states that bone's internal and external architecture changes in response to stress placed on it.

Bone Types

The five types of bones are designated as long, short, flat, irregular, and sesamoid (figure 5.7). **Long bones** have a greater length than width. The humerus and the femur are long bones. **Short bones** have a width that is approximately the same as their length. The carpals and tarsals are considered short bones.

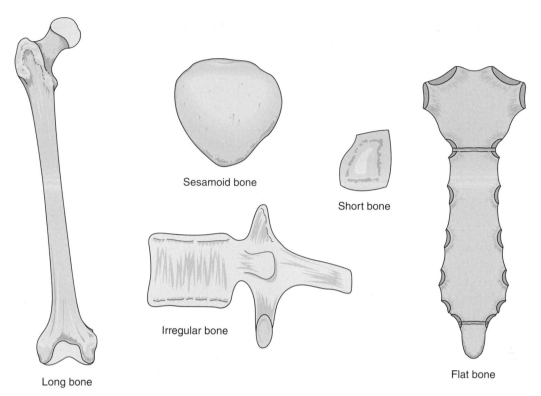

Sesamoid bone

Short bone

Irregular bone

Long bone

Flat bone

▶ **FIGURE 5.7** Types of bones.

Flat bones provide a large surface area for muscular attachments. They are usually broad and thin. The scapula is a flat bone. **Irregular bones** include the vertebrae and facial bones. These bones do not fall into one of the other categories because of their irregular features. **Sesamoid bones** are small round bones of the first metacarpal, the first metatarsal, and the patella.

Cartilage

Cartilage is a specialized tissue that unites bones of cartilaginous joints, covers the ends of long bones of synovial joints, and provides a wear-resistant surface that allows smooth gliding of healthy joint surfaces across one another.

Fibrocartilage

A very abundant material in the human body, **fibrocartilage** contains predominantly type I collagen but is actually a mixture of connective tissue and cartilage. It is found in the meniscus of the knee, the annulus fibrosus of the intervertebral discs of the spine, the labrum of the shoulder and hip, and the disc of the temporomandibular joint. These interposing structures provide stability and support and may also help guide arthrokinematic movement. Elastic cartilage contains mostly type II collagen and is more elastic in nature. It is the structural component of the ear.

Fibrocartilage lacks pain sensory fibers except in some locations near the insertions to bone (meniscal horns). It also in most instances exhibits a limited blood supply and is therefore dependent on synovial diffusion for nourishment, demonstrated by its maintenance of healthy status through intermittent compressive loads, which can easily be seen in the spine and knee. Specific properties of the knee fibrocartilage known as the meniscus are described in Clinical Correlation 5.3.

Hyaline Articular Cartilage

Hyaline cartilage is a very complex, highly ordered structure that is relatively thin and covers the ends of long bones in the synovial joints. Articular cartilage provides a smooth,

CLINICAL CORRELATION 5.3

The fibrocartilage in the knee is known as a meniscus. Although both meniscuses (medial and lateral) are fully vascularized at birth, each have vascular penetration of only 25% around the periphery in adulthood. This can impair the healing response. If an injury does not extend into the periphery (the vascularized zone), the potential for healing is limited. Symptomatic tears near the center of the knee will require debridement and thus a loss of meniscal tissue. Clinically one would like to have the tear into the periphery because healing through conservative or surgical means is possible. This will result in maintaining more meniscal tissue and decreasing the risk of early advanced osteoarthritis.

low-friction surface for joints to glide on and helps dissipate stress to the underlying subchondral bone. Hyaline articular cartilage is primarily type II collagen (90 to 95%). Chondroblasts are cells located in the articular cartilage that produce collagen and hyaluronic acid. This tissue contains very high levels of the protein proteoglycan. Proteoglycans constitute 12% of the total weight of articular cartilage and occupy the interstices within collagen fibers. The primary proteogylcan in articular cartilage is aggrecan, which contains chondroitin sulfate and keratan sulfate. When bound together, these tissues can resist compressive forces.

Water binds to proteoglycan and contributes about 65 to 80% of the total weight depending on the presence or absence of degeneration (Noyes et al. 1980). Chondrocytes occupy about 2% of the total volume of normal cartilage. Chondrocytes are living tissues that are metabolically active and synthesize extracellular matrix macromolecules as well as secrete degradative enzymes directing cartilage remodeling and regeneration (Jackson et al. 2001).

Articular cartilage structure varies over four distinct zones (figure 5.8). The superficial tangential zone (10 to 20%) is composed of a highly organized, packed, hydrated layer of cells that run parallel to the cartilage surface. This zone is thought to resist tensile forces on the surface. The second zone, or middle layer (40 to 60%), contains what look like random large-diameter collagen and rounded chondrocytes. The deep zone, or third layer (30%) is low in water concentration but rich in proteoglycan. The collagen is organized in

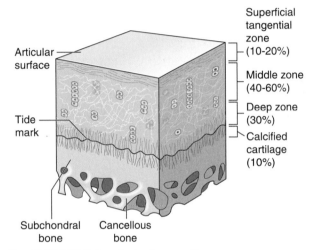

FIGURE 5.8 Articular cartilage.

a highly structured vertical fashion thought to resist compressive forces. The deepest zone (10%) is where the cartilage transitions from collagen to bone. This deepest layer, called calcified cartilage, is separated from the subchondral bone and cancellous bone that lies underneath the cartilage.

Because of this structure, healing of cartilage injuries is precarious at best. Since articular cartilage is isolated from marrow cells by the subchondral bone, it therefore is not vascularized. Injuries to other areas of the body rely on a vascular supply to create a hemorrhage; allow fibrin clot formation; and mobilize cells, mediators, and growth factors. Because of the lack of blood supply, injured articular cartilage has a very limited ability to heal (see Clinical Correlation 5.4).

▶ **KEY POINT**
Cartilage comes in two forms: hyaline articular or fibrocartilage

CLINICAL CORRELATION 5.4

Damage to hyaline articular cartilage creates an invariable catch twenty-two situation. Minor cartilage injuries rely on chondrocytes for repair. Chondrocytes can create limited cell clusters around the site of injury but are incapable of proliferating into a larger lesion. A deep, full-thickness lesion can initiate a reparative response if the injury is severe enough to penetrate the subchondral plate; this creates a vascular insult, resulting in hemorrhage and fibrin clot formation and associated vascular ingrowth. The catch twenty-two is simply that to create a healing response, you have to have a more severe injury.

Ligaments

Ligaments are highly organized structures that connect bone to bone. They play an especially important role when they become taut at the end of a range of movement. Ligaments and capsular structures are important for maintaining joint stability and are considered dense connective tissues. These types of tissues have few cells (fibroblasts), relatively low to moderate proportions of proteoglycan and elastin, and an abundance of type I collagen fibers (Neumann 2010).

▶ **KEY POINT**

Ligaments are highly organized tissues that connect bone to bone.

The fibers in isolated ligaments are considered regular dense connective tissue because they are developed in a parallel arrangement and function best when stressed along the long axis of the ligament. Capsular ligaments are different in that they have more of a haphazard arrangement that may allow resistance of tension from multiple directions. Collagen fibers in ligaments have a wavy, or crimped, configuration when unloaded. At even low levels of strain, this crimp will disappear. As stress is continued beyond physiological limits, a ligament will either partially or completely tear. A partial ligament tear is considered a sprain. Sprains are graded from I to III according to severity. A minor ligament tear with no loss of stability in the joint is considered a grade I sprain; in a grade II sprain, there is moderate tearing and mild to moderate ligament instability; a grade III sprain involves a complete rupture, resulting in a dramatic loss of stability.

Ligaments can be inside the joint proper (intra-articular ligaments), such as the anterior and posterior cruciate ligaments within the knee. They are also found outside the joint proper (extra-articular ligaments), such as the lateral collateral ligament of the knee. Finally, ligaments may actually be a part of the capsule (capsular ligaments). This occurs in the glenohumeral joint (glenohumeral ligament) and the deeper portions of the medial collateral ligament of the knee. Regardless of their type or location, ligaments are a fundamental means of maintaining joint stability.

Joint Positions and Movements

Joints in the human body are designed for stability, mobility, or a combination of the two. The terms *closed packed* and *open packed* describe positions of the joint and its congruency to the opposing joint structure. To obtain unrestricted movements, a joint must have accessory motion that occurs at the surface. These accessory movements required for full unrestricted motion are rolls, glides, and spins.

Closed Packed Versus Open Packed Positions

To allow for full range of movement, every joint must have an open and closed packed position. Both of these joint positions are important for movement and stability. In the **closed packed position**, the two joint surfaces fit congruently. The capsule and ligaments surrounding the joint are on maximal tension and taut, and the joint surfaces cannot be separated. The open packed position, or

loose packed position, is the point in the joint's passive motion when the joint capsule has its greatest capacity. The resting position is the maximum loose packed position, where ligaments and the capsule are at their greatest laxity and the joint is the most separated.

Joint Movements

To allow for full unrestricted mobility, joints go through several forms of movement types. Physiological movements are those that are performed actively or passively through a range of motion. An active movement of a joint is performed exclusively by the patient; passive physiological movements are performed exclusively by the clinician (Cook 2007). These motions, commonly termed *osteokinematic* because they are the planar movements such as flexion, extension, abduction, and adduction, are clearly visible to the naked eye and are used to measure and quantify motion via goniometry on a daily basis during rehabilitation.

Another form of movement that is not as easily seen or quantified is arthrokinematic movement. These are considered passive accessory joint movements involving motions specific to the two articulating surfaces (Barak et al. 1990; Wooden 1989). A passive accessory movement is any movement mechanically or manually applied to a body with no voluntary muscular activity by the patient (Grieve 1998). These accessory motions—rolls, glides, and spins—are required for full unrestricted motion and occur during both active and passive ROM of any joint. A roll occurs as a joint is moved through its ROM, and the first surface of the joint comes in contact with new points on the second surface. This is similar mechanically to a car tire rolling on pavement. As the car moves forward, there is a new contact point on both the tire and the pavement. During a glide, or slide, one point on the resting surface comes in contact with two or more points on the second surface. Mechanically, this would be more analogous to the car tire in which the brakes have been applied while on ice. One spot on the tire is gliding across

different spots on the ice. The motion of spin implies that one portion of the joint spins or rotates around a stationary mechanical axis, very similar to a top. These motions can occur in combination but do not necessarily occur in equal proportions. These are all considered accessory motions. Determining the amount of accessory motion available is defined as accessing "joint play." Joint play is any motion within the joint that is outside of the patient's available control. Therefore, the force used to impart joint play comes from a passive movement via another person.

▶ **KEY POINT**
An osteokinetic movement is one that can be seen by the naked eye, such as when the humerus moves through flexion, extension, abduction, or adduction. An arthrokinetic movement is one that occurs at the joint level. These accessory movements—rolls, glides, and spins—are much harder to see with the naked eye.

Convex–Concave Rules

Synovial structures all have some degree of convexity or concavity. If the two ends of a joint are incongruent, there is usually an additional cartilaginous meniscus that helps make the joint more congruent. When discussing joint movement, gliding follows a set pattern described as the convex–concave rules (Kaltenborn 1980).

When a convex surface is moving on a stationary concave joint surface, the gliding movement will occur opposite to the rolling motion—that is, opposite the visualized bone movement (figure 5.9). This can be seen at the knee during closed-chain knee flexion: The femur is rolling posteriorly but gliding anteriorly. When the concave surface is moving on a stationary convex surface, the rolling and gliding occur in the same direction, which is the direction of bone movement. During open kinetic chain movements at the knee, as the tibia moves into flexion, the tibial surface of the joint glides in a posterior direction (figure 5.10).

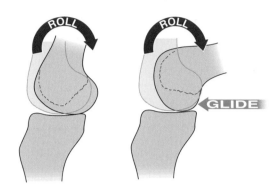

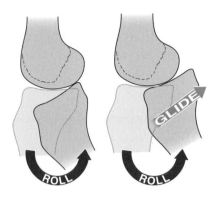

▶ **FIGURE 5.9** Convex-on-concave arthrokinematics.

▶ **FIGURE 5.10** Concave-on-convex arthrokinematics.

Conclusion

Human joints make up the mechanical systems that allow movement and transmission of forces in the musculoskeletal system. Joints vary in size, shape, and type depending on location and need. They are highly organized structures designed for a specific purpose and degree of motion. Joints are named according to the type of connective tissue used to maintain stability or mobility, and the amount of mobility and stability at a given joint is dependent on its function. Joints are further classified as either simple or complex depending on how many bones make up their structure. Components of joints include biological tissues such as bones, fibrocartilage, hyaline cartilage, and ligaments. The amount of and type of movement created by a joint is determined by its arthrokinematic pattern. Joint arthrokinematic movements are rolls, glides, and spins. These movements are required for full unrestricted functional activities.

REVIEW QUESTIONS

1. Which of the following is an example of a diarthrosis?

   a. pubic symphysis

   b. femoral head and acetabulum

   c. coronal sutures

   d. lumbar spine discs

2. Which form of nerve receptors are sensitive to stretch at the extremes of range of motion?

   a. Golgi tendon organs

   b. Pacinian corpuscles

   c. Ruffini endings

   d. Unmyelinated endings

3. Which joint allows biplanar motion with little to no rotation?

   a. ball-and-socket joint

   b. complex joint

   c. condyloid joint

   d. ellipsoidal joint

4. Which form of bone cell is responsible for synthesis, deposition, and mineralization of bone material?

   a. osteoclasts

   b. osteoblasts

   c. osteocytes

   d. osteotones

5. Bones that are thin and broad and have ample places for muscular attachments are known as which kind of bones?

    a. short bones

    b. long bones

    c. irregular bones

    d. flat bones

Regional Anatomy and Kinesiology

From the previous parts of this book, you have gained a good understanding of the general principles of biomechanics and joint and muscle structure and function. Now we turn our attention to how these principles are applied to specific joints. Part III includes 10 chapters, each dedicated to a particular joint complex. Each chapter follows a consistent format. The osteology section starts every chapter with a detailed review of the bones that make up the joints of the anatomical region. After this section is a thorough discussion of each joint articulation and joint anatomy. Explicit discussion of the joint capsules, ligaments, and other soft-tissue structures that are imperative to the stability and function of the joint is offered. Joint function is discussed in the next section and describes the axes of movement, arthrokinematics, range of motion, closed and loose packed positions, end feel, and capsular pattern of each joint. The final portion of these regional chapters is a discussion on the muscles that act on the described joints. The muscle origin, insertion, innervation, and action are included. An emphasis is placed on the functional action of the muscle.

The first four chapters of part III, chapters 6 through 9, include the cervical, thoracic, and lumbar spines, and the craniomandibular joints. Chapters 10 through 12 cover the joints of the shoulder, elbow and forearm, and wrist and hand. Chapters 13 through 15 are dedicated to the lower extremity: the hip, knee, and foot and ankle joints. Several clinical correlations are contained within all chapters to bridge theory to practice. The clinical correlations include common clinical examples that support the didactic material within the text.

At the completion of part III, the student should have a good understanding of the major joints of the human body. The last part of this text takes this information and applies it to whole-body movements such as walking, running, and jumping.

Cervical Spine

This is the first in a series of chapters covering the entire vertebral column. This chapter covers the general information common to all areas of the vertebral column to acquaint the reader with the framework of the spine. It is important to recognize that there are characteristics common to all sections of the spinal column as well as unique characteristics of each section. Specific characteristics of the cervical spine, how they are unique to this area of the spine, and so forth are all detailed in this chapter.

Vertebral Column

The entire vertebral column is composed of 33 vertebrae and 23 intervertebral discs. Primary regions of the vertebral column are cervical, thoracic, lumbar, sacral, and coccygeal (figure 6.1). Vertebrae in each region generally have characteristics and functional demands specific to that particular region. Vertebrae size increases from the cervical spine to the sacrum and then decreases from the sacral to the coccygeal regions. Seven vertebrae are present in the cervical spine, 12 in the thoracic region,

OBJECTIVES

After reading this chapter, you should be able to do the following:

> Describe the primary functions of the spine.

> Describe the unique characteristics of the upper and lower cervical spines as well as how they differ from each other.

> List the primary features (components) of the cervical spine.

> List the major ligaments in the cervical spine.

> List the components of the intervertebral discs in the cervical spine.

> Describe the orientation of the zygapophyseal joints in the cervical spine.

> Describe the arthrokinematics in all three planes of movement in the cervical spine.

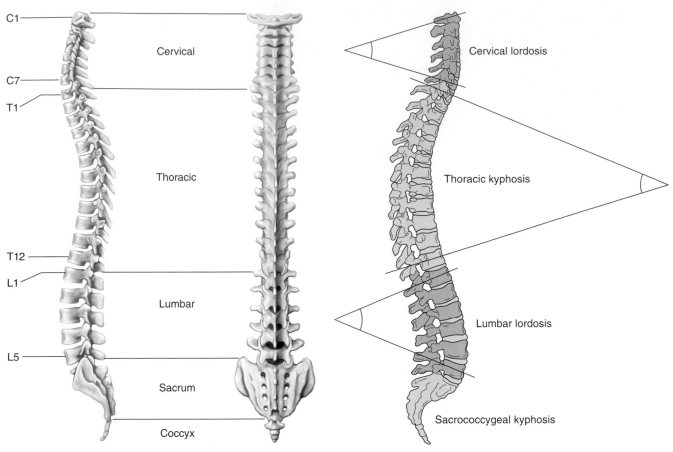

C1

Cervical

C7

T1

Thoracic

T12

L1

Lumbar

L5

Sacrum

Coccyx

Cervical lordosis

Thoracic kyphosis

Lumbar lordosis

Sacrococcygeal kyphosis

▶ **FIGURE 6.1** Vertebral regions of the spine.

▶ **FIGURE 6.2** Primary and secondary curves of the spine.

and 5 in the lumbar region. Five fused vertebrae form the sacrum, and 4 fused vertebrae form the coccyx.

The human spine has one **kyphotic** curve at birth. That is, the entire spine is one large C-shaped curve. The cervical **lordotic** curve is developed during childhood as a result of lifting the head into extension, while sitting and standing in the young child creates lordosis in the lumbar spine. Therefore, the two **primary spinal curves** are the kyphotic curves present at birth in the thoracic and sacral regions. **Secondary curves** in the cervical and lumbar spines are typically the result of adaptation to surroundings (figure 6.2).

Following are the primary functions of the spine:

- Provide a structural base of support for the rest of the body. The spine provides a base of support for the head and trans-

mits the entire weight of the head, arms, neck, and trunk to the pelvis.

- Allow for movement of the body. Each region of the spine has different amounts of movement availability (discussed later in each separate region), but as a unit the spine can move in all three planes of motion.

- Provide a stable base for attachment of ligaments, tendons, bones, muscles of the extremities, rib cage, and pelvis. The various areas of the spine provide multiple attachment sites for each of these types of tissue.

- Provide a link between upper and lower extremities. The spine as a unit is a common link between the upper and lower extremities.

- Protect the spinal cord and internal organs. The spinal cord is directly protected by the spinal vertebrae, and the internal organs are protected by the spinal column and rib cage.

- Absorb shock for the entire body. The forces initially absorbed via the upper or lower extremities are ultimately absorbed by the spinal column.

Vertebrae in each region of the spine have specific common characteristics (figure 6.3). Each vertebra consists of two major parts: a cylindrically shaped anterior **vertebral body** and an irregularly shaped posterior neural or **vertebral arch**. The vertebral body is the major weight-bearing structure of the spinal column. Trabeculae in the cancellous bone of the vertebral body are primarily responsible for providing resistance to compressive forces. A cortical bone shell reinforces the vertebral body.

The vertebral arch consists of the bilateral **pedicles, laminae, transverse processes**, and a spinous process. Pedicles are the region of the vertebral arch that lies anterior to the transverse processes on either side. These pedicles transmit tension and force from the posterior elements and the vertebral bodies. They are short, stout pillars with thick walls

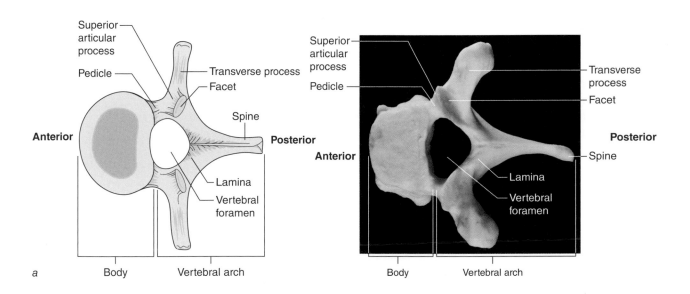

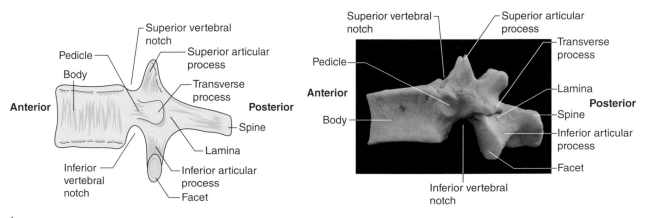

▶ **FIGURE 6.3** A typical vertebra: (a) superior view and (b) lateral view.

and therefore are able to transmit forces and tension quite well.

Osteology

The cervical spine has characteristics that differ from other areas of the spinal column. In particular is the osteology of this region of the spine. The upper cervical spine (atlas and axis) has several unique characteristics that differ not only from the thoracic and lumbar spine but even from the lower cervical spine. These unique characteristics are vital to the function of the entire cervical spine. The lower cervical spine also has unique characteristics as compared with other spinal column regions. Understanding the osteological features of the cervical spine is therefore paramount to understanding the function of the cervical spine.

Made up of 36 joints, the cervical spine allows for more motion than any other spinal region. The multiple joints provide increased motion with a sacrifice in stability. The cervical spine is vulnerable to both direct and indirect trauma.

The cervical spine consists of seven vertebrae and is divided into four anatomical units: the atlas, the axis, the C2-C3 junction, and the rest of the cervical vertebrae (Mercer & Bogduk 1993). The third through seventh vertebrae follow a typical morphology, with minor variations. These lower cervical vertebrae are often described as a functional unit.

▶ **KEY POINT**
The cervical spine consists of seven vertebrae and is divided into four anatomical units: the atlas, the axis, the C2-C3 junction, and the rest of the cervical vertebrae.

Upper Cervical Spine

The **craniovertebral junction** is a collective term that refers to the occiput, atlas, axis, and supporting ligaments. Twenty-five percent of the vertical height of the cervical spine is accounted for by this region. The occiput articulation with the atlas is via paired occipital condyles situated in a posterolateral to anteromedial orientation with the superior condyles of the concave atlas.

Atlas

The **atlas** (figure 6.4) is a ringlike structure formed by two lateral masses; it sits like a washer between the skull and lower cervical spine. The two lateral masses are interconnected by anterior and posterior arches, transforming the atlas into a ring and allowing the lateral masses to act in parallel (Mercer & Bogduk 2001). Anterior and posterior tubercles are present where each lateral mass connects with the other side. Functions of the atlas include cradling of the occiput, transmitting forces from the head to the cervical spine, and serving as an attachment for ligaments and muscles. Although the atlas is considerably wider than any of the other cervical vertebrae,

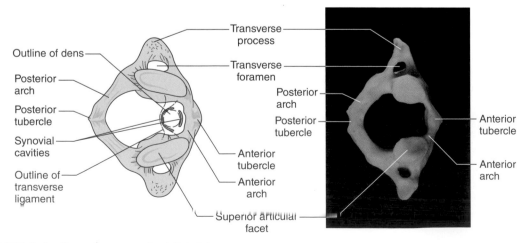

▶ **FIGURE 6.4** Superior aspect of the atlas.

it has a smaller vertical dimension than the rest. Since this vertebra has only a posterior tubercle and not a spinous process, there is no palpable bony process posteriorly between the occiput and the spinous process of C2. Therefore, the first spinous process palpable distally from the occiput is that of C2, or the axis. This lack of a spinous process on C1, or the atlas, allows for increased craniovertebral extension.

The lateral masses include four articulations: two superior and two inferior. The superior aspect of the atlas has a superior articular (zygapophyseal) facet on each side that articulates with the convex occipital condyles on their respective sides. These facets are concave anteroposteriorly and mediolaterally and provide an ideal articulation with the respective convex occipital condyles so that the skull rests securely on the atlas. The size and shape of the facet sockets can vary greatly, but generally they are directed upward and laterally, with their outer margins projecting more superiorly (Van Roy et al. 1997). These superior facets on C1 are elongated from anterior to posterior, with the anterior surfaces being closer together and more upwardly curved than their posterior counterparts (Gray 1995). This arrangement also allows for greater extension than flexion at the atlantooccipital joint (Panjabi et al. 1988). There is a right–left asymmetry in the atlantal sockets (Gottleib 1994; Singh 1965). The inferior facets are slightly convex and directed inferiorly for articulation with the superior articular facet of the axis (C2). The atlas also possesses a small, smooth facet on the internal surface of the anterior arch for articulation with the dens (odontoid process) of the axis. This bony articulation provides a solid resistance to posterior translation of the atlas.

The large transverse processes of C1 serve as a primary site of muscle attachment for this vertebra. The length of each transverse process increases the moment arms of the muscles attached to it. Between the superior articular facet and the most lateral portion of each transverse process is a transverse foramen through which the **vertebral artery** passes (see Clinical Correlation 6.1).

The central foramen of the atlas has two distinct parts. The smaller anterior part partially encircles the **odontoid** process, or dens, while the larger posterior portion is the vertebral foramen. The **transverse ligament** is the dividing structure between these two distinct parts.

Axis

The **axis** is a transitional vertebra between the craniovertebral region and the traditional vertebrae of the cervical spine. This vertebra, like the atlas, is unique compared with C3-C7. The axis is considerably less wide than the atlas (figure 6.5). The axis has a true spinous process, and it is the most cranial spinous process, which is palpable as the first midline structure below the occiput (Williams 1995).

One unique feature of the axis is the odontoid process (dens), which extends superiorly from the body to just proximal to the C1 vertebra, at which point it tapers to a blunt point. Dense, thick trabecular bone is present in the center of the dens, with uniformly thick cortical bone at the anterior base of the body of C2 (Sasso 2001). Beneath the odontoid process at the upper portion of the body of C2 is an area of hypodense bone that is susceptible to fracture (Heggeness & Doherty 1993). The dens has an anterior facet for articulation with the anterior arch of the atlas and a posterior

CLINICAL CORRELATION 6.1

The vertebral artery is significant in respect to providing blood flow to the brain. The course this artery takes as it winds around the C1 vertebral level suggests that rotation of the cervical spine could potentially compromise this artery. Assessment of the vertebral artery is often recommended in patients presenting with signs and symptoms suggestive of its involvement. Such signs and symptoms include dysarthria, diplopia, numbness around the mouth, ataxia, nystagmus, and nausea.

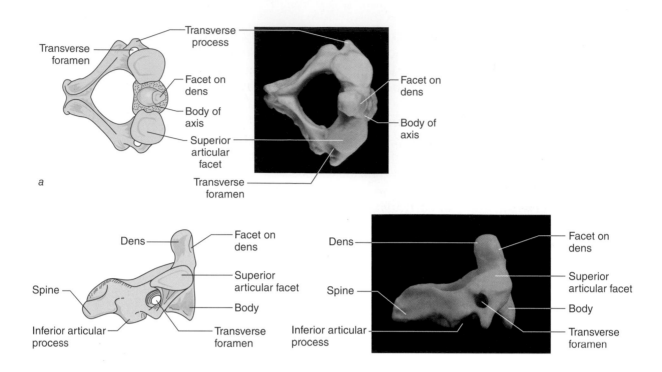

▶ **FIGURE 6.5**   The axis: *(a)* superior aspect and *(b)* right lateral aspect.

groove for articulation with the transverse ligament. The dens functions as a pivot for the upper cervical joints and as the center of rotation for the atlantoaxial (AA) joint (see figure 6.4).

The laminae of the axis are broad and join at the spinous process. Similar to the transverse processes of the atlas and the lower cervical spine, the axis also has transverse foramen for the vertebral artery.

Lower Cervical (C3-C7) Spine

The lower five cervical vertebrae have load-bearing, stability, and mobility functions. These vertebrae are similar in their characteristic osteology, function, and so on. The vertebral body, as well as the superior and inferior **end plates**, typically has a greater transverse diameter than anteroposterior diameter and height.

Because of the presence of the uncinate processes along the posterolateral edges (figure 6.6), the superior surface of the body of a lower cervical vertebra is concave in the frontal plane, while the superior surface slopes

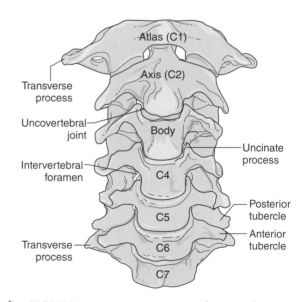

▶ **FIGURE 6.6**   Anterior view of cervical spine.

forward and downward in the sagittal plane. These uncinate processes are present at birth and gradually enlarge from 9 to 14 years of age (Bland & Boushey 1990). The inferior surface of the vertebra is concave with an anterior lip that projects anteroinferiorly toward the anterior superior edge of the vertebra below (Bland & Boushey 1990).

Just posterior to the vertebral body is the relatively large, triangular **vertebral foramen** formed by the borders of the vertebral body anteriorly, the two pedicles and transverse processes anterolaterally, and the two pedicles posterolaterally (figure 6.7). These pedicles project posterolaterally and are located halfway between the superior and inferior surfaces of the vertebral body. The transverse processes have a groove for the spinal nerves exiting the spinal cord. Two parts of each transverse process have been described: (1) the anterior tubercle, which is an attachment site for the longus capitis, scalenus anterior, and longus colli, and (2) the posterior tubercle, which is considered the true transverse process. The muscles that attach to the posterior tubercle include the iliocostalis cervicis, scalenus medius, scalenus posterior, and levator scapulae. The transverse processes have an inferolateral orientation.

Long, narrow laminae project posteromedially on each side just posterior to the pedicles and transverse processes. The pedicles and laminae form the neural arch, which encloses the vertebral canal. The laminae meet and join together to form the spinous process.

The vertebral canal, or foramen, in the cervical spine contains the entire cervical part of the spinal cord as well as the upper part of the first thoracic spinal cord segment. This is due to the fact that there are eight cervical spinal nerves on each side but only seven cervical vertebrae (see figure 6.8 and Clinical Correlation 6.2).

The **intervertebral foramina** (figure 6.9) are present in all the lower cervical vertebrae, with each one 4 to 5 mm long and 8 to 9 mm high. The pedicles of the vertebrae form the superior and inferior boundaries, while the anterior boundary is formed by the **intervertebral disc** (IVD) and portions of both vertebral bodies. The posterior boundaries are the respective zygapophyseal joints.

The spinous processes extend inferiorly and are typically short and slender. These spinous processes are bifid at their tip. Their length decreases slightly from C2 to C3, remains

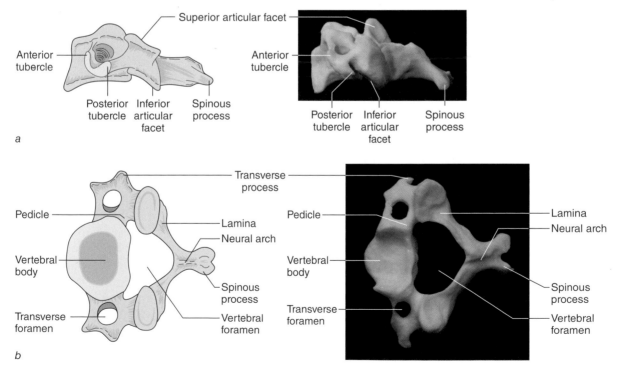

▶ **FIGURE 6.7** Typical cervical vertebra: (a) left lateral aspect and (b) superior aspect.

CLINICAL CORRELATION 6.2

Since there are eight cranial nerves, there are then eight dermatomal levels of the cervical spine. The clinician should be cognizant of the dermatomal levels and the orientation of the exiting nerve roots in relation to the cervical spinal level when examining patients with potential nerve root problems. Understanding this anatomical nature of the cervical spine will assist the clinician in properly determining nerve root level involvement.

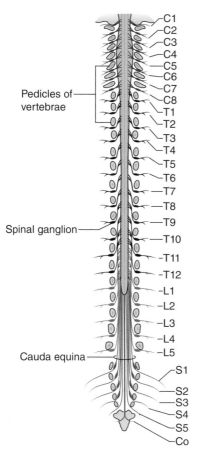

▶ **FIGURE 6.8**   Spinal nerves with corresponding cervical levels.

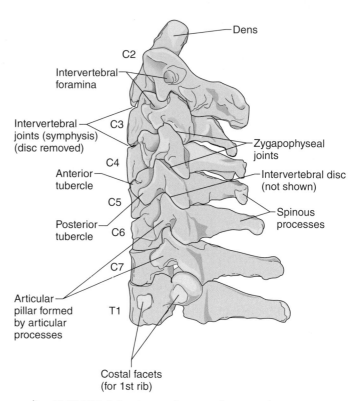

▶ **FIGURE 6.9**   Lateral view of cervical spine.

constant from C3 to C5, and significantly lengthens at C7 (Panjabi, Duranceau, et al. 1991). The seventh vertebra also has a wider transverse process, no inferior uncinate facet, and no transverse foramen (although there is some disagreement regarding the lack of a transverse foramen).

Each vertebra has a superior and inferior articular facet that articulates with the vertebra either above or below. Each articular facet is teardrop shaped, with the superior facet facing superiorly and posteriorly, while the inferior facet faces inferiorly and anteriorly. The superior articular facets change from a posteromedial orientation at C2-C3 to a posterolateral orientation at the C7-T1 level. The transition typically occurs at the C5-C6 level (Pal et al. 2001). The average horizontal angle of the joint planes of the middle cervical segments is approximately 45° between the horizontal/transverse and vertical/frontal planes (Williams 1995; White & Panjabi 1990). Upper cervical levels are closer to 35° and lower levels approximately 65°, although clinically the joint planes are thought of as passing through the patient's nose.

Joint Articulations

Once the reader has a good understanding of the osteology of the different regions of the cervical spine, the specific joint articulations are easier to appreciate. The distinct nature of each joint articulation in the cervical spine allows for complex interaction among each of these joints for ideal cervical spine function. The specific articulations, how they relate to each other, and how they differ from each other are described in this section.

Atlantooccipital Joint

The atlantooccipital (AO) joints (one each on the right and left side) represent the most superior zygapophyseal joints of the vertebral column and the only vertebral articulations that have a convex surface (occipital condyle) moving on a concave surface (superior articular facets of the atlas) (Walsh & Nitz 2001). These joints are true synovial joints with intra-articular fibroadipose meniscoids and lie nearly in the horizontal plane.

Atlantoaxial Joint

Four synovial joints make up the atlantoaxial (AA) joint. The two medial joints are (1) between the anterior surface of the dens and the anterior surface of the atlas and (2) between the posterior surface of the dens and the anterior hyalinated surface of the transverse ligament (Williams 1995). These joints are synovial pivot (or trochoid) joints, where the dens rotates in an osteoligamentous ring. There are also two lateral joints between the superior zygapophyseal facets of the axis and the inferior zygapophyseal facets of the atlas.

The relatively large superior articular facets of the axis lie lateral and anterior to the dens. They appear to be plane synovial joints based on bony structure; however, the articular cartilages of both the atlantal and axial facets are convex, resulting in biconvex zygapophyseal facet joints (Bogduk & Mercer 2000). These facets slope considerably downward from medial to lateral, in line with the zygapophyseal facets of the middle to lower cervical spine (see figure 6.9) (Ellis et al. 1991). These superior articular facets of the axis also face superiorly and laterally. The inferior articular facets are located posterior to the superior articular facets in a position similar to the articular processes of the lower cervical vertebrae. Therefore, the C2 vertebra serves as a transition from the upper cervical spine to the lower cervical spine (see Clinical Correlation 6.3).

Lower Cervical (C3-C7) Spine

Each pair of vertebrae in the lower cervical region is connected via a pair of zygapophyseal joints, two uncovertebral joints, and an interbody joint with the intervertebral disc (IVD). Since each vertebra has two superior and two inferior articular facets, there are then 14 facet or zygapophyseal joints from the occiput to the first thoracic vertebra. These zygapophyseal joints are typical synovial joints, with hyaline cartilage covering their articulating surfaces.

The uncovertebral joints are present from C3 to T1 and are saddle-shaped diarthrodial joints. These articulations are formed between the uncinate process on the lateral aspect of the superior surface of the caudal, or inferior,

CLINICAL CORRELATION 6.3

Transitional zones of the spine, such as C2-C3, are often clinically thought to be potential areas of dysfunction. The reason is that the spine is transitioning not only anatomically but also biomechanically and functionally from one area of the spine to another. This change in anatomy and therefore biomechanics and function can put increased stress on these regions of the spine. The clinician should be conscious of other such areas of the spine as well, namely C7-T1, T12-L1, and L5-S1.

vertebra and the beveled inferolateral aspect of the superior, or cranial, vertebra. These joints develop within the first 12 years of life and become fully developed by about 33 years of age (Orofino et al. 1960).

Interbody joints of the lower cervical region include the superior vertebra, the inferior vertebra, and the IVD between these two vertebral bodies. These joints have been traditionally described as amphiarthrodial, with the capability of contributing to motion in all three planes of motion, although others (Mercer & Bogduk 2001) have argued that these joints between the bodies of cervical vertebrae are essentially saddle joints and allow movements in only two planes.

▶ **KEY POINT**

Functionally, the cervical spine can be divided into upper and lower regions, with each having unique anatomical and arthrokinematic characteristics.

Joint Anatomy

The anatomy of the joints of the cervical spine, like the osteology and joint articulations, has distinctive features for each joint. The atlantooccipital, atlantoaxial, zygapophyseal, and intervertebral joints each have distinctive anatomical features that are described in this section.

Atlantooccipital Joint

The AO joint capsule and corresponding ligaments allow for motion because they are relatively lax. This joint is typically described as the nodding joint, as it contributes to head-nodding motion. The amount of motion in this joint is greater in the sagittal plane than any other, yet relatively small compared with other cervical spine joints. For a list of ligaments for the AO joint, see table 6.1.

Atlantoaxial Joint

A primary function of the atlantoaxial joint is head rotation. Because of the amount of motion required at this joint, the capsule has a large amount of extensibility and is therefore relatively lax. For a list of ligaments for the AA joint, see table 6.1.

▶ **KEY POINT**

The cervical spine allows for more motion than any other region of the spine.

TABLE 6.1   Ligaments of the Upper Cervical Spine

Ligament	Attachments	Function
Atlantooccipital ligaments		
Joint capsule and accessory capsular ligaments	Run obliquely from the base of the occiput to the transverse process of C1	Naturally quite lax to permit maximal motion and therefore only moderately support the joints during contralateral head rotation
Anterior AO membrane	Superior continuation of anterior longitudinal ligament/anterior atlantoaxial membrane	Connects anterior arch of atlas to anterior aspect of foramen magnum and therefore assists with stability of the anterior atlantooccipital joint
Posterior AO membrane	Superior continuation of ligamentum flavum/posterior atlantoaxial membrane	Interconnects posterior arch of atlas and posterior aspect of foramen magnum; forms part of posterior boundary of vertebral canal (Pick & Howden 1995) and therefore assists with stability of the posterior atlantooccipital joint

Ligament	Attachments	Function
Atlantoaxial ligaments		
Anterior atlantoaxial membrane (continuation of anterior longitudinal ligament)	Connects atlas to axis anteriorly	Limits extension
Posterior atlantoaxial membrane (continuation of ligamentum flavum)	Broad and thin; attaches to posterior ring of atlas and axis; along with posterior AO membrane is anatomically analogous to ligamentum flavum, which is first present at C2-C3	Limits flexion
Lateral atlantoaxial joint capsule	Connects atlas to axis laterally on each side	Generally described as loose and thin to allow for axial rotation in each direction; contributes to stability of the joint at end range of each of these movements
Cruciform ligament	Has superior, inferior, and transverse (transverse ligament) portions	See details of each ligament listed below
Transverse ligament	Stretches between tubercles on the medial aspects of lateral masses of atlas	Maintains the position of the dens relative to the anterior arch of the atlas; often considered a separate ligament; see text for additional details
Superior band of cruciform ligament	Runs from superior portion of transverse ligament to anterior edge of foramen magnum	Minimal importance in controlling physiological motion; checks inferior–superior displacement of transverse ligament
Inferior band of cruciform ligament	Runs from inferior portion of transverse ligament to body of axis	Minimal importance in controlling physiological motion; checks inferior–superior displacement of transverse ligament
Ligaments connecting axis with occiput		
Apical ligament	Extends from the apex of the dens of C2 to the anterior rim of the foramen magnum	Appears to be only a moderate stabilizer against posterior translation of dens relative to both the atlas and the occipital bone (Pick & Howden 1995)
Alar ligament	Connects superior part of either side of dens to fossae on medial aspect of occipital condyles, although can also attach to lateral masses of the atlas (Dvorak & Panjabi 1987; Dvorak, Panjabi, et al. 1987)	Limits rotation of the head. Also limits lateral flexion to the opposite side; see text for detailed description
Tectorial membrane	Most superficial of the three membranes and interconnects the occipital bone and axis; it is the superior continuation of the posterior longitudinal ligament; runs from body of C2 up over the posterior portion of dens and then makes a 45° angle in anterior direction as it attaches to anterior rim of foramen magnum	Important limiter of upper cervical flexion, extension, and vertical translation; holds the occiput off the atlas (Pal & Sherk 1988)

Upper Cervical Nonskeletal Muscular Soft Tissue

The nonskeletal muscular soft-tissue structures in the upper cervical spine include the ligaments, spinal nerves, and blood supply to this region. This region of the cervical spine is quite complex anatomically and biomechanically, and injuries can affect multiple structures and have serious consequences. Ligaments are therefore important for overall cervical spine stability. The stabilization these ligaments provide supports all the structures in this area, including the nervous and blood supply systems.

Upper Cervical Spine Ligaments

The transverse ligament (see figure 6.4) resists posterior translation of the dens into the vertebral foramen and, therefore, the spinal cord. It also limits the amount of flexion between the atlas and axis (White et al. 1975). These limiting functions are extremely important. Excessive movement of either type could result in the dens compressing not only the spinal cord but also the epipharynx, vertebral artery, or superior cervical ganglion. Transverse ligament integrity is also vital to the stability of atlas fractures; degenerative, inflammatory, and congenital disorders; and any other abnormalities affecting the craniovertebral junction (see Clinical Correlation 6.4).

Spontaneous or isolated traumatic tears of the transverse ligament are extremely rare. The ligament is primarily composed of collagen and has a parallel orientation close to the atlas and dens. There are 30° obliquities at other points in the ligament. The transverse ligament is almost twice as strong as the alar ligament, with a tensile strength of 330 N (Dvorak, Panjabi, et al. 1987; Dvorak, Schneider, et al. 1987). In fact, the transverse ligament is so strong that the dens will fracture before the ligament will tear (Williams 1995). Rheumatoid arthritis and other conditions such as Down syndrome can compromise the transverse ligament. For a list of other primary upper cervical ligaments, refer to table 6.1.

The paired **alar ligaments** arise on either side of the superior portion of the dens and extend superiorly and laterally to attach to the respective sides of the occipital condyles and lateral masses of the C1. They consist mainly of collagen fibers arranged in parallel and are approximately 1 cm in length (Panjabi, Duranceau, et al. 1991). These paired ligaments are relaxed with the head in neutral position and are taut in flexion (Panjabi, Duranceau, et al. 1991). The left upper and right lower portions limit right lateral flexion of the head and neck (Crisco, Oda et al. 1991; Crisco, Panjabi, et al. 1991). Axial rotation of the head and neck tightens both alar ligaments (Panjabi, Oxland, et al. 1991). They also help prevent distraction of C1 on C2.

The primary muscles of the upper cervical spine include the anterior suboccipital muscles (rectus capitis anterior and rectus capitis lateralis) and the posterior suboccipital muscles (rectus capitis posterior major and minor, superior and inferior obliques). More detail on these muscles is given later in the chapter under muscle function.

Upper Cervical Spine Nerves

The posterior, or dorsal, ramus of spinal nerve C1, along with the vertebral artery, exits the spinal canal posteriorly between the posterior

---

CLINICAL CORRELATION 6.4

Transverse ligament integrity is vital to the stability of the upper cervical spine. Damage to the transverse and alar ligaments is primarily due to trauma, although degenerative disease, inflammatory conditions, and congenital disorders are significant risk factors for upper cervical instability. Specific conditions the clinician should be aware of are rheumatoid arthritis and Down syndrome. Both these conditions have the potential to compromise the integrity of one or more of these ligaments and contribute to craniovertebral junction instability

arch of the atlas and the rim of the foramen magnum, entering the suboccipital triangle. It supplies most of the muscles that form the triangle and usually has no cutaneous distribution. The dorsal ramus of spinal nerve C2, also known as the greater occipital nerve, is the largest of the dorsal rami and is primarily a cutaneous nerve. It supplies most of the posterior aspect of the scalp.

Upper Cervical Spine Blood Supply

The intradural vertebral artery supplies the most superior segments of the cervical spinal cord. The cervical cord is supplied by two discrete but overlapping arterial systems: central and peripheral. Both these systems supply blood to the upper cervical spine.

Lower Cervical (C3-C7) Spine

Synovial intra-articular structures have been observed in the zygapophyseal joints. These fat-filled vascular structures have been described as fibroadipose meniscoids, synovial folds, and capsular rims. The meniscoids consist of connective and fatty tissue that is highly innervated and vascularized (Dvorak 1998). These meniscoids function as space fillers in the cervical spine for the uneven articular surfaces, especially where the elasticity of the relatively thin articular cartilage is not sufficient (Dvorak 1998). These inclusions are also prone to entrapment between the two articulating surfaces. They can contribute to intra-articular fibrosis and cervical spine pain (Giles & Taylor 1987).

The joint capsules of the zygapophyseal joints are relatively lax and allow for a large range of motion (ROM). The anterior capsule is strong but is lax in neutral and extension (Lysell 1969). The posterior capsule is thin and weak, allowing for translation between facets but making it vulnerable to injury (Bogduk & Mercer 2000). The posterior capsule often integrates with a meniscoid and cervical multifidus musculature (Bogduk & Mercer 2000). The anterior joints do not articulate with meniscoids (Bogduk & Mercer 2000). Elastic fibers of the medial aspect are oriented similar to those of the ligamentum flavum and may join with the ligamentum flavum in spanning from one articular process to the other. The ligaments of the vertebral column and the IVD are the major supports of these joints.

The presence of a synovial cavity in the uncovertebral joints has been controversial. Panjabi, Duranceau, et al. (1991) suggest that these joints maintain a synovial compartment and create the posterior lateral border of the IVD. These articulations, also known as joints of Luschka, could then potentially be pain-generating structures.

Intervertebral Disc

The intervertebral disc (figure 6.10) provides a separation of the adjacent vertebral bodies. Each IVD is composed of three parts: (1) the **nucleus pulposus** (NP), (2) the **annulus fibrosus** (AF), and (3) the end plate (EP). The NP is typically, except in the cervical spine, the gelatinous mass found in the center of the disc, whereas the AF is the fibrous outer ring. The vertebral EP is the cartilaginous layer covering the superior and inferior surfaces of the IVD. All three of these structures are composed of water, collagen, and proteoglycans (PGs). Fluid and PG concentrations are highest in the NP and lowest in the outer regions of the AF and EP. Collagen fibers, on the other hand, are highest in the EP and outer AF and lowest in the NP. There is little distinction between the NP and the AF where they merge, while greater distinction between these structures

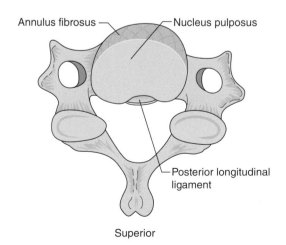

Annulus fibrosus — — Nucleus pulposus

Posterior longitudinal ligament

Superior

▶ **FIGURE 6.10** Superior view of cervical disc.

is more readily apparent where they are farthest apart.

Both type I and type II collagen are present in the NP, although type II predominates. Type II collagen is most prevalent in the NP because of its ability to resist compressive loads. When the IVD is compressed and deforms, the increased pressure stretches the walls of the IVD, similar to a balloon being compressed. Tensile forces are created along the walls of the IVD in such instances. Resistance to these tensile forces reflects the need of the AF to be composed of predominantly type I collagen, although type II collagen fibers are also present in the AF (Bogduk 1997; Lundon & Bolton 2004). The AF fibers are attached to the cartilaginous EP via Sharpey's fibers on the inferior and superior vertebral plateaus of adjacent vertebrae as well as the epiphyseal ring region.

The EP covers the entire NP and is strongly attached to the AF but only weakly attached to the vertebral body. The EP is considered to be a component of the disc rather than the vertebral body (Bogduk 1997; Lundon & Bolton 2004). The cartilage of the vertebral EP is both hyaline cartilage and fibrocartilage. Hyaline cartilage is present closest to the vertebral body, and fibrocartilage is present closest to the NP. Fibrocartilage becomes the major component of the EP with increasing age. Little or no hyaline cartilage remains, reflecting the need to tolerate high compressive forces. Hyaline cartilage is found primarily in young discs.

The form and function of the IVDs of the cervical spine are distinctly different from those of the lumbar spine (Mercer & Bogduk 1999). The first cervical disc is located between C2 and C3. Several unique features distinguish the cervical disc from the lumbar disc (Mercer & Bogduk 1999):

- The AF in the cervical spine is a discontinuous structure surrounding a fibrocartilaginous core compared with a fibrous ring enclosing a gelatinous NP like the AF in the lumbar region. At birth, the NP makes up less than 25% of the disc, whereas in the lumbar disc, it makes up at least 50% (Taylor 1971).

- No region of the AF in the cervical spine has successive lamellae exhibiting alternating orientations as in the lumbar region. Only the anterior portion of the AF has a cruciate pattern.

- The NP is contained only by the alar fibers of the **posterior longitudinal ligament** (PLL) in the posterolateral region of the cervical disc. Protection against disc herniation in this region of the disc is afforded via these fibers and the uncovertebral joints. The NP must pass through or under these alar fibers if it is to herniate.

- There is an absence of an AF over the uncovertebral region. Clefts are present that progressively extend across the back of the disc as a result of collagen fibers being torn during the first 7 to 15 years of life. These clefts have been reported as either enabling (Kokubun et al. 1996) or resulting from rotary motions of the cervical vertebrae.

- Axial rotation of a typical cervical vertebra occurs around an oblique perpendicular to the plane of its facets.

The inferior surface of the cervical IVD is concave, and the NP sits in, or near, the center of the disc (slightly more posterior than anterior). The small vertebral body size affords the greatest ratio of IVD height to body height (2:5) in the cervical spine. In fact, the IVD make up approximately 25% of the superior to inferior height of the cervical spine (Walsh & Nitz 2001). This ratio, along with the facet joint orientation, allows for the greatest possible ROM of the typical lower cervical vertebrae.

Compared with the lumbar spine, the previously described morphology might put into question the etiology and mechanism of cervical discogenic pain. The lack of a posterolateral AF would seem to lend credibility to other possible sources of disc-related pain in the cervical spine, as there would not be any posterolateral fissures of the AF as in the lumbar spine. Other possible sources of disc-related pain of the spine include strain or tears of the AF, nerve compression, altered pH levels around the nerve root, inflammatory mediators in the herniated disc tissue, neovascularization, and autoimmune reactions. As in the lumbar spine, the pain associated with cervical disc lesions probably occurs from an inflammatory process initiated by nerve root compression, resulting in nerve swelling. Relief of this type of pain with nonsteroidal anti-inflammatory drugs (NSAIDs) and controlled low-level exercise designed to increase blood flow, therefore limiting chemical mediate stagnation, would seem to lend credibility to this hypothesis.

Lower Cervical Spine Ligaments

The major ligaments of the lower cervical spine are outlined in table 6.2 and pictured in figure 6.11. Some are continuations of the

TABLE 6.2 Ligaments of the Lower Cervical Spine

Ligament	Attachments	Function
Continuous ligaments		
Anterior longitudinal ligament	A strong band extending along the anterior surfaces of the vertebral bodies and IVDs, from the anterior aspect of the sacrum to the anterior aspect of C2; firmly attached to the superior and inferior EPs of the cervical vertebrae but not to the cervical IVDs	Narrower in the upper cervical spine but well developed in the lower cervical, lower thoracic, and lumbar regions; limits spinal extension and reinforces the anterolateral portion of AF and anterior aspect of intervertebral joints
Posterior longitudinal ligament	Extends from the sacrum to the body of C2, where it is continuous with the tectorial membrane; firmly attached to the posterior aspect of the IVDs, the laminae of hyaline cartilage, and adjacent margins of vertebral bodies	Broad in the cervical and thoracic regions; narrow in the lumbar region; expands laterally over the posterior portion of the discs to give a saw-tooth appearance; prevents disc protrusions; acts as a restraint to segmental spinal flexion
Ligamentum nuchae	Spans the entire cervical spine from the external occipital protuberance to the spinous process of C7	Controversial functions
Segmental ligaments		
Ligamentum flavum	Extends from the sacrum to C2, where it is continuous with the posterior atlantoaxial ligament; connects the laminae of successive vertebrae from the zygapophyseal joint to the root of the spinous process	Thickest in the lumbar region; limits forward spinal flexion via resisting separation of the laminae, especially in the lumbar spine
Interspinous ligament	Thin ligaments interconnecting successive spinous processes	Poorly developed in the upper cervical spine but well developed in the lower cervical spine (Johnson et al. 1975); thought to limit spinal flexion
Supraspinous ligament	C7 to L3 or L4; continuation of ligamentum nuchae in cervical spine	Weak in lumbar spine; limits spinal flexion
Intertransverse ligament	Primarily in the lumbar spine	Limits contralateral lateral spinal flexion

ligaments of the upper cervical spine (table 6.3). The continuous ligaments are those that attach to the peripheral aspects of all the vertebrae, whereas the segmental ligaments are those short ligaments that interconnect adjacent vertebrae.

The **anterior longitudinal ligament** (ALL) has two layers consisting of thick bundles of collagen fibers (Putz 1992; Maiman & Pintar 1992). The superficial fibers are long and bridge several vertebrae, while the deep fibers are short and run between adjacent vertebrae. The deep fibers blend with the AF. The ligament is well developed in areas of lordosis in the spine (cervical and lumbar spine) but is much less developed in the kyphotic thoracic

spine. The ALL also increases in thickness and width from the lower thoracic to the L5-S1 vertebrae (Putz 1992). The greatest tensile strength of the ALL is in the lumbar region, followed by the upper cervical and lower thoracic spines (Myklebust et al. 1988). The ALL has been reported to be twice as strong as the posterior longitudinal ligament (PLL) (Myklebust et al. 1988).

The PLL is broader and considerably thicker in the cervical region than in the thoracic and lumbar regions (Johnson et al. 1975). The PLL also has two layers, a superficial and deep layer. Similar to the ALL, the superficial fibers span several levels, while the deep fibers run only between adjacent vertebrae. The deep layer interlaces with the outer layers of the AF and attaches to the margins of the EP in a manner that varies from segment to segment (Putz 1992). The PLL is stretched in flexion and slack in extension. It significantly narrows in the lumbar spine and provides only one-sixth the resistance to axial tension as compared with the PLL in this region (Myklebust et al. 1988).

The **ligamentum nuchae** is considered by some to be an extension of the supraspinous and interspinous ligaments (Hollinshead 1982; Johnson et al. 2000), while others (Allia & Gorniak 2006) suggest it is a distinct ligamentous structure made up of four portions. It is commonly described as being a strong and thick ligament that aids in the suspension of the head and neck (Johnson et al. 2000), although the function of this ligament is controversial.

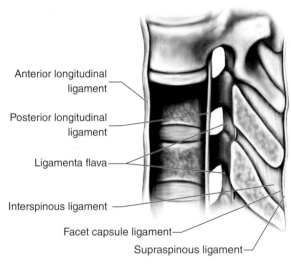

Anterior longitudinal ligament

Posterior longitudinal ligament

Ligamenta flava

Interspinous ligament

Facet capsule ligament

Supraspinous ligament

▶ **FIGURE 6.11** Ligaments of the lower cervical and thoracic spine.

TABLE 6.3   Major Ligament Continuations of the Upper Cervical Ligaments

Upper cervical ligament	Continuation of ligament in lower cervical spine	Level of continuation
Atlantooccipital ligaments		
Anterior AO membrane	Anterior atlantoaxial membrane	C1
Posterior AO membrane	Posterior atlantoaxial membrane	C1
Atlantoaxial ligaments		
Anterior atlantoaxial membrane	Anterior longitudinal ligament (ALL)	C2
Posterior atlantoaxial membrane	Ligamentum flavum	C2
Ligament connecting axis with occiput		
Tectorial membrane	Posterior longitudinal ligament (PLL)	C2

The **ligamentum flavum** spans from the anterior surface of one vertebral lamina to the adjacent vertebral lamina. It forms the smooth posterior surface of the vertebral canal (Olszewski et al. 1996), while some fibers extend laterally to cover the articular capsules of the zygapophyseal joints (Maiman & Pintar 1992). This ligament is strongest in the lower thoracic spine and weakest in the midcervical spine (Myklebust et al. 1988).

The ligamentum flavum is often referred to as the yellow ligament on account of its formation from collagen and yellow elastic tissue. Because of its elastic quality, it differs from other ligaments of the cervical spine. This highly elastic quality serves two primary purposes: (1) It creates a continuous compressive force on the discs, which maintains high intradiscal pressure, and (2) it prevents ligament buckling. Scarring or fatty infiltration of the ligament in the cervical spine can make the ligament lax, particularly with cervical extension. This laxity increases the potential for the ligament to buckle and compress the contents of the vertebral canal (Penning 1978). The spinal nerve can also become impinged by enlargement of this ligament (Watson & Trott 1993).

The interspinous and supraspinous ligaments are the first to be damaged with excessive flexion (Adams et al. 1980). The function of the **interspinous ligament** is controversial, although most authors agree that it resists flexion. It may also resist end-range extension and posterior shear of the superior vertebra on the inferior one because of the oblique nature of the fiber orientation (McGill 2007).

The **supraspinous ligament** is a cord-like structure that connects the spinous processes from C7 to L3 or L4. In the lumbar spine, these fibers merge with the thoracolumbar fascia and become indistinct.

The **intertransverse ligaments** are paired ligaments demonstrating extreme variability. These ligaments connect adjacent transverse processes on each side of the vertebral body. Only a few fibers of the ligaments are found in the cervical spine. These ligaments are alternately stretched or compressed with lateral bending, depending on the side.

Lower Cervical Spine Nerve Supply

The zygapophyseal joints are innervated by the medial branches of the cervical posterior rami from C2 to C8 and the recurrent meningeal (sinuvertebral) nerve. The IVDs are innervated in the outer one-third to one-half of the fibers of the AF (Bogduk 1997). The cervical and lumbar IVDs are innervated by the **sinuvertebral nerve**[1], formed by branches from the anterior, or ventral, nerve root and the sympathetic plexus (Bogduk et al. 1988).

Lower Cervical Spine Blood Supply

The vertebral artery (figure 6.12) arises from the first part of the subclavian artery and passes upward along the longus colli muscle to enter the transverse foramen of C6. Occasionally it may enter at the fourth or fifth transverse foramen. The vertebral arteries on either side then travel through the transverse foramen on each side of the cervical vertebrae through C1, pierce the posterior AO membrane, and meet at the foramen magnum.

The IVD does not receive blood supply from any major arterial branches. The outer surfaces of the AF do receive a blood supply from small branches from metaphyseal arteries, which also form a dense capillary plexus at the base of the EP cartilage and subchondral bone deep to the EP (Bogduk 1997). The remainder of the disc receives nutrition via diffusion through these sources.

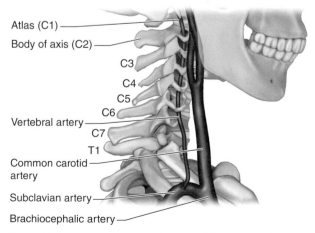

FIGURE 6.12 Sagittal view of the cervical spine demonstrating the vertebral artery.

Joint Function

Joint function in the cervical spine is quite complex because there are distinct units in this region of the spine, each with unique anatomical and therefore biomechanical and functional characteristics. Understanding a joint's anatomy and biomechanics is quite helpful in understanding the joint's function.

Atlantooccipital Joint

The atlantooccipital joint is the most proximal joint in the cervical spine. In this section, we describe the joint axes of motion, arthrokinematics of the joint, range of motion, closed and loose packed positions, end feel, and capsular patterns of this most proximal cervical spine joint.

Axes of Motion

Flexion and extension occur in the sagittal plane around a mediolateral axis. Axial rotation takes place in the transverse plane around a vertical axis, and side bending takes place in the frontal plane around an anteroposterior axis.

Arthrokinematics

The deep atlantal sockets of the atlas (concave articulating joint surfaces) (see figure 6.4) facilitate flexion–extension (or nodding) movements but impede other motions (Bogduk & Mercer 2000). The nodding motion during flexion of the head is a result of rolling and gliding of the occipital condyles in their sockets (figure 6.13a). As the head nods forward, the occipital condyles roll forward, rolling up the anterior wall of the atlantal socket. This results in a compression loading of the mass of the head in the socket. The flexor musculature, the tension in the joint capsule, or both, causes the occipital condyles to concomitantly translate downward and backward (Bogduk & Mercer 2000). Therefore, forward nodding of the head (or anterior rotation of the occipital condyles) is coupled with downward and posterior gliding of the condyles. The occipital condyles essentially stay settled in the floor of the atlantal sockets, ensuring maximum stabil-

ity of the head on the neck. During extension of the head on the atlas, the converse occurs (figure 6.13b).

The concept of coupled motion, or coupling, has been defined as "a phenomenon of consistent association of one motion (translation or rotation) about an axis with another motion about a second axis" (Blauvelt & Nelson 1994). In essence, one motion cannot be produced without the other. In many schools of thought, examination and treatment of the spine are thought to be based on this very concept. Biomechanical descriptions of spinal movement also follow this concept.

Recently, the concept of coupling has come under some scrutiny. Although the understanding of biomechanical movement of the spine is commonly accepted, inconsistencies in recent literature reviews (Cook 2003; Legaspi & Edmond 2007) have led to the suggestion of using caution when applying the concepts of coupled motion to the evaluation and treatment of patients with pain in the low back or lumbar spine (Legaspi & Edmond 2007). This text describes the coupling patterns as initially described by the literature and suggests the reader independently investigate this concept.

▶ **KEY POINT**

The concept of coupled motion, or coupling, describes the fact that one motion of the spine cannot be produced without the other. Coupling involves consistent association of one motion (translation or rotation) about an axis with another motion about a second axis. Although the understanding of biomechanical movement of the spine is commonly accepted, sufficient literature has suggested that clinicians utilize caution in universal adoption of this concept.

The normal coupling pattern of the AO joint appears to vary. Although it is generally accepted that rotation and side bending at this joint occur to opposite sides when they are combined, current literature does not support these defined coupled patterns (Mercer 2004). Occipital rotation and, to some degree,

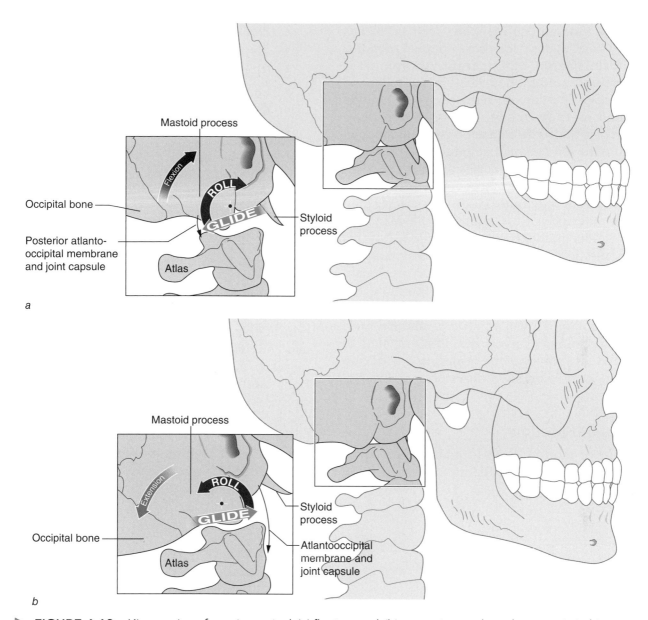

FIGURE 6.13 Kinematics of craniocervical *(a)* flexion and *(b)* extension at the atlantooccipital joint.

anteroposterior translation of the occiput on C1 are limited by the alar ligaments (O'Brien & Lenke 1997).

There is controversy regarding lateral flexion and rotation of the AO joint. Although, as just pointed out, some believe that coupling is inconclusive (Van Roy et al. 1997), Steindler (1955) describes lateral tilt as being associated with a contralateral rotation at this joint. Jirout (1973) contends that lateral flexion of the AO joint occurs together with rotation of the atlantoaxial (AA) joint. Still others

(Penning 1978; Panjabi et al. 1993; Panjabi & Crisco et al. 2001) propose that it depends on which motion occurs first, side bending or rotation. Each of the authors, through three-dimensional in vitro analyses, determined axial rotation was coupled to the ipsilateral side of side bending when side bending was initiated first, while axial rotation was coupled to the contralateral side of side bending when rotation was the initial movement (Penning 1978; Panjabi et al. 1993; Panjabi & Crisco et al. 2001).

The mnemonic "MIA has nice LiPS" has been used to describe these coupled motions. Lateral flexion of the right AO joint, for example, could be described in the following manner: The convex right occipital condyle would move **M**edially, **I**nferiorly, and **A**nteriorly in the right concave occipital condyle, while the left convex occipital condyle would move **L**aterally, **P**osteriorly, and **S**uperiorly in the left concave atlas facet. The concave atlas facet is generally more lateral posteriorly than it is anteriorly, thus allowing the MIA and LiPS movements (figure 6.14). The use of this mnemonic and description of this coupled motion in opposite directions is, as previously mentioned, controversial, and therefore the clinician must carefully consider its clinical applicability. However, several manual therapy techniques have been implemented to correct dysfunction of AO joint lateral flexion based loosely on such descriptions of movement.

Range of Motion

Range of motion (ROM) for the entire cervical spine has been listed as follows (Nilsson et al. 1996):

- Flexion: 50°
- Extension: 60°
- Right side bending: 45°
- Left side bending: 45°
- Right rotation: 80°
- Left rotation: 80°

A large variability exists in the reported ranges of motion for the AO joint. The total ROM for combined flexion and extension varies between a mean of 14 and 35°. Most of the total ROM values reported have been less than 15°. The ROM for AO joint lateral flexion has been reported to range from 2 to about 11°. Axial rotation ROM of this joint is also minimal and has been reported to range from about 0 to 7°. Lateral flexion and rotation of this joint have often been described as a coupled motion, with each motion in the opposite direction of the other. Therefore, right lateral flexion of the AO joint would couple with left rotation and vice versa. Since the pattern of coupling depends on the shape of the joint surfaces, and there has been some description of asymmetry of these joints, Van Roy et al. (1997) concluded that no single rule for a pattern of coupling can be applied.

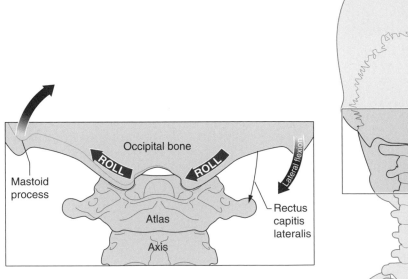

▶ **FIGURE 6.14**   Kinematics of craniocervical lateral flexion at the atlantooccipital joint.

Closed and Loose Packed Positions

These are not described.

End Feel

The end feel of this joint is generally firm.

Capsular Pattern

The capsular pattern of this joint has been described as extension and side bending being equally limited. Rotation and flexion are not affected (Magee 2002). Forward bending restricted more than backward bending has also been described as the capsular pattern for this joint. Cyriax described the capsular pattern for the entire cervical spine as follows: side bending and rotation are equally limited, and extension is more limited than flexion (Cyriax 1982).

Atlantoaxial Joint

The atlantoaxial joint is the next most proximal joint in the cervical spine after the atlantooccipital joint. In this section, we describe the axis of motion, arthrokinematics, range of motion, closed and loose packed positions, end feel, and capsular pattern of this unique cervical spine joint.

Axis of Motion

Rotation is the primary motion at this joint. This motion occurs in the transverse plane around a vertical axis.

Arthrokinematics

The C2 vertebra is designed to allow axial rotation of the head and C1. Since both articulating surfaces are convex, the apex of the cartilage on the inferior facet of C1 balances on the apex of the superior articular cartilage of C2 at rest. Axial rotation of C1 requires anterior displacement of one lateral mass and a reciprocal posterior displacement of the opposite lateral mass. The inferior articular cartilages of C1 must glide down the respective slopes of the convex superior articular cartilages of C2. Therefore, C1 essentially screws down onto C2 as it rotates (Koebke & Brade 1982). During this movement, the C1 articular facets take up the space occupied by the intra-articular meniscoids. Since the lateral mass draws the capsule of the joint anteriorly or posteriorly on each side, these meniscoids are withdrawn to this space. On reversal of the rotation movement, the meniscoids return passively to their normal position. A small amount of side bending may accompany this rotation if the articular cartilages are asymmetrical. This coupling motion may be ipsilateral or contralateral depending on the bias of the asymmetry (White et al. 1975). The alar ligaments provide the principal restraint to axial rotation at this joint, with the lateral AA joint capsule providing a minor role (Dvorak, Panjabi et al. 1987). The first 45° of rotation of the head on either side occurs at the C1-C2 level before initiation of movement of the lower cervical segments (see Clinical Correlation 6.5) (Mercer & Bogduk 2001).

CLINICAL CORRELATION 6.5

A patient presents to your clinic with the complaint of difficulty turning his head to the right. Onset of symptoms was insidious and has been present for about 3 weeks. As part of his clinical examination, you notice normal range of motion for all cervical motions except cervical rotation to the right. This motion is restricted to about 20°. Since you are aware that normal ROM for this movement is approximately 60°, and your patient demonstrates that much motion to the left, you conclude that at least part of his motion restriction is at the atlantoaxial joint. The fact that the first 45° of rotation of the head occurs at this joint (Mercer & Bogduk 2001) helped you determine this conclusion.

Posterior translation of C1 is limited by bony impact of the anterior arch of C1 against the odontoid process of C2, while anterior translation is limited by the transverse and alar ligaments (Dvorak, Schneider et al. 1987). Either ligament alone provides sufficient integrity of the AA joint.

Lateral translation is not a physiological movement at the AA joint, although it is assessed by some manual therapists. Because of the inferior and lateral slope of the superior articular facets of C2, lateral translation of C1 must be accompanied with ipsilateral side bending.

Essentially, the movement of the upper cervical spine occurs between the occiput and C2 and is regulated by C1 (Penning 1978). In all motions, whether proximal or distal to the upper cervical spine, C1 is mobile, and the movement of the occiput and C2 predicates on the initiation of the movement (Penning 1978).

Side bending of the head is a complex phenomenon (figure 6.14). Side bending exerts an axial compression force along the entire ipsilateral side of the vertebral column. As a result of this vertical compressive load, the inferior articular process of C2 is driven posteriorly and inferiorly along the sloping superior articular process of C3. This posterior displacement causes C2 to rotate in the direction of the side bending. Since this would result in the head facing the same direction as side bending and not directly forward, a compensatory contralateral rotation at the lateral AA joints and the atlas must occur. The result is therefore contralateral atlas rotation with ipsilateral side bending.

Range of Motion

Total ROM of flexion–extension of the AA joint is approximately 10° (Werne 1958). Extension of C1 is possible because the odontoid process curves slightly posteriorly, allowing the anterior arch of C1 to glide up and slightly backward (Werne 1958). Flexion occurs by reciprocal motion but also involves anterior translation of C1, during which the anterior arch separates from the odontoid process.

Clinical descriptions indicate rotation ROM at this joint is 50% of total cervical spine ROM, and therefore approximately 40°.

Closed and Loose Packed Positions

These are not described.

End Feel

The end feel of this joint is firm.

Capsular Pattern

There is restriction with rotation. Cyriax described the capsular pattern for the entire cervical spine as follows: side bending and rotation are equally limited, and extension is more limited than flexion (Cyriax 1982).

▶ **KEY POINT**

Clinical descriptions indicate rotation ROM of the atlantoaxial joint is 50% of total cervical spine ROM, and therefore approximately 40°. The C2 vertebra is designed to allow axial rotation of the head and C1. Since both articulating surfaces are convex, the apex of the cartilage on the inferior facet of C1 balances on the apex of the superior articular cartilage of C2 at rest.

Lower Cervical (C3-C7) Spine

The lower cervical spine demonstrates a consistent and predictable coupling pattern regardless of initiation of motion; side bending and rotation occur to the same side (Penning 1978; Panjabi et al. 1993; Panjabi & Crisco et al. 2001). Again, these findings were determined with three-dimensional in vitro analyses.

In this section, we describe the axes of motion, arthrokinematics, range of motion, closed and loose packed positions, end feel, and capsular pattern of these joints (C3-C7), all of which function in the same manner.

Axes of Motion

Flexion–extension occurs in the sagittal plane around a mediolateral axis. Axial rotation takes place in the transverse plane around a vertical axis, and side bending takes place in the frontal plane around an anteroposterior axis.

Arthrokinematics

Pure anterior translation cannot occur in the lower cervical spine because of the impact of the inferior articular processes of the superior vertebra against the superior articular processes of the lower vertebra. Flexion in the lower cervical spine, therefore, is always a combination of anterior translation and anterior rotation in the sagittal plane (with the inferior articular process of the superior vertebra gliding up the superior articular process of the vertebra below). Extension would therefore involve coupling of posterior sagittal rotation and posterior translation.

The basis for the coupling motion of side bending and rotation in the lower cervical spine lies in the morphology of the articular processes. During axial rotation, the contralateral inferior articular process impacts the superior articular process of the inferior vertebra. Axial rotation continues only if the inferior articular process glides up the superior facet, resulting in ipsilateral side bending of the moving superior vertebra. Therefore, rotation and side bending of the lower cervical spine are always coupled ipsilaterally.

During side bending, the ipsilateral inferior articular process moves down the slope of the superior articular process of the inferior vertebra. The inferior articular process must then move posteriorly, resulting in the vertebra rotating toward the side of the side bending. Therefore, again, side bending and rotation of the lower cervical spine are always coupled ipsilaterally, regardless of which movement is initiated first.

Simplifying the movements of flexion and extension would require minimizing the role of this translation and rotation in the sagittal plane, but it might assist the reader to describe these movements in a less complex manner. During flexion of a particular segment of the lower cervical spine, the motion could be described as movement "up and forward" of the superior vertebra on the inferior vertebra (and movement "down and back" for extension). Each segment of the spine will include the inferior articular process of the superior vertebra articulating with the superior articulating process of the inferior vertebra on each respective side. The movement of each one of these particular segments could better be described as the superior articulating process of the segment (the inferior articular process of the superior vertebra) gliding on the inferior articular process (the superior articular process of the inferior vertebra). Therefore, flexion of the spine at C4-C5 would be described as C4 gliding "up and forward" on C5 since it is generally agreed to describe movement of the superior vertebra on the inferior vertebra unless otherwise noted. Extension of C4-C5 (and for any other localized segment) would be described as C4 gliding "down and back" on C5. Right side bending of C4-C5 would involve the right side of C4 gliding down and back on C5 and the left side of C4 gliding up and forward on C5. Therefore, the right C4-C5 zygapophyseal joint would be in extension (down and back) and the left C4-C5 zygapophyseal joint would be in flexion (up and forward).

Range of Motion

The middle to lower cervical ROM values demonstrate variability, with C4-C5 and C5-C6 exhibiting the highest values (White & Panjabi 1990). Unilateral side bending progressively decreases from cranial to caudal, with a peak of 10 to 11° at C2-C3, C3-C4, and C4-C5 to the low of 4° at C7-T1. Unilateral rotation is greatest from C3-C4 to C6-C7, with a lowest value of 0 to 7° reported at C7-T1.

Closed and Loose Packed Positions

The close packed position of the typical cervical spine zygapophyseal joint is full extension, with a resting position of midway between flexion and extension.

End Feel

The end feel of the typical cervical spine zygapophyseal joint is firm for lateral flexion, rotation, extension, and flexion.

Capsular Pattern

Cyriax (1982) described the capsular pattern for the entire cervical spine as follows: side bending and rotation are equally limited, and extension is more limited than flexion. Others (Hertling & Kessler 1996) describe the capsular pattern for C2-C7 as recognizable by pain and equal limitation of all motions except flexion, which is usually minimally restricted. They also describe the capsular pattern for unilateral facet joint involvement as a greater restriction of movement in side bending to the contralateral side and in rotation to the ipsilateral side (Hertling & Kessler 1996).

▶ **KEY POINT**

The middle to lower cervical ROM values demonstrate variability, with C4-C5 and C5-C6 exhibiting the highest values. Unilateral side bending progressively decreases from cranial to caudal, with a peak of 10 to 11° at C2-C3, C3-C4, and C4-C5 to the low of 4° at C7-T1. Unilateral rotation is greatest from C3-C4 to C6-C7, with a lowest value of 0 to 7° reported at C7-T1.

Muscles

Two of the major functions of the musculature of the cervical spine are supporting the weight of the skull and optimally positioning the sensory organs of the skull to respond to stimuli. It is often necessary for people to move their heads in a specific direction rather rapidly to optimally position their ears or eyes. The muscles of this region can have many classifications. The classifications covered in this text are listed in table 6.4. Refer to table 6.5 for specific details of each muscle.

▶ **KEY POINT**

Two of the major functions of the musculature of the cervical spine are supporting the weight of the skull and optimally positioning the sensory organs of the skull to respond to stimuli.

The anterior neck muscles consist of the sternocleidomastoid (SCM), the scalene muscles (anterior, medial, and posterior) (see Clinical Correlation 6.6), and the platysma. The SCM primarily produces ipsilateral side bending and contralateral rotation. Torticollis is the most common clinical condition involving the SCM. Whether this condition is congenital or spasmodic, the resultant position of the person's head is an ipsilateral side bend and contralateral rotation. Treatment therefore would involve stretching in the opposite directions. The scalene muscles are a group of three muscles in the lateral region of the neck

TABLE 6.4   Muscle Groups of the Cervical Spine

Muscle group	Muscles
Anterior neck muscles	SCM, anterior scalene, middle scalene, posterior scalene, platysma
Prevertebral and deep neck flexor muscles	Longus colli, longus capitis, rectus capitis anterior, rectus capitis lateralis
Posterior neck muscles	Upper trapezius, levator scapulae, semispinalis capitis and cervicis, splenius capitis and cervicis, longissimus capitis and cervicis, spinalis cervicis, rectus capitis posterior major and minor, obliquus capitis superior and inferior
Suprahyoid muscles	Mylohyoid, geniohyoid, stylohyoid, digastric
Infrahyoid muscles	Sternohyoid, omohyoid, sternothyroid, thyrohyoid

TABLE 6.5 Primary Muscles of the Cervical Spine

Muscle	Origin	Insertion	Action	Innervation
Anterior neck muscles				
SCM	Mastoid process and lateral superior nuchal line	Sternal head: anterior manubrium Clavicular head: superior medial clavicle	Extension of head, ipsilateral side bending (SB), and contralateral rotation	Spinal root of accessory nerve
Anterior scalene	Anterior tubercles of transverse processes of C4-C6	1st rib	Elevation of 1st rib, ipsilateral SB, and contralateral rotation	C4-C6
Middle scalene	Transverse processes of C2-C7	1st rib	Elevation of 1st rib, ipsilateral SB, and contralateral rotation	Ventral rami of cervical spinal nerves (C3-C8)
Posterior scalene	Transverse processes of C5-C6	2nd rib	Elevation of 2nd rib, ipsilateral SB, and contralateral rotation	Ventral rami of cervical spinal nerves C3, C4 (C6-C8)
Platysma	Inferior mandible	Fascia of pectoralis major and deltoid	Draws skin of neck superiorly with clenched jaw, draws corners of mouth inferiorly	Cervical branch of fascial nerve
Splenius muscles				
Splenius capitis	Ligamentum nuchae; spinous process of C7-T3 or T4	Mastoid process and occipital bone of skull	Rotation of head and cervical vertebral column to ipsilateral side Bilateral action: extension of head	Dorsal rami of middle cervical spinal nerves
Splenius cervicis	Spinous processes of T3-T6	Transverse processes of upper 2 to 4 cervical vertebrae	Rotation of head and cervical vertebral column to ipsilateral side Bilateral action: extension of head and vertebral column	Dorsal rami of lower cervical spinal nerves
Serratus posterior muscles				
Serratus posterior superior	Ligamentum nuchae; spinous processes of C7-T2 or T3	Upper ribs	Elevation of ribs	Ventral rami of upper 3 or 4 thoracic spinal nerves (intercostals)
Serratus posterior inferior	Spinous process of T11-L2	Lower 3 or 4 ribs	Pulls lower ribs inferiorly	Ventral rami of lower thoracic (9th to 12th) spinal nerves (intercostals)

> continued

TABLE 6.5 > *continued*

Muscle	Origin	Insertion	Action	Innervation
Erector spinae muscles				
Erector spinae	Common tendon of origin; posterior surface of sacrum, iliac crest, spinous processes of T11-L5 (specific origins given below)	As described for each muscle	Bilateral action: extension of vertebral column Unilateral action: bending vertebral column toward ipsilateral side (lateral flexion) *Note:* These are the actions for all muscles of this group.	Dorsal rami of spinal nerves in area of muscle *Note:* This is the innervation for all muscles of this group.
Iliocostalis lumborum	Iliac crest; sacrum	Lower borders of lower 6 or 7 ribs		
Iliocostalis thoracis	Upper borders of lower 6 or 7 ribs	Lower borders of upper 6 ribs		
Iliocostalis cervicis	Angles of upper 6 ribs	Transverse processes of C4-C6		
Longissimus thoracis	Intermediate part of common tendon	Lower 9 or 10 ribs and adjacent transverse processes of vertebrae		
Longissimus cervicis	Transverse processes of upper 4 to 6 thoracic vertebrae	Transverse processes of 2nd to 6th cervical vertebrae		
Longissimus capitis	Tendons of origin of longissimus cervicis; articular processes of C4-C7	Mastoid process		
Spinalis thoracis	Spinous processes of T11-L2	Spinous processes of upper thoracic vertebrae (varies from 4 to 8)		
Spinalis cervicis	Ligamentum nuchae; spinous processes of C7-T2	Spinous process of C2 (and possibly C3 & C4)		
Spinalis capitis	Considered the medial part of semispinalis capitis			
Transversospinalis muscles				
Semispinalis thoracis	Transverse processes of T7-T12	Spinous processes of C6-T4	Extension of vertebral column	Dorsal rami of cervical and thoracic spinal nerves
Semispinalis cervicis	Transverse processes of T1-T6	Spinous processes of C2-C5	Extension of vertebral column	Dorsal rami of cervical and thoracic spinal nerves
Semispinalis capitis	Transverse processes of T1-T6; articular processes of C5-C7	Occipital bone (between superior and inferior nuchal lines)	Extension of head	Dorsal rami of cervical and thoracic spinal nerves

Muscle	Origin	Insertion	Action	Innervation
Multifidus	Sacrum and transverse processes of lumbar through lower cervical vertebrae	Spinous processes of lumbar through C2; fascicles span 2 to 4 segments of the column	Extension, lateral flexion, and rotation (to opposite side) of vertebral column	Dorsal rami of spinal nerves
Rotators	Sacrum and transverse processes of lumbar through lower cervical vertebrae	Spinous processes of lumbar through C2; fascicles span 1 or 2 segments of the column	Rotation (to opposite side) and extension of vertebral column	Dorsal rami of spinal nerves
Segmental muscles				
Interspinalis	Spinous processes of vertebrae (absent in much of thoracic region)	Spinous processes of vertebrae (span between adjacent vertebrae)	Extension of cervical column	Dorsal rami of cervical spinal nerves
Intertransversarii	Transverse processes of vertebrae (absent in most of thoracic region)	Transverse processes of vertebrae (span between adjacent vertebrae)	Lateral flexion of vertebral column (unilateral action)	Dorsal and ventral rami of spinal nerves
Suboccipital and deep neck flexor muscles				
Obliquus capitis inferior	C2 spinous process	C1 transverse process	Rotation of atlas (turn head to same side)	Dorsal ramus of C1
Obliquus capitis superior	C1 transverse process	Occipital bone	Extension and lateral flexion of head	Dorsal ramus of C1
Rectus capitis posterior major	C2 spinous process	Occipital bone	Extension of head; rotation of head to same side	Dorsal ramus of C1
Rectus capitis posterior minor	Posterior tubercle of C2	Occipital bone	Extension of head	Dorsal ramus of C1
Longus colli	Bodies of T1-T3; transverse processes of C3-C5; bodies of C5-T3	Transverse processes of C5 & C6; anterior surface of C1; bodies of C2-C4 (respectively with listed origins)	Flexion; possibly lateral flexion of neck	Ventral rami of C2-C6
Longus capitis	Transverse processes of C3-C6	Occipital bone	Flexion of head and upper cervical vertebrae	Ventral rami of C1-C3
Rectus capitis anterior	Lateral mass of atlas	Occipital bone	Stabilization of AO joint; flexion of head	Ventral rami of C1 & C2
Rectus capitis lateralis	Transverse process of C1	Occipital bone	Stabilization of AO joint; lateral flexion of head	Ventral rami of C1 & C2

(figure 6.15). The anterior and medial scalenes attach to the first rib, while the posterior scalene attaches to the second rib. Together they function to laterally flex the neck ipsilaterally, assist with neck stabilization, and elevate the

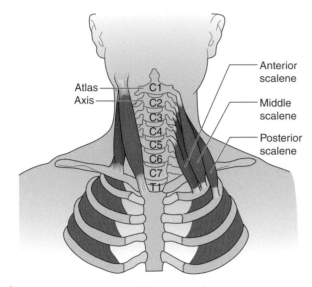

▶ **FIGURE 6.15**   Scalene muscles.

ribs as accessory muscles of respiration. The platysma does not have a primary influence on head or neck movement.

The splenius muscles consist of the splenius capitis and cervicis. These muscles originate on the spinous processes of the cervical and upper thoracic spine and pass laterally and superiorly to attach to the cervical transverse processes and the skull. The suboccipital and deep neck flexor muscle groups include several muscles.

Other muscles of primary importance for the cervical spine and its function include the levator scapulae and the upper trapezius. More detailed descriptions of their functions are given in chapter 10 (shoulder complex). Regarding the cervical spine, the levator scapulae produces extension of the cervical spine when the scapula is fixed (therefore increasing lordotic posture), and unilaterally it also produces ipsilateral lateral flexion and contralateral rotation of the cervical spine with the scapula relatively fixed.

CLINICAL CORRELATION 6.6

Thoracic outlet or inlet is a collection of signs and symptoms caused by compression or irritation of the brachial plexus and subclavian vessels, including at least six compression sites. One of these sites, the interscalene triangle, can be decreased for several reasons, including the scalene muscles themselves (Atasoy 2004). Adson's test (Malanga et al. 2003) (figure 6.16) is a provocation test that is commonly used to implicate scalene muscle involvement in this condition. The test involves the patient rotating her head to the side of symptoms, tucking her chin, and extending the head slightly as her ipsilateral upper extremity is externally rotated and extended. The clinician monitors the radial pulse as the patient inhales and holds her breath for 10 seconds. A positive test is noted if the radial pulse disappears and there is a reproduction of symptoms.

▶ **FIGURE 6.16**   Adson's test for thoracic outlet and inlet.

Conclusion

Like other regions of the spine, the cervical spine has several unique characteristics and functions. This portion of the spine has the smallest vertebral bodies and also the greatest amount of motion than any other region of the spine. The cervical spine has a large amount of variability in its structure, anatomy, and function. The upper cervical spine, with its large availability of motion, also has several ligaments providing stability.

REVIEW QUESTIONS

1. Which joint in the cervical spine is commonly referred to as the nodding joint?

   a. atlantoaxial joint (C1-C2)

   b. atlantooccipital joint (C0-C1)

   c. cervical 7 to thoracic 1 (C7-T1)

   d. cervical 2 to cervical 3 (C2-C3)

2. Which vertebral levels in the cervical spine do not have a disc between them?

   a. atlantooccipital joint (C0-C1)

   b. atlantoaxial joint (C1-C2)

   c. C0-C1 and C1-C2

   d. none

3. Which joint, unique to the cervical spine, can significantly restrict cervical side bending, especially as one ages?

   a. uncovertebral joint (joint of Luschka)

   b. zygapophyseal joint

   c. intervertebral joint

   d. atlantooccipital joint (C0-C1)

4. Which joint provides the largest amount of motion for cervical rotation?

   a. zygapophyseal joint

   b. intervertebral joint

   c. atlantooccipital joint (C0-C1)

   d. atlantoaxial joint (C1-C2)

5. Which artery does the vertebral artery arise from?

   a. internal carotid artery

   b. subclavian artery

   c. basilar artery

   d. epigastric artery

Craniomandibular Complex

Sue Klein

The craniomandibular complex consists of the cranial articulation at the temporal bone on each side of the head in front of the ear and the two condyles of the mandible. Each joint is referred to as the temporomandibular joint (TMJ). The craniomandibular complex allows for functional movement of the jaw during **mastication**, speech, and various facial expressions. Each TMJ is one of the most frequently moved joints in the body, and its design reflects its variety of functional demands. A person with healthy TMJs can eat, laugh, talk, and carry on other activities of daily living without notice or care. Conversely, someone with dysfunction in the TMJs may have pain at rest, have difficulties or pain with many orofacial activities, and may even experience challenges related to proper nutrition because of dietary restrictions. This chapter discusses the mechanics and kinesiology of the TMJ in normal function and in dysfunction and provides practical applications for the clinician.

Developmentally, the mandible arises from the ventral portion of the **first branchial arch** of the embryo. From the dorsal aspect of the first branchial arch, the bones of the inner ear and the tensor tympani develop.

OBJECTIVES

After reading this chapter, you should be able to do the following:

> List the bones associated with the craniomandibular complex.

> Describe the two portions of the craniomandibular complex.

> Explain the ligamentous structure that supports the craniomandibular complex.

> List the primary muscles that control motion of the craniomandibular complex and name their origins, insertions, and actions.

> Discuss a variety of injuries associated with the craniomandibular complex.

By 8 weeks of gestation, portions of the TMJ are present and by 21 weeks the superior portion of the lateral pterygoid attaches to the anterior and medial intra-articular disc, and the area is considered to be prenatally developed. Bony development of the maxilla may not be completed until as long as 2 years after full skeletal height is achieved (Rocabado & Iglarsh 1991).

The function of the TMJs is closely related to that of the cervical spine and can be influenced by postural adaptations associated with childhood habits. These habits include thumb sucking; prolonged use of a bottle or pacifier; and open-mouth breathing associated with asthma, allergies, or other **upper respiratory obstructions**.

Osteology

The bones of the TMJ are the mandible and the temporal bone (figure 7.1). The mandible is essentially suspended from the temporal bones on either side of the skull by ligaments.

The mandible is a U-shaped bone with condyles on each end. The body of the mandible is the more horizontal component; it receives the lower teeth at the alveolar process at the upper margin of the body. Notable markings on the interior of the body of the mandible are the **digastric fossa**, site of the attachment of digastric muscles on the inferoanterior aspect

of the anterior portion of the body, and the **mylohyoid line**, the area of attachment of the flat mylohyoid muscle on the interior surface of the body of the mandible (Neumann 2010).

▶ **KEY POINT**
The mandible is a U-shaped bone with condyles on each end. The ramus is the more vertical component of the mandible.

The ramus is the more vertical component of the mandible. At the superior end of the ramus are two projections, the **coronoid process** and the **mandibular condyle**. The coronoid process is the anterior bony projection that provides attachment for the temporalis and the masseter muscles. The most superior portion of the coronoid process rests deep to the zygomatic arch, and its anterior aspect can be palpated posterior to the third molars high in the intraoral cavity. The mandibular notch is the area between the coronoid process and the mandibular neck.

The angle of the mandible is the posterior inferior portion of the mandible where the body and the ramus meet. This bony area is sandwiched in muscle, with the internal surface providing attachment for the medial pterygoid muscle and the lateral surface covered by the masseter.

The mandibular neck is the posterior projection of the ramus that supports the con-

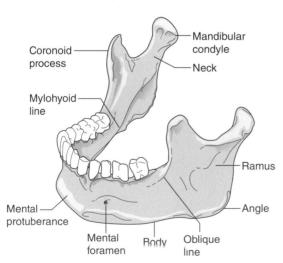
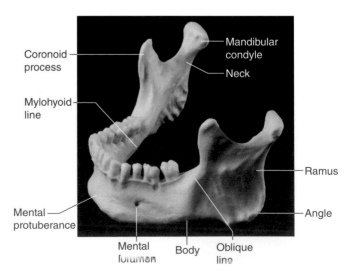

▶ **FIGURE 7.1** TMJ anatomy.

dyle. Located on the anterior aspect of the neck is the **pterygoid fossa**, which provides attachment for the inferior head of the lateral pterygoid. The mandibular condyle is a continuation of the neck of the mandible. This convex condyle has a lateral and a medial pole. The lateral pole is not as long as the medial pole and rests anterior to the medial pole. An extended line transecting the anterior and posterior halves of each condyle would meet at a point just anterior to the **foramen magnum**. This is notable from a clinical standpoint when mobilizing the joint (see Clinical Correlation 7.1).

The mandibular condyle articulates with the disc, and together with the disc and the temporal bone form the TMJ, the unilateral component of the craniomandibular complex.

At the inferior and lateral aspect of the temporal bone is the mandibular fossa. The mandibular fossa rests above the mandibular condyle, but only a portion of the bone is suitable for the loading and weight bearing required for proper functioning of the TMJ. The most superior bony portion of the mandibular fossa forms the roof of the TMJ. This region of bone is quite thin, and there may be only 3 mm of bone thickness between the superior portion of the mandibular fossa and the cranial cavity. The bony roof is covered by a thin layer of fibrocartilage that makes the area a poor choice for weight-bearing activities (Levangie & Norkin 2011).

It is notable that the cartilaginous covering of the temporal bone in the TMJ is fibrocartilage rather than the hyaline cartilage typically associated with synovial joints. Compared with hyaline cartilage, fibrocartilage allows for increased tolerance of repetitive or high loads as well as improved ability to repair damage sustained during micro- or macrotrauma (Krause 1994). When microtrauma to the joint exceeds the tissue's ability to repair itself, tissue breakdown occurs. **Parafunctional activities** are activities that contribute to microtrauma of the TMJ, including fingernail biting, lip chewing, gum chewing, tooth grinding (**bruxism**), and jaw clenching. These parafunctional activities repetitively load the joint and over time may produce a sequence of degenerative changes to the joint such as joint laxity, disc dysfunction, and changes to the articular cartilage (Rocabado & Iglarsh 1991). For example, during normal functional use of the craniomandibular complex, the teeth come into full occlusion only during swallowing and to a lesser degree, during chewing (Rocabado & Iglarsh 1991). Even during chewing, food creates varying degrees of separation of the upper and lower arches of the teeth. Consider the increased repetitive loading of the TMJs when the patient chews gum several hours during the day compared with eating at normal intervals. In general, the clinician should counsel the patient to avoid parafunctional activities.

▶ **KEY POINT**

The cartilaginous covering of the temporal bone is fibrocartilage rather than the hyaline cartilage typically associated with synovial joints. The fibrocartilage allows for tolerance of higher loads and for improved repair of damage sustained during micro- or macrotrauma.

The more vertical posterior border of the mandibular fossa is called the **posterior glenoid tubercle** and rests just anterior to the external auditory meatus. Occasionally, this area of the mandibular fossa can be fractured with a blow to the anterior mandible. The fibrocartilage covering the posterior glenoid tubercle is thin.

CLINICAL CORRELATION 7.1

When joint mobilization of the TMJ is indicated, the clinician will mobilize the mandibular condyle in order to separate the bone/disc/bone surfaces and elongate the joint capsule. The lateral mobilization force in this case is not purely lateral but is directed laterally and slightly anteriorly, along a line parallel to the mandibular condyle.

The **articular eminence** is the downward-sloping anterior border of the fossa. The fibrocartilage lining the trabecular bone of the articular eminence is thick, especially at the middle third of the eminence, and is well suited for repetitive and high-force loading. There is no bony block on the anterior eminence that might prevent the mandibular condyle from translating forward off of the eminence. Infrequently, in hypermobile joints the mandibular condyle may dislocate anteriorly and lodge anterior to the articular eminence. This occurs most often with excessive opening of the mouth, such as during yawning (Ugboko et al. 2005).

Other bones related to the TMJ complex include the styloid process, a thin bony projection from the tympanic plate that provides attachment for ligaments and muscles. The zygomatic process is the squamous portion of the temporalis bone that articulates with the zygomatic bone to form the arch of the cheek bone. The maxillae on each side fuse to form the hard palate of the mouth and the maxillary arch that receives the upper teeth.

The sphenoid bone contributes to the bony base of the skull anterior to the foramen magnum. The medial and **lateral pterygoid plates** descend from the sphenoid bone, and the medial and inferior portion of the lateral pterygoid muscles arise from the lateral pterygoid plates, while the superior portion of the lateral pterygoid arises from the greater wing of the sphenoid bone.

▶ **KEY POINT**

The sphenoid bone contributes to the bony base of the skull anterior to the foramen magnum. The medial and lateral pterygoid plates descend from the sphenoid.

The **hyoid bone** is the superficial U-shaped bone that floats in the soft tissue in the superior anterior area of the throat. The body of the hyoid bone rests horizontally above the thyroid cartilage, usually at the level of the body of C3. It provides attachments for the suprahyoid and intrahyoid musculature and acts as a type of balance point for the musculature, the point from which one group sta-

bilizes while the other group acts to produce movement. For example, during the opening of the mandible, the infrahyoids stabilize the hyoid and prevent it from elevating with the contraction of the digastrics and other suprahyoid muscles.

Joint Articulations

The mandible is a bicondylar bone with interdependent movement of the TMJs on the right and the left. Each TMJ is divided into an upper joint space and a lower joint space by the articular disc. Each joint space has a separate joint capsule with its own synovial lining. The lower joint is formed by the mandibular condyle and the inferior surface of the articular disc. This joint functions as a ginglymus (or hinge) joint, with the primary movement of rotation. The articular eminence of the temporal bone and the superior surface of the articular disc form the upper joint of the temporomandibular complex. The upper joint is classified as an amphiarthrodial joint or a plane joint; translation between the disc and the articular eminence is the primary movement available.

▶ **KEY POINT**

The TMJ is divided into an upper joint space and a lower joint space by the articular disc. The lower joint functions as a ginglymus (or hinge) joint with the primary movement of rotation. The upper joint is classified as an amphiarthrodial joint or a plane joint, where translation occurs between the disc and the articular eminence.

Joint Anatomy

The articular disc of the TMJ is a biconcave fibrocartilagenous structure that rests between the mandibular condyle and the mandibular fossa (figure 7.2). The mediolateral length of the disc is greater than the anteroposterior distance, consistent with the shape of the mandibular condyle. The disc has been described as having three distinct bands: an anterior band, an intermediate band, and a posterior

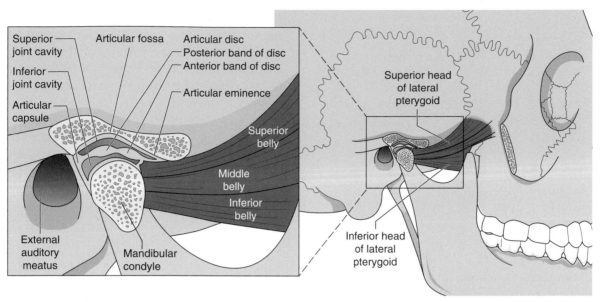

FIGURE 7.2 Anatomy of the TMJ joint.

band (Rees 1954). The intermediate band of the disc is thinner than the anterior or posterior band, and the shape of the disc improves the congruence between the convex surfaces of the condyle and the articular eminence. The disc is innervated and vascular at the thicker outer margins but is without blood or nerve supply at the thinner center. The thinner center portion is therefore better designed for accepting load.

Posteriorly, the disc is attached to the posterior ligament, a loose, highly vascularized and innervated tissue also known as the **bilaminar zone**. The upper portion of the bilaminar zone is a fibroelastic tissue thought to provide assistance in retracting the disc as it returns to a closed position from opening. The lower portion of the bilaminar zone is composed of fibrous tissue.

The superior joint capsule attaches to the margins of the mandibular fossa as far forward as the articular eminence. At the anterior attachment, the capsule also connects to the tendon of the superior head of the lateral pterygoid. The disc, the capsule, and the superior head of the lateral pterygoid are derived from the same embryological tissue, and it can be difficult to differentiate distinct boundaries between these tissues (Porter 1970). Posteriorly, the capsule connects to the borders of the posterior glenoid spine. As the capsule extends inferiorly to the articular disc,

it forms collateral ligaments that attach firmly to the disc. The upper joint capsule is loose, allowing for translation to occur in the upper joint. The capsule is slightly thicker laterally, reinforcing the lateral aspect of the joint.

The inferior joint capsule attaches to the inferior portion of the articular disc and reaches inferiorly to the neck of the condyle. The inferior capsule is tighter and shorter compared with the upper capsule, allowing rotation to occur but limiting translation between the mandibular condyle and the articular disc.

The capsules are highly innervated and vascularized and, in the healthy joint, provide a high degree of proprioceptive feedback for speech, expression, and mastication. The capsule thickens at its medial and lateral aspects to provide support to the joint and is sometimes referred to as the capsular ligament.

The largest ligament of the TMJ is the **lateral ligament** or the temporomandibular ligament, which has two portions: an outer oblique portion and an inner horizontal portion (figure 7.3). The oblique portion attaches to the posterior neck of the condyle and travels in a superior anterior direction and to the lateral margins of the articular eminence. The inner horizontal portion also attaches to the articular eminence but runs horizontally to attach to the mandibular condyle. The ligament suspends the mandible from the

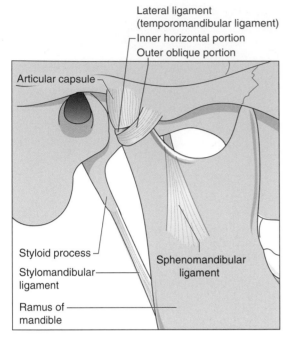

Lateral ligament
(temporomandibular ligament)
Inner horizontal portion
Outer oblique portion

Articular capsule

Styloid process

Stylomandibular
ligament

Sphenomandibular
ligament

Ramus of
mandible

▶ **FIGURE 7.3**   Ligaments of the TMJ.

temporal bone, supports the lateral aspect of the joint, and prevents excessive rotation and posterior displacement of the mandible. The integrity and pain response of the temporomandibular lateral ligament is tested by moving the mandibular condyle on the ipsilateral side posteriorly and inferiorly, stretching the fibers of the ligament, monitoring the patient's complaints of pain, and assessing the end feel of the movement.

▶ **KEY POINT**

The largest ligament of the TMJ is the lateral ligament, which consists of two portions: an outer oblique portion and an inner horizontal portion.

Accessory ligaments include the stylomandibular and sphenomandibular ligaments. The **stylomandibular ligament** originates from the styloid process and inserts on the internal angle of the mandible. The **sphenomandibular ligament** originates at the spine of the sphenoid bone and inserts on the internal surface of the ramus of the mandible to the small bony projection called the lingula. These ligaments, along with the lateral liga-

ment, are thought to aid in suspending the mandible. When the ligamentous system is compromised, joint dysfunction can occur. For example, patients with systemic hypermobility present with generalized ligamentous laxity. Systemic hypermobility is a predisposing factor in the development of TMJ dysfunction. Buckingham et al. (1991) found that of the patients who suffered from severe signs and symptoms of TMD, 54% met the criteria for systemic hypermobility. The inference is that patients whose joints are hypermobile are less likely to tolerate the excessive joint stresses if they engage in parafunctional activities.

Nerve Supply

The craniomandibular system is innervated primarily by the trigeminal and facial nerves, and cranial nerves V and VII, respectively. The fibers of the trigeminal nucleus exit the brain at the anterolateral surface of the pons with separate sensory and motor roots. The nerves pass anteriorly to reach the trigeminal ganglion, which is located on the internal surface of the skull at the lateral wall of the cavernous sinus. The nerves exit the ganglion in three distinct divisions: the ophthalmic division (V1), the maxillary division (V2), and the mandibular division (V3). The ophthalmic division and the maxillary division are sensory nerves only, with V1 supplying sensation to the ipsilateral scalp, forehead, and nose and V2 supplying sensory nerves to the ipsilateral cheek, upper lip, and temple area. V3 is a mixed nerve, with the sensory component innervating the anterior chin, the skin over the zygoma, and the posterior temple area and the motor portion innervating the ipsilateral muscles of mastication (i.e., the temporalis, the masseter, and the lateral and medial pterygoids). The motor nerve to the medial pterygoid also supplies the tensor tympani and tensor palati. Nerve fibers from the sympathetic nervous system are also associated with the divisions of the trigeminal nerve (Green & Silver 1981).

The trigeminal nerve, or cranial nerve V, presents with nuclei in the pons and a large sensory root that extends from the midbrain down to the second cervical nerve segment.

This large nucleus is known as the **trigemino-cervical nucleus** and has important clinical relevance (Grieve 1988). Changes in mechanical pain sensitivity to tissues in the trigeminal distribution in patients with chronic mechanical cervical pain are thought to be the result of the convergence of the neurological inputs from both the cervical and trigeminal areas that occurs in the trigeminocervical nucleus (La Touche & Fernández-de-las-Penas 2009a).

Sensitivity of tissues innervated by the trigeminal nerve (the masseter and temporalis muscles) decrease with treatment of manual therapy and exercise to the cervical spine (La Touche & Fernández-de-las-Penas 2009b). Studies indicate that the health care provider should evaluate and treat the craniofacial and craniovertebral areas as interrelated and interdependent neurological and functional units.

A nerve important to the TMJ is the **auriculotemporal nerve**. This sensory branch of cranial nerve V runs posterior and deep to the lateral pterygoid and provides sensory innervation to the TMJ as well as sympathetic and parasympathetic fibers to the area. Occasionally, patients with TMJ dysfunction experience ringing or fullness in the ear. In addition to the auriculotemporal nerve passing close to the joint, the common embryological origins of the TMJ and the ear may allow for some common neural pathways that interfere with the patient's ability to differentiate between pain arising from the ear and pain arising from the TMJ. During the subjective examination, the patient may relate a negative ear examination but continued ear-type symptoms. The clinician can implicate the TMJ by asking if the ear fullness relates to orofacial activities such as chewing, eating, or other functional activities of the jaw.

The facial nerve supplies structures derived from the second branchial arch and is the motor nerve for the muscles for facial expression and some of the suprahyoid muscles. The sensory component of the nerve carries the fibers that are responsible for transmitting taste from the anterior two-thirds of the tongue. Although the facial nerve does not directly supply the TMJ, its proximity to the joint puts the nerve at risk during facial trauma and it may be affected during surgical repair of condylar and subcondylar fractures.

Blood Supply

The maxillary artery is the larger of the two terminal branches of the external carotid artery. The maxillary artery arises behind the neck of the mandible in the parotid gland. Relative to this discussion, the artery supplies the buccinator, temporalis, pterygoids, and masseter (Green & Silver 1981). Behind the condyle, in the bilaminar zone of the retrodiscal tissues, is a plexus of venous tissue that empties and fills relative to healthy mandibular movement (Alomar et al. 2007). This vascular pump may contribute to the health of the intra-articular structures; however, in the presence of an anteriorly displaced disc, the superficial portion of the retrodiscal area becomes avascular (Heffez & Jordan 1992).

Joint Function

The function of the craniomandibular complex is unique in that movement in each of the joints is interdependent, simultaneous, and influenced significantly by each joint as well as by the occlusal surfaces of the teeth, the cranium, and the cervical spine. Concurrent care by a dental temporomandibular disorder (TMD) specialist may be warranted for management of oral appliances, occlusal contacts, and other aspects of patient management (Grieve 1988). All patients who are seen for care of the TMJ and for craniomandibular pain should receive a careful evaluation of the cervical spine. Emphasis on the restoration of the normal relationship of the head on C1 and of C1 to C2 (craniovertebral angle) is important, as is the normalization of the cervical lordosis. These areas are critical since a forward head posture will disrupt the length–tension relationship of the anterior and posterior cervical muscle groups, creating a stretch to the infrahyoids that will lead to a traction force on the mandible. This force tends to move the mandible into a retruded position and potentially moves the condyle posterior to the disc and onto the retrodiscal area and the vascularized, innervated bilaminar zone. In addition, the cranium is posteriorly rotated on C1 in a forward head posture, decreasing the suboccipital space, with potential ramifications to the soft tissue and neurological tissue in the area (see Clinical Correlation 7.2).

The joint function of the craniomandibular system is described with both joints starting in a neutral head and neck posture, with the resting position of teeth slightly apart. This section covers the axes of motion, arthrokinematics, range of motion, closed and loose packed positions, end feel, and capsular pattern of the region.

Axes of Motion

Arthrokinematically, the TMJ has three degrees of freedom and moves around a different axis for each of the three planes of movement. The axes of motion of the TMJ are through the mandibular condyle in the sagittal, frontal, and transverse planes. The axis of rotation changes throughout opening and is somewhat variable depending on the person's anatomy (e.g., degree of steepness of the articular eminence). Arthrokinematics may also change slightly as a result of those anatomical variabilities.

Osteokinematically, the axis of motion during opening moves inferiorly and has been described as occurring in an axis connecting the rami near the mandibular foramen.

Arthrokinematics

The arthrokinematics of the TMJ are complex because of the number of degrees of freedom

as well as the interdependent nature of the bicondylar joint structure. The possible movements of the mandible are depression (or mouth opening), elevation (or mouth closing), protrusion, retrusion, and lateral excursion to each side.

Mandibular depression (figure 7.4) is thought to occur in two phases, a rotation phase and a translation phase. The rotation phase begins as the mandible depresses and the condylar head rotates under the inferior surface of the articular disc. This usually occurs in the first 35 to 50% of available opening. Once the capsular ligaments restrict further rotation of the mandibular head on the disc, the condyle and disc (condyle–disc complex) move together and translate anteriorly

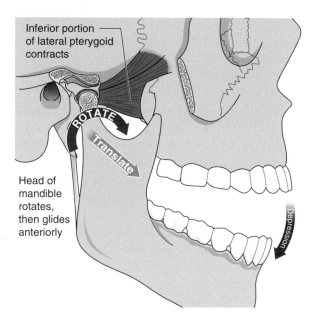

Inferior portion of lateral pterygoid contracts

ROTATE

Translate

Head of mandible rotates, then glides anteriorly

Depression

▶ **FIGURE 7.4** Mandibular depression.

CLINICAL CORRELATION 7.2

Temporomandibular disorders (TMDs) are a group of interrelated disorders that may result in pain and dysfunction in the craniofacial and craniovertebral regions. The pain patterns are varied but can include headache, toothache, facial pain, neck pain, jaw pain (with or without joint noise), and ear pain (Travell & Simons 1983). The clinician must be aware of the variety of TMD presentations in order to target treatment to the appropriate area. Dysfunction in the region can include limitations in active and passive range of motion in the jaw and the cervical and thoracic spines, joint locking, painful bite, changes in initial occlusal contacts, and ringing or buzzing in the ears (Grieve 1988). Parafunctional activities are often a contributing factor.

and inferiorly on the articular eminence to complete the final 50 to 65% of opening (see Clinical Correlation 7.3).

Mandibular elevation arthrokinematics occur in reverse order to mandibular depression. The condyle–disc complex translates posteriorly and superiorly on the articular eminence for the first 50 to 65% of closing. The remainder of closing is accomplished by the posterior rotation of the condyle on the articular disc. Full elevation of the mandible occurs when the teeth are fully approximated. A bony end feel is present as a result of the dental approximation.

Mandibular protrusion is the movement of sliding the lower teeth anteriorly relative to the upper teeth (figure 7.5). During protrusion, the condyle and disc translate anteriorly and slightly inferiorly until the complex abuts the

articular eminence. Protrusion is more limited with mandibular depression because the condyle and disc are in closer approximation to the articular eminence during depression compared with the rest position of the mandible.

Mandibular retrusion occurs when the lower teeth slide posteriorly relative to the upper teeth (figure 7.6). During this movement, the mandibular condyle and the articular disc translate posteriorly and increase the space between the anterior condyle–disc complex and the articular eminence. The posterior aspect of the complex approximates the posterior glenoid spine and compresses the soft tissue between the bony components of the posterior joint.

Lateral excursion is the lateral movement of the mandible from side to side, sliding the lower teeth laterally relative to the upper teeth

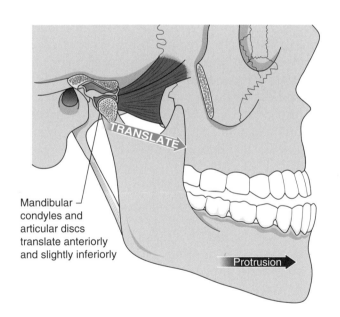

▶ **FIGURE 7.5**   Mandibular protrusion.

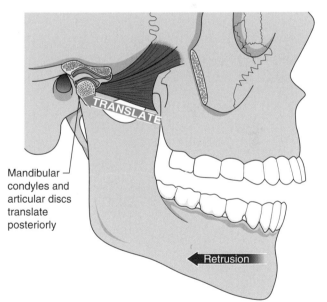

▶ **FIGURE 7.6**   Mandibular retrusion.

CLINICAL CORRELATION 7.3

During rehabilitation, the patient may be instructed to limit the amount of translation allowed during opening. Opening may be limited to only the rotation phase if the disc has been recaptured and is likely to displace again with excessive opening force. For example, the patient is instructed to yawn with the tongue in the rest position of the soft "no," with the lips wrapped gently around the teeth. This prevents the patient from overopening as well as limits the translation at the joint, therefore decreasing the compression in the upper aspect of the joint.

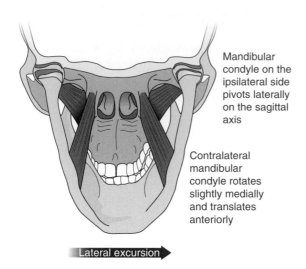

Mandibular condyle on the ipsilateral side pivots laterally on the sagittal axis

Contralateral mandibular condyle rotates slightly medially and translates anteriorly

Lateral excursion

▶ **FIGURE 7.7**   Mandibular lateral excursion.

(figure 7.7). The direction of lateral deviation is named for the side that the mandible is gliding toward. Lateral excursion occurs with small multiplanar movements because of the sloping of the articular eminence. A pivot point is established through the mandibular condyle in the transverse plane on the side to which the excursion is occurring. The mandibular condyle on the ipsilateral side pivots laterally on the sagittal axis, while the contralateral mandibular condyle rotates slightly medially and translates anteriorly. This most likely occurs in the upper portion of the joint because of the laxity of the joint capsule.

Chewing is a biomechanically complex activity that combines the previously defined movements, with some degree of variation between people (Okeson 2003).

Range of Motion

The normal range of motion of the TMJ during opening is between 40 and 55 mm. Opening is measured using a millimeter scale, placed in front of the upper and lower **central incisors**, by observing the distance between the distal aspect of each tooth. Because many patients' maxillary incisors overlap the incisors of the lower arch both vertically and horizontally, the measurements should take the overlap into account. For example, if the patient's opening

is 42 mm, and the vertical overlap in 2 mm, the patient's total opening measurement is 44 mm. Functional opening may be easily measured by requesting that the subject place the width of his second and third proximal interphalangeal joints between the front teeth. Difficulty in placing two knuckles between the upper and lower front teeth indicates restricted range of motion in the TMJ, whereas easily placing three or more fingers in the space may suggest excessive mobility. The normal range of motion of the TMJ during closing is full occlusion of the teeth.

▶ **KEY POINT**

The normal range of motion in the TMJ is between 40 and 55 mm and is measured by the distance between the maxillary and mandibular central incisors.

Protrusion is measured in a similar manner as opening, accounting for horizontal overlap of the teeth. Normal range of motion of protrusion is 3 to 6 mm, measured from the patient's resting mandibular position (not from full retrusion) to the end range when the condyle–disc complex contacts the articular eminence. The end feel is firm.

The normal range of motion in retrusion is 3 to 4 mm, measured from the patient's resting mandibular position (not from full protrusion) to the end range when contact with the posterior glenoid spine occurs. The end feel is firm.

Normal range of motion in lateral excursion is 10 to 12 mm and, in normal subjects, should be in approximate 1:4 ratio relative to opening. Lateral excursion is measured as the distance from the center of the maxillary incisors to the center of the mandibular incisors at the end of full excursion. A decrease in joint play and abnormal joint mechanics associated with a tight joint capsule and decreased ROM in lateral excursion make passive or forceful depression of the mandible inadvisable because of higher compressive forces during the movement. Restoration of lateral excursion should be attempted concomitantly with restoration of opening in a 1:4 ratio (Rocabado & Iglarsh 1991).

Closed and Loose Packed Positions

The closed packed position of the TMJ occurs in full occlusion. In the loose packed position, the teeth are separated by approximately 2 to 3 mm.

End Feel

The normal end feel during opening is a soft-tissue stretch, a bony end feel during full occlusion of the teeth, and a firm end feel during protrusion and retrusion. During lateral excursion, the end feel is capsular.

Capsular Pattern

The capsular pattern of the TMJ is a restriction in inferior glide.

Muscles

Although the muscles of the cervical spine influence the craniomandibular system, this section addresses the primary and secondary muscles related to movement of the mandible and the TMJ.

Primary Muscles of Mastication

The muscles of mastication contract in a coordinated and complex fashion in a healthy, asymptomatic person. Figure 7.8 illustrates these muscles, and table 7.1 describes the muscles in detail.

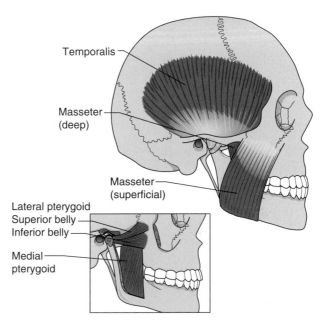

▶ **FIGURE 7.8** Muscles of the TMJ involved in mastication.

TABLE 7.1 Primary Muscles of the TMJ

Muscle	Origin	Insertion	Action	Innervation
Temporalis	Temporal fossa superior to the zygomatic arch	Anterior fibers: coronoid process Middle fibers: coronoid notch Posterior fibers: anterior condyle	Elevation of mandible (all fibers contracting together) Bilateral action: posterior fibers are active in retrusion Unilateral action: temporalis deviates the mandible to the same side	Anterior and posterior deep temporal nerves from the mandibular portion of cranial nerve V
Masseter	Superficial layer: anterior and inferior border of the zygomatic arch Deep layer: medial portion of the zygomatic arch	Distally on the lateral surface of the coronoid process, the ramus, and the angle of the mandible	Elevation of mandible (primary function)	Masseteric nerve from the anterior branch of the mandibular division of cranial nerve V

> continued

TABLE 7.1 > *continued*

Muscle	Origin	Insertion	Action	Innervation
Lateral pterygoid	Superior portion: infratemporal crest of the temporal bone and the inferior lateral greater wing of the sphenoid bone Inferior portion: lateral surface of the lateral pterygoid plate of the sphenoid bone	Superior portion: anterior and medial aspect of the disc Middle head of the inferior portion: condyle of the mandible Inferior head of the inferior portion: neck of the condyle	Superior portion: electrically silent during opening and active during closing Inferior portion: active during opening and silent during closing Bilateral action: protraction of mandible	Lateral pterygoid nerve from anterior division of the mandibular branch of cranial nerve V
Medial pterygoid	Lateral aspect of the lateral pterygoid plate	Medial aspect of the lower border of the mandibular ramus near the angle	Bilateral action: assists the masseter and temporalis muscles in closing; assists in protrusion Unilateral action: deviation of mandible to contralateral side	Medial pterygoid nerve from the mandibular division of cranial nerve V

Temporalis

The temporalis muscle attaches proximally to the temporal fossa superior to the zygomatic arch. The muscle fills the temporal fossa completely. The myotendinous junction is located above the zygomatic arch before the tendon dives medial to the zygomatic arch. The anterior fibers of the tendon attach to the coronoid process, the middle fibers attach to the coronoid notch, and the posterior fibers attach to the anterior aspect of the mandibular condyle. The temporalis muscle, contracting all portions simultaneously, will elevate the mandible. The posterior fibers are active during retrusion when acting bilaterally and will glide the mandible posteriorly from a protruded position. During unilateral activity, the temporalis deviates the mandible to the same side. The anterior and posterior deep temporal nerves from the mandibular portion of cranial nerve V (trigeminal) innervate the temporalis muscle.

Masseter

The superficial layer of the masseter attaches to the anterior and inferior border of the zygomatic arch and is angulated slightly posteriorly as it attaches to the lateral surface of the coronoid process, the ramus, and the angle of the mandible. The deep layer attaches to the medial portion of the zygomatic arch and is the more vertical layer of the two portions of this muscle. The deep portion of the masseter attaches to the superior half of the internal ramus and, in some cases, to the inferior border of the lateral ramus. The primary function of the masseter is elevation of the mandible. It also assists the lateral and medial pterygoid in protrusion. The innervation of the masseter muscle is from the masseteric nerve from the anterior branch of the mandibular division of cranial nerve V (trigeminal).

Lateral Pterygoid

The lateral pterygoid muscle is typically divided into two portions, superior and inferior. Rocabado (2012) describes a division of the inferior portion of the lateral pterygoid into two portions, therefore further delineating the muscle into superior, middle, and inferior heads. Anteriorly, the superior head attaches to the intratemporal crest of the temporal bone and the inferior lateral aspect of the greater wing of the sphenoid bone, traveling posteriorly and slightly laterally to insert into the medial portion of the disc. Some fibers from the superior head also attach to the anterior portion of the disc. The inferior portion attaches to the lateral surface of the lateral pterygoid plate of the sphenoid bone. As the muscle travels posteriorly and laterally, the middle head of the inferior portion of the lateral pterygoid attaches to the condyle of the mandible, and the inferior head attaches to the neck of the condyle, although some sources report the inferior head also attaches to the anterior capsule and the disc (Dutton 2008; Rocabado & Iglarsh 1991). The superior head of the muscle is electrically silent during opening and active during closing. The activity recorded during closing is thought to be a result of the superior head acting eccentrically to guide the disc on the condyle during closing. The inferior portion is active during opening and silent during closing. Bilaterally, all heads of the muscles act to protract the mandible. Innervation is from CN V, mandibular branch.

▶ **KEY POINT**
The lateral pterygoid muscle is divided into two portions, superior and inferior. The superior head of the muscle is active during closing. The inferior portion is active during opening. Bilaterally, both portions of the muscle act to protract the mandible.

Medial Pterygoid

The medial pterygoid arises from the medial aspect of the lateral pterygoid plate, and as it runs laterally, posteriorly, and inferiorly, it inserts on the lower medial aspect of the ramus of the mandible near the angle. Working bilaterally, the medial pterygoid assists the masseter and temporalis muscles in closing. It also assists the lateral pterygoid in protrusion while acting bilaterally. Acting unilaterally, the medial pterygoid deviates the mandible to the contralateral side. Innervation to the medial pterygoid is provided by the medial pterygoid nerve from the mandibular branch of cranial nerve V.

Secondary Muscles of Mastication

The secondary muscles of mastication assist the primary muscles of mastication by stabilizing the hyoid bone and also assist in resisted opening. The suprahyoid muscles stabilize the hyoid and provide a firm floor for the tongue during oral activities. The infrahyoids are thin, strap-like muscles that insert on the inferior surface of the hyoid bone; they depress the hyoid and stabilize it for action of the suprahyoids. The muscles are superficial to the longus colli muscles and deep to the sternocleidomastoid muscles. Table 7.2 describes the muscles in detail.

TABLE 7.2 Secondary Muscles of Mastication

Muscle	Origin	Insertion	Action	Innervation
Suprahyoids				
Anterior digastric	Inferior border of the anterior mandible near the symphysis	Joined to the posterior digastric by a common tendon that attaches to the hyoid bone through a fibrous loop	Assists in opening the mandible on a fixed hyoid bone; with a fixed mandible, elevates the hyoid bone. Active during coughing, swallowing, and retrusion (Krause 1994)	Inferior alveolar branch of cranial nerve V

> continued

TABLE 7.2   > *continued*

Muscle	Origin	Insertion	Action	Innervation
Suprahyoids				
Posterior digastric	Mastoid notch deep to the other muscle attachments	Joined to the anterior digastric by a common tendon that attaches to the hyoid bone through a fibrous loop	Assists in opening the mandible on a fixed hyoid bone; with a fixed mandible, elevates the hyoid bone Active during coughing, swallowing, and retrusion	Cranial nerve VII
Geniohyoid	Mental spine of mandible	Body of hyoid bone	Elevates hyoid bone	Ventral ramus of C1 via hypoglossal nerve
Mylohyoid	Medial surface of mandible	Body of hyoid bone	Stabilizes tongue on a fixed hyoid, raises floor of the mouth, elevates hyoid bone	Mylohyoid branch of cranial nerve V and mandibular division
Stylohyoid	Styloid process of temporal bone	Body of hyoid bone	Elevates hyoid bone	Cranial nerve VII
Infrahyoids				
Omohyoid	Superior angle of scapula	Inferior body of hyoid bone	Depresses and stabilizes hyoid bone	Ansa cervicalis
Sternohyoid	Manubrium and medial portion of clavicle	Body of hyoid bone	Depresses and stabilizes hyoid bone	Ansa cervicalis
Sternothyroid	Posterior aspect of manubrium	Thyroid cartilage	Depresses larynx	Ansa cervicalis
Thyrohyoid	Thyroid cartilage	Inferior body of hyoid bone	Depresses hyoid bone and elevates larynx	C1 via hypoglossal nerve

Conclusion

The craniomandibular complex, comprised of individual temporomandibular joints and the teeth, form a unique functional unit that is integral for daily functions such as talking, eating, and breathing. This chapter discusses the anatomy of the region, the arthrokinematics of the TMJ, and its relationship with the cranium and the cervical spine. In addition, the available range of motion of the mandible and the muscles necessary to produce the movements are reviewed. Clinical correlations may help the reader find practical applications of the material to clinical scenarios and problems.

REVIEW QUESTIONS

1. Describe the articulation of the TMJ.

2. What is the main action of the temporalis muscle?

3. Name the main parts of the mandible.

4. Describe the shape and location of the articulating TMJ disc.

5. Describe the arthrokinematics of mandibular depression.

Thoracic Spine

The thoracic spine may be best divided into three separate and related regions: upper, middle, and lower. Ligamentous support throughout the thoracic spine includes ligaments of the cervical and lumbar spine. Therefore, it is advised that the reader refer to relevant chapters, particularly chapter 6 (cervical spine), for a more detailed description of these ligaments. The overall injury rates for the thoracic spine are not well understood but are generally accepted to be lower than the injury rates for the cervical and lumbar spine regions. The thoracolumbar region, though, has been cited as the most frequently injured region of the thoracic spine (Tawackoli et al. 2004). The thoracic spine often does not get the attention that the cervical and lumbar spines do with respect to pathology or musculoskeletal dysfunction, but relevant dysfunction can be found here. Compression fractures, burst fractures, and flexion-distraction injuries as well as idiopathic and acquired scoliosis are common in this region of the spine (Singh et al. 2004; White et al. 1977). Other less prevalent pathology of the thoracic spine includes herniated discs, with a prevalence ranging from 0.2 to 5.0% of all intervertebral disc injuries (Benson & Burnes 1975; Stone et al. 1994).

OBJECTIVES

After reading this chapter, you should be able to do the following:

> Describe the variability of spinous process orientation throughout regions of the thoracic spine.

> Describe the differences between true, floating, and false ribs.

> Understand the correlation between muscle fiber type in the thoracic spine, posture-related dysfunction, and training emphasis to improve dysfunction.

> Name the major articulations of the anterior thorax.

> Name the major articulations of the posterior thorax.

This chapter describes the anatomy and biomechanical components of the thoracic spine, emphasizing how the anatomical structure and biomechanics of this region differ from the cervical and lumbar spines.

Osteology

Osteology of the thoracic spine, similar to the cervical and lumbar spines, includes distinct components. Detailed description of the osteological components of the thoracic spine is given, with descriptive figures to enrich the reader's understanding of this section of the spine.

Thoracic Spine

The upper thoracic spine is often thought to mimic the movement and, to some extent, the anatomy of the cervical spine. The lower thoracic spine is often considered somewhat analogous with the lumbar spine in the same respect.

The thoracic vertebral body (figure 8.1) is primarily made of cancellous bone. These vertebral bodies are progressively wider from the upper to the lower thoracic segments (Panjabi et al. 1991). A concavity is present in the horizontal plane of each segment (superior and inferior border of the vertebral body). This may reduce the prevalence of disc herniation in the thoracic spine (Panjabi et al. 1981). The anterior surface of the body is convex from side to side, while the posterior surface is deeply concave (Panjabi et al. 1991). The posterior height of the vertebral body increases slightly with caudal progression, lending to the normal kyphotic posture of the thoracic spine (see Clinical Correlation 8.1). The inclination of the end plates remains constant throughout the thoracic spine despite this posterior height increase with resultant kyphosis (Panjabi et al. 1991). It is generally believed that the typical thoracic vertebra has equal mediolateral and anteroposterior diameters, although others describe the anteroposterior diameter as slightly larger (Panjabi et al. 1991; Reoniok et al. 1997).

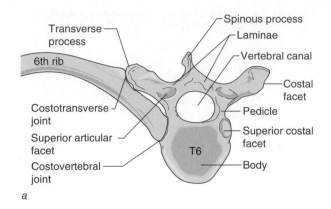

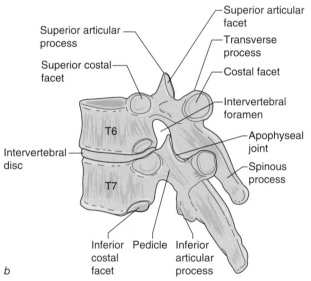

▶ **FIGURE 8.1** *(a)* Superior and *(b)* lateral view of structures of the thoracic spine.

Facets for rib articulation (**costal facets**) are present on most of the thoracic vertebrae (figure 8.1*a*). All of these, except the 1st and the 10th through 12th vertebral bodies, have **demifacets** (sometimes also called costal facets of the vertebral body) along their posterolateral aspects on the superior and inferior borders (Panjabi et al. 1981; Williams & Bannister 1995; Vanichkachorn & Vaccaro 2000). The **head of the rib** articulates with the respective superior and inferior borders of the vertebral bodies of thoracic 2 through 9 (T2-T9). T1 and T10 through T12 vertebral bodies provide the entire attachment for the rib.

As in the rest of the spine, just posterior to the vertebral body is the vertebral arch, consisting of the pedicles and laminae bilater-

CLINICAL CORRELATION 8.1

Progressive wedging of the thoracic vertebral bodies occurs in the majority of people with increasing age. Disc space narrowing at multiple levels has been shown to occur from the third decade of life (Romanes 1981). As this progressive wedging and disc space narrowing continue, excessive kyphosis of the thoracic spine is noted. Relevant pathology with this excessive kyphosis during aging can lead to thoracic vertebral fractures, decreased intrathoracic cavity space, and balance difficulty, among other things. Decreased intrathoracic cavity space is a potential contributing factor to decreased air exchange capability as one ages. Therefore, the clinician should try to minimize these deleterious effects. Possible interventions might include erector spinae muscle strengthening and anterior chest muscle (pectoralis) stretching. The client should take care to avoid excessive stress when performing these activities.

ally. The pedicles are composed primarily of cancellous bone for structural support. The outer shell of the pedicles has different thicknesses of cortical bone as well as variations in trabeculae. The pedicles generally face posteriorly with minimal to no lateral projection, resulting in a small vertebral canal. As in all other regions of the spine, the convergence of the laminae bilaterally forms the vertebral arch.

The vertebral spinal canal in the thoracic spine is largest at T1 and smallest in the midthoracic area, and it increases again in the lower thoracic region. The spinal canal is generally smaller in the thoracic spine than in the cervical or lumbar spine. The larger canal size in these two regions is necessary for the increased movement demands as well as the formation of the brachial (cervical spine) and lumbosacral (lumbar spine) plexuses. The spinal cord occupies approximately 25% of the area of the canal in the cervical spine but about 40% in the thoracic spine (Panjabi et al. 1981; Williams & Bannister 1995). Unlike in the cervical and lumbar regions, the spinal canal in the thoracic area is oval in cross section and becomes triangular at the upper and lower levels.

Thoracic vertebral level 7 (along with C6 and L4) is considered one of the tension points along the spine. These are often areas of decreased gliding capability of the nervous system or decreased vertebral canal area. This decrease in vertebral canal area can be due to a true decrease in anatomical structure or the fact that these tension points occur at areas of transition in the shape of the spine (e.g., peak of the kyphotic curve). Manual therapy intervention to T7 is a suggested treatment approach for people with symptoms in the upper or lower extremities with a potentially adverse neurodynamic mobility component (Butler 1992; Shacklock 2005).

▶ **KEY POINT**
The spinal canal in the thoracic spine is largest at the first thoracic spinal level and smallest in the midthoracic spine. The canal is generally smaller in the thoracic spine than in the cervical or lumbar spine. The spinal cord occupies approximately 25% of the area of the canal in the cervical spine but about 40% in the thoracic spine.

The transverse processes of the thoracic spine differ from those of the cervical spine for a couple of reasons. There are no longer anterior and posterior tubercles with the spinal nerve running between them, as in the cervical spine. The transverse processes of the thoracic and lumbar spines are solid structures. Length of the transverse processes varies, with the greatest length noted in the upper thoracic spine and the shortest length in the lower thoracic spine (Panjabi et al. 1981). They are located directly between the inferior articulating process and the superior articulating process of the zygapophyseal joints of each level. The transverse processes of the first 10 thoracic vertebrae have on the lateral aspect of their anterior surfaces a facet for articulation with the tubercle of the rib.

Spinous processes in the thoracic spine also differ from those of the cervical and lumbar spine regions. These spinous processes slope inferiorly (figure 8.2) and are the longest of any region of the spine. There is a lack of complete agreement regarding the extent of this inferior slope. Typically, the upper thoracic spinous processes have a more horizontal orientation similar to the cervical spine, and the lower thoracic spinous processes (particularly T11 and T12) are shorter than the other thoracic vertebrae with more posterior projections, similar to the lumbar spine. The first three spinous processes and the last three are almost horizontal, while those in the midthoracic spine are steeply inclined. Thoracic level seven (T7) has the greatest spinous process angulation.

Facet (zygapophyseal) joints in this region of the spine are plane synovial joints with fibroadipose meniscoids. The orientation of these joints is approximately 10 to 20° from the frontal/vertical plane or approximately 70 to 80° from the transverse/horizontal plane, with the lower thoracic facets slightly more vertical than the upper thoracic facets (Panjabi et al. 1991). The superior articular facet of a vertebra faces posteriorly and slightly lateral, while the inferior articular facet of each vertebra faces anteriorly and slightly medial. The frontal plane orientation of the thoracic vertebrae is similar to those of the

cervical spine (10 to 20°), while the facets of the lumbar spine lie closer to the sagittal plane (Panjabi et al. 1991).

The facet joint orientation dictates the availability of motion in the respective regions of the spine. The frontal plane orientation of the thoracic and cervical spines allows for mobility with rotation and side bending. The vertical orientation of the thoracic vertebrae limits flexion mobility in the thoracic spine to an extent. In the lower thoracic spine, because the facet joint orientation is more similar to that of the lumbar spine, sagittal plane motion is most favorable, with limitations for rotation and side bending.

Ribs

The primary function of the **rib cage** is to protect the heart and lungs. Although the ribs share common characteristics, they each have unique characteristics. They all differ from each other in size, width, and curvature. The ribs increase in length from the first rib until the seventh and then progressively shorten again. The first rib is the shortest of all the ribs.

The ribs (figure 8.3) are often classified into three general categories: true, false, and floating. Ribs 1 through 7 are considered **true ribs** since their respective cartilages directly attach to the **sternum** anteriorly. Ribs 8 through 10 are considered **false ribs**, as their anterior attachment is to the costochondral cartilage of the rib superior to them. All the false ribs articulate with the sternum via the seventh rib.

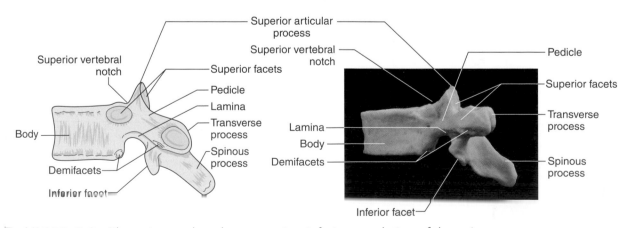

► FIGURE 8.2   Thoracic vertebra demonstrating inferior angulation of the spinous process.

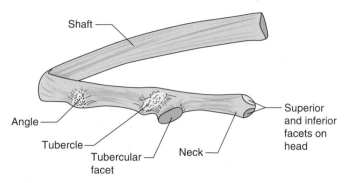

FIGURE 8.3 View of a typical rib.

Ribs 11 and 12 are considered **floating ribs** since they do not have an anterior attachment.

Posteriorly, ribs 2 through 9 have a superior costal facet that attaches to the costal demifacet of the vertebra above and an inferior costal facet articulating with the costal demifacet of the vertebra of the same level (Panjabi & White 1980) (figures 8.4, 8.5, and 8.6). The 1st and 9th through 12th ribs articulate only with their respective vertebra via one facet (Panjabi & White 1980).

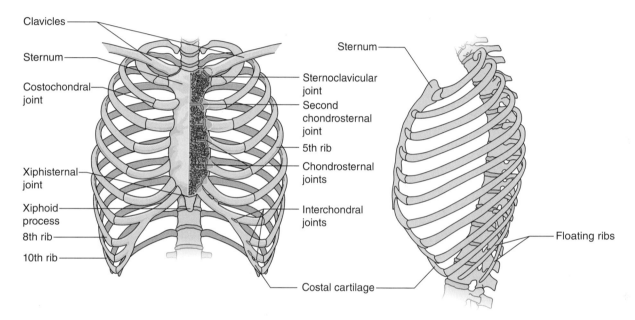

FIGURE 8.4 Anterior and lateral view of rib cage demonstrating true and false ribs.

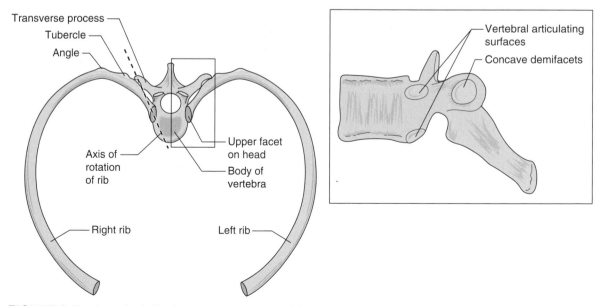

FIGURE 8.5 A typical rib demonstrating costal facet articulation with the demifacet on the vertebral body.

The head and neck of the rib are the most posterior, or dorsal, portions and directly articulate with the vertebral body. The **body of the rib** extends laterally from the neck and curves anteriorly from the rib angle. The

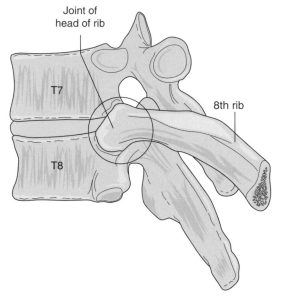

FIGURE 8.6 A typical rib articulating with vertebrae of the same level, vertebrae of the level above, and the interposed intervertebral disc.

bodies of the ribs serve as common muscle attachment sites, including the abdominal, erector spinae, and intercostal muscles.

Sternum

The sternum (figure 8.7) is generally a flat bone consisting of three segments (**manubrium**, body of sternum, and xiphoid process); it is convex anteriorly and concave posteriorly. The manubrium is the widest and most proximal portion of the sternum; its proximal border is at approximately the third thoracic vertebra. The manubrium's most superior, or proximal, edge is the **sternal notch**, or jugular. Just lateral to this are the facets for the sternoclavicular joints. The body of the sternum spans the fifth through ninth thoracic vertebrae and is the longest portion of the sternum (White & Panjabi 1990). Laterally along the body are the facets for the costal cartilages of the second through seventh pairs of ribs. The **manubriosternal joint** is easily palpable since these structures join at an angle. This angle is often referred to as the sternal angle, or angle of Louis. This joint is cartilaginous and does have a tendency to ossify with

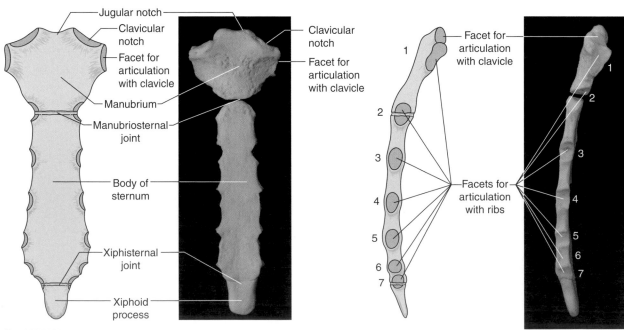

FIGURE 8.7 The sternum.

CLINICAL CORRELATION 8.2

With aging, both the xiphisternal and manubriosternal junctions have a tendency to fuse, resulting in a less pliable rib cage. The manubriosternal joint especially has a tendency to bend a few degrees with respiration, especially if it is forced respiration (Kapandji 1974; Williams & Bannister 1995). This bending capability, and therefore the pliability of the rib cage, would therefore be decreased with the normal aging process. The clinician should recognize that the aging client may therefore have limitations in breathing. When designing training programs for these clients, the clinician should understand that gains may be limited because of these structural changes.

age. The **xiphisternal junction joint** is also cartilaginous and tends to fuse with age (see Clinical Correlation 8.2). The **xiphoid** is the smallest portion of the sternum.

Joint Articulations

There are several joint articulations in the thoracic spine region at the rib cage. Each type of joint articulation, its specific articulation, and its relationship with the thoracic spine are discussed here. The corresponding figures are intended to help the reader understand the complex relationship of these joint articulations.

Zygapophyseal (Facet) Joints of the Thoracic Spine

The facet joints of the thoracic spine differ in angulation from those in the cervical spine. The zygapophyseal joints of the upper thoracic spine show morphological features similar to the cervical spine, and the lower thoracic spine joints progressively approximate those of the upper lumbar spine (Singer et al. 2004). These joints, as with the cervical and lumbar spines, are gliding synovial joints with a synovial capsule and synovial fluid.

Intervertebral Joints of the Thoracic Spine

These joint articulations are between each respective superior vertebra and inferior vertebra, with an interposed intervertebral disc.

The characteristic properties of the intervertebral disc in the thoracic spine differ from those of the cervical spine and are detailed later in the Joint Anatomy section.

Rib and Rib Cage Articulations

The major joint articulations involving the ribs and rib cage include (from posterior to anterior): costovertebral, costotransverse, costochondral, interchondral, chondrosternal, xiphisternal, and manubriosternal.

- **Costovertebral joints** (figure 8.8) are hyalinated synovial joints formed by the head of the rib, two adjacent vertebral bodies, and the interposed intervertebral disc. The vertebral bodies of T2 through T9 have demifacets for articulation with two separate ribs.

- **Costotransverse joints** (figure 8.8) are pairs of synovial joints. These joints consist of vertebrae T1 through T10 articulating with the rib of the same number. Ribs 11 and 12 do not articulate with their respective transverse processes of T11 and T12.

- **Costochondral joints** are formed by articulation of the 1st to 10th ribs anterolaterally with the costal cartilages. These joints are both synchondroses and synovial joints.

- **Interchondral joints** (figure 8.9) are synovial joints and are supported by a capsule and interchondral ligament. This joint also has a tendency to become fibrous and fuse with age.

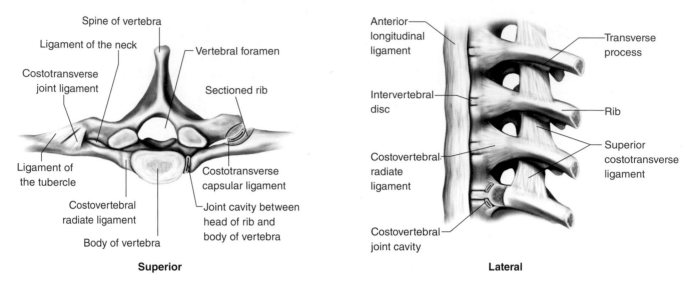

FIGURE 8.8 Articulations and ligaments of the rib cage.

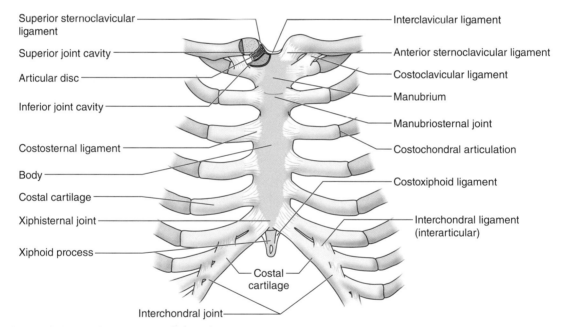

FIGURE 8.9 Anterior view of the rib cage.

- **Chondrosternal joints** are formed by the articulation of costal cartilages of ribs 1 through 7 anteriorly with the sternum. Chondrosternal joints of ribs 1, 6, and 7 are synchondroses, while chondrosternal joints of ribs 2 through 5 are synovial joints.

- Xiphisternal joints are also synchondrosis joints that have a tendency to fuse after the age of 40 years. This articulation is between the main body of the sternum and the xiphoid process.

- Manubriosternal joints are the joint articulations between the most superior portion of the sternum (manubrium) and the largest portion of the sternum (the sternal body). These are synchondrosis joints that have a tendency to fuse as one ages.

▶ **KEY POINT**

There is a significant amount of variability between the various joint articulations in the thoracic spine and thorax.

Joint Anatomy

The joint anatomy in the thoracic spine is also complex because of the rib cage, multiple joint articulations, and variation between the different regions of the thoracic spine. This section describes the various components of the thoracic spine and rib cage. Again, the figures are intended to assist the reader in understanding the complexity of the anatomy of this region of the spine.

Zygapophyseal Joint

The superior facets of the thoracic spine are on the superior aspect of the lamina–pedicle junction and face posteriorly, superiorly, and laterally, while the inferior articulating facet faces anteriorly, inferiorly, and medially. The articulating surface of the superior facet lies on the posterior aspect, while the articulating surface of the inferior facet is on the anterior aspect. The facet surfaces are concave anteriorly and convex posteriorly.

The angle of inclination of the zygapophyseal joints changes depending on the level. The upper segments are inclined at 45 to 60° from the horizontal, similar to the cervical spine; the middle segments incline at approximately 90° to the horizontal, in typical thoracic spine form; and the lower segments are inclined as in the lumbar spine (predominantly in the sagittal plane).

The capsules in the thoracic and lumbar spine areas are tauter than those in the cervical spine and therefore help limit flexion and anterior translation of a superior vertebra on an inferior vertebra (Panjabi et al. 1981; Panjabi & White 1980). The common spinal ligaments are present in all thoracic vertebrae. Ligament reinforcement is similar to the cervical spine, although ligaments are not as relevant here as in the cervical spine because of other stabilizing structures in the thoracic spine (e.g., rib cage, other joints). They do perform similar functions as in the rest of the spine. The anterior longitudinal ligament is narrower but thicker compared with elsewhere in the spine, while the posterior longitudinal ligament is wider at the level of the intervertebral disc (IVD) and narrower at the vertebral body than in the lumbar spine (Gray 1995). Please refer to table 6.2 for other continuous and segmental ligaments of the lower cervical and thoracic spine.

Fibroadipose meniscoids, thought to originate medially from the ligamentum flavum or laterally from the joint capsule, are present in each joint (Bogduk & Engle 1984). Each joint demonstrates asymmetry from side to side at nearly all levels (Bogduk & Engle 1984), although these differences are small in most cases (Boszczyk et al. 2001).

Intervertebral Joint

Thoracic IVD to vertebral body height ratio in the thoracic spine (1:5) is smaller than the ratios in the cervical (2:5) and lumbar (1:3) spines (Kapandji 1978). This lack of IVD height is one reason for decreased mobility in the thoracic spine as compared with other regions of the spine.

▶ **KEY POINT**
Decreased intervertebral disc height, rib articulation with the thoracic spine, and facet joint orientation are three primary reasons for decreased thoracic spine mobility as compared with other regions of the spine.

As in other portions of the spine, the IVD consists of the nucleus pulposus, annulus fibrosus, and end plate. More detail about these specific structures is given in chapter 6 and chapter 9. Unlike the IVDs of the cervical spine, those in the thoracic and lumbar spines have concentric cartilaginous rings of the annulus around the entire periphery of the nucleus. The nucleus is also relatively small, especially in comparison with its counterpart in the lumbar spine IVDs. Thoracic spine discs are also narrower mediolaterally than the cervical and lumbar IVDs. Additionally, the thoracic disc has a relatively small nucleus pulposus (Galante 1967).

The compliance of the thoracic disc is typically lost sooner than that of the cervical or lumbar disc (Edmondston & Singer 1997). This is due, in part, to the fact that there is much less movement in this area of the spine as compared with the other two. Although thoracic disc herniations are much less prevalent (see Clinical Correlation 8.3), degenerative changes, including disc degeneration, osteophytes, and disc space narrowing, are common in the midthoracic spine from the third decade of life (Horton 2002).

Rib and Rib Cage Articulations

Articulations at the ribs and rib cage, like the rest of the thoracic spine region, are complex and require understanding when prescribing exercise or implementing manual therapy interventions. The articulations of this region are described here so the reader can appreciate these factors when working with patients presenting with dysfunction in this area.

Costovertebral Joint

As previously mentioned, these joints are synovial joints formed by the head of a rib, the demifacets of two adjacent vertebral bodies, and the interposed IVDs for ribs 2 through 9. The 1st, 10th, 11th, and 12th pairs of ribs are considered atypical since they articulate with only one vertebral body. The 1st rib is small; it is the most curved and the most inferiorly oriented (posterior to anterior) rib. The 1st rib attaches to the manubrium just under the sternoclavicular joint. Posteriorly it attaches only to the T1 vertebra. Occasionally a cervical rib articulates with the C7 vertebra. This cervical rib, if present, or the 1st rib can be implicated in thoracic outlet syndrome (see Clinical Correlation 8.4). The 2nd rib articulates with the sternum at the manubriosternal junction.

The demifacets of each vertebral body are small, oval, and slightly concave. These demifacets are also referred to as the superior and inferior costovertebral facets.

A costovertebral joint is reinforced by the intra-articular ligament extending medially from the tip of the head of each rib to the annulus fibrosus of the IVD and the synovial capsule. The intra-articular ligament essentially divides the typical costovertebral joint into two cavities. This ligament is absent in the atypical costovertebral joints of ribs 1, 10, 11, and 12. The capsule is supported by the radiate ligament. This ligament is within the joint capsule and is so named because it radiates from the rib to the bodies of the superior and inferior vertebrae and to the interposed

---

CLINICAL CORRELATION 8.3

Thoracic disc herniations constitute only 0.25 to 0.75% of all disc herniations (Arce & Dohrmann 1985), partly because the intervertebral foramina are quite large. Therefore, osseous contact with the nerve roots is seldom encountered in the thoracic spine (Lyu et al. 1999). Central disc protrusions are more common in the thoracic spine despite a smaller spinal cord size and the more oval shape of the vertebral foramen (Lyu et al. 1999). Therefore, the clinician should consider the increased likelihood of central disc protrusions versus direct pressure on the nerve root as it exits the intervertebral foramen as the cause of radiculopathy related to the thoracic spine.

CLINICAL CORRELATION 8.4

An elevated 1st rib has been implicated in thoracic outlet syndrome. Treatment of an elevated 1st rib is often performed manually by joint mobilization, with and without breathing techniques. The clinician must be aware of the possibility of a cervical rib versus an elevated 1st rib off of the T1 vertebra. Cervical ribs are a rare anatomical variant, with an overall prevalence of 0.74%, and a higher rate in females compared with males (1.09 and 0.42%, respectively) (Brewin et al. 2009). Assessment techniques for thoracic outlet syndrome are covered in chapter 6 (cervical spine).

IVD (White & Panjabi 1990). The radiate ligament has three bands (superior, intermediate, and inferior) (figure 8.10). Radiate ligaments are fibrous bands that surround the entire articulation of, and are present in, each costovertebral joint.

Costotransverse Joint

The costotransverse joints have a thin, fibrous capsule surrounding the articulation of the rib tubercle and the thoracic vertebral transverse process. The articulation of the ribs with the transverse processes yields two synovial capsules, one above and one below the articulations, with an interarticular ligament that provides stability (Williams & Bannister 1995).

The costotransverse joint is supported by a joint capsule reinforced by three major ligaments: lateral costotransverse ligament, costotransverse ligament, and superior costotransverse ligament (figure 8.10). The lateral costotransverse ligament provides support most laterally between the transverse pro-

cess tip and the same-level rib (see Clinical Correlation 8.5). The costotransverse ligament lends support between the body of the transverse process and the articulating rib, while the superior costotransverse ligament articulates between the same-level rib and the superior transverse process. The superior costotransverse ligament is wider and runs more obliquely than the intertransverse ligament (articulating between corresponding levels of transverse processes.

Costochondral Joint

The costochondral joint is the most lateral anterior joint articulation of the rib cage (figure 8.9). These joint articulations lack true ligamentous support because they are a direct attachment of the lateral-most rib and costal cartilage. These joints have been described as being both synchondrosis and synovial type joints.

Interchondral Joint

Interchondral joints are the synovial articulations between corresponding ribs (figure 8.9). The costal cartilages of ribs 7 through 10 articulate with the cartilage immediately above them. This joint is the only connection to the sternum for ribs 8 through 10. As previously mentioned, these joints are supported by a capsule and interchondral ligament.

Chondrosternal Joint

These joints are continuous with the sternum in that the periosteum of the sternum is continuous with the perichondrium of the costal cartilage. The upper seven joints have a relatively thin fibrous capsule blending with the sternocostal ligaments. The joint surface is supported by capsular, radiate sternocostal, and intra-articular ligaments.

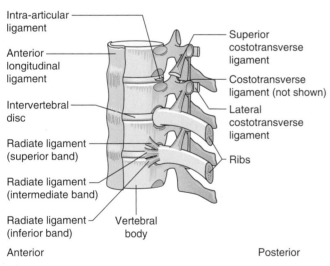

Intra-articular ligament
Anterior longitudinal ligament
Intervertebral disc
Radiate ligament (superior band)
Radiate ligament (intermediate band)
Radiate ligament (inferior band)
Vertebral body
Superior costotransverse ligament
Costotransverse ligament (not shown)
Lateral costotransverse ligament
Ribs
Anterior
Posterior

▶ **FIGURE 8.10** Ligament support of the costovertebral joint.

CLINICAL CORRELATION 8.5

The lateral costotransverse ligament is a short (as it runs from the tip of the transverse process to the rib), thick, and generally strong ligament. It is, however, often damaged by direct blows to the chest. Mechanisms most likely to damage this ligament include a punch or kick to the chest. The clinician is advised to perform a comprehensive examination of any client complaining of vague rib cage pain as a result of such mechanisms.

Xiphisternal Joint

The xiphisternal joint is initially a cartilaginous joint early in a person's life, but it ossifies in adulthood. This joint does not have any direct ligamentous support.

Manubriosternal Joint

The articulation between the manubrium and sternum is lined with hyaline cartilage. There is no direct ligamentous support for this ligament either.

Nerve Supply

As with other regions of the spine, there are posterior and anterior rami braches of the spinal nerves in this region. The branches of the posterior rami supply the skin of the medial two-thirds of the back and neck, the deep muscles of the back and neck (lateral branches), the zygapophyseal joints (the medial branches) (Bogduk & Marsland 1988), and the ligamentum flavum.

The anterior or ventral rami (the anterior branches) from T2 through T11 become intercostal nerves. These nerves supply the intercostal muscles, costotransverse joints, and part of the abdomen. The anterior rami above T2 and below T11 form the somatic plexuses that innervate the extremities.

The peripheral nerves of this region include the posterior scapulae, thoracodorsal, and long thoracic nerve. These nerves also course through the chest wall.

The sympathetic nervous system has an important role in the thoracic spine (see Clinical Correlation 8.6). Twelve sympathetic ganglia exist within the thoracic region. The thoracolumbar sympathetic fibers arise from the dorsolateral region of the anterior column of the spinal cord gray matter and pass with the anterior roots of all the thoracic and upper two or three lumbar spinal nerves (Williams & Bannister 1995).

Blood Supply

The thoracic spine and rib cage region has a poor blood supply. Vascularization is mainly accomplished via the posterior, or dorsal, branches of the posterior intercostal arteries, while the venous drainage occurs via the anterior and posterior venous plexuses. There is poor vascularization of the spinal cord region between T4 and T9 (Dommisse 1974).

Joint Function

Understanding the biomechanical aspects of any portion of the body is necessary to implement appropriate intervention. Thoracic spine biomechanics is less understood and therefore has little consensus of agreement. Most of the understanding of thoracic spine biomechanics is based largely on ex vivo studies (Panjabi et al. 1981; Panjabi et al. 1976; White 1969) and clinical models (Lee 1988; Flynn 2001).

The upper thoracic vertebrae zygapophyseal joints function similar to the lower cervical facet joints, and the lower thoracic vertebrae zygapophyseal/facet joints function similar to the lumbar spine zygapophyseal joints. The middle segments of the thoracic spine vertebrae are designed for less overall mobility, since the thoracic cage articulations limit sagittal plane motions while accommodating axial displacements (Gregersen & Lucas 1967; Singer et al. 2004). These joints restrain the amount of flexion and anterior translation as well as facilitate rotation of the vertebral segment (White 1969). They appear

---

CLINICAL CORRELATION 8.6

Manual therapy treatment of the cervical and thoracic spine has demonstrated a sympathetic nervous system effect. Stimulation of these areas causes upper extremity changes in pain response and pressure pain as well as measurable improvement in strength (Cleland et al. 2004; Vicenzino et al. 2001; Reiman et al. 2008). Mobilization and manipulation of the thoracic spine are thought to affect the sympathetic trunk since it runs directly anterior to the costovertebral joint.

to have minimal influence on lateral flexion (White 1969).

This section covers the axes of motion, arthrokinematics, range of motion, closed and loose packed positions, end feel, and capsular pattern of the thoracic spine.

Axes of Motion

Flexion–extension occurs in the sagittal plane around a mediolateral axis. Axial rotation takes place in the transverse plane around a vertical axis, and side bending takes place in the frontal plane around an anteroposterior axis.

Arthrokinematics

Coupling characteristics of the thoracic spine are location dependent. The upper thoracic spine tends to couple the same as the lower cervical spine (ipsilateral side bending and rotation) (White & Panjabi 1990). The middle thoracic spine demonstrates variability in regard to a specific coupling pattern (White & Panjabi 1990; Panjabi et al. 1976; Buchalter et al. 1988). The lower thoracic spine, in general, mimics the coupling pattern of the lumbar

spine (White & Panjabi 1990). As mentioned in chapter 6 (cervical spine), the concept of coupling has been questioned. Inconsistencies in previous literature reviews investigating coupling of the lumbar spine (Cook 2003; Legaspi & Edmond 2007) have led to the suggestion of using caution when applying the concepts of coupled motion to the evaluation and treatment of patients with low back pain (Legaspi & Edmond 2007).

Flexion (figure 8.11a) of the thoracic spine is initiated by the abdominal muscles and eccentrically controlled by the erector spinae muscles. As described in the cervical spine, physiological flexion in the thoracic spine is a combination of anterior translation and anterior rotation in the sagittal plane (superior and anterior glide of the superior facet of the segment on the inferior facet of the segment). The superior facet of the respective zygapophyseal joint glides up and forward on the inferior facet of the joint or segment. Extension (figure 8.11b) involves posterior sagittal rotation and posterior translation (superior facet glides down and back). End-range extension is resisted by the posterior portion of the disc annulus and impaction of the zygapophyseal joints.

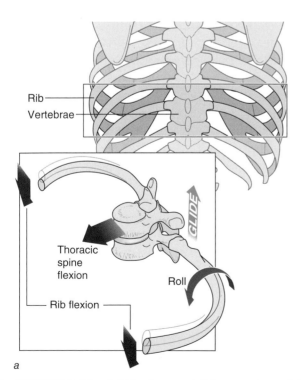

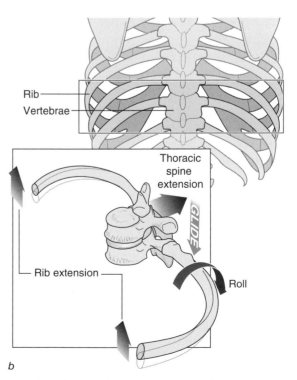

▶ **FIGURE 8.11**   *(a)* Flexion and *(b)* extension of the thoracic spine and corresponding rib movement.

Unique to the physiological movement of the thoracic spine is the movement of the rib cage and respective ribs. During flexion, the rib rotates (rolls) superiorly with respect to the transverse process. During extension, the opposite occurs, with inferior rotation (roll) of the rib with respect to the transverse process (Lee 1996). Posterior translation of the vertebra with extension pushes the superior aspect of the head of the rib posteriorly at the costovertebral joint, producing a posterior rotation of the rib (anterior aspect of rib travels superiorly, while the posterior aspect travels inferiorly) (Lee 1994).

Extension of the thoracic spine is primarily produced by the lumbar erector spinae muscles. Extension is resisted by stiffness in the anterior intervertebral disc, stiffness in the anterior longitudinal ligament, and bony contact of the zygapophyseal joint and between the spinous processes (Edmondston & Singer 1997; Panjabi et al. 1981).

Joint Function During Rotation

Approximately 35 to 50° of total rotation motion is available in the thoracic spine (White 1969). Segmental axial rotation in the thoracic spine averages about 7° in the upper thoracic spine, about 5° in the midthoracic spine, and 2 to 3° in the last two or three segments (White 1969; Singer et al. 1989). Pure axial rotation is thought to occur at the cervicothoracic and thoracolumbar junctions, and thoracic segmental rotation has been described as coupled with contralateral side bending (Singer et al. 1989) in other regions of the thoracic spine. The reader is again reminded of the controversy regarding coupling mechanics of the spine.

Joint Function During Side Bending

Side bending in the thoracic spine is initiated by the ipsilateral erector spinae and abdominal muscles. Twenty-five to 45° of side bending motion has been described, with an average of 3 to 4° to each side per segment and 7 to 9° per lower segment (White 1969).

Side bending movement at the zygapophyseal joint can be described similar to flexion and extension. During side bending to the right, the right superior facet of the segment will glide inferiorly on the inferior facet of the segment. In other words, the ipsilateral zygapophyseal joint is performing extension, or down and back movement. The contralateral left side is performing flexion, or up and forward movement, since the superior facet of the segment is gliding superiorly on the inferior facet of the segment.

Side bending is stopped by soft-tissue stretch contralaterally or rib approximation ipsilaterally. Additional side bending is modified by the fixed ribs (Lee 1994). The ipsilateral transverse process glides inferiorly on the rib at the same level, resulting in a relative anterior rotation of the neck of the rib, while the contralateral transverse process glides superiorly, producing a posterior rotation of the rib (Lee 1994). Contralateral rotation of the superior vertebra of the segment is suggested as the result of such movement.

Joint Function During Respiration

The main movement of the upper six ribs during respiration and other movements is rotation of the neck of the rib, with minimal amounts of superior and inferior motion. The primary motion of ribs 7 to 10 is superiorly, posteriorly, and medially during inspiration, with the opposite occurring during expiration (Lee 1988).

Since the anterior portion of the ribs are more inferior than the posterior, inspiration causes the anterior portion to rise superiorly, or upward, while the posterior portion drops downward. There is an anterior elevation of the upper ribs (described as a pump-handle motion) and a lateral movement in the middle and lower ribs (a bucket-handle motion). The upper ribs expand the anteroposterior diameter of the rib cage, while the lower rib movement results in an increase in the transverse diameter of the rib cage. The last two floating ribs perform a caliper type of motion as a result of a superior and lateral movement of the nonfixed anterior portion of the rib compared with the posterior portion.

During normal inspiration, with the second rib being longer than the first, the superior portion of the manubrium tilts posteriorly while the inferior portion moves anteriorly. This posterior tilt of the manubrium results in an anterior roll of the clavicle. The inferior portion of the sternum also moves farther anteriorly than the superior portion, again due to the increasing length of the ribs distally. The manubriosternal junction is the hinge for this sternal movement. As previously mentioned, this joint can stiffen, or ossify, affecting rib cage expansion (refer to Clinical Correlation 8.2).

Normal expiration is passive. Forced expiration is a result of several muscle groups being involved. Forced inspiration and expiration are discussed in more detail in the muscle section later in the chapter.

Motion of the thorax and rib cage is primarily described as occurring in the previously described joints. Stability is afforded to the thoracic spine and rib cage via several of these joints, although sometimes not until adolescence. A large amount of thoracic rotation and side bending can be demonstrated in younger people who perform movements requiring such flexibility (e.g., gymnasts). The rib head does not ossify until about age 13 (Feiertag et al. 1995), thus affording such a large range of motion (ROM). There is very little transverse or anteroposterior translation available at the costotransverse joint. The chondrosternal joint, though, is capable of about 2° of motion from full inspiration to full expiration and allows the full excursion of the sternum.

▶ **KEY POINT**
Arthrokinematics of the thoracic spine are complex, accounting for rib movement with all motions.

Range of Motion

The range of motion of the entire thoracic and lumbar spine is as follows (American Medical Association 1988):

- Flexion: 60°
- Extension: 25°
- Right side bending: 25°
- Left side bending: 25°
- Right rotation: 30°
- Left rotation: 30°

Combined flexion and extension ROM in the thoracic spine is bimodal, superior to inferior, with greater ROM available in flexion than extension (White & Panabi 1990). The upper thoracic spine demonstrates a combined 3 to 5° of flexion or extension, while only 2 to 7° are available at T5-T6, increasing to 6 to 20° at T12-L1 (White & Panjabi 1990). Combined side bending is also bimodal in the thoracic spine. In the upper thoracic spine, 5° of motion is available, with 3 to 10° for T7 through T11, and 5 to 10° at T12-L1. Combined rotation is reported to be 14° at T1-T2, progressively declining to 2 to 3° combined at T12-L1, similar to the lumbar spine.

Closed and Loose Packed Positions

The closed packed position of the thoracic spine zygapophyseal joints is full extension, with a resting position of midway between flexion and extension.

End Feel

End feel is generally described as firm for side bending and rotation and as tissue stretch for extension and flexion.

Capsular Pattern

The capsular pattern for the thoracic spine zygapophyseal joints is a greater limitation of extension, side bending, and rotation than of forward bending.

Muscles

The primary muscles of the thorax and thoracic spine region include the latissimus dorsi, erector spinae, trapezius, and rhomboid major and minor posteriorly and the external oblique, internal oblique, rectus abdominis, and pectoralis major and minor anteriorly. Many of these muscles are most relevant in other areas of the body, and more detailed discussion of those muscles is given in other chapters. Also, please refer to these respective chapters (shoulder, cervical spine, and lumbar spine particularly) for the tables listing these muscles' origins, insertions, actions, and innervations.

The latissimus dorsi, trapezius, and rhomboids have all been reported to produce side bending and rotation of the spine with unilateral contraction. These motions require fixation of the distal attachment of each respective muscle.

All these muscles also act on other areas of the body and therefore have multiple functions that may or may not be relevant to the thoracic spine. Most likely, though, those muscles with direct attachment to the thoracic spine will act on and produce tension across this region of the spine.

Erector Spinae Muscles

The erector spinae includes (from medial to lateral) the spinalis, longissimus, and iliocostalis lumborum muscles. The longissimus consists of the longissimus thoracis, cervicis, and capitis. The longissimus thoracis has both a thoracic portion (pars thoracis) and a lumbar portion (pars lumborum) (Loring & Woodbridge 1991; Macintosh & Bogduk 1987). The iliocostalis consists of segments in the cervical, thoracic, and lumbar spines. Additionally, the iliocostalis lumborum has a thoracic component (pars thoracis) and a lumbar component (pars lumborum).

The longissimus thoracis pars thoracis consists of 12 pairs of fascicles arising from the ribs and the transverse processes of T1 through T12. Caudally the tendons join in the erector spinae aponeurosis together with the tendons of the iliocostalis lumborum pars thoracis. This broad sheet of tendinous fibers is attached to the ilium, the sacrum, and the spinous processes of the lumbar spine. When contracting bilaterally, the longissimus thoracis pars thoracis acts indirectly on the lumbar spine and uses the erector spinae aponeurosis to increase lumbar lordosis.

The longissimus thoracis pars lumborum arises from the dorsal surface of the transverse processes and inserts into the medial aspect of the posterior superior iliac spine (PSIS). Unilateral contraction produces ipsilateral side bending; bilateral contraction produces extension of the spine.

The fascicles of the iliocostalis lumborum pars thoracis arise from the rib angle of the lower 7 ribs via a ribbon-like tendon, contributes to the erector spinae aponeurosis, and inserts caudally into the ilium and sacrum. The muscle does not have any attachments into the lumbar vertebrae. Bilateral contraction causes a bowstring effect, causing an increase in lumbar lordosis.

Unilateral contraction of the iliocostalis lumborum pars lumborum produces side bending as the fibers arise from the tip of the transverse processes of L1 through L4 and insert on the iliac crest. This muscle is reasonably disposed to produce trunk rotation.

As a group, the erector spinae muscles perform trunk extension with bilateral contraction, side bending with unilateral contraction (ipsilateral), and ipsilateral rotation of the trunk (unilateral contraction).

Type I, or slow-twitch, muscle fibers are predominant in the erector spinae muscles. The percentage of Type I muscle fibers in the erector spinae is more prevalent in the thoracic spine (about 75%) than in the lumbar

CLINICAL CORRELATION 8.7

The erector spinae muscles in the thoracic region of the spine are predominantly Type I (slow-twitch) muscles. Type I muscles are physiologically favored to perform endurance-type activity. Therefore, these muscles play a primary role in postural support and in stabilizing the costovertebral joint (Saumarez 1986). Endurance training, with higher repetitions, increased sets, and lighter loads, would provide more benefit when training the postural thoracic spine erector spinae muscles than strength training with fewer repetitions, decreased sets, and heavier loads. The clinician must therefore take this information into account, along with other information discussed in chapter 4, when prescribing exercise for a patient requiring thoracic spine intervention.

CLINICAL CORRELATION 8.8

Thoracic paraspinal muscle strength is of paramount importance in posture-related dysfunction. Trunk extensor strength is negatively correlated with degree of kyphosis (Itoi & Sinaki 1994; Sinaki et al. 1996; Wang et al. 1999). For all people with posture-related dysfunction, the clinician should implement a program to strengthen the thoracic paraspinal musculature (starting with isometric strengthening and progressing with increasing ranges of movement isotonics as tolerated).

spine (approximately 57%) (Sirca & Kosteve 1985) (see Clinical Correlations 8.7 and 8.8).

Muscles of Respiration

The primary muscles of respiration include the diaphragm, the sternocostals, and the intercostals (figure 8.12). All these muscles act on the rib cage to promote inspiration. The secondary muscles include the anterior and medial scalenes, serratus posterior, pectoralis major and minor, latissimus dorsi, trapezius, and when the head is fixed, sternocleidomastoid (Gray 1995; Kendall et al. 1952). These are typically muscles that attach the rib cage to the shoulder girdle, head, vertebral column, or pelvis.

The diaphragm is attached around the thoracoabdominal junction circumferentially. Contraction of the diaphragm pulls the central tendon inferiorly, producing diaphragmatic inspiration. The phrenic nerve (C3-C4) provides the motor innervations, and the lower six intercostal nerves provide the sensory supply to the diaphragm.

The intercostal muscles have been described primarily as the internal and external intercostals. These muscles lie between each rib, and the neurovascular bundle lies just below the rib. The internal intercostals are deep to the external intercostals, and their posterior fibers pull the upper rib down, but only during forced expiration (DeTroyer & Sampson 1982; Taylor 1960). The external intercostals pull

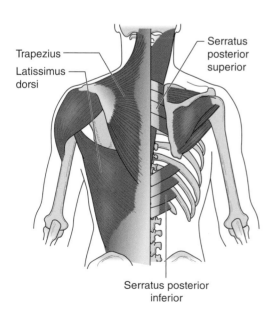

Posterior

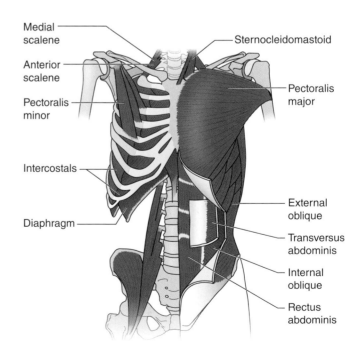

Anterior

▶ **FIGURE 8.12** Musculature of the trunk demonstrating muscles of respiration.

the lower rib toward the upper rib, resulting in inspiration.

The secondary muscles of respiration assist with inspiration and expiration in situations of stress, such as increased activity or disease. The sternocleidomastoid, along with the trapezius stabilizing the head, moves the rib cage superiorly with a bilateral action, therefore expanding the upper rib cage in a pump-handle motion. The pectoralis major can elevate the upper rib cage when the shoulders and humerus are stabilized. The pectoralis minor can help elevate the 3rd, 4th, and 5th ribs during forced inspiration. The subclavius is between the clavicle and the 1st rib, and it can also assist in raising the upper chest for inspiration. The abdominal muscles (transverse abdominis, internal and external obliques, and rectus abdominis) function principally as expiratory muscles, assisting with forced expiration. Intra-abdominal pressure (IAP) created by active abdominal contraction during forced expiration pushes the diaphragm cranially, optimizing its length–tension relationship for expiration. Increased IAP created by a lowered diaphragm in inspiration is countered by tension in abdominal musculature.

Several musculoskeletal changes occur as one ages. Among these are changes in rib cage mobility (see Clinical Correlation 8.9), muscle endurance, and posture, all of which affect respiration. Changes in posture will also affect muscle length, with the posterior thoracic spine musculature elongating and the chest musculature typically shortening. These postural changes affect respiration in people without pathology but to an even greater extent in those with pathology. This is especially relevant in people with respiratory pathology, such as chronic obstructive pulmonary disease. These normal adaptive changes require careful monitoring and progression during therapeutic exercise programs. Systematic progression of strength and endurance training, gait training, and so on will be affected. Modifications to include longer and more frequent rest periods than typically employed in a rehabilitation program most likely will be necessary.

CLINICAL CORRELATION 8.9

Many changes happen as one ages, including compliance and mobility of the rib cage. Some of the major changes that occur in elderly persons include the following:

- Decreased compliance of the bony rib cage due to ossification of different joints (e.g., manubriosternal, xiphisternal), restricted joint mobility, and so on
- Overall decreased compliance of the respiratory system
- Decreased effectiveness of ventilatory muscles
- Decreased ventilatory reserve

Because of these changes, elderly people will have less air exchange capacity. This will be especially prevalent with increasing activity levels. The practicing clinician must account for these changes and modify rehabilitation programs accordingly.

Conclusion

The thoracic spine has the least mobility of all the spinal regions. Rib articulations, the smaller ratio of disc to vertebral body height, the number of joint articulations, and facet joint orientation are contributory factors to this decreased mobility. The function of the thoracic spine is therefore unique. The different portions of the rib cage itself are variable.

REVIEW QUESTIONS

1. What type of muscle fiber is predominant in the erector spinae muscles, thereby determining the primary type of training program to improve their muscular performance?
   a. Type I, endurance training emphasis
   b. Type IIa, combination of strength and endurance training
   c. Type IIb, strength training emphasis
   d. Type IIc, strength training emphasis

2. What variables are least likely to account for limited motion in the thoracic spine when compared with the cervical and lumbar spine?
   a. facet joint orientation
   b. soft-tissue elasticity
   c. intervertebral disc to body height ratio
   d. rib articulation with the spine

3. Which ribs are considered floating ribs?
   a. ribs 11 and 12
   b. ribs 10 through 12
   c. ribs 1 and 2
   d. ribs 8 through 10

4. Which of the following statements about facet joint orientation in the thoracic spine compared with facet joint orientation in the cervical spine is correct?
   a. Orientation in the thoracic spine is more favorable for increased motion than in the cervical spine.
   b. Orientation in the thoracic spine is a larger angle (a greater number of degrees) from the horizontal plane than in the cervical spine.
   c. Orientation of the facets in both the thoracic and cervical spines is the same number of degrees from the horizontal.
   d. Orientation of the thoracic spine facets favors only sagittal plane (flexion–extension) motion.

5. The primary muscles of respiration are the
   a. diaphragm, sternocostals, and intercostals
   b. diaphragm and obliques
   c. diaphragm only
   d. intercostals only

Lumbar Spine and Pelvic Girdle

Low back pain (LBP) is the most prevalent of all musculoskeletal conditions. It afflicts nearly everyone at some point (Woolf & Pfleger 2003). Low back pain is the leading cause of injury and disability for those under the age of 45 and the third most prevalent impairment for those 45 and older (Truchon 2001). Physical load on the low back has commonly been described as a risk factor for future LBP. Certain dysfunctions related to the lumbar spine and pelvis have specific anatomical relationships. A detailed understanding of anatomy and biomechanics of the lumbar spine and pelvis is essential for proper examination and treatment.

The lumbar spine consists of five lumbar vertebrae. The sacroiliac (SI) joint, argued to be of prime importance in understanding vertebral joint problems (Grieve 1981), is a large diarthrodial joint connecting the spine to the pelvis. The pelvic girdle consists of bilateral innominate bones that articulate anteriorly to form the pubic symphysis and posteriorly with the sacrum on either side to form the SI joints. The bilateral hip joints are also often included as part of the pelvic girdle.

OBJECTIVES

After reading this chapter, you should be able to do the following:

> Detail the motion available in the lumbar spine as it relates to the zygapophyseal joint orientation.

> Describe the differences in the intervertebral discs of the lumbar spine as compared with the cervical spine.

> Describe the mechanisms via which the lumbar spine intervertebral discs receive nutrition.

> Detail the differences in sacral nutation versus sacral counternutation.

> List the components of the pelvic girdle.

Osteology

The osteology of the lumbar spine has some characteristics that are similar to and dissimilar to the cervical and thoracic spines. The details of lumbar spine osteology, as well as similarities with these other regions of the spine, are described here.

Lumbar Spine

The lumbar spine consists of five lumbar vertebrae that gradually increase in size from lumbar vertebra one (L1) to lumbar vertebra five (L5). The fifth lumbar vertebra articulates with the sacroiliac joint. The specific anatomical structures of the lumbar spine—vertebral body, spinous and transverse processes, and so on—are described here so the reader can appreciate how the anatomical structure relates to joint structure and function (described later). A failure of the junction between the first and second sacral vertebrae to fuse is known as lumbarization. The result is six lumbar vertebrae that are mobile versus the normal five. When the lumbosacral junction fuses during growth, it is known as sacralization of L5. The result is only four mobile lumbar vertebrae. Neither lumbarization nor sacralization appears to increase the risk of LBP (van Tulder et al. 1997). Therefore, the clinician should recognize that both these conditions may or may not be relevant to the reason the patient is being seen.

> ▶ **KEY POINT**
> The lumbar spine consists of five lumbar vertebrae that gradually increase in size from the first lumbar vertebra to the fifth lumbar vertebra. The fifth lumbar vertebra articulates with the sacroiliac joint.

Vertebral Body

The vertebral body in the lumbar spine is larger than in any other spinal region. As in the other regions of the spine, intervertebral discs (IVDs) separate the vertebral bodies. Therefore, the lumbar spine consists of five lumbar vertebrae with five interposed IVDs, 12 zygapophyseal joints (including T12-L1 through L5-S1), multiple ligaments and muscles, and complex neurovascular contributions.

The lower vertebrae and IVDs of the lumbar spine are wedge shaped, promoting the natural postural lordosis of this region (figure 9.1). The

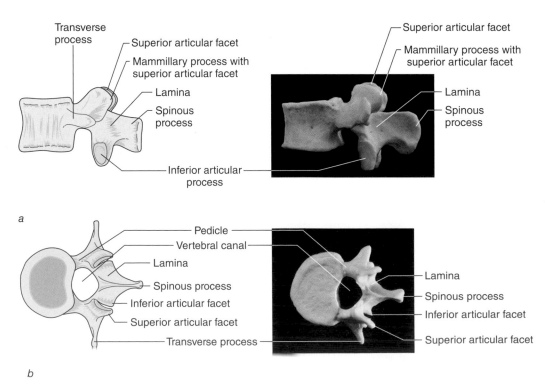

▶ **FIGURE 9.1** (a) Lateral and (b) superior view of a lumbar vertebra.

anterior aspect of each vertebra is generally concave, while the posterior aspect is flattened and stable (Bogduk 1997). The anterior portion of the vertebral body is flat on the superior and inferior surfaces and provides contact points for the intervertebral disc (Bogduk 1997). The vertebral bodies are composed of cancellous bone that has a good blood supply. They are slightly narrowed in their midsection, giving them a unique biconcave arrangement and providing a deep passageway for neurovascular structures. This biconcave shape, along with vertical and transverse arrangement of the bony trabeculae, creates a system well designed to tolerate compressive loads (see Clinical Correlation 9.1). Because of its high trabecular bone content, the vertebral body is more susceptible to fractures, end plate damage, degenerative changes, and so on than if it were cortical bone. Osteoporosis is one such degenerative change. Osteoporosis is a condition of decreased bone density in the spine. It primarily affects trabecular bone and therefore causes damage in the vertebral body more so than in other spinal structures. The principle concern of osteoporosis involving the lumbar spine is compressive loads.

The lumbar vertebral bodies widen slightly at the superior and inferior margins. This widening corresponds to the epiphyseal ring and forms the site of the strong peripheral attachments of the IVDs.

The vertebral bodies in the lumbar spine are also covered by cartilaginous vertebral end plates. The end plates are relatively flat and composed of hyaline and fibrocartilage approximately 0.6 to 1.0 mm thick (Bogduk & Twomey 1997). They are located within the inner margin of the epiphyseal ring on the superior and inferior surfaces of the vertebral bodies and are perforated by numerous small holes that allow the passage of metabolites from bone to the central regions of the avascular discs (Roberts et al. 1993). These holes most likely weaken the end plate, potentially a reason why the end plate is the most easily damaged structure in the lumbar spine.

The cartilaginous end plate is more strongly attached to the disc than the vertebral body. It is responsible for nutrient transfer to the disc (figure 9.2) and becomes thicker and less permeable with increasing age. This structure also contributes to confining the annulus and nucleus (Humzah & Soames 1988). The IVD is discussed in the joint anatomy section.

Vertebral end plates are usually flat in young adults but develop a marked concavity with increasing age. This may reflect repeated minor injuries to the end plates themselves or to the vertically oriented trabeculae that support them

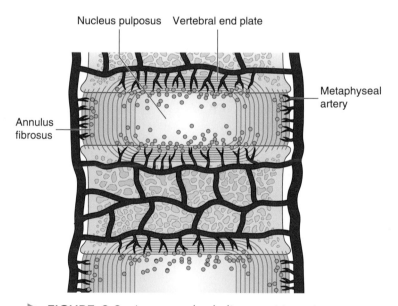

FIGURE 9.2 Intervertebral disc nutrition via cartilaginous end plate.

CLINICAL CORRELATION 9.1

In upright postures, the vertebral bodies of the lumbar spine assume 80 to 90% of the compressive load bearing (Bogduk & Twomey 1991; Humzah & Soames 1988). This capacity is further enhanced by spaces in the cancellous bone occupied by blood and hematopoietic tissue, reinforcing the bony trabeculae. Damage to the lumbar spine, including any injury to the vertebral body, will affect the person's ability to tolerate loads in the upright position. Rehabilitation exercises may have to be modified to include positions other than standing (e.g., supine, prone) in initial phases of rehabilitation.

(Twomey & Taylor 1987; Vernon-Roberts & Pirie 1973) (see Clinical Correlation 9.2).

Vertebral Arch

Posterior to the vertebral body is the vertebral foramen, or canal. The anterior wall of the vertebral foramen is formed by the posterior surfaces of the lumbar vertebrae, and the posterior wall is formed by the laminae and ligamentum flava of the same vertebrae (figure 9.3, *a-b*) (Bogduk 1997). The anterior wall of the foramen is flattened.

The pedicles make up the middle section of the lumbar vertebra, while the posterior portion of the vertebra includes the superior and inferior articular processes and the spinous process (see figure 9.1). The vertebral, or neural, arch is a bony ring posterior to the vertebral body. It encloses the vertebral foramen. This arch consists of paired pedicles binding the vertebral body and the paired laminae together. Arising from the neural arch are seven bony projections: two superior and two inferior articular processes, two transverse processes, and one spinous process.

The pedicles in the lumbar spine, composed of strong cortical bone, are stout and roughly cylindrical. Because of their shape and cortical bone makeup, they are well suited to sustain high compressive and tensile loads that occur during spinal flexion, extension, and rotation. These pedicles (one from the superior and one from the inferior vertebra) also form the superior and inferior boundaries of each intervertebral foramina through which the spinal nerve runs.

Laminae in the lumbar spine are relatively flat, blade-shaped bones projecting posteriorly from lateral to medial and converging at the midline posteriorly to give rise to the spinous

---

CLINICAL CORRELATION 9.2

A Schmorl's node is an intraosseous herniation of the nucleus pulposus, with a calcified shell that forms around it, passing vertically through the damaged end plate (Adams et al. 2000). In older spine specimens with reduced bone density, nodes tend to be more irregular (Hansson & Roos 1983). Adolescent specimens demonstrate posterior-edge fractures running from the end plate down to the cartilage growth plate, which appears to be a zone of weakness before skeletal maturity (Lundin et al. 1998). Many end plate pathologies may go undetected, as currently there is no clinical examination able to detect them. Therefore, the clinician should be cognizant of the patient's having increased pain with increased weight-bearing activities. These increased loading activities would more likely stress the end plate.

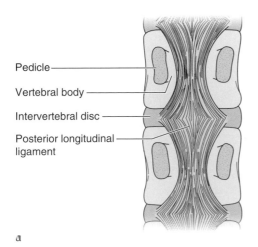

Pedicle
Vertebral body
Intervertebral disc
Posterior longitudinal ligament

a

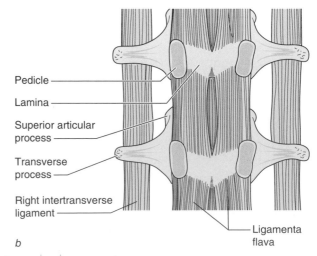

Pedicle
Lamina
Superior articular process
Transverse process
Right intertransverse ligament
Ligamenta flava

b

▶ **FIGURE 9.3** *(a)* Anterior and *(b)* posterior view of vertebral segments.

process (see figure 9.1). The laminae form the posterior bony border of the neural arch. Although they have less load-bearing demand than the pedicles, they are responsible for shunting forces between the spinous process and the articular processes, such that may occur with forceful lumbar rotation (Lucas & Bresler 1961).

The neural arch is mostly cortical bone, with only a small amount of trabecular bone; therefore, it is unlikely to weaken as much with age as the vertebral body (which has a high trabecular bone content). It therefore is not affected as much by osteoporosis as the vertebral body. The spinal canal, however, is involved in lumbar spinal stenosis (see Clinical Correlation 9.3).

Spinous and Transverse Processes

As in the rest of the spine, these processes (see figure 9.1) are sites for muscle attachments and increase the lever arm for muscles of the vertebral column. Additionally, ligaments and fascia attach to these structures. The spinous processes are not bifid in the lumbar spine as they are in the cervical spine. Unlike in the thoracic spine, the spinous processes project mostly straight posteriorly and are on the same transverse plane as the vertebral body.

Transverse processes are typically long and flat. The transverse processes of L3 are the widest of the lumbar spine, while the transverse processes of L5 are the thickest. Several stabilizing structures for the lumbar spine are attached to the transverse processes, including thoracolumbar fascia, the quadratus lumborum, and the iliolumbar ligament.

Articular Processes

The superior and inferior articular processes arise from the junction of the posterior pedicles and lateral laminae. As throughout the rest of the spine, there is a superior and inferior articular process on each vertebra (see figure 9.1). On the posterior edge of each of the superior articular processes is a small bump called the mamillary process. It functions as an attachment for the lumbar musculature.

▶ **KEY POINT**

The lumbar spine varies from the cervical spine in several ways: It has a much higher tolerance for weight bearing because of a larger vertebral body size; zygapophyseal joint orientation to a much greater extent in the sagittal plane; decreased motion due to a larger vertebral body to intervertebral disc size; distinct transverse processes; nonbifid spinous processes; and more consistency in characteristics compared with the variability of the cervical spine.

Sacrum

Three bones make up the **sacroiliac (SI) joint**: two **innominates** and the **sacrum**. The sacrum is a strong triangular bone located between the two innominates (figure 9.4). The **sacral base** is superior and anterior, with its

CLINICAL CORRELATION 9.3

Lumbar spinal stenosis (LSS), a focal narrowing of the spinal canal, nerve root canals, or intervertebral foramina (Arnoldi et al. 1976; Penning et al. 2005), is a common and disabling condition in the older adult (Tomkins et al. 2007; Deyo et al. 2005; Weinstein et al. 2006). A high depression score has been associated with more severe symptoms, poorer walking capacity, and less treatment satisfaction (Katz et al. 1995), as well as poorer postoperative treatment satisfaction (Katz et al. 1999). These physical and mental impairments may continue to increase in prevalence; it has been estimated that approximately 4% of patients visiting their primary care physicians and 14% of patients seeking assistance from a specialist for low back pain (LBP) present with LSS (Fanuele et al. 2000; Hart et al. 1995; Long et al. 1996). The clinician should therefore be cognizant of the multifactorial nature of LBP and LSS.

apex inferior and posterior. Five vertebrae fuse to form the central part of the sacrum, which contains remnants of the IVD enclosed by bone. The sacrum has four pairs of pelvic sacral foramina for transmission of the anterior (ventral) primary rami of the sacral nerves and four pairs of posterior (dorsal) sacral foramina for transmission of the posterior (dorsal) primary rami.

The **sacral ala** forms the superolateral portions of the sacral base. The **sacral superior articular processes**, which are concave and

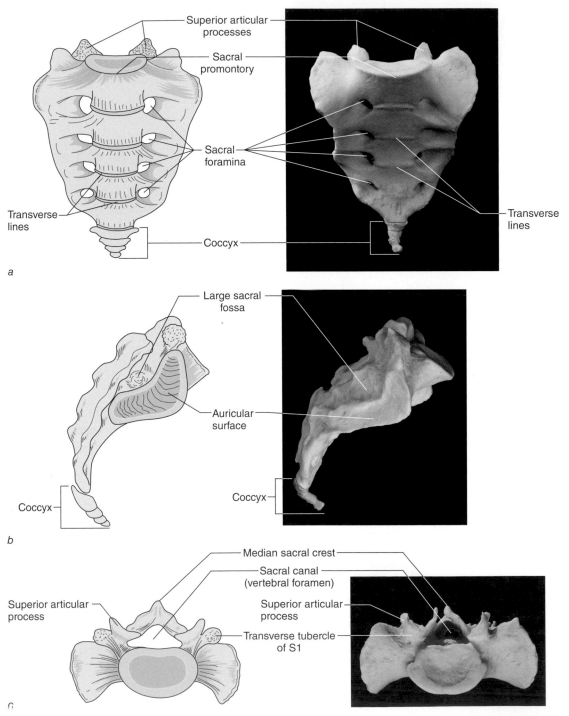

▶ **FIGURE 9.4**   The sacrum and coccyx. (a) Anterior aspect of the sacrum and coccyx. (b) Right lateral aspect of the sacrum and coccyx. (c) Superior aspect of the sacrum.

oriented posteromedially, extend upward from the base to articulate with the inferior articular processes of L5.

The **median sacral crest** is a midline ridge of bone on the posterior surface of the sacrum. It represents the fusion of the sacral spinous processes of S1 to S4.

The **sacral cornua** are bilateral downward projections off the sacral hiatus that are connected to the coccyx via the intercornual ligaments. On the inferolateral borders of the sacrum, about 2 cm to either side of the sacral hiatus, are the inferior lateral angles.

Pelvis

The pelvic girdle consists of the two innominate bones, the sacrum, and the bones of the coccyx (figure 9.5). The only other joint in the pelvic girdle, besides the two SI joints posteriorly, is the pubic symphysis. This joint is the articulation between each respective innominate anteriorly. The pelvic girdle should be considered one functional unit. Interrelated movement between the hip joints, **pubic symphysis**, bilateral SI joints, and lumbar spine (especially L4 and L5) can affect movement

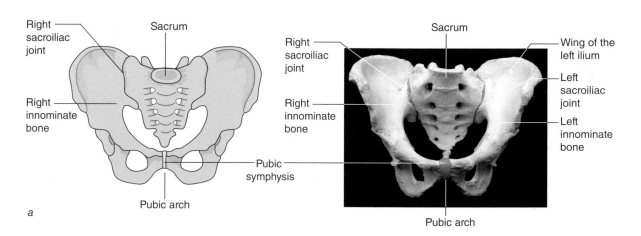

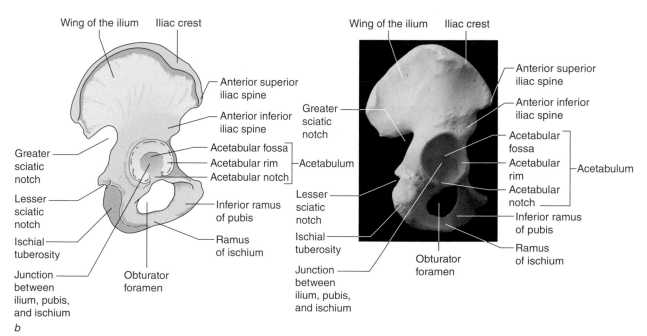

▶ **FIGURE 9.5** *(a)* Anterior aspect of the pelvis. The right innominate bone showing the *(b)* lateral aspect. *(continued)*

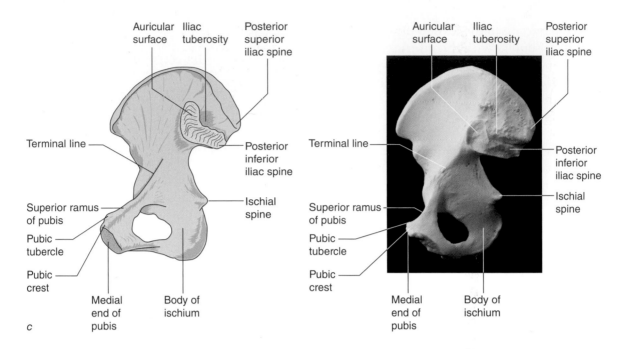

▶ **FIGURE 9.5** *(continued)* (c) Medial aspect.

throughout the pelvic girdle. Dysfunction in one area should be investigated in other areas as well. As described by Erhard (1977), "Dysfunction in any unit of the system will cause delivery of abnormal stresses to other segments of the system with the development of a subsequent dysfunction here as well."

▶ **KEY POINT**

The pelvic girdle consists of bilateral innominate bones, articulating anteriorly to form the pubic symphysis and with the sacrum posteriorly to form the sacroiliac joints, as well as bilateral hip joints. The pelvic girdle functions as one unit since motion of this area is interdependent on components of the pelvic girdle. Additionally, many muscles have common attachment areas on the pelvic girdle.

Joint Articulations

The multiple joint articulations in the lumbar spine and pelvic girdle primarily include the zygapophyseal joints on each respective side, the intervertebral joints, the SI joints, and the pubic symphysis. Each of these joints has unique characteristics that are described here.

Zygapophyseal (Facet) Joint

Each pair of vertebrae in the lumbar region is connected via a pair of zygapophyseal joints and an interbody joint with the IVD. Since each vertebra has two superior and two inferior articular facets, there are then 12 facet, or zygapophyseal, joints (T12-L1 to L5-S1). These zygapophyseal joints are typical synovial joints with hyaline cartilage covering their articulating surfaces.

Joint orientation from an anteroposterior perspective appears straight, but a superior view reveals a J- or C-shaped curve. There is variability in orientation, respective to both the level of the lumbar spine and the individual subject. Both the C- and J-shaped curves resist rotation (figure 9.6). It is thought that the C-shaped joints are better at resisting anterior displacement than the J-shaped joints because of the curvature (Bogduk & Twomey 1997; Ahmed et al. 1990). Coronal (frontal) plane orientation is present in isthmic spondylolisthesis (IS), while sagittal plane orientation of the facet joint is prevalent in degenerative spondylolisthesis (DS) (Boden et al. 1996; Grobler et al. 1993; Sato et al. 1989). The increase in the sagittal orientation of the facet joints is not just at the level of the degenera-

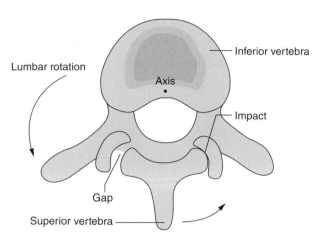

▶ **FIGURE 9.6**   Lumbar spine rotation to the left, resulting in impaction of right zygapophyseal joint and gapping of left zygapophyseal joint.

tive slip but also at the levels above the slip (Sato et al. 1989). The area most involved in resisting this shear force is the anteromedial part of the superior zygapophyseal joint. It is this area that is most vulnerable to fibrillation (Bogduk & Twomey 1997). The vertical tearing of the cartilage that occurs with aging may be part of the normal degeneration of the joint (Bogduk & Twomey 1997).

In the normal lumbar spine, the primary function of the zygapophyseal joint is to protect the motion segment from anterior shear forces, excessive rotation, and flexion (Tulsi & Hermanis 1993). These joints best resist forces acting perpendicular to their broad surfaces, approximately in the plane of the disc. The range of axial rotation is severely limited in the lumbar spine by these joints (Ahmed et al. 1990). They protect the disc from excessive strain and keep the segment stable. Additional functions include the following:

- Production of spinal motions including coupling movements (see chapter 6 regarding the controversy over spinal joint coupling)
- Minimal restriction of physiological movements of extension and side bending (Abumi et al. 1990)

Intervertebral Joint

Interbody joints of the lumbar region include the superior vertebra, inferior vertebra, and the IVD between these two vertebral bodies. These joints have been described as amphiarthrodial, with the capability of contributing to motion in all three planes.

Sacroiliac Joint

There is considerable controversy regarding the sacroiliac (SI) joint, including the type of joint it is. It has been classified as a diarthrodial (synovial) joint (Morris 1925; Cunningham 1925; Sashin 1930), as an amphiarthrosis (cartilaginous, less than 10° of movement) type of joint (Heisler 1923), and as a diarthrosis–synarthrosis (syndesmosis) combination joint (Gray 1980). The SI joint is most recently and generally described as a synovial joint (sacral surface covered with hyaline cartilage; iliac surface covered with a type of fibrocartilage) connecting the spine with the pelvis (Lee 1996).

Pubic Symphysis Joint

As previously mentioned, this joint is the articulation between the left and right innominate anteriorly. It is classified as a symphysis because it has no synovial tissue or fluid, and it contains a fibrocartilaginous lamina, or disc (see figure 9.5a). Some have also classified this joint as an amphiarthrodial joint.

Joint Anatomy

The joint anatomy of the lumbar spine and **pelvic girdle** is critical for the biomechanics and function in this region. This section describes the joint anatomy of the zygapophyseal and intervertebral joints of the lumbar spine as well as the joints of the pelvic girdle, including the SI joints. The discrete anatomical structure of each of these joints contributes in similar and differing manners to the function of this region of the body.

Zygapophyseal Joint

The joint capsule is fibrous and surrounds the entire zygapophyseal joint except anteriorly, which consists of the ligamentum flavum. The deep layers of the lumbar multifidus muscle reinforce the posterior facet joint capsule (Lewin 1964). Some of these fibers blend with the posterior capsule fibers and appear to keep the capsule taut (Macintosh & Bogduk 1986) (see Clinical Correlation 9.4). The capsule is quite loose superiorly and inferiorly. Both the superior and inferior poles of the capsule have a very small hole that allows passage of fat from within the capsule to the extracapsular space (Lewin et al. 1962).

Localized thickenings of the joint capsule are sometimes considered distinct ligaments. In full flexion, these ligaments provide 39% of a motion segment's resistance to bending (Adams et al. 1980). They also resist hyperextension (Adams et al. 1988; Hedtmann et al. 1989) but offer little resistance to axial rotation, at least in the lumbar spine (Adams & Hutton 1981; Panjabi et al. 1983).

The articular surfaces are approximately vertical in the upper lumbar spine but are more oblique at L4-L5 and L5-S1. This explains why the lower lumbar zygapophyseal joints resist approximately 20% of the compressive force acting perpendicular to the midplane of the discs, while at the upper levels they resist only half as much (Adams & Hutton, 1980). Even in healthy spines, extremely lordotic postures and extension movement can cause substantial compressive forces to be transmitted through these joints.

Three types of intra-articular meniscoids have been described within the zygapophyseal joints (Bogduk & Twomey 1997):

1. **Connective tissue rim**—a wedge-shaped thickening of the internal surface of the capsule that fills the space left by the curved margin of the articular cartilage. It is thought to be simply a space filler, although it might also increase the surface area of the joint, helping to transfer loads.

2. **Adipose tissue pad**—found at the superior and inferior poles of the joint, consisting of synovial folds. They project 2 mm into the joint.

3. **Fibroadipose meniscoids**—project approximately 5 mm into the joint cavity from the inner surface of the superior and inferior capsules.

The function of these structures is thought to be filling the joint cavity, increasing the articular surface area without reducing the flexibility in the joint, and protecting the articular surfaces. During lumbar spine flexion, the superior facet of the joint glides 5 to 8 mm up and forward, exposing the articular cartilage on both facets. The meniscoids remain in contact with the exposed cartilage, thereby maintaining a film of synovial fluid on the cartilage, ensuring proper lubrication against friction as the joint moves back into its resting position. Little and Khalsa (2005) describe these menisci as invaginations of the joint capsule that may occasionally project into the joint space rather than true intra-articular structures.

Morphological configuration of the zygapophyseal joints at the thoracolumbar junction is quite variable. In general, there is a change from a relatively coronal orientation at T10-T11 to a more sagittal orientation between L1 and L3, before returning to a more coronal orientation at L5 and S1.

The zygapophyseal joints are innervated by the medial branches of the posterior (dorsal) rami (Bogduk et al. 1982; Bogduk 1983). Each

CLINICAL CORRELATION 9.4

Posterior fibers of the lumbar multifidus blend with the posterior portion of the facet joint capsule, keeping it taut. Lumbar multifidus muscle contraction is not only beneficial when training for trunk stabilization (Richardson et al. 1999) and prevention of reoccurrence of LBP but also has been theorized to distract the facet joint when contracted segmentally. Clinically, distracting the facet joint either by lumbar multifidus contraction or manual therapy is thought to "unblock" a stuck facet joint caused by an entrapped meniscoid (Bogduk & Jull 1985).

joint receives its nerve supply from the corresponding medial branch above and below the joint (Bogduk et al. 1982; Bogduk 1983). For example, the L4-L5 joint receives its nerve supply from the corresponding medial branches of L3 and L4.

Intervertebral Joint

The intervertebral joint is composed of the articulation between the intervertebral disc and the vertebral body. As previously mentioned, the cartilaginous end plate is strongly attached to the intervertebral disc and is a primary mechanism for disc nutrition.

The intervertebral disc is the major load-bearing and motion-control element. The intervertebral disc guides the motion of rotation, whereas the facet restricts motion beyond disc boundaries (Tencer et al. 1982; Gracovetsky et al. 1985). The three primary components of the IVD are the (1) annulus fibrosus, (2) nucleus pulposus, and (3) cartilaginous end plate (figure 9.7). The major constituents of the IVD include proteoglycans, collagen, and water. The IVD functions as a shock absorber and a deformable space, and it resists compressive forces of the spine (Frymoyer & Cats-Baril 1987).

The annulus fibrosus is predominantly composed of rings of fibrocartilage forming the outer portion of the IVD. The typical lumbar annulus consists of between 10 and 20 layers of collagen fibers that are obliquely oriented to one another (Taylor 1990). These concentric-oriented lamellar rings lie at a 30° plane from the horizontal (Peng et al. 2005). Lamellae are designed to counter compression, side bending, shear, and distraction forces. Nerve endings in the outer portion of the annulus are responsible for pain generation and somatic referral of symptoms.

Strong attachments exist between the annulus and the outer portion of the vertebral bodies and vertebral end plates as well as the anterior longitudinal ligament (ALL). The inner two-thirds of the disc attach to the cartilaginous end plate.

The annulus blends into the nucleus in a gradual transition rather than an abrupt transition between two separate structures (Pope & Panjabi 1985). Cadaveric examination has shown the differentiation between the annulus and nucleus is difficult to discern, especially as one ages. For other changes in the IVD with aging, refer to Clinical Correlation 9.5. The lumbar spine disc, unlike the cervical spine disc, has consistent annulus fibers around the entire periphery of the disc. The lumbar spine disc undergoes greater torsional and compressive loads than the cervical spine disc.

▶ **KEY POINT**
The intervertebral discs of the lumbar spine are distinctly different from those of the cervical spine, due primarily to the differences in the annulus fibrosus.

The nucleus pulposus is the inner portion of the IVD and accounts for up to 50% of the disc. The nucleus is responsible for nutrient transport via osmosis of the middle cartilaginous end plate and articulation with the disc.

The IVD is avascular, except for the very periphery of the annulus fibrosus. Therefore, the hydration by diffusion of the IVD is mediated by mechanical forces and osmotic gradients. Cyclic loading creates movement of tissue fluid in and out of the IVD. This mechanism, along with diffusion from the vessels in the outer annulus and the capillary plexuses beneath the vertebral end plate, provides nutrition to the disc.

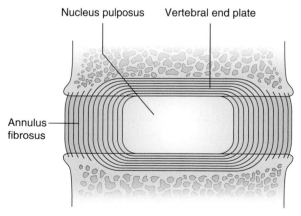

▶ **FIGURE 9.7** Intervertebral disc demonstrating annulus fibrosus, nucleus pulposus, and vertebral end plate.

Because of the low blood supply, the oxygen concentration in the center of the nucleus is only 2 to 5% of that at the periphery. Therefore, the cells must rely on anaerobic metabolism. As a result, an acidic environment is created from the large amounts of lactic acid produced. The metabolism of the disc is very sensitive to changes in pH. The disc functions properly at a pH range of 6.9 to 7.2. If the pH drops below 6.3 the metabolism activity falls below 15%. Impaired nutrition, inflammatory mediates, or changes in pH can lead to major changes in matrix status.

The outer half of the IVD is innervated by the **sinuvertebral nerve**[2] (Bogduk 1983) and the gray rami communicantes. The posterolateral aspect innervated by both the sinuvertebral nerve (Bogduk 1983) and the gray rami communicantes. The lateral aspect of the IVD receives only sympathetic innervation.

Sacroiliac Joint

The sacral joint surfaces are formed from hyaline cartilage and the iliac joint surface from fibrocartilage (Schunke 1938). The hyaline cartilage is three to five times thicker than the fibrocartilage (MacDonald & Hunt 1951). The auricular articular surface of the sacrum is described as an inverted L shape (figure 9.8). The short (superior) arm lies in a craniocaudal plane and corresponds to the depth of the sacrum. It is widest superiorly and anteriorly. The long (inferior) arm lies in an anteroposterior plane and represents the

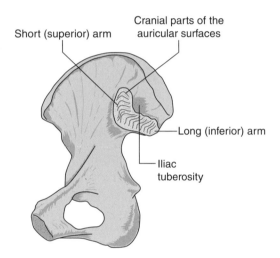

Short (superior) arm

Cranial parts of the auricular surfaces

Long (inferior) arm

Iliac tuberosity

▶ **FIGURE 9.8**   Inverted L-shaped orientation of the sacral articulation.

length of the sacrum from superior to inferior. It is widest inferiorly and posteriorly.

The articular surfaces of the sacrum and **ilium**[1] have extensive elevations and depressions (Vleeming et al. 1990; Weisl 1954), with a central depression on the sacral side. The sacral concavity and iliac convexity, and other elevations and depressions, increasingly develop over time (Bowen & Cassidy 1981).

The SI joint capsule is extensive and very strong, consisting of two layers. It attaches to both articular margins of the joint and is thickened inferiorly. Patterns of extracapsular extravasation from the SI joint have been observed on postarthrography computed tomography (CT) scans (Fortin et al. 1999; Fortin & Tolchin 2003). This capsular disruption could allow inflammatory mediates to leak from the SI joint to nearby neural structures and therefore could explain radicular pain in certain patients (Fortin et al. 1999; Fortin et al. 2003).

Pubic Symphysis Joint

The bone surfaces of the pubic symphysis are covered with hyaline cartilage but are kept apart by the presence of the disc. Ligamentous support of this joint includes the following (Kapandji 1991):

- Superior pubic ligament, a thick, fibrous band
- Inferior arcuate ligament attaching to the inferior pubic rami bilaterally and blending with the disc
- Posterior pubic ligament, a membranous structure that blends with the adjacent periosteum
- Anterior ligament, a thick band with both transverse and oblique fibers

Ligaments of the Lumbar Spine and Pelvic Girdle

A primary function of the lumbar spine ligaments is to restrain motion. Biomechanically, ligaments of all regions of the spine, with the exception of the ligamentum flavum, are relatively inelastic and exhibit a viscoelastic response to loading (Bogduk & Twomey 1991). The ligaments of the lumbar spine are traditionally thought to be primary spinal stabilizers. The spine has been shown to buckle under as little as 2 kg (5 lb) of loading without muscular support (Lucas & Bresler 1961). This would therefore imply that the ligaments of the lumbar spine contribute only a small portion to spinal stability.

The lumbar spine ligaments have demonstrated large densities of sensory end organs, including free nerve endings and mechanoreceptors (Bogduk 1983; Kellgren 1938; Rhalmi et al. 1993; Yahia et al. 1988). Solomonow and colleagues (1998) have suggested that these ligaments, along with the lumbar musculature, provide information regarding the position of the motion segment of the lumbar spine, thereby influencing lumbar muscle tension.

The SI ligaments include both intra- and extra-articular ligaments and are some of the strongest ligaments in the body. The ligaments of the lumbar spine (figure 9.9 and table 9.1) are primarily the same ligaments of the rest of the spine except for some additional ones that are discussed here (e.g., iliolumbar ligament) (see chapter 6, cervical spine).

The anterior longitudinal ligament (ALL) is under tension in a neutral spine and primarily resists overextension. It also provides a minor limitation to anterior translation and vertical separation of the vertebral bodies. The space between the ALL and the bone is filled with fatty tissue, blood vessels, and nerves. The ligament is only loosely attached to the front of the disc. The recurrent branch of the gray rami communicantes provides the nerve supply to the ALL (Bogduk et al. 1981).

The posterior longitudinal ligament (PLL; see figure 9.3a) forms a narrow band over the posterior portion of the lumbar vertebral bodies. It expands laterally over the posterior portion of the discs, giving it a saw-tooth appearance. The primary function of the PLL in the spine is resistance to flexion. Traction and posterior vertebral body shearing tend to make the PLL taut. The PLL also limits flexion over a number of segments, although it is less a restraint than the ligamentum flavum because of its proximity to the center of rotation (Willard 1997). The PLL is innervated by the sinuvertebral nerve.

The ligamentum flavum (see figure 9.3b) is a short but thick ligament that connects laminae of consecutive vertebrae. It consists of 80% elastin and 20% collagen (Yahia et al. 1990). Therefore, it is stretched in flexion and returns to normal length in neutral position or extension. This ligament is often referred to as the yellow ligament because its elastin content—which differentiates it from other ligaments of the spine—gives it a yellow appearance. Because of its elastic properties, the ligament does not buckle when the spine extends (Schonstrom et al. 1989). Elastin also enables this ligament to be stretched by 80% without failure (Nachemson & Evans 1968). The primary function of the ligamentum flavum is to resist flexion. It appears to prevent the anterior capsule from becoming trapped

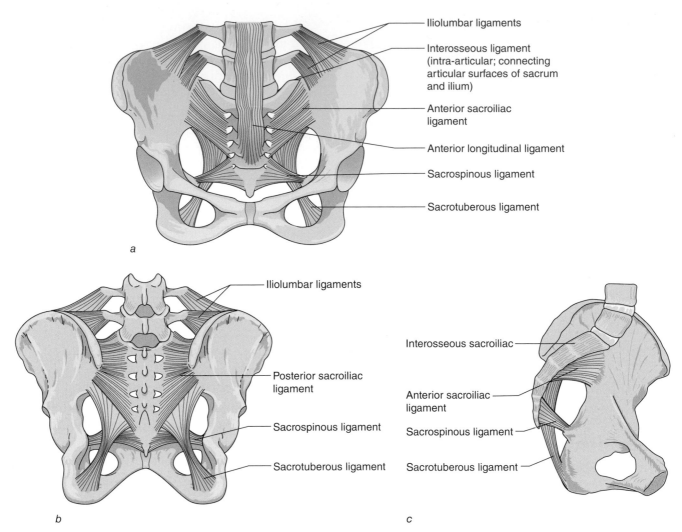

- Iliolumbar ligaments
- Interosseous ligament (intra-articular; connecting articular surfaces of sacrum and ilium)
- Anterior sacroiliac ligament
- Anterior longitudinal ligament
- Sacrospinous ligament
- Sacrotuberous ligament

a

- Iliolumbar ligaments
- Posterior sacroiliac ligament
- Sacrospinous ligament
- Sacrotuberous ligament

b

- Interosseous sacroiliac
- Anterior sacroiliac ligament
- Sacrospinous ligament
- Sacrotuberous ligament

c

▶ **FIGURE 9.9**   Supporting ligaments of the sacroiliac articulations: *(a)* anterior aspect, *(b)* posterior aspect, and *(c)* left aspect of medial section through the pelvis.

between the articular margins as it recoils during extension (Bogduk & Twomey 1997). Additionally, it is thought to pretension the intervertebral disc (IVD), although the significance of this remains elusive. The ligamentum flavum is innervated by the medial branch of the posterior (dorsal) rami (Bogduk et al. 1982).

The interspinous ligament lies deep between and connects adjacent spinous processes. Although this ligament most likely resists separation of the spinous processes during flexion (Hukins et al. 1990), it may provide little contribution to clinical stability of the lumbar spine (Panjabi et al. 1983). The interspinous ligament is innervated by the medial branch of the posterior (dorsal) rami (Bogduk et al 1982)

The supraspinous ligament lies in the midline and runs posterior to the posterior edges of the spinous processes, to which it is attached. It is broad, thick, and cord-like but is only well developed in the upper lumbar region (Bogduk & Twomey 1997; Gray 1995). The supraspinous ligament is not considered a true ligament by some because part of it is derived from the posterior part of the interspinous ligament and also merges with the insertions of the lumbar dorsal muscles (Heylings 1978). The supraspinous ligament is also innervated by the medial branch of the dorsal rami (Bogduk et al. 1982).

Intertransverse ligaments consist of sheets of connective tissue extending from the upper border of one transverse process to the lower

TABLE 9.1 Principal Ligaments of the Pelvic Girdle

Ligament	Attachments	Function
Extra-articular ligaments		
Sacrotuberous	Composed of three large fibrous bands, attached broadly to the PSIS and lateral sacrum, and blended with the dorsal sacroiliac ligament Descends obliquely to attach to the medial margin of the ischial tuberosity; also has attachments to the coccyx, lowest fibers of the gluteus maximus, and piriformis Superficial fibers on the inferior aspect of the ligament can continue into the tendon of the biceps femoris	Stabilizes against nutation or flexion of the sacrum as well as counteracting against posterior and cranial migration of the sacral apex during weight bearing (Van Wingerden et al. 1993; Vleeming et al. 1989)
Sacrospinous	Triangular in shape, extends from the ischial spine to the lateral margins of the sacrum and coccyx and laterally to the spine of the ischium[1] Runs deep to the sacrotuberous ligament, blending with it and attaching to the capsule of the SI joint (Willard 1997)	Converts the lesser sciatic notch into the lesser foramen Along with the sacrotuberous ligament, opposes forward tilting of the sacrum on the innominates (nutation or flexion) during weight bearing of the vertebral column
Iliolumbar	Connects transverse process of L5 and occasionally L4 to the ilium	Prevents anterior translation of L5; resists side bending and, to a lesser degree, twisting and forward and backward bending of L5; appears to resist sagittal nutation and counternutation of the SI joint
Intra-articular ligaments		
Anterior sacroiliac	An anteroinferior thickening of the fibrous capsule that is relatively weak and thin compared with other SI ligaments Extends between the anterior and inferior borders of the iliac auricular surface and the anterior border of the sacral auricular surface (Bowen & Cassidy 1981)	Prevents anterior distraction of the joint Often injured and described as a source of pain; stressed with any distraction of the anterior portion of the joint
Interosseous sacroiliac	Located deep to the dorsal SI ligament; very strong	Resists anterior and inferior movement of the sacrum Forms the major connection between the sacrum and the innominate
Posterior (dorsal) sacroiliac, also known as the long ligament	Easily palpable directly distal to the PSIS Fibers are multidirectional and blend laterally with the sacrotuberous ligament; also attaches medially to the erector spinae (Vleeming 1996), multifidus muscles (Willard 1997), and thoracodorsal fascia	Nutation or flexion of the sacrum appears to slacken this ligament, while counternutation or extension of the sacrum stresses the ligament (Vleeming 1996). Resists sacral extension or counternutation

border of the transverse process above. Rather than being a true ligament, they form part of a complex fascial system. They form a septum that divides the anterior and posterior muscles of the lumbar spine.

A ligament unique to the lumbar spine is the **iliolumbar ligament**. Along with the **sacrotuberous ligament** and sacrospinous ligament, it is one of the three vertebropelvic ligaments. On each side of the vertebra, they connect the transverse process of the fifth lumbar vertebra to the ilium. Kapandji (1991) also recognizes a superior part that runs from the tip of the L4 transverse process to the iliac crest. It is variously believed to be a degenerate part of the quadratus lumborum or the iliocostalis, not developing fully until approximately 30 years of age (Luk et al. 1986). Regardless of what its structure and development are reported to be, it is generally accepted to form a strong bond between L5 and the ilium in the adult. It restrains flexion, extension, axial rotation, and side bending of L5 on S1 (Chow et al. 1989).

Thoracolumbar Fascia

The **thoracolumbar fascia** is a complex array of dense connective tissue traveling from the spinous process of T12 to the posterosuperior iliac spine and iliac crest. The thoracolumbar fascia serves as an attachment for several upper limb and trunk muscles, including the latissimus dorsi, gluteus maximus, transverse abdominis, and internal oblique musculature (Sturesson et al. 1989). It also stabilizes the spine against anterior shear and flexion moments; resists segmental flexion; and provides stability to the posterior aspect of the lumbar spine, especially with lifting and pulling activities in trunk flexion, as it reinforces the posterior ligaments and muscular system (Gracovetsky 1986; Vleeming et al. 1995). This fascia transmits forces between the spine, pelvis, and legs and may contribute to trunk stabilization during rotation- and flexion-based activities (Pool-Goudzwaard et al. 1998). The thoracolumbar fascia consists of three layers that envelop the lumbar musculature and separate them into anterior, middle, and posterior layers (figure 9.10). The large posterior

layer arises from the spinous processes of the thoracic, lumbar, and sacral vertebrae and covers the erector spinae muscles. It blends with the outer layers of the fascia along the lateral border of the iliocostalis lumborum in a dense thickening of the fascia called the lateral raphe (Bogduk & Macintosh 1984).

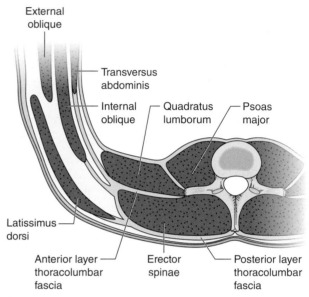

▶ **FIGURE 9.10** Layers of the thoracolumbar fascia.

Nerve Supply

Because of the complexity of the sacroiliac joint, the precise innervations of this joint remain somewhat unclear. The anterior portion likely receives innervation from the posterior rami of the L2 through S2 roots (Bogduk 1997), although their contribution is highly variable and may differ from person to person (Dreyfuss et al. 1996). Additional innervations to the anterior joint may be from the obturator nerve, superior gluteal nerve, or lumbosacral trunk (Pitkin & Pheasant 1936). The posterior portion of the joint is most likely innervated by the posterior rami of L4 through S3, especially S1 and S2 (Grob et al. 1995). An additional autonomic component of the joint's innervations further increases the complexity (Pitkin & Pheasant 1936). Fortin and colleagues (1999) concluded that, based on an anatomical study on adult cadavers with fetal correlation, the joint is predominantly, if not entirely, innervated by sacral posterior rami.

Blood Supply

The lumbar spine blood supply is provided by the lumbar arteries, and its venous drainage occurs via the lumbar veins.

Joint Function

This section explains the function of the various joints in the lumbar spine. For each type of joint, we describe its axes of motion, arthrokinematics, range of motion, closed and loose packed positions, end feel, and capsular pattern, where applicable, with corresponding figures to supplement the reader's understanding of lumbar spine joint function.

Zygapophyseal Joint

The zygapophyseal joints are the principle guiding and restraining mechanism of the segment. They protect the disc from excessive strain and keep the segment stable.

Axes of Motion

Flexion–extension occurs in the sagittal plane around a mediolateral axis. Axial rotation takes place in the transverse plane around a vertical axis, and side bending takes place in the frontal plane around an anteroposterior axis.

Arthrokinematics

The arthrokinematics of the zygapophyseal joint have unique characteristics dependent on the motion being performed. The joint arthrokinematics for flexion, extension, rotation, and side bending are detailed here.

Joint Function During Flexion and Extension As previously mentioned for the cervical spine (chapter 6), up-and-forward movement and down-and-back movement occur with flexion and extension, respectively. The inferior facet of the superior vertebra will glide up and forward on the superior facet of the inferior vertebra. Along with this motion will be opening of a small gap between facets.

Segmentally, the amount of motion at each respective segmental level increases from L1-L2 distally to L4-L5 and then decreases slightly at L5-S1. The total motion is variable but has been described as ranging from 35 to 52° (Grieve 1981; Pearcy et al. 1984).

Extension of the facet joint involves the inferior facet of the superior vertebra gliding down and back on the superior facet of the inferior vertebra (figure 9.11). Along with this posterior and inferior motion of the superior

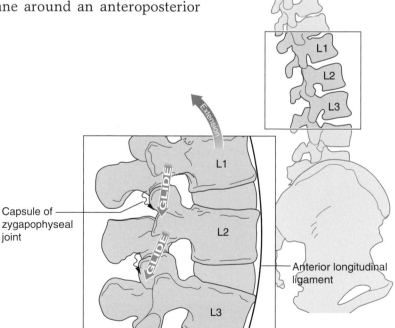

▶ **FIGURE 9.11**   Facet joint movement during extension. Bilateral superior facets glide inferiorly on the inferior facets of the segment.

facet is an impaction of that zygapophyseal joint and buckling of the supraspinous ligament.

The amount of motion at each segmental level is variable but is typically largest at L5-S1 and L1-L2 for a total of 15 to 29° (Grieve 1981; Pearcy et al. 1984). Again, variability exists among sources.

As mentioned in chapters 6 and 8 (cervical spine and thoracic spine), the concept of coupling has been questioned. Also, as previously mentioned in these chapters, inconsistencies in previous literature reviews investigating spinal coupling (Cook 2003; Legaspi & Edmond 2007) has led to the suggestion of using caution when applying the concepts of coupled motion to the evaluation and treatment of patients with low back pain (Legaspi & Edmond 2007).

Joint Function During Rotation   Because of the facet joint orientation, the amount of rotation is significantly reduced in the lumbar spine. Each segment demonstrates 1 to 4° for a total of 5 to 16° (Grieve 1981; Pearcy et al. 1984). As a result of the joint orientation, rotation results in gapping of the ipsilateral and compression of the contralateral facet joint (see figure 9.6).

Joint Function During Side Bending   Facet joint orientation in the sagittal plane also restricts side bending motion in the lumbar spine. Each segment generally contributes approximately 3 to 6° except L5-S1, which has only 1 to 2°, for a total range of 16 to 25° (Grieve 1981; Pearcy et al. 1984).

The motion description is the same as for the other regions of the spine. Right side bending is caused by the right superior facet of that joint gliding down and back on the inferior facet of that segment (i.e., extension of the right zygapophyseal joint), while the left zygapophyseal joint undergoes up and forward movement (flexion of the left zygapophyseal joint) (figure 9.12).

The facet joints also serve other important roles in the load bearing of the lumbar

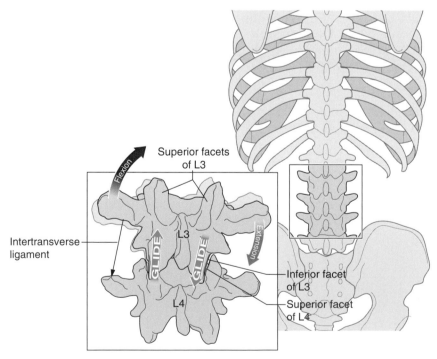

▶ **FIGURE 9.12**   Facet joint movement during right side bending. The right superior facet glides inferiorly on the inferior facet of the same segment, while the left superior facet glides superiorly on the inferior facet of the same segment.

spine. They resist anterior shear forces and, along with the IVD, resist torsion (Sharma et al. 1995). They also assist with resisting compressive forces. During upright posture, approximately 18 to 20% of the compressive load acting on the lumbar spine is exerted at the facets (Bogduk & Twomey 1997). This percentage of compressive loading will vary depending on lumbar posture.

Range of Motion

Range of motion of the entire thoracic and lumbar spine per the American Medical Association (1988) is as follows:

- Flexion: 60°
- Extension: 25°
- Right side bending: 25°
- Left side bending: 25°
- Right rotation: 30°
- Left rotation: 30°

Closed and Loose Packed Positions

The closed packed position of the lumbar spine zygapophyseal joints is full extension, with a resting position of midway between flexion and extension.

End Feel

The end feel is generally described as firm for side bending and rotation and as tissue stretch for extension and flexion.

Capsular Pattern

The capsular pattern for the lumbar spine is a marked and equal restriction of side bending followed by restriction of flexion and extension (Cyriax & Cyriax 1983).

▶ **KEY POINT**
  The lumbar spine is the most distal and largest portion of the spine. Motion of the lumbar spine is primarily in the sagittal plane (flexion and extension) because of the orientation of the zygapophyseal (facet) joints.

Intervertebral Joint

The intervertebral joint of the lumbar spine is quite different anatomically and biomechanically from the zygapophyseal joints. Please note that closed and loose packed positions, end feel, and capsular pattern of the intervertebral joints are not described because they are not synovial joints.

Axes of Motion

Flexion–extension occurs in the sagittal plane around a mediolateral axis. Axial rotation takes place in the transverse plane around a vertical axis, and side bending takes place in the frontal plane around an anteroposterior axis.

Arthrokinematics

Segmentally, lumbar flexion produces a combination of an anterior roll and glide of the vertebral body. During this anterior rocking motion, the inferior facets of the superior vertebra glide up and forward (as mentioned previously). Lumbar extension is just the reverse—a combination of posterior roll and glide of vertebral bodies and facet movement of the superior vertebra down and back on the inferior vertebra.

The spinous process can make firm contact in full extension, or after a pathological disc narrowing, so that a proportion of the compressive forces acting on the spine can be resisted by these "kissing spines" (Adams et al. 1988).

The IVD allows joint displacement to occur by maintaining a separation between the vertebral bodies and by being capable of deformation in all planes of motion. The IVD can act as a ball between the vertebral bodies, with a larger IVD allowing for more potential motion in each direction. Gliding motions can also occur at this joint. Anterior gliding of the vertebral body is resisted by contact between the inferior articular processes of the superior vertebra abutting against the superior articular processes of the lower vertebra (Bogduk & Twomey 1997).

External forces that approximate the vertebral bodies exert compressive load on the IVD. The IVD tolerates these loads by converting vertically applied compression into circumferentially applied tension via hoop stress. In other words, compression loading on the nucleus pulposus causes it to exert radial stresses on the annulus fibrosus.

Range of Motion

Range of motion of the entire thoracic and lumbar spine as per the American Medical Association (1988) is as follows:

- Flexion: 60°
- Extension: 25°
- Right side bending: 25°
- Left side bending: 25°
- Right rotation: 30°
- Left rotation: 30°

The flexion–extension ROM in the lumbar spine that occurs between vertebral segments is approximately 12° in the upper lumbar spine, reaching a maximum ROM of 20 to 25° between L5 and S1 (Bogduk & Twomey 1997).

Sacroiliac Joint

Because of the tight articulation at the SI joint, it was thought for several decades that the SI joint was immobile. It has been demonstrated that mobility of the SI joint is not only possible (Zheng et al. 1997; Harrison et al. 1997; Miller et al. 1987; Sturesson et al. 2000; Egund et al. 1978; Smidt et al. 1995; Frigerio et al. 1974; Weisl 1955) but essential for shock absorption during weight-bearing activities (Lee & Vleeming 2004). The pelvis must absorb the majority of lower extremity rotation, particularly during bipedal gait (Basmajian & Deluca 1985), as the lumbar spine does not tolerate rotation well.

A recent systematic review has questioned the utilization of clinical examination methods based on the amount of motion available at the SI joint. The authors' findings were that rotation ranged between −1.1 and 2.2° along

the x-axis, −0.8 and 4.0° along the y-axis, and −0.5 and 8.0° along the z-axis. Translation ranged between −0.3 and 8.0 mm along the x-axis, −0.2 and 7.0 mm along the y-axis, and −0.3 and 6.0 mm along the z-axis. Since the motion of the SI joint was found to be limited to minute amounts of rotation and of translation, the authors suggested that clinical methods utilizing palpation for diagnosing SI joint pathology may have limited clinical utility (Goode et al. 2008).

▶ **KEY POINT**
More recent evidence suggests limited mobility of the sacroiliac joint. The clinician must consider this when examining people with low back pain.

Many biomechanical interpretations of SI and pelvic girdle movement exist, and again, there is considerable controversy regarding this movement (even within a particular discipline). Although there are other models of description for pelvic girdle movement (e.g., osteopathic model, chiropractic model), the anatomical–biomechanical model is described here. This model has been influenced by Vleeming et al. (1997), Snijders et al. (1997), and Richardson et al. (1999). It is considered the most contemporary and describes joint mechanics as being potentially influenced by articular, neuromuscular, and emotional factors.

Axes of Motion

Five different axes of motion have been described, amid much controversy. Further detail regarding these motions is given in the section on arthrokinematics.

Arthrokinematics

During **sacral nutation** (or flexion), the superior base of the sacrum moves anteriorly and the inferior border of the sacrum moves posteriorly (figure 9.13a). This motion is resisted by various factors: the wedge shape of the sacrum, ridges and depressions on the articular surfaces, the friction coefficient of the joint surface, and the integrity of the soft-tissue structures surrounding the joint (figure 9.14*a*). **Sacral counternutation** (or extension) (figure 9.13*b*) is accompanied by sacral superior base movement posteriorly and inferior border movement anteriorly. This motion is primarily resisted by the **long dorsal ligament** (figure 9.14*b*).

Commonly described axes of motion are demonstrated in figure 9.15. Different models of description for movement often use differ-ent axes of motion. The oblique axes of motion are commonly considered more contemporary. Sacral torsion—rotation and side bending of the sacrum in the same direction—is often described as occurring in an oblique axis. Therefore, the sacrum is thought to have motions of nutation, counternutation, side bending, and rotation.

The innominate motion of posterior rotation is resisted by the sacrotuberous ligament, and anterior rotation can potentially be resisted

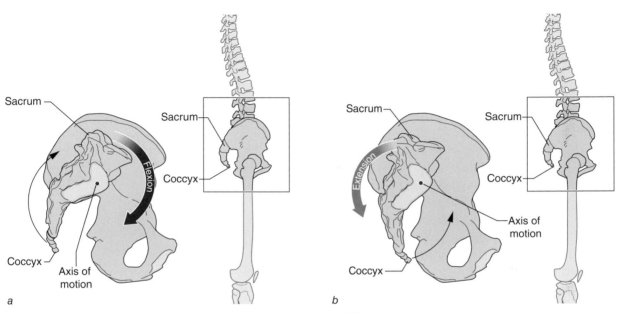

▶ **FIGURE 9.13**   *(a)* Nutation and *(b)* counternutation of the sacrum.

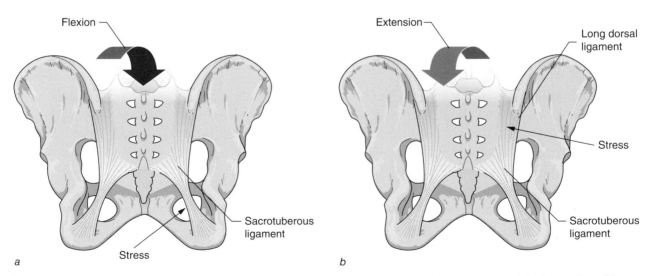

▶ **FIGURE 9.14**   Stress on the *(a)* sacrotuberous ligament due to sacral nutation and *(b)* long dorsal ligament due to sacral counternutation.

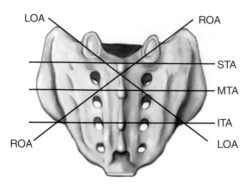

▶ **FIGURE 9.15** Commonly described axes of motion about the SI joint. LOA = left oblique axis, ROA = right oblique axis, STA = superior transverse axis, MTA = middle transverse axis, ITA = inferior transverse axis.

by the long dorsal ligament. The direction of the innominate rotation in normal arthrokinematics depends on the initiating movement (table 9.2).

During trunk forward bending, as the lumbar spine flexes (superior facet glides up and forward on the bottom facet), the innominates move in an anterior and outward rotation. This results in an approximation of bilateral PSISs, while the sacrum nutates. The sacrum stops this nutation motion at about 60° of trunk flexion (Lee 2004).

With backward bending of the trunk, the spine extends (superior facet glides down and back on the bottom facet), involving an inferior motion of both PSISs. Minimum posterior rotation of the innominate occurs, and the sacrum remains nutated.

Range of Motion

Rotation ranges between –1.1 and 2.2° along the x-axis, –0.8 and 4.0° along the y-axis, and –0.5 and 8.0° along the z-axis. Translation ranges between –0.3 and 8.0 mm along the x-axis, –0.2 and 7.0 mm along the y-axis, and –0.3 and 6.0 mm along the z-axis (Goode et al. 2008).

Closed and Loose Packed Positions

The closed packed and resting positions of the SI joint and pubic symphysis are typically not described, in part due to the unique nature of this joint.

End Feel

The end feel is not described.

Capsular Pattern

The capsular pattern for both the sacroiliac and pubic symphysis joints is described as pain when the joints are stressed.

Pubic Symphysis Joint

The amount of movement at the pubic symphysis is also relatively small. Walheim and Selvik (1984) described 3° of rotation and 2° of translation. During a single-leg stance, 2.6 mm of vertical and 1.3 mm of sagittal movement has been demonstrated on the weight-bearing side. During walking, the pubic symphysis can piston up and down up to 2.2 mm vertically and 1.3 mm sagitally (Meissner et al. 1996).

Muscles

The primary trunk extensors include (as discussed in previous chapters) the longissimus, iliocostalis, and multifidus groups. The longissimus and iliocostalis also have thoracic components. Origins, insertions, innervations, and muscle actions of all relevant musculature of the lumbar spine and pelvis are listed in table 9.3. Several muscles attach to the lumbar–pelvic–hip complex, particularly posteriorly;

TABLE 9.2   Interrelated Hip, Lumbar Spine, and Innominate Motions

Joint of motion initiation	Innominate motion	Sacral motion
Hip flexion	Posterior rotation	n/a
Hip extension	Anterior rotation	n/a
Lumbar spine flexion	Anterior rotation	Nutation, then counternutation
Lumbar spine extension	Posterior rotation	Nutation

TABLE 9.3 Muscles of the Lumbosacral Region

Muscle	Origin	Insertion	Action	Innervation
Posterior trunk musculature				
Erector spinae muscles				
Erector spinae	Common tendon of origin; posterior surface of sacrum, iliac crest, spinous processes of T11-L5 (specific origins given below)	As described for each muscle	Bilateral action: extension of vertebral column Unilateral action: bend vertebral column toward same side (lateral flexion)	Dorsal rami of spinal nerves in area of muscle *Note:* This is the innervation for all muscles of this group.
Iliocostalis lumborum	Iliac crest; sacrum	Lower borders of lower 6 or 7 ribs	*Note:* This is the action for all muscles of this group.	
Iliocostalis thoracis	Upper borders of lower 6 or 7 ribs	Lower borders of upper 6 ribs		
Iliocostalis cervicis	Angles of upper 6 ribs	Transverse processes of C4-C6		
Longissimus thoracis	Intermediate part of common tendon	Lower 9 or 10 ribs and adjacent transverse processes of vertebrae		
Longissimus capitis	Tendons of origin of longissimus cervicis; articular processes of C4-C7	Mastoid process		
Spinalis thoracis	Spinous processes of T11-L2	Spinous processes of upper thoracic vertebrae (varies from 4 to 8)		
Transversospinalis muscles				
Semispinalis thoracis	Transverse processes of T7-T12	Spinous processes of C6-T4	Extension of vertebral column	Dorsal rami of cervical and thoracic spinal nerves
Rotators	Sacrum and transverse processes of lumbar through lower cervical vertebrae	Spinous processes of lumbar through C2; fascicles span 1-2 segments of column	Rotation (to opposite side) and extension of vertebral column	Dorsal rami of spinal nerves
Segmental muscles				
Interspinalis	Spinous processes of vertebrae (absent in much of thoracic region)	Spinous processes of vertebrae (span between adjacent vertebrae)	Extension of cervical column	Dorsal rami of cervical spinal nerves
Intertransversarii	Transverse processes of vertebrae (absent in most of thoracic region)	Transverse processes of vertebrae (span between adjacent vertebrae)	Lateral flexion of vertebral column (unilateral action)	Dorsal and ventral rami of spinal nerves

> *continued*

TABLE 9.3 > *continued*

Muscle	Origin	Insertion	Action	Innervation
Deep trunk musculature				
Transversus abdominis	Deep surfaces of the costal cartilages of lower 6 ribs, interdigitating with the diaphragm, thoracolumbar fascia between iliac crest and 12th rib, anterior 2/3 of inner lip of iliac crest, lateral 1/3 of inguinal ligament, and fascia over iliacus muscle	Pubic crest and aponeurosis that fuses with posterior layers of aponeurosis of internal oblique; the innermost of the three muscles of the abdominal wall	Provides segmental stabilization to the lumbar spine and pelvic girdle	Ventral rami of lower 6 thoracic and 1st lumbar spinal nerves
Multifidus	At each level, several fascicles arise via a common tendon from the spinous process	Diverge caudally to assume separate attachments to the mamillary process, iliac crest, and sacrum Some of the deeper fibers attach to facet joint capsule	Opposes flexion effect of abdominal muscles as they produce rotation Increases intra-abdominal pressure and assists with spinal stabilization	Dorsal rami of spinal nerves
Anterior trunk musculature				
Rectus abdominis	Crest of **pubis**[1]; ligaments covering ventral surface of pubic symphysis	By 3 portions into cartilages of 5th, 6th, and 7th ribs	Trunk flexion; assists with trunk rotation; rib depression	Ventral rami of lower 6 or 7 thoracic spinal nerves
External oblique	8 digitations from external surfaces and inferior borders of lower 8 ribs	Anterior 1/2 of outer lip of iliac crest; aponeurosis to pubic tubercle and pectineal line in middle interlaces with aponeurosis of opposite muscle forming linea alba, extending from xiphoid process to pubic symphysis	Trunk rotation in contralateral direction; trunk flexion; increases intra-abdominal pressure; rib depression; spinal stabilization	Ventral rami of lower 6 thoracic spinal nerves
Internal oblique	Lateral 1/2 of inguinal ligament; anterior 2/3 of middle lip of iliac crest; posterior layer of thoracolumbar fascia near the crest	Crest of pubis and medial part of pectineal line; linea alba; cartilages of 7th, 8th, and 9th ribs; inferior borders of cartilages of the last 3 or 4 ribs	Trunk rotation in ipsilateral direction	Ventral rami of lower 6 thoracic and 1st lumbar nerves

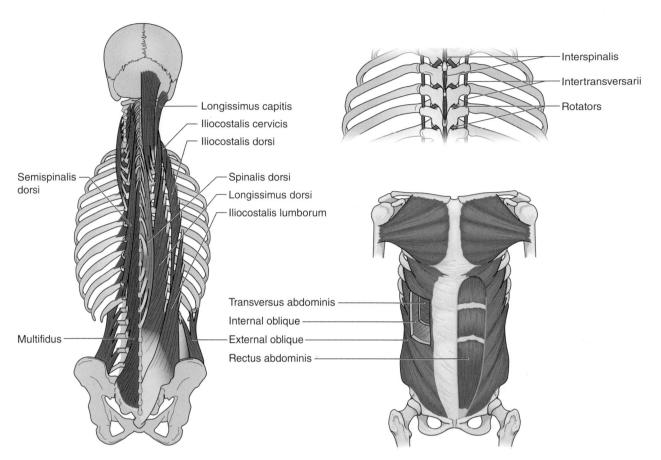

FIGURE 9.16   Muscles of the posterior lumbar–pelvic–hip complex.

these are illustrated in figure 9.16. Refer to chapter 13 for the musculature of the hip joint.

The iliocostalis lumborum pars lumborum consists of four overlying fascicles arising from L1-L4. Each fascicle arises from the tip of the transverse process and from the thoracolumbar fascia 2 to 3 cm lateral to the transverse process. This muscle group inserts on the iliac crest. Unilateral contraction of this muscle group produces side bending as well as some rotation.

The longissimus thoracis pars lumborum arises from the accessory process and the dorsal surface of the transverse process to insert into the medial aspect of the PSIS. Unilateral contraction of this muscle produces side bending, while bilateral contraction produces extension.

Proper muscle function, as with other areas of the spine and the body in general, is imperative for normal function. Muscle defi-

cits in strength, endurance, power, or motor control will result in suboptimal function and potential dysfunction.

Panjabi (1992a, 1992b) has described three systems that contribute to active spinal stabilization:

1. Passive subsystem—the inert tissue of the spine including the spinal column and ligaments
2. Active subsystem—the spinal musculature
3. Control subsystem—neural motor control

The passive subsystem, including the spinal structure, anatomy, and ligamentous tissue, provides passive restraint and stability. The active subsystem involves the spinal musculature. This muscular control is often described in terms of the local and global muscle systems. The control subsystem involves interaction between the nervous and muscular systems (refer to chapter 3).

The local muscle system, or inner muscle unit, consists of muscles predominantly involved in joint support or stabilization. These muscles are not movement specific but rather provide stability of the lumbar spine to allow movement of a peripheral joint (Bergmark 1989; Richardson et al. 1999). These muscles either originate or insert (or both) into the lumbar spine (Fritz et al. 2007) and include the transversus abdominis, lumbar multifidus, internal oblique, diaphragm, and muscles of the pelvic floor (Bergmark 1989; Richardson et al. 1999).

The global muscle system, or outer muscle unit, is predominantly responsible for movement and consists of more superficial musculature that attaches from the pelvis to the rib cage or lower extremities. Major muscles in this group include the rectus abdominis, external oblique, erector spinae, hamstrings, gluteus maximus, latissimus dorsi, adductors, and quadriceps (Bergmark 1989; Richardson et al. 1999). These muscles transfer and absorb forces from the upper and lower extremities to the pelvis (Richardson et al. 1999). The global muscle system has been divided into different force couples: the deep longitudinal, posterior oblique, anterior oblique, and lateral subsystems (Lee 1996).

The deep longitudinal global muscle system consists of the erector spinae muscles, thoracolumbar fascia, sacrotuberous ligament, and ipsilateral biceps femoris (lateral hamstring) muscle. The deep longitudinal system, when working properly, can provide SI joint stability.

The posterior oblique system (POS) consists of the contralateral gluteus maximus and latissimus dorsi with the interposed thoracolumbar fascia (figure 9.17). The POS works synergistically with the deep longitudinal system. This system can also be recruited to create lumbar spine and SI joint stability (see Clinical Correlation 9.6). The POS is also a significant contributor to load transference through the pelvic girdle during rotational activities of gait.

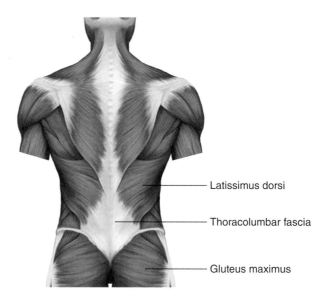

— Latissimus dorsi

— Thoracolumbar fascia

— Gluteus maximus

▶ **FIGURE 9.17** Posterior oblique system contraction producing thoracolumbar compression with corresponding lumbar and SI joint stabilization.

Conclusion

The lumbar spine differs anatomically, arthrokinematically, and functionally from the thoracic and cervical spines. The articulation of the lumbar spine and pelvis results in an interdependent biomechanical relationship between these two anatomically variable areas of the body. Motion initiating in either the lumbar spine or pelvis affects motion in the other, as well as motion in the hip joint.

Because of multiple muscular connections to the thoracolumbar fascia, contraction of these muscles can create increased tension in this fascia. The contralateral latissimus dorsi and gluteus maximus have muscle fibers running in the same oblique plane. Therefore, contraction of both these muscle groups at the same time can pull the thoracolumbar fascia taut, compressing the lumbar and SI joints, thereby providing stabilization to these structures. Examples of a lower-level exercise and higher-level exercise employing this mechanism are demonstrated in figures 9.18 and 9.19 respectively.

▶ **FIGURE 9.18** Lower-level exercise utilizing (a) contralateral latissimus dorsi and (b) gluteus maximus (left) contraction to tighten the thoracolumbar fascia.

Photos courtesy of Michael Reiman.

▶ **FIGURE 9.19** Higher-level exercise utilizing (a) contralateral latissimus dorsi and (b) gluteus maximus (right) contraction to tighten the thoracolumbar fascia.

Photos courtesy of Michael Reiman.

REVIEW QUESTIONS

1. The lumbar spine zygapophyseal joints favor motion in which plane of movement?
   a. sagittal
   b. frontal
   c. transverse
   d. lateral

2. Which of the following ligaments is much thicker and broader in the lumbar spine than in the cervical spine?
   a. supraspinous ligament
   b. interspinous ligament
   c. posterior longitudinal ligament
   d. anterior longitudinal ligament

3. Which shape of the zygapophyseal joints in the lumbar spine is most likely to resist sagittal plane translation of one vertebral body on the other?
   a. C-shaped
   b. J-shaped
   c. I-shaped
   d. L-shaped

4. Which of the following are components of the posterior oblique system?
   a. contralateral gluteus medius and latissimus dorsi, thoracolumbar fascia
   b. ipsilateral gluteus medius and latissimus dorsi, thoracolumbar fascia
   c. contralateral gluteus maximus and latissimus dorsi, thoracolumbar fascia
   d. ipsilateral gluteus maximus and latissimus dorsi, thoracolumbar fascia

5. Sacral flexion is also known as
   a. nutation
   b. counternutation
   c. side bending
   d. rotation

6. Sacral nutation is resisted by which of the following ligaments?
   a. long dorsal
   b. sacrotuberous
   c. iliolumbar
   d. anterior sacroiliac

Shoulder

The shoulder joint, composed of five bones and four different articulations, allows an incredible range of motion. This increased motion afforded by the shoulder sets it up to have certain pathological consequences that lead to significant impairments such as loss of motion and loss of muscular strength. Knowledge of the kinesiology of the shoulder is imperative for understanding normal and abnormal forces, force couples, and appropriate movement patterns of the shoulder.

The shoulder is the most proximal link of the upper extremity to the axial skeleton. It must work in cooperation with the elbow and hand for efficient human function. A complex joint, the shoulder has three true joints in the glenohumeral, acromioclavicular, and sternoclavicular joints and a pseudojoint in the scapulothoracic joint. Articulations of this complex include the humerus, clavicle, scapula, and dorsal surface of the ribs. The shoulder complex is the most mobile of any joint in the human body. The trade-off of this multiplane joint with the large degree of motion is a relative laxity that can lead to multiple injuries.

OBJECTIVES

After reading this chapter, you should be able to do the following:

> Name the five major articulations in the human shoulder.

> Differentiate between normal and pathological scapulohumeral rhythm.

> Apply alterations in anatomy that lead to pathological changes in the shoulder.

> Define capsular patterns of the major articulations of the shoulder.

> Generalize major actions of muscles or groups of muscles surrounding the shoulder.

Osteology

The skeletal system of the shoulder girdle provides attachment sites for muscles, tendons, and capsular structures. Bones also provide stability and a means for muscles to exert force to create movement. With a joint structure like no other, the shoulder exhibits the greatest overall range of motion availability in the human body. The bones of the shoulder and the various facets, protuberances, and tuberosities are discussed in this section.

Sternum

Although in its entirety the sternum (figure 10.1) is not actually considered one of the osseous pieces of the shoulder joint, its attachment at the sternoclavicular joint is a very important proximal component of the shoulder. The attachment of the clavicle to the sternum is the only true articulation of the upper extremity to the axial skeleton. The upper lateral portions of the manubrium (the cranial portion of the sternum) form a shallow concavity known as the clavicular notch, which is the attachment site of the medial (proximal) end of the clavicle. This notch is relatively small in comparison to the end of the clavicle. The medial clavicle is so much larger than the manubrium that it extends both superiorly and posteriorly from the manubrium. In between the two clavicular notches is the most superior portion of the sternum, the sternal (or jugular) notch. The sternoclavicular joint is discussed in greater detail later in the chapter. A last important landmark on the anterior sternum is where the manubrium meets the sternum, known as the sternal angle, the manubriosternal joint, or the angle of Louis.

► **KEY POINT**

The attachment of the clavicle to the sternum is the only true articulation of the shoulder to the axial skeleton.

Scapula

On the posterior portion of the shoulder is the large, flat shoulder blade, or scapula (figure 10.2). The anterior surface is concave forward. The scapula's primary function is to provide attachments for the 15 major muscles that act upon the shoulder (Lucas 1973; Williams et al. 1995), including the rotator cuff and scapular

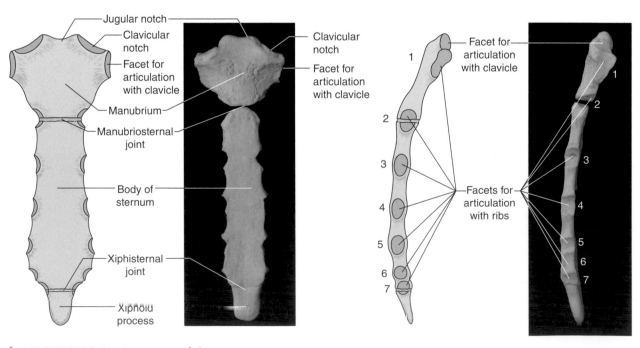

► **FIGURE 10.1** Anatomy of the sternum.

stabilizer muscles. There are notable areas in which the bone of the scapula is thicker than others. Areas of thickness include the superior and inferior angles and the lateral border, where more of the shoulder's power muscles are attached. Various thickened processes include the coracoid process, the acromion process, the glenoid, and the scapular spine (Rockwood & Wirth 1998).

▶ **KEY POINT**
The primary function of the scapula is for muscular attachment for rotator cuff and scapular stabilizer muscles.

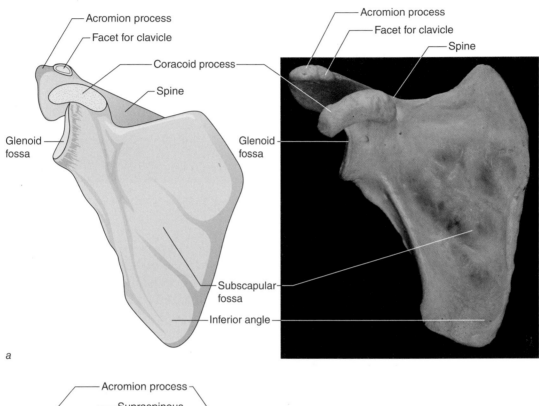

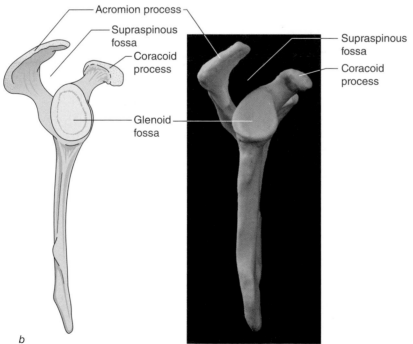

▶ **FIGURE 10.2** *(a)* Anterior and *(b)* lateral aspects of the right scapula. *(continued)*

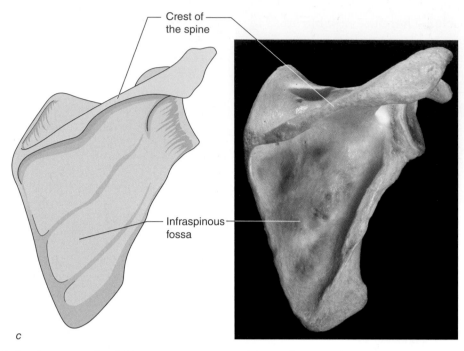

Crest of
the spine

Infraspinous
fossa

c

▶ **FIGURE 10.2** *(continued)* (c) Posterior aspects of the right scapula.

The costal surface (anterior) is separated from the thorax by the serratus anterior muscle. Additionally, there is a concavity that is the origin of the subscapularis muscle. The dorsal surface (posterior) is divided into two smaller cavities where the supraspinatus and infraspinatus muscles originate. These two cavities may appear hollowed out in those with shoulder pathology such as chronic tendon tears or peripheral nerve injuries.

With such a large surface area, the scapula is perfectly formed for muscular attachments on either the three angles or the three processes. The medial border is referred to as the vertebral border because of its close relationship and proximity to the spine. The medial border can be palpated easily up to the **root of the scapula**, at which point the scapula angles anteriorly. This anterior angulation allows for the scapula to more closely mimic the angulation of the posterior upper thorax. Superior to the root of the scapula on the medial surface is the superior medial angle. This superior border is the common attachment of the trapezius and the levator scapulae and associated bursae. The lateral border has attachments for the teres minor and major.

When following the superior border laterally or the lateral border superiorly, you

ultimately come to the glenoid fossa (see figure 10.2, *a* and *b*). The **glenoid fossa**, the proximal component of the glenohumeral joint, is located on the exposed lateral portion of the scapula. The average anteroposterior depth of the glenoid is only 2.5 mm, whereas the average superoinferior depth is 9.0 mm (Howell & Galinat 1989). The glenoid fossa is pear shaped and looks like an inverted comma, with its cephalad (superior) portion narrower than the bulbous caudal (inferior) portion. This fossa is slightly retroverted about 7.4° in approximately 75% of the population (Saha 1971; Saha 1973). The fossa is also tilted upwardly about 5° relative to the scapula's medial border (Basmajian & Bazant 1959) and tilted from superomedial to inferolateral an average of 15°. Many believe the superior tilt of the glenoid provides an avenue for joint stability by counteracting an inferior and anterior displacement of the humeral head (Doos et al. 1966; Saha 1971; Saha 1973; Warner et al. 1992). Radiographic studies, however, have demonstrated that in the unloaded, abducted shoulder, the glenoid faces slightly downward (Freedman & Munro 1966; Poppen & Walker 1978; Warner et al. 1991).

The glenoid has an average height of 35 mm and an average width of 25 mm in the adult

(Saha 1983). At the very superior margin of the glenoid is the supraglenoid tubercle, from which the long head of the biceps arises. The infraglenoid tubercle, an analogous tubercle, exists at the inferior margin of the glenoid, where the long head of the triceps attaches.

One last important projection from the scapula should be noted: the coracoid process. This projection occurs on the anterior surface of the scapula (figure 10.2, a and b). From its base, the coracoid projects upward before turning downward anteriorly and laterally. The coracoid process is a roughened fingerlike projection that allows for numerous attachments of both muscles and ligaments including the pectoralis minor muscle, the short head of the biceps brachii, the coracobrachialis muscle, and the coracoacromial ligament. The undersurface is smooth at the base, forming a floor for the subscapularis muscle. A recent anatomical study defined the spatial relationship of the tip of the coracoid process based on a clock face of the glenoid. The tip in all shoulders was between 1:24 and 2:18, while the distance of the tip to the nearest portion of the glenoid labrum was 21.5 mm (Tham et al. 2009).

The anterior, or costal, surface of the scapula is large and slightly concave. The two muscles that attach anteriorly are the subscapularis, a portion of a collection of muscles known as the rotator cuff, and the serratus anterior, which attaches to the anterior vertebral, or medial, border. The costal, or anterior, surface is concave and smooth over the lateral one-third of the surface, whereas the medial portion has oblique ridges for attachment of the subscapularis muscle. Approximately one-quarter of the distance from the posterior top surface of the scapula is the scapular spine. This is a very prominent thickening that is an attachment site for multiple muscles. The scapular spine also separates the posterior surface into two depressions known as fossas, which serve as attachment sites for two more of the rotator cuff muscles. The smaller superior fossa is known as the supraspinatus fossa, which is the attachment of the supraspinatus muscle, while the larger inferior fossa is known as the infraspinatus fossa, the attachment of the infraspinatus muscle. The supraspinatus fossa is wider at the medial border than it is toward the lateral side. The origin of the muscle is along the medial two-thirds of the bone, whereas the lateral third has no real attachments. Careful inspection of either of these fossas can give obvious signs of pathology (see Clinical Correlation 10.1).

The lateral, or axillary, border of the scapula exhibits a flattened area along its course for attachments of the teres major, the teres minor, and the long head of the triceps. The medial, or spinal, border of the scapula provides attachment sites for the levator scapulae and the two rhomboid muscles on the dorsal surface and the serratus anterior on the costal surface. The scapular spine runs the entire length of the posterior scapula from its medial, or vertebral, border to the far lateral end. The far lateral end of the spine of the scapula projects past the lateral border of the scapula to form what is known as the acromion process. This bone is roughly a triangle in shape, yet oblong. The acromion is thicker at its base and thinner toward its terminal portion. The superior surface is convex and roughened to provide attachments for the deltoid muscle. The inferior underside is smooth and in contact with the subacromial bursa. On the medial end is a concave facet that accepts the lateral portion of the clavicle, forming the acromioclavicular joint. The distal clavicle is

CLINICAL CORRELATION 10.1

Hollowing in the area of the supraspinatus and infraspinatus fossas is often seen clinically as a chronic long-standing rotator cuff tear causing muscle wasting. In younger patients, this wasting may be the result of a supraglenoid cyst compressing the suprascapular nerves along the spine of the scapula, resulting in a loss of innervation to the supraspinatus and infraspinatus muscles. If hollowing or wasting is seen in the supraspinatus or infraspinatus fossa, the clinician should suspect either tendon rupture or nerve compression.

oriented posteriorly and laterally, while the acromion faces medially and anteriorly, forming the acromioclavicular joint. The posterior surface has roughened areas for attachment of the trapezius muscle.

This entire acromion process projects laterally, superiorly, and anteriorly and forms somewhat of a "roof" over the glenoid fossa and the head of the humerus (see figure 10.1b). Although having a roof over the glenoid fossa and humeral head sounds useful, it is a double-edged sword. This bony projection protects against trauma to the humeral head and superior dislocation of the humerus, but it is a common site of spur formation that can lead to a very painful condition called impingement syndrome, which results in a superior migration of the humeral head, pinching the rotator cuff and bursa between the humeral head and the acromion. The anterior medial portion of the acromion is covered with fibrocartilage and attaches to the clavicle at the acromioclavicular joint, which is discussed later. Additionally, the lateral acromion has its own ossification center that sometimes does not completely fuse (figure 10.3). This os acromiale can be a source of superior shoulder pain during overhead movements.

▶ **KEY POINT**

Impingement syndrome is a painful condition in which the rotator cuff and the humeral head are impinged into the undersurface of the acromion or coracoacromial arch.

Three variations of the orientation of the undersurface of the acromion exist. A type I acromion has a flat undersurface, a type II has a curved undersurface, while a type III acromion contains a hooked undersurface (Bigliani et al. 1986). Certain types of acromion

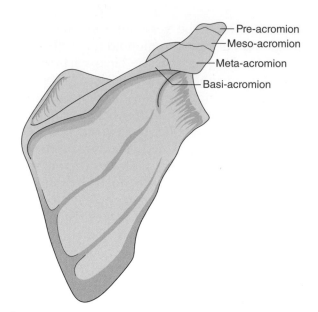

▶ **FIGURE 10.3** Regions of the acromion in which bony union may fail to occur, resulting in os acromiale.

may be more likely to result in impingement syndromes (see Clinical Correlation 10.2).

Humerus

The humerus is the second longest bone in the human body, second only to the femur. Key parts of the humerus include the head, neck, and body (figure 10.4). The humeral head is almost half of a perfect sphere (Iannotti et al. 1992; Soslowsky et al. 1992; van der Helm et al. 1992; Williams et al. 1995) and forms the convex portion of the glenohumeral joint. The head has a radius of curvature of about 2.25 cm (Perry 1988). A bisection of the humeral head will show it is positioned to face medially, superiorly, and posteriorly with respect to the shaft and the medial and lateral humeral condyles. There is a better orientation of the humeral head in the plane of the scapula for articulation with the glenoid fossa.

CLINICAL CORRELATION 10.2

Although any type of acromion (I to III) can contribute to impingement syndromes, the type III hooked acromion is the most resistant to conservative care because of excessive encroachment into the subacromial space. That is not to say that conservative care will not help any patients with a type III acromion; studies have shown that symptoms will resolve in 64% of cases of impingement syndrome in those with a hooked acromion (Morrison et al. 1997).

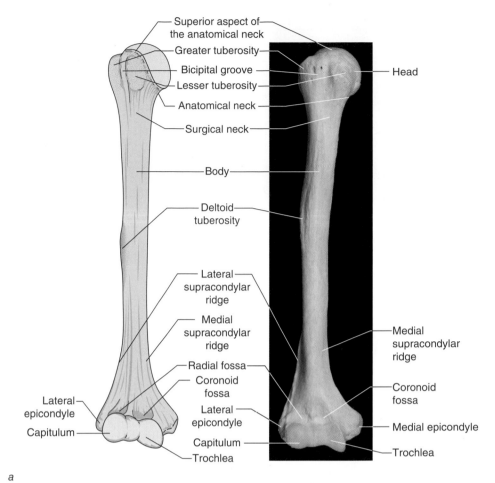

Superior aspect of the anatomical neck
Greater tuberosity
Bicipital groove
Lesser tuberosity
Anatomical neck
Surgical neck
Head

Body

Deltoid tuberosity

Lateral supracondylar ridge
Medial supracondylar ridge
Radial fossa
Coronoid fossa
Lateral epicondyle
Capitulum
Trochlea

Lateral epicondyle
Capitulum

Medial supracondylar ridge
Coronoid fossa
Medial epicondyle
Trochlea

a

Greater tuberosity
Greater tuberosity
Head
Anatomical neck
Surgical neck

Body

Olecranon fossa
Medial epicondyle
Medial epicondyle

b

▶ **FIGURE 10.4** (a) Anterior and (b) posterior views of the right humerus.

The frontal plane of the head of the humerus forms an angle with the shaft of about 135°, known as the **angle of inclination**[1] (figure 10.5). This head-to-shaft angle is less for small humeral heads and greater for larger humeral heads (Iannotti et al. 1992). Most shoulders exhibit some degree of **humeral retroversion**, or posterior rotation of the humeral head. In the transverse plane the humeral head is retroverted 30° posteriorly in the horizontal plane (figure 10.6), although Boileau and Walch, using three-dimensional analysis and computerized modeling, found cadaveric retroversion to be in a range from –6.7 to 47.5° (Boileau & Walch 1997), and Saha noted an average of 7° of retroversion (Saha 1961).

Two well-known areas on the head of the humerus are the surgical and anatomical necks (see figure 10.4). The **surgical neck of the humerus** is the area just slightly below the tuberosities and is a frequent site of fractures of the humerus. The anatomical neck is the space where the shaft meets the articular surface of the humeral head. It is around this location where the shoulder capsule inserts. Just distal to the anatomical neck is where the tendons attach to the humeral head. The tendon attachments of the rotator cuff are the lesser and greater tubercles surrounding the anterior and lateral portions of the proximal end of the humerus.

The lesser tubercle is the distal attachment of the subscapularis of the rotator cuff and also forms the medial border of the bicipital groove. The greater tubercle is the larger, more lateral tubercle that is the medial border of the bicipital groove and the attachment site of several rotator cuff muscle tendons, including the supraspinatus, infraspinatus, and teres minor. The greater tubercle has three facets: upper, middle, and lower. From each of these tubercles are crests that run distally down the humeral shaft. These crests increase the surface area for tendon attachment. Between these two crests is a sulcus called the **intertubercular groove**, offset about 30° in relation to a line drawn from the shaft of the humerus through the center of the head. The long head of the biceps tendon runs through this groove to its superior origin at the supraglenoid tubercle of the scapula. Covering the superior portion of this groove is the intertubercular ligament, also known as the transverse humeral ligament.

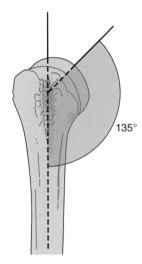

▶ **FIGURE 10.5** Humeral angle of inclination. The angle formed between the shaft and head of the humerus in the frontal plane.

Clavicle

The clavicle (collarbone) is located on the anterior part of the shoulder. When looking at the clavicle from the superior surface, it is shaped like an S or a crank because of its double curve (Pratt 1994) (figure 10.7). The clavicle is shaped concave forward laterally and convex

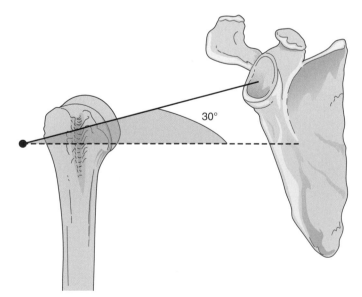

▶ **FIGURE 10.6** Humeral retroversion. Posterior rotation of the head in relation to the distal humerus.

forward medially. When this bone is viewed from the anterior position, it appears to be relatively straight. A fracture of the clavicle can generally be palpated, as the superficial skin in this region is relatively thin (see Clinical Correlation 10.3). The medial two-thirds of the clavicle are triangular in shape in its cross section. The bone then takes a more flattened cross section lateral to the coracoid. When the entire arm is in resting position at the side, the clavicle is in a position slightly posterior to the frontal plane and above the horizontal by about 20°.

Because the clavicle is located between two bones, it has two ends. The medial end is flared and rounded; it articulates somewhat poorly with the sternum and is sometimes called the sternal end of the clavicle. The inferior medial surface has a costal facet, or tuberosity, that rests against the first rib. Additionally, on the inferior medial side there are ligamentous attachments for the costoclavicular ligaments. The flattened lateral side of the clavicle, or the acromial end, is oval distally and attaches to the acromion process of the scapula to form the acromioclavicular joint. The acromial facet is a flat-shaped portion of the lateral end of the clavicle. On the inferior lateral portion of the clavicle is the conoid tubercle and trapezoid line. The conoid tubercle is the attachment of the conoid ligament, while the trapezoid ligament attaches to the trapezoid line. The superior surface is roughened for attachments of the deltoid and trapezius muscles.

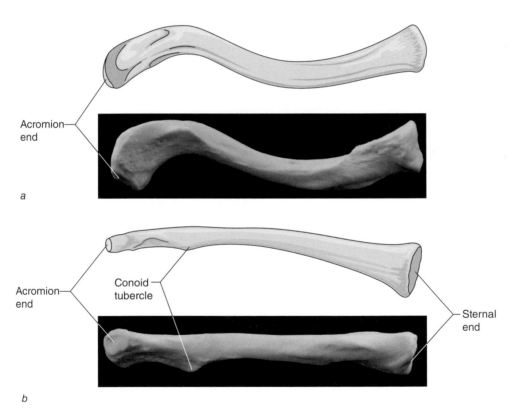

Acromion end

a

Acromion end

Conoid tubercle

Sternal end

b

▶ **FIGURE 10.7** *(a)* Superior and *(b)* anterior views of the clavicle.

CLINICAL CORRELATION 10.3

Clavicle fractures are the most common of all pediatric fractures. Because of the limited soft-tissue covering in the area of the clavicle, these fractures are easily seen if there is a deformity or palpated if a deformity is not present. These are very painful fractures, and any motion of the shoulder or upper extremity will reproduce discomfort.

Joint Articulations

The joints that make up the shoulder include the glenohumeral joint, the acromioclavicular joint, the sternoclavicular joint, and the scapulothoracic joint.

Glenohumeral Joint

The glenohumeral joint is a synovial joint formed by the articulation between the large humeral head and the shallow glenoid fossa (figure 10.8). This joint has a large capsule and several associated ligaments. Its ball-and-socket configuration provides the shoulder with more mobility than any other joint in the human body, allowing rotation and three translational degrees of freedom (Curl & Warren 1965). In addition to the bones just listed, the following structures are integral components of the shoulder.

The glenoid **labrum** is the fibrocartilaginous structure that encircles the glenoid fossa of the scapula. The labrum is thought to be composed of either fibrocartilaginous tissue (Bost & Inman 1942; Codman 1934), dense cartilaginous fibrous tissue with chondrocytes (Prodromos et al. 1990; Williams et al. 1995; Nishida et al. 1996), or dense fibrous collagen tissue with a small transitional zone between the hyaline cartilage and fibrous labral tissue (Cooper et al. 1992; Moseley & Overgaard 1962; Huber & Putz 1997). This structure surrounds the glenoid fossa and increases shoulder stability by deepening the glenoid. The labrum provides about 50% of the overall depth of the fossa (Howell & Galinat 1989). The depth of the glenoid fossa has been shown to double (2.5 to 5.0 mm) when the labrum is present (Perry 1978). This deepening of the fossa may also allow for increased surface area contact between the glenoid and the humeral head (Cooper et al. 1992). Hertz and colleagues (1986) and Levine and Flatow (2000) have reported that humeral head contact is increased by one-third via the labral complex.

▶ **KEY POINT**
The labrum nearly doubles the depth of the glenoid fossa.

Finally, the labrum acts as a valve with the humeral head, creating a seal against atmospheric pressure, thus maintaining negative intra-articular pressure and helping to limit excessive translation (Bahk et al. 2007). Breaking this seal via capsular or labral injury could increase joint instability (Habermeyer et al. 1992; Ferguson et al. 2001; Howell & Galinat 1989) (see Clinical Correlation 10.4).

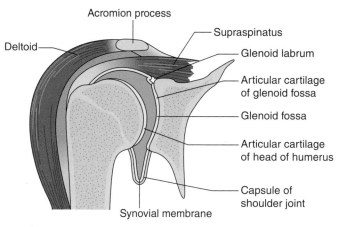

Acromion process
Supraspinatus
Deltoid
Glenoid labrum
Articular cartilage of glenoid fossa
Glenoid fossa
Articular cartilage of head of humerus
Capsule of shoulder joint
Synovial membrane

▶ **FIGURE 10.8**   Glenohumeral joint.

CLINICAL CORRELATION 10.4

The labrum can be torn away from the glenoid in cases of trauma such as a shoulder dislocation or a fall on an outstretched hand (FOOSH). The tearing of the anterior inferior portion of the glenoid labrum is known as a Bankart tear. This is thought to be the primary reason for recurrent shoulder dislocations. If someone has had an anterior shoulder dislocation and has not had surgical repair, odds are that a Bankart lesion exists. A posterior shoulder dislocation will result in a reverse Bankart lesion. Without some form of stabilization, the patient with a Bankart lesion will continue to have chronic dislocations or subluxations of the shoulder. When dislocations become chronic, they often occur during simple activities of daily living such as putting on clothes or rolling over in bed.

Kumar and Balasubramaniam (1985) have demonstrated that piercing the capsule of an intact shoulder will equalize atmospheric pressure, resulting in an inferior subluxation of the humeral head on the glenoid fossa. Habermeyer and colleagues (1992) reported that the glenohumeral joint should act like a valve block, sealing the joint from atmospheric pressure. They found that applying traction to the arm increases negative intra-articular pressure, while the same phenomenon does not occur in those with labral tears.

▶ **KEY POINT**

A Bankart lesion occurs when the glenoid labrum is torn away from the glenoid fossa during a glenohumeral dislocation.

Acromioclavicular Joint

The acromioclavicular joint (ACJ) is located at the lateral end of the clavicle and the medial end of the anterior portion of the acromion, which is part of the scapula (figure 10.9). The ACJ is a plane diarthrodial synovial joint with three degrees of freedom. Because it is a synovial joint, it is susceptible to degenerative conditions and inflammatory, septic, and

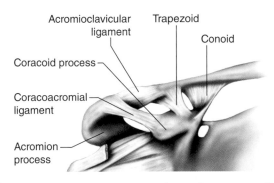

▶ **FIGURE 10.9**   Acromioclavicular joint and surrounding ligaments.

crystalline arthropathies (Blotter & Bruckner 1995; Cooper et al. 1993; Lahtinen et al. 1999; Oppenheimer 1943). Each side of the joint exhibits a facet that varies greatly in configuration (flat to reciprocally convex/concave or concave/convex) and size (Williams et al. 1995). The distal portion of the clavicle faces posteriorly and laterally, while the articular surface of the acromion faces medially and anteriorly.

Within the joint capsule, each end of the ACJ is covered with articular fibrocartilage. Additionally, a wedge-shaped articular disc is interposed between the two surfaces. This articular disc is not yet fully understood, as it commonly degenerates early in life. In children under the age of 2, the disc is merely a fibrocartilage bridge between the acromion and clavicle, with no true joint cavity. The first sign of an actual joint cavity occurs at 3 to 5 years (DePalma et al. 1949). Although the disc is fibrocartilage through the first several years of life, it undergoes rapid degeneration with upper extremity use, typically shows a high rate of degenerative changes by the second decade of life (DePalma 1959), and is not even functional by the fourth decade (McCluskey & Todd 1995) (see Clinical Correlation 10.5).

▶ **KEY POINT**

Osteoarthritis and osteolysis of the acromioclavicular joint are very common in weightlifters or those with labor-intensive vocations.

The actual ACJ itself is about 9 mm (vertical), 19 mm (anteroposterior), and only 1 to 3 mm thick (Bosworth 1949; Bonsell et al. 2000; Oppenheimer 1943). As a result, joint forces across the ACJ are distributed over a small area of contact, thus increasing pressure greatly

CLINICAL CORRELATION 10.5

Acromioclavicular joint arthritis and osteolysis are very common problems in weightlifters or those in labor-intensive vocations. These patients will commonly present with discomfort during overhead loaded activities and at times even simple reaching activities to full elevation. Full elevation of the shoulder is the closed packed position of the acromioclavicular joint and requires maximal approximation of the two joint surfaces.

over any one area of the joint surface. Exact ACJ contact forces are not known.

In more than 50% of people, the clavicle slightly overrides the acromion, although other variations do exist (DePalma 1957). Edelson believed that those with a more vertical facet orientation of the ACJ may be at risk for developing arthritic changes (Edelson 1996). Pitchford and Cahill (1997) agree, claiming that those with a more vertical joint orientation (under- or overriding) are more susceptible to osteolysis of the ACJ.

Sternoclavicular Joint

The sternoclavicular joint is the articulation of the medial end of the clavicle with the manubrium. Although described by Romanes (1981) as a ball-and-socket joint, most consider it a saddle-shaped diarthrodial synovial joint (Steindler 1955; Williams et al. 1995). Because of its saddle-shaped arrangement, the sternoclavicular joint is inherently unstable. To further this instability, less than half of the medial end of the clavicle articulates with the lateral end, or upper angle, of the manubrium. Because of this limited congruence, this joint relies heavily on ligamentous support.

The sternoclavicular joint contains a fibrocartilaginous articular disc, or meniscus (figure 10.10). This meniscus divides the joint into two separate compartments. The strong articular disc is attached superiorly to the upper medial end of the clavicle and passes downward between the articular surfaces to the sternum and first costal cartilage (Warrwick & Williams 1973). This anatomical arrangement allows the disc to act as a hinge between the clavicle and sternum. It also serves as a stabilizing structure, resisting forces applied to the shoulder medially through the clavicle and sternum; reduces incongruities between the articulating surfaces; and guards against medial translations of the clavicle over the sternum. The disc itself is approximately 3 to 5 mm thick in the upper posterior portion (Brossmann et al. 1996). During motion of the shoulder, various stresses are placed on the disc. These concepts are discussed further in the biomechanical section related to the sternoclavicular joint.

The sternoclavicular joint is the only skeletal articulation between the upper limb and the axial skeleton (Hollinshead 1982). The larger medial portion of the clavicle stands slightly superior and more anterior and posterior than the sternum. It is exactly this anatomical arrangement that develops the suprasternal notch. The medial end of the clavicle is concave in the anteroposterior direction, while it is convex superiorly and inferiorly. Both surfaces, despite having a slight convexity or concavity, are really fairly flat.

Stability of the sternoclavicular joint is provided by soft-tissue structures such as the articular disc; the anterior, posterior, costoclavicular, and interclavicular ligaments; and the sternoclavicular joint capsule that surrounds the joint and attaches beyond the articular surfaces. The capsule itself is relatively weak but is reinforced by several well-defined ligaments or accessory ligaments that are thickenings of the capsule itself (figure 10.11). These ligaments appear to be robust, as the sternoclavicular ligament is the least constrained joint in the human body (Rockwood & Wirth 1998). Despite this, sternoclavicular joint dislocations are a rare occurrence, representing only 1% of dislocation in the human body and only 3% in the entire upper extremity (Cave 1958).

Anterior and posterior clavicle displacement is resisted by the anterior and posterior sternoclavicular ligaments. Additionally, they provide some support for the limits of motion through protraction and retraction. These ligaments run superiorly from their attachment to the sternum to their superior attachment on the clavicle. Probably the most important of these ligaments is the posterior

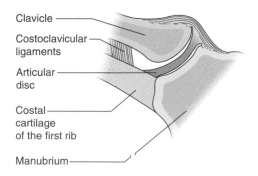

FIGURE 10.10 Cross section of sternoclavicular joint.

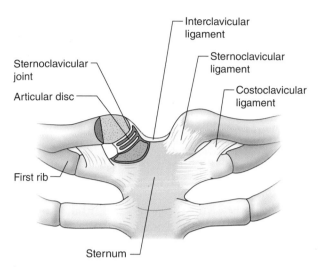

FIGURE 10.11 Anatomy of the sternoclavicular joint.

sternoclavicular ligament, which limits posterior translations and is also important for resisting inferior depression of the lateral end of the clavicle (Bearn 1967). More recently, Spencer et al. (2002) found that sectioning the posterior capsule caused significant increases in anterior and posterior translations, more so than the sectioning of any other structure. Cutting the anterior capsule resulted in increased translation in only the anterior direction. Cutting the costoclavicular ligament complex had little effect on sternoclavicular joint translations. From this study, it appears that the most important restraint is the posterior capsule, as a posterior dislocation can have grave consequences because vital nerve and vascular structures lie closely behind this joint (Gangahar & Flogaites 1978; Gardner & Didstrup 1983; Jougon et al. 1996; Noda et al. 1997; Ono et al. 1998; Rayan 1994).

▶ **KEY POINT**
A posterior dislocation of the sternoclavicular joint is an immediate emergency, as vital nerve and vascular structures lie closely behind this joint.

Scapulothoracic Joint

The scapulothoracic joint is the articulation between the broad, flat triangular scapula and the posterolateral aspect of the thorax and ribs. It is usually considered a pseudojoint because it does not have all the characteristics of a true synovial joint. The top of the scapula typically rests near the second thoracic vertebra, the lower border around the seventh thoracic vertebra. The resting position is with the medial border of the scapula approximately 2 in. (5 cm) from the spine. This joint is angled so that the glenoid fossa faces anteriorly about 30 to 45° anterior to the coronal (frontal) plane, a position known as the **plane of the scapula**, or scaption (Codman 1934; Johnston 1937; Saha 1961) (figure 10.12). Additionally, the scapula is tipped anteriorly 10 to 20° from vertical and upwardly rotated 10 to 20° from vertical (Ludewig & Cook 2000). However, it is very common to have a substantial amount of variability in resting scapular position from person to person. Variations of resting scapular position can occur for a multitude of reasons, including pain, muscle fatigue, or compensation, or as a consequence of postural habits, hand dominance, occupation muscle tone, and even age (Kelley & Clark 1995) (see Clinical Correlation 10.6).

▶ **KEY POINT**
The plane of the scapula is the position in which the scapula faces anteriorly 30 to 45° from the frontal plane.

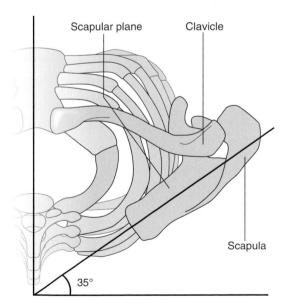

FIGURE 10.12 Plane of the scapula, or scaption—the orientation of the scapula deviated approximately 35° anterior to the frontal plane.

CLINICAL CORRELATION 10.6

According to Burkhart and colleagues (2003), what was originally thought to be a dysfunction (asymmetrical malposition of the scapula) may actually be a normal compensation in those who perform repetitive overhead actions, such as throwing athletes. This condition is known as the SICK scapula (**S**capular malposition, **I**nferior medial border prominence, **C**oracoid pain and malposition, and dys**K**inesis of scapular movement). This finding implies clinically that not all abnormal positions of the scapula should be considered pathological. Any abnormality must be correlated with clinical signs and symptoms.

Joint Anatomy

A clear understanding of shoulder joint anatomy is needed because multiple joints make up the shoulder complex. Important joint anatomy includes the previously mentioned bony anatomy as well as the capsule and ligaments, bursae, nerves, and blood supply.

Shoulder Capsuloligamentous Complex

The glenohumeral joint has a large capsule and associated ligaments that cover approximately twice the surface area of the humeral head (figure 10.13). This fibrous structure isolates the intra-articular joint from the external surrounding tissues. The large capsule is synovial lined and extends from the glenoid neck, or labrum, to the proximal shaft, or anatomical neck. Inferiorly this capsule is quite loose, allowing for great ranges of movement. On the external capsular surface, the capsule blends with ligamentous structures to help reinforce the passive restraint. The extracapsular ligaments surrounding the capsule anteriorly include the superior, middle, and inferior glenohumeral ligament complex. The coracohumeral ligament reinforces the glenohumeral joint superiorly. These ligaments may all function differently depending on collagenous integrity, attachment sites, and the position of the arm.

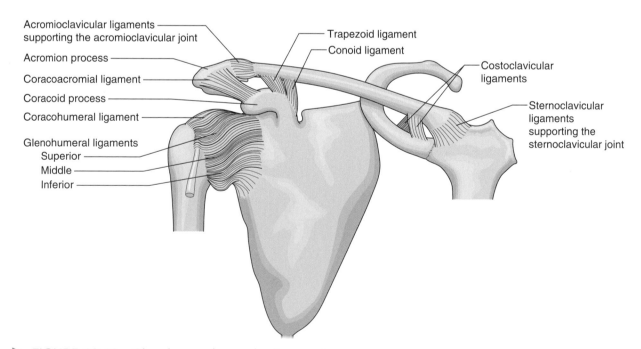

▶ **FIGURE 10.13** Glenohumeral capsular ligaments.

Superior Glenohumeral Ligament

The superior glenohumeral ligament (SGHL) runs from the superior portion of the labrum, near the attachment of the long head of the biceps, and from the base of the coracoid process to the superior aspect of the humeral neck. The SGHL can arise in several variations including a common attachment with the long head of the biceps, from the labrum between the long head of the biceps and the middle glenohumeral ligament, and from a common attachment to the labrum with the middle glenohumeral ligament (DePalma et al. 1949). This ligament has a varied thickness, from robust tissue to a wispy thickening. The SGHL is not very useful in the abducted shoulder (Schwartz et al. 1987) but is taut when the arm is fully adducted near the side or when an inferior translation of the humerus occurs with the arm in a dependent position (Warner et al. 1992).

▶ **KEY POINT**

The superior glenohumeral ligament is under tension when the arm is in a dependent position at the side hanging against gravity or when inferior force is applied to the humerus.

Middle Glenohumeral Ligament

The middle glenohumeral ligament (MGHL) originates from just beneath the supraglenoid tubercle and the anterosuperior portion of the labrum, and it inserts just medial to the lesser tuberosity, blending with fibers of the subscapularis tendon (Turkel et al. 1981). It displays a large range of variety in its structure and presence. In some cases, the MGHL is very well developed, while in others it may be poorly defined or even absent.

The MGHL can resist an anteriorly directed force with the arm in an adducted position up to approximately 45°, and it limits the extremes of external rotation with the arm at the side (Turkel et al. 1981; O'Brien et al. 1990). It can additionally function as a secondary stabilizer for anterior translation if the anterior portion of the inferior glenohumeral ligament is injured (O'Brien et al. 1995).

Inferior Glenohumeral Ligament Complex

The **inferior glenohumeral ligament complex** (IGHLC) is, as its name implies, an extensive complex of ligaments compared with the previously discussed ligaments, which are singular in origin. The IGHLC is attached proximally along the anterior inferior rim of the glenoid fossa and to the glenoid labrum. Its distal attachments include the anterior and posterior inferior margins of the **anatomical neck of the humerus**. This complex has both an anterior and a posterior band, with a "hammock" of tissue interposed between that acts as a pouch. Insertions of the anterior band are typically between the 2 and 4 o'clock positions, while the posterior band is between the 7 and 9 o'clock positions (O'Brien et al. 1990).

The ability of the IGHLC to restrict motion depends on the position of the humerus in regard to elevation and rotation. The IGHLC becomes the primary resistance to anterior and posterior instability when the humerus is abducted to 90° (Ticker et al. 1996; Warner et al. 1992). At 90° of shoulder abduction, the anterior band becomes the primary stabilizer when the humerus is abducted (even in neutral) (Terry et al. 1991; Burkhart & Debski 2002). When external rotation is added to the abducted shoulder, the IGHLC tightens as the anterior band elevates around the front of the humeral head, limiting anterior translation (O'Brien et al. 1990; Terry et al. 1991; Turkel et al. 1981; O'Brien et al. 1995). The opposite occurs with glenohumeral internal rotation and abduction (when abducted 90°) as the posterior band elevates around the posterior portion of the humeral head, limiting posterior translation. Any condition that causes immobility or pain could be due to tightening of the shoulder capsular ligaments. This is especially true of adhesive capsulitis, or frozen shoulder (see Clinical Correlation 10.7).

▶ **KEY POINT**

The inferior glenohumeral ligament complex restricts motion and has tension that is dependent on the amount of elevation and rotation of the humerus.

A common condition that affects the glenohumeral joint capsule is adhesive capsulitis, or frozen shoulder. In adhesive capsulitis, the capsule becomes inflamed and thickened, with adhesions forming between the synovial folds. This condition generally progresses very slowly, leading to pain and discomfort as the shoulder's range of motion gradually diminishes, resulting in a loss of both passive and active range of motion. Frozen shoulder should be considered part of the differential diagnosis in any patient with slow to no improvement of persistent shoulder pain that appears to be resistant to common treatment methods.

Bursa

The glenohumeral joint has an extensive **bursa**, which is a fluid-filled sac surrounding a synovial joint to help reduce friction between tendons, muscles, ligaments, and bones. The subacromial bursa—one of the largest in the body—separates the supraspinatus tendon from the acromion above it (figure 10.14). When healthy, these layers of bursal tissue are very thin, with minimal fluid inside. When the bursa becomes irritated, excessive synovial fluid is produced, causing a swelling of the bursa that results in pain with active movement of the shoulder.

Nerve Supply

The nerve supply of the shoulder includes the brachial plexus, cranial nerve XI, and the supraclavicular nerves. A large number of terminal nerves supply the shoulder, and some of the more important are listed here.

The subscapular nerves, upper (C5) and lower (C5 and C6), supply two-thirds to four-fifths of the upper portion of the subscapularis and the lower portion of the subscapularis, respectively. The axillary nerve (C5 and C6) divides; its motor portion branches posteriorly to supply the teres minor and the posterior one-third of the deltoid, and an anterior branch travels to the anterior two-thirds of the deltoid. A lateral brachial cutaneous nerve supplies the area of superficial sensation to the skin covering the deltoid. The suprascapular nerve (C5 and C6) innervates the infraspinatus muscle and provides two additional articular branches: one in the supraspinatus fossa to the acromioclavicular and superior glenohumeral joints, and one in the infraspinatus fossa to the posterior superior glenohumeral joint. The spinal accessory nerve, or cranial nerve XI, supplies motor innervation to the upper trapezius. The supraclavicular nerves originate from spinal nerves C3 and C4 and supply sensation to the shoulder in the area

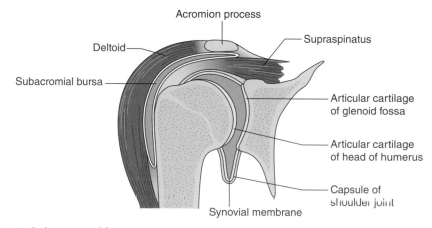

▶ FIGURE 10.14 Subacromial bursa.

above the clavicle and anteriorly to the skin overlying the acromion and deltoid.

Blood Supply

Arterial and venous supply to the shoulder and arm is provided by the axillary artery and vein. The axillary artery is a continuation of the subclavian artery. Past the shoulder, the axillary artery becomes the brachial artery. The axillary artery is further described by its three portions. The first portion gives off the superior thoracic artery. The second portion gives off the thoracoacromial artery, which is further subdivided into the deltoid and pectoral arteries. The last, or third, portion is the subscapular artery, which divides into the circumflex scapular artery, the posterior humeral circumflex artery, and the thoracodorsal arteries.

Joint Function

The shoulder joint has an extremely large range of available motion yet is still required to maintain stability. This creates quite a paradox because usually a more stable joint has less motion, and thus achieving both stability and mobility is not totally possible for any joint in the human body. In this section, we describe the axes of motion, arthrokinematics, range of motion, closed and loose packed positions, end feel, and capsular patterns of the joints of the shoulder complex.

Glenohumeral Joint

The glenohumeral joint provides the greatest range of motion of any joint in the body. This is due to a complex set of joint interactions of bony structures rotating in a specific plane and position. When these motions can occur unimpeded by pathology, many amazing movements can occur, such as those used during throwing a baseball or swimming.

Axes of Motion

The glenohumeral joint is a ball-and-socket synovial joint with three available degrees of freedom. Movements in the sagittal plane include shoulder flexion and extension and occur around a mediolateral axis of rotation. Shoulder abduction and adduction occur in the frontal plane around an anteroposterior axis. Finally, internal and external shoulder rotation occurs in the transverse plane around a superoinferior axis.

Arthrokinematics

Arthrokinematics describe the actual joint movements occurring during shoulder motion. The arthrokinematics at the shoulder are extremely important because they identify normal movement patterns and help the clinician discern when abnormal movement patterns are occurring.

Shoulder Flexion Shoulder flexion occurs through the coronal axis and the sagittal plane as a spinning of the joint along the face of the glenoid (figure 10.15). Rolling or gliding motions, which are required for abduction and rotational motions, are not necessary for motions of flexion and extension as long as the motion occurs in the sagittal plane. Although some report axial rotations through the long axis of the humerus during motions of elevation, controversy exists regarding rotation of the humerus through its long axis during isolated shoulder flexion. Traditional views are that the humerus undergoes external axial rotation during shoulder abduction (Blakely & Palmer 1984; McClure et al. 2001; Steindler 1955; Stokdijk et al. 2003), and some feel that medial rotation occurs with flexion (Blakely & Palmer 1984; Saha 1983; Steindler 1955). These rotations are very subtle and may not even be detectable with the naked eye. It is thought that tension in the coracohumeral ligament may create a slight internal rotation torque as the arm is elevated past 90° of flexion (Palmer & Blakely 1986). Saha (1983) reports that little axial rotation occurs with elevation in the plane of the scapula, while others are in disagreement (An 1991; Soslowsky et al. 1992).

Shoulder Extension Shoulder extension occurs through the coronal axis and the sagittal plane as a spinning along the face of the glenoid.

Shoulder External Rotation   Shoulder external rotation occurs through a superoinferior, vertical, or longitudinal axis, while the motion occurs in the transverse plane. As the humerus moves into external rotation, a posterior rolling of the convex humeral head occurs on the smaller shallow glenoid fossa. As this occurs, a simultaneous anterior gliding has to occur (figure 10.16). These arthrokinematics seem to work very well; during external shoulder rotation, the normal humerus translates only 1 to 2 mm posteriorly (Harryman et al. 1990).

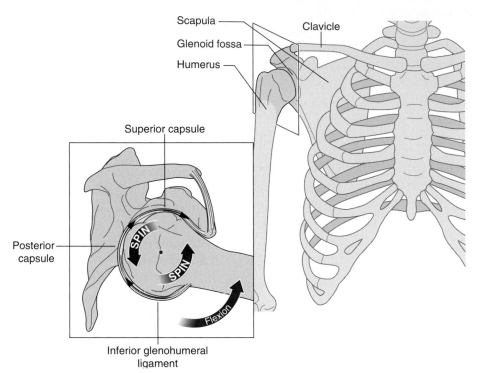

▶ **FIGURE 10.15**   Sagittal plane movement of glenohumeral flexion. The axis of rotation is seen spinning about a point on the glenoid fossa.

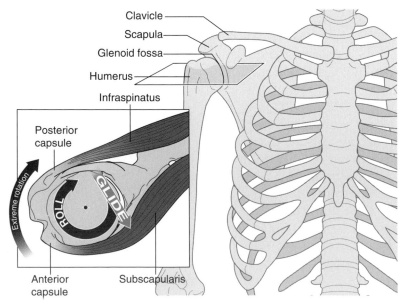

▶ **FIGURE 10.16**   Superior view of active shoulder external rotation, creating a posterior roll of the humeral head and a concomitant anterior gliding.

Approximately 70 to 80° of internal rotation can occur, and around 60 to 70° of external rotation is possible with the arm at the side. As the shoulder is brought up to 90° of elevation, a much larger rotation is allowed, reaching nearly 90° for each. Theoretically, the 90° position places increased tension on the inferior glenohumeral ligament but allows some amount of slack in both the anterior and posterior capsules. Extremes of motion in both internal and external shoulder rotation probably occur with help from the scapulothoracic joint. Scapular protraction and retraction assist maximal efforts into internal and external rotation, respectively.

Shoulder Internal Rotation   Shoulder internal rotation occurs through a superoinferior, vertical, or longitudinal axis, while the motion occurs in the transverse plane. As the humerus moves into internal rotation, an anterior rolling of the convex humeral head occurs on the smaller shallow glenoid fossa. As this occurs, a simultaneous posterior gliding has to occur concomitantly to keep the joint from dislocating.

Shoulder Abduction and Adduction   Shoulder abduction and adduction occur in the frontal plane around an anteroposterior axis of motion. The center of rotation occurs through the center of the humeral head. Joint arthrokinematics for these motions involve the convex–concave rules, as the convex humeral head will roll superiorly and simultaneously glide inferiorly during abduction. Since full shoulder abduction is 180°, it is easy to see that motion from some other joint is needed to achieve full elevation. In a normal-functioning glenohumeral joint, 60° of elevation comes from upward scapular rotation.

Clinically, when the humerus is placed in extreme internal glenohumeral rotation, it is unable to fully elevate into either flexion or abduction because of contact between the greater tuberosity and the acromion. Multiple important structures lie in the space between the humeral head and the acromion—known as the **subacromial space**. Structures in the subacromial space include the supraspinatus muscle and tendon, the subacromial bursa, the

superior joint capsule, and the intra-articular portion of the long head of the biceps brachii. During glenohumeral elevation with concomitant external rotation of approximately 35 to 40° (An 1991; Soslowsky et al. 1992), the greater tuberosity moves behind the acromion, sparing the subacromial contents compressive stress.

▶ **KEY POINT**
During glenohumeral elevation, a concomitant external rotation occurs so that the greater tuberosity will move behind the acromion, sparing the subacromial contents excessive compression.

Regardless of terminology, for the humerus to have full unrestrained motion, complex arthrokinematics at the joint must occur. Because the humeral head is larger than its counterpart, the glenoid fossa, a certain amount of inferior gliding of the humerus must occur during humeral elevation. As the humerus moves into elevation, the head rolls superiorly. If the humeral head rolled only superiorly, it would soon run out of room. If this occurred, the humeral head would impinge underneath the acromion (figure 10.17). For additional information regarding rotator cuff dysfunction, see Clinical Correlation 10.8. When the glenohumeral capsule is of sufficient length and the rotator cuff muscles are working functionally as dynamic stabilizers, the humerus is drawn inferiorly as the humerus concomitantly rolls superiorly.

Range of Motion

The glenohumeral joint affords the shoulder a tremendous amount of motion. Motions occur in all three planes around three different axes (AAOS 1965):

- Flexion: 0 to 180°
- Extension: 0 to 60°
- Abduction: 0 to 180°
- Medial rotation: 0 to 70°
- Lateral rotation: 0 to 90°

Isolated motions at the glenohumeral joint do occur but are probably fairly rare, as most

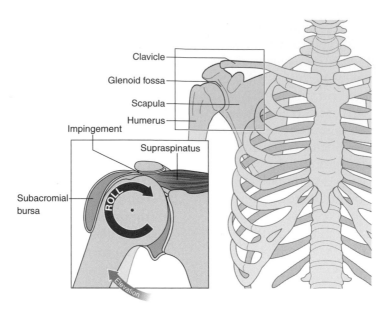

Clavicle
Glenoid fossa
Scapula
Humerus
Impingement
Suprasprinatus
Subacromial
bursa
ROLL
Elevation

▶ **FIGURE 10.17**   Humerus translating superiorly without an inferior glide component during shoulder abduction, creating an impingement of soft tissues between the rotator cuff and the acromion.

CLINICAL CORRELATION 10.8

With a rotator cuff dysfunction, structures in the subacromial space can be pinched, a condition clinically known as impingement syndrome. A very common condition, rotator cuff impingement requires a thorough evaluation to determine its exact cause. This condition may be the result of several factors including weakness of the rotator cuff muscles, weakness of the scapulothoracic muscles, instability, and tightness of posterior shoulder tissues. Failure to adequately address any of these problems or deficiencies could ultimately result in a progression of rotator cuff disease and subsequent rotator cuff rupture.

motion incorporates the rest of the shoulder including the scapulothoracic joint, the acromioclavicular joint, and the sternoclavicular joint. Movements into either flexion or abduction are commonly termed *elevation*. When distinguishing any individual specific movement, one should describe the motion as either flexion or abduction, as the specific plane of motion should be identified.

The motions of glenohumeral flexion and extension occur in the sagittal plane around a coronal axis. Isolated glenohumeral flexion ROM is still being debated. Classical reports describe motion from 100 to 120° as common (Inman et al. 1944; Poppen & Walker 1976; van der Helm & Pronk 1995). Recently Rundquist and colleagues (2003) reported flexion measurements of 97° in patients with

an average age of 50. This most recent finding may indicate an age-related loss of isolated glenohumeral motion. The ability to fully flex 180° requires accessory motions of the scapulothoracic joint.

Scapulohumeral Rhythm   Inman (Inman et al. 1944) was one of the first to write about what he termed **scapulohumeral rhythm**. Despite the fact that numerous studies have assessed scapulohumeral rhythm, Inman's study continues to stand the test of time and is historically used to describe contributions made by multiple joints in the shoulder to perform elevation. Inman has reported that for every 2° of glenohumeral abduction or flexion, a corresponding 1° of upward rotation will occur at the scapulothoracic joint.

Essentially, for every 3° of shoulder elevation, two-thirds comes from the humerus and one-third comes from the scapula (figure 10.18). Based on these numbers, for the shoulder to transverse 180° of elevation into flexion or abduction, 120° will come from humeral elevation, while another 60° will come from the scapulothoracic joint. This 2:1 ratio is not always consistent because during early stages of elevation, the setting phase occurs. During the setting phase, there is an undulation of the scapula that does not always equate the total motion to a 2:1 ratio. Usually after 30 to 60° of elevation, the scapula finds a position of stability and the 2:1 ratio is closely maintained. Multiple authors have reported scapulohumeral rhythms that range from 1.25:1 to 4:1 (Bagg & Forrest 1988; Freedman & Munro 1966; Graichen et al. 2000; McQuade et al. 1998; McClure et al. 2001; Poppen & Walker 1976).

▶ **KEY POINT**
During scapulohumeral rhythm, for every 2° of glenohumeral motion there is 1° of scapulothoracic motion.

Sternoclavicular and Acromioclavicular Joint Contributions to Scapulohumeral Rhythm Neumann (2002) has taken data from Inman (Inman et al. 1944) and described contributions of the sternoclavicular and acromioclavicular joints to scapulohumeral rhythm. Neumann has divided the full 180° of elevation into an early and a late phase, each consisting of 90° of shoulder elevation. In the early phase during the first 90° of elevation, approximately 60° comes from the humerus

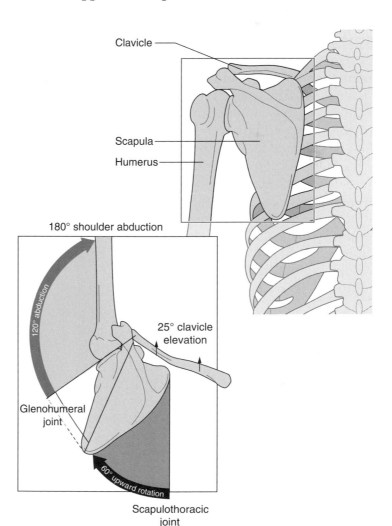

▶ **FIGURE 10.18** Scapulohumeral rhythm. With shoulder elevation of 180°, 60° is obtained by the scapulothoracic joint while the other 120° will come from the glenohumeral joint. With scapulohumeral rhythm, there is also a concomitant posterior clavicle roll of about 25°.

while 30° comes from scapular rotation. The first 30° of scapular rotation occurs in part from the 20 to 25° of clavicular elevation at the sternoclavicular joint and about 10° of upward rotation at the acromioclavicular joint.

During the late phase, another 90° of elevation of total shoulder motion occurs via several joints. Just as in the early phase, about 60° comes from the humerus, while another 30° comes from the scapula. Only another 5° of clavicle elevation is possible from the sternoclavicular joint, but rotation at the acromioclavicular joint can go through another 20 to 25°. Inman originally reported that about 50% of the motion of the scapula comes from clavicle elevation, while the other 50% occurs from acromioclavicular joint rotation (Inman et al. 1994).

As the clavicle elevates initially, tension slowly increases in the coracoclavicular ligament. As elevation continues, tension increases as the coracoid process of the scapula gets pulled inferiorly. The coracoclavicular ligaments have been found to be approximately three times stiffer than the acromioclavicular ligaments (Dawson et al. 2009). Failure loads of the trapezoid at 312 N and the conoid at 266 N demonstrate significant strength (Costic et al. 2003). Because the coracoclavicular ligament does not significantly elongate, tension is placed on the attachment of the conoid ligament at the conoid tubercle, which lies posterior to the longitudinal axis of the clavicle (Inman et al. 1994). This creates a posterior rotation about the long axis of the clavicle of 30 to 55° (Pratt 1994).

Closed and Loose Packed Positions

The closed packed position, one of stability for the glenohumeral joint, is full abduction followed by lateral rotation. The loose packed position, the one of mobility, is 40 to 50° of abduction and 30° of horizontal adduction in the scapular plane.

End Feel

The end feel for shoulder motions are unique to this joint. During shoulder flexion, there should be a firm end feel because of tension in the posterior band of the coracohumeral ligament, posterior joint capsule, posterior deltoid, teres minor, teres major, and infraspinatus. Shoulder extension should also be firm because of tension in the anterior band of the coracohumeral ligament, anterior joint capsule, clavicular portion of the pectoralis major, coracobrachialis, and anterior deltoid. Shoulder abduction is firm from tension in the middle and inferior band of the inferior glenohumeral ligament complex, teres minor, and clavicular portion of the pectoralis major. Internal (medial) rotation is firm because of tension in the posterior joint capsule, infraspinatus, and teres minor. Finally, external (lateral) rotation is firm from tension in the anterior joint capsule, all bands of the glenohumeral capsule, the coracohumeral ligament, the subscapularis, the teres major, and the clavicular fibers of the pectoralis major muscle.

Capsular Pattern

The capsular pattern of the shoulder is described in a typical pattern of restriction, with the greatest restriction listed first: lateral rotation, abduction, internal (medial) rotation. In this case, lateral humeral rotation is limited more than abduction, which is limited more than medial rotation.

Scapular Plane Movements

It is often a moot point to describe flexion and abduction as anything other than elevation. Part of the reason for this movement has come from arguments from Codman, Johnston, McGregor, and Saha, who strongly believe in measuring elevation in the scapular plane (Codman 1934; Johnston 1937; McGregor 1937; Saha 1961). The scapular plane is described as being 30 to 45° anterior to the coronal plane. For most people, this is a much easier plane to move through since the bones of the articulating joints maintain optimal alignment. Additionally, the capsuloligamentous structures remain looser than in flexion or abduction, the rotator cuff muscles remain relatively relaxed because humeral rotation is not required, and the deltoid and supraspinatus are in an optimal length–tension position (Johnston 1937; McGregor 1937).

These contentions are not without dispute; An (1991) has reported that maximum elevation actually occurs 10 to 37° anterior to the scapular plane. Additionally, researchers have disputed the fact that the scapular plane does not require external humeral rotation (An 1991; Soslowsky et al. 1992).

Acromioclavicular Joint

The acromioclavicular joint is formed between the acromion of the scapula and the lateral end of the clavicle. This joint allows independent motion between both the scapula and clavicle. Without this integral joint, the scapula and clavicle would be required to move as one solid unit.

Axes of Motion

The acromioclavicular joint is a plane synovial joint that allows primary motions of upward and downward rotation. These motions occur in an oblique anteroposterior axis around the frontal plane. Other accessory motions include anterior and posterior tipping, which occur in an oblique coronal axis around the sagittal plane. Finally, internal and external rotation occur at the acromioclavicular joint around a vertical axis in the transverse plane.

Arthrokinematics

The true arthrokinematics of the acromioclavicular joint are not usually discussed, as there are few studies and most are inconsistent.

Range of Motion

The motions of upward and downward rotation refer to movement of the glenoid fossa so that upward rotation causes an upward movement of the glenoid and a downward rotation creates the opposite. Conway is one of the few to describe acromioclavicular motions and has reported 30° of upward rotation and approximately 17° of downward rotation (Conway 1961).

Anterior tipping can be seen in the posterior shoulder as a prominent inferior angle (backward tipping) of the scapula and a forward tipping of the anterior acromion. Posterior tipping results in elevation of the acromion and is thought to increase the subacromial space. Some amount of tipping may be necessary because the posterior thorax is not flat. The scapula must adjust to this posterior convexity (Dempster 1965; McClure et al. 2001).

To function properly, the acromioclavicular joint will also anteriorly tip up to 60° in cadaveric specimens (Dempster 1965). This amount is reportedly increased with in vivo studies to 40° or more during maximal range of motion shoulder flexion and extension (Ludewig et al. 2009; Sahara et al. 2007). Posterior tipping generally brings the acromioclavicular joint back to its original resting position.

Acromioclavicular movements are described in reference to the glenoid fossa, so that internal rotation implies movement of the glenoid fossa anteriorly and medially, while lateral rotation means movement of the glenoid posteriorly and laterally. Motions at the acromioclavicular joint are thought to maintain the scapula against the thorax; therefore internal and external rotation are similar in description to motions at the scapula of protraction and retraction. Combined motions of internal and external rotation have been described as being from 0 to 20° (Dempster 1965; Conway 1961) and from 0 to 60° of motion (McClure et al. 2001; Fung et al. 2001). A recent study among healthy young adults revealed that during elevation to 90°, internal rotation of the acromioclavicular joint averaged 68° (Teece et al. 2008).

Closed and Loose Packed Positions

The closed packed position of the acromioclavicular joint is 90° of shoulder abduction, while the loose packed position is with the arm at the side.

End Feel

End feels are all firm because of tension in the acromioclavicular joint ligaments and capsule.

Capsular Pattern

The capsular pattern for the acromioclavicular joint is described as having pain at extremes of range of motion.

Sternoclavicular Joint

The sternoclavicular joint is the articulation between the manubrium of the sternum and the medial end of the clavicle.

Axes of Motion

The sternoclavicular joint is a saddle-shaped synovial joint, yet it allows a full three degrees of freedom. Typically a saddle-shaped joint is considered biaxial, but the sternoclavicular joint exhibits a small degree of rotation, which gives it a third degree of freedom. Elevation and depression of the clavicle occur around an anteroposterior axis in the coronal plane. Protraction and retraction of the clavicle occur around a vertical axis around the transverse plane. Finally, anterior and posterior rotations occur around a coronal axis in the sagittal plane.

These motions are sometimes hard to visually comprehend because movement at the sternoclavicular joint is described as motion of the distal end of the clavicle, not the proximal end that is being viewed as going through the motions. Therefore motions of protraction, retraction, elevation, and depression should reference the lateral end of the clavicle. Rotation movements of the clavicle can be viewed as through the long axis. During shoulder motion, varying degrees of stress are placed on the sternoclavicular disc. During elevation and depression, most motion occurs between the clavicle and the articular disc, whereas during protraction and retraction, the greatest amount of movement occurs between the articular disc and the sternal surface (Dempster 1965).

Arthrokinematics

Arthrokinematics of the sternoclavicular joint require a unique pattern of accessory movement to allow the motion that occurs from this joint.

Elevation of the sternoclavicular joint occurs between a convex medial clavicle and a concave sternum. With elevation of the sternoclavicular joint, the medial end of the clavicle rolls superiorly but depresses (glides) inferiorly, while the lateral end of the clavicle elevates (figure 10.19). Depression occurs as the medial end of the clavicle actually rolls inferiorly and glides superiorly at the joint proper. This motion is seen as a depression of the lateral portion of the clavicle (figure 10.20).

Protraction occurs around a vertical axis along the transverse plane. During these motions, the medial saddle-shaped end of the clavicle takes a concave form. During protraction, the concave medial clavicle glides anteriorly on the convex sternum.

Retraction occurs around a vertical axis along the transverse plane. During retraction, the concave medial end of the clavicle glides posteriorly on the convex sternum (figure 10.21).

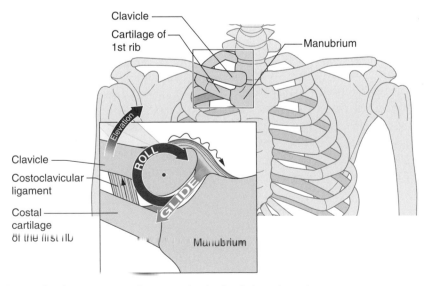

▶ **FIGURE 10.19** Arthrokinematics of sternoclavicular joint elevation.

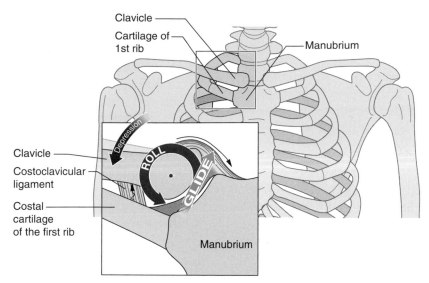

FIGURE 10.20 Arthrokinematics of sternoclavicular joint depression.

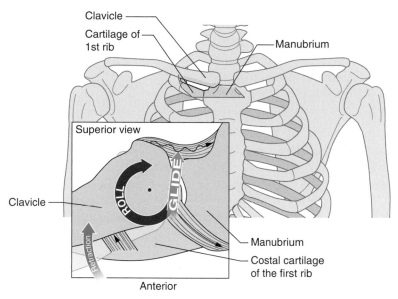

FIGURE 10.21 Arthrokinematics of sternoclavicular joint retraction.

Finally, anterior and posterior rotation of the sternoclavicular joint is required during any elevation of the shoulder of more than approximately 90° of rotation. This rotation occurs around the clavicle's long axis as a spinning between the saddle-shaped sternum and the manubrium.

Range of Motion

Despite the fact that the sternoclavicular joint is the only real attachment of the upper extremity complex to the axial skeleton and requires a great deal of stability, it also requires some degree of mobility to allow movements of the upper quadrant to occur.

Elevation and Depression Elevation occurs up to about 45° and depression only about 5° at the sternoclavicular joint (Moseley 1958). Ludewig and colleagues (2004) reported 10 to 15° of elevation in healthy subjects. This lack of depression is useful to offset the weight of the upper extremity at the sternoclavicular joint. The interclavicular ligament may support some inferior depression of the medial end of the clavicle on the sternum, as does the bony contact between the inferior portion of the clavicle and sternum.

Protraction and Retraction   Approximately 15 to 30° of both protraction and retraction are reported at the sternoclavicular joint (Conway 1961; Moseley 1958; Steindler 1955).

Anterior and Posterior Rotation   Posterior rotation occurs from neutral and is reported to be as high as 50° (Inman et al. 1944). As the shoulder is brought back to neutral, the rotation returns to its original position. Anterior rotation occurs from neutral but is limited to only about 10°. Ludewig and colleagues (2004) have reported 15 to 31° of posterior rotation during overhead shoulder movements.

Closed and Loose Packed Positions

The closed packed position of the sternoclavicular joint is when the shoulder is in full elevation and protraction. The loose packed position occurs with the arm at the side.

End Feel

The end feel is firm because of tension in the anterior and posterior sternoclavicular ligaments, the costoclavicular ligament, the interclavicular ligament, and the joint capsule.

Capsular Pattern

Pain can be present at extremes of range of motion, especially with horizontal adduction and full elevation.

Scapulothoracic Joint

The scapulothoracic joint is a pseudojoint formed between the scapula and the posterior thorax. Therefore it does not possess a capsular pattern, an end feel, or closed and loose packed positions. The scapulothoracic joint works in concert with the other joints of the upper extremity: the glenohumeral, acromioclavicular, and sternoclavicular joints.

Axes of Motion

The scapulothoracic joint is a pseudojoint with motions that occur in three planes and three axes. Elevation and depression of the

scapula are gliding motions that occur along the frontal plane. Abduction, adduction, and upward (lateral) and downward (medial) rotations are all gliding motions that also occur along the frontal plane. Upward and downward rotations typically occur around an anteroposterior axis. Internal and external rotations occur around a vertical axis along the transverse plane, whereas anterior and posterior tilting occurs around a mediolateral axis in the sagittal plane.

Arthrokinematics

Since the scapulothoracic joint is not a true synovial joint, it does not have joint arthrokinematics.

Range of Motion

Motions of the scapulothoracic joint are referenced to the glenoid fossa. There are five groups of motions (figure 10.22): elevation and depression, upward and downward rotation, abduction and adduction, internal and external rotation, and anterior and posterior tipping.

Elevation and Depression   Two of the most common primary motions of the scapulothoracic joint are elevation and depression (figure 10.22a). Movement of the entire scapula in a superior direction is termed *elevation*. This occurs when someone shrugs his shoulders. Depression occurs when the scapula is brought back to its neutral position. As you will soon see, these motions of the scapula do not occur in isolation and require motion at the sternoclavicular and acromioclavicular joints. Even simple elevation and depression require minor adjustments of tilting and winging. Most scapulothoracic motions are difficult to measure clinically and are commonly observed for differences or for faulty patterns. Movements of elevation have been reported to be between 2 and 10 cm, while those for depression are significantly less at 2 cm (Kapandji 1982; Kelley 1995).

Upward and Downward Rotation   Upward and downward rotations are defined as rotations in the anteroposterior axis, along the

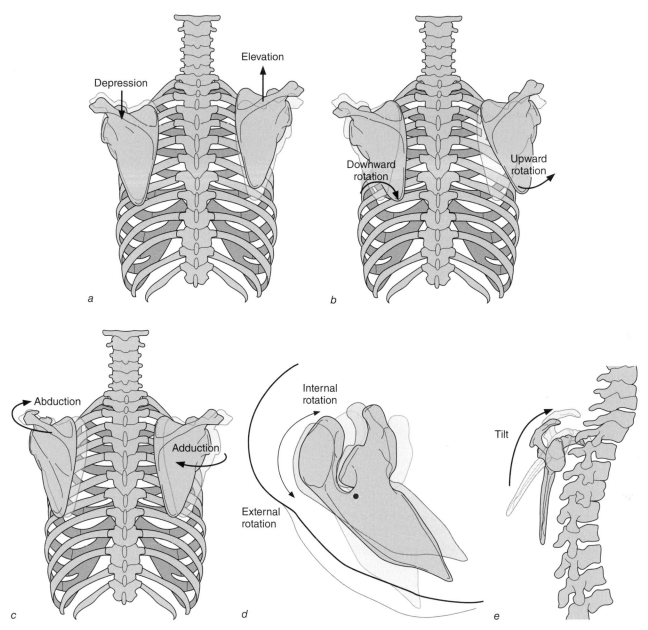

▶ **FIGURE 10.22**   Primary motions that occur at the scapulothoracic joint: *(a)* elevation and depression; *(b)* upward and downward rotation; *(c)* abduction and adduction; *(d)* internal and external rotation; *(e)* anterior and posterior tipping.

coronal plane, resulting in the glenoid moving upward (upward rotation) or downward (downward rotation) (figure 10.22b). The axis of motion in the anteroposterior plane with upward and downward rotation of the scapula actually changes as the scapular motion progresses into higher levels of elevation. At the initiation of movement, the axis is at approximately the root of the scapular spine. As the arm is elevated, the axis moves laterally to near the acromioclavicular joint at the end of elevation. Movements of upward (lateral) rotation are around 10 cm, while those for downward (medial) rotation are approximately 4 to 5 cm (Kapandji 1982; Kelley 1995).

Abduction and Adduction   Abduction and adduction of the scapulothoracic joint occur as the scapula moves away from the spine or midline (abduction) or toward the spine or midline (adduction) (figure 10.22c). Because the thorax is not flat, this is not a pure translation and generally occurs during protraction or retraction movements of the scapula.

Internal and External Rotation   Internal and external rotations are at times described in reference to the vertebral border of the scapula. As the scapula moves through protraction, it either rotates internally (glenoid moves anteriorly, vertebral axillary border posteriorly) or retracts (glenoid moves posteriorly, vertebral axillary border anteriorly) (figure 10.22d). These motions occur at the clavicle, the sternoclavicular joint, and the acromioclavicular joint. When the vertebral border moves excessively from the thorax, it is commonly known as **scapular winging**. This can be seen either in the resting position or with active movement of the shoulder.

▶ **KEY POINT**
Scapular winging occurs when the medial posterior border of the scapula moves posteriorly, either at rest or during movement of the shoulder.

Anterior and Posterior Tipping   Anterior and posterior **scapular tipping**, or tilting, occur around a coronal or mediolateral axis. During anterior tilting, the superior portion of the scapula and the glenoid fossa move anteriorly, while the inferior angle of the scapula moves posteriorly (figure 10.22e). Posterior tilting occurs as the superior portion of the scapula and the glenoid fossa both move posteriorly, which means the inferior angle moves in the anterior direction. These motions

actually occur at the acromioclavicular joint and the sternoclavicular joints. Similarly, as during protraction and retraction, the thorax is shaped convexly in a superior and inferior direction. Clinical Correlation 10.9 looks at abnormalities in tipping and winging.

▶ **KEY POINT**
Anterior scapular tipping occurs when the inferior angle of the scapula moves posteriorly.

Protraction and Retraction   Other translation motions at the scapulothoracic joint are protraction and retraction of the scapulothoracic joint toward or away from the spine. Because the posterior thorax is convex, the motion is not pure translation. The translation of the scapula moves outward and anteriorly, following the thorax during protraction. The motion is coupled with internal and external scapular rotation in which the glenoid fossa moves anteriorly (internal rotation) and posteriorly (external rotation).

Muscles

Because of the dynamic nature of shoulder movement and function, the musculature is of paramount importance. The muscular system is the major component that drives movements of the shoulder needed for full functional activity. The muscles in the shoulder are numerous and include powerful large muscles and smaller, less powerful stabilizers. Please see table 10.1 for shoulder muscle origin, insertion, action, and innervation.

CLINICAL CORRELATION 10.9

A pathological situation can exist in which either tipping or winging occurs excessively. This can be caused by weakness of the scapular muscles or by limited flexibility of the musculature around the shoulder due to habitual postural positions. Excessive motion at the scapulothoracic joint can lead to encroachment of the acromion near the humeral head during activities of elevation, creating a scapular-related impingement. If tipping or winging is occurring in excess on the injured shoulder, the clinician should adequately examine the scapular musculature to determine which muscle weakness is most likely causing this faulty movement pattern. Failure to address this weakness or motor pattern could result in a progression of rotator cuff disease and ultimately rupture of the rotator cuff tendons.

TABLE 10.1 Muscles of the Upper Extremity

Muscle	Origin	Insertion	Action	Innervation
Sternocleido-mastoid	Tendinous head from sternum; muscular head from medial third of clavicle	Mastoid process of skull	One muscle: flexion of neck toward same side (face to opposite side, ear of same side toward clavicle)	Accessory nerve
Subclavius	Rib 1	Undersurface of clavicle	Possibly depresses clavicle; maintains sternoclavicular joint	Nerve to subclavius
Trapezius	External occipital protuberance; ligamentum nuchae; spinous processes of C7 and all thoracic vertebrae	Spine of scapula; acromion; lateral 1/3 of clavicle	Upper fibers: elevation of scapula Middle fibers: retraction of scapula Inferior fibers: depression of scapula Rotation of glenoid cavity upward	Accessory nerve (C3 and C4 sensory)
Latissimus dorsi	Spinous processes of lower 6 thoracic and all lumbar and sacral vertebrae; posterior part of iliac crest	Medial lip (crest of lesser tubercle) and floor of intertubercular groove of humerus	Extension, adduction, and medial rotation of arm	Thoracodorsal nerve
Levator scapulae	Transverse processes of upper 4 cervical vertebrae	Superior angle and upper part of medial border of scapula	Elevation of scapula	C3 and C4 dorsal scapular nerve
Rhomboid minor	Lower part of ligamentum nuchae; spinous processes of C7 and T1	Medial border of scapula at base of spine	Elevation and retraction of scapula; downward rotation of glenoid cavity	Dorsal scapular nerve
Rhomboid major	Spinous processes of T2-T5	Medial border of scapula below rhomboid minor	Elevation and retraction of scapula; downward rotation of glenoid cavity	Dorsal scapular nerve
Serratus anterior	Ribs 1-8 on anterolateral thoracic wall	Medial border of scapula; heaviest insertion to inferior angle	Protraction of scapula; upward rotation of glenoid cavity; holds medial border against thoracic wall	Long thoracic nerve
Deltoid	Lateral 1/3 of clavicle; acromion; spine of scapula	Deltoid tuberosity on shaft of humerus	Middle fibers: abduction Anterior fibers: flexion and medial rotation Posterior fibers: extension and lateral rotation	Axillary nerve
Supraspinous	Supraspinous fossa of scapula	Greater tubercle of humerus	Abduction of the arm	Suprascapular nerve
Infraspinatus	Infraspinous fossa of scapula	Greater tubercle of humerus below supraspinatus	Lateral rotation of arm	Suprascapular nerve

> *continued*

TABLE 10.1 > *continued*

Muscle	Origin	Insertion	Action	Innervation
Teres minor	Upper 2/3 of lateral border of scapula	Greater tubercle of humerus below infra-spinatus	Lateral rotation of arm	Axillary nerve
Teres major	Inferior angle of scapula	Medial lip of intertu-bercular groove of humerus	Adduction, medial rotation, and exten-sion of arm	Lower subscapular nerve
Subscapularis	Subscapular fossa	Lesser tubercle and crest of humerus	Medial rotation of arm	Upper and lower subscapular nerves
Pectoralis minor	Ribs 3-5	Medial portion of cora-coid process	Scapular protraction	Medial and lateral pectoral nerve
Pectoralis major	Medial 1/4 of clavicle, sternum, costal carti-lage of ribs 2-6	Crest of greater tuber-cle	Humerus adduction, medial rotation	Medial and lateral pectoral nerve

Scapulothoracic Muscles

The scapulothoracic muscles (figure 10.23) provide dynamic support for the scapula. These are especially critical because the scapula has no direct attachment to the remainder of the skeleton.

Levator Scapulae

The levator scapulae is a small muscle located on the posterior lateral aspect of the cervical spine. It originates from a series of slips extending from the C1 (atlas) to C4 transverse processes (see figure 10.23*a*). These slips form a flat conjoined tendon that inserts onto the posterior border of the scapula at the superior medial angle and the root of the scapula. This muscle is innervated by small branches of the anterior division of the third and fourth cervical nerve roots and a branch from the dorsal scapular nerve. The spinal accessory nerve crosses laterally in the middle section of this muscle. The levator scapulae elevates the superior angle of the scapula. It is also a functional portion of the scapular force couples that upwardly rotate the scapula.

Trapezius

The largest and most superficial of the scapulothoracic muscles is the trapezius. The trapezius is a large, broad, flat trian-gular muscle that originates on the spinous processes of the C7 through T12 vertebrae (see figure 10.23*a*). Both the right and left trapezius together form a trapezoid. The trapezius is commonly divided into upper, middle, and lower thirds. The upper trapezius fibers (above C7) originate on the ligamentum nuchae, some as high as the external occipital protuberance. The upper fibers run in an outward and downward direction to insert on the posterior border of the distal third of the clavicle. Some of the lower cervical and upper thoracic (middle trapezius) fibers attach along the acromion and the spine of the scapula. The lower trapezius fibers run outward and upward, originating at the thoracic spine and inserting on the lower border of the scapular spine.

As an entire unit, the trapezius acts as a scapular retractor, while the upper fibers elevate the lateral border. The trapezius also assists with upward scapular rotation. Inman found the upper trapezius muscles to be active during all upward rotations of the scapula (Inman et al. 1994). The middle trapezius has an excellent line of pull to retract the scapula when functioning concentrically. The middle trapezius may also function eccentrically when the arm is forcefully taken into pro-traction such as when punching or throwing overhead.

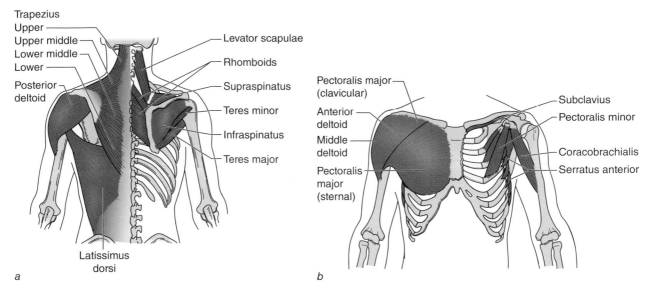

► **FIGURE 10.23** *(a)* Posterior and *(b)* anterior scapulothoracic muscles.

Rhomboid Minor and Rhomboid Major

The rhomboid minor is the more superior of the two muscles and originates from the lower ligamentum nuchae at C7-T1, while the rhomboid major is the lower muscle and originates from T2 to T5 (see figure 10.23a). The rhomboid muscle fibers run in an outward and upward direction; the minor inserts on the posterior portion of the medial base of the scapular spine, while the major inserts into the posterior surface of the medial border, from the attachment of the minor to the inferior angle of the scapula. These two muscles are innervated by the dorsal scapular nerve.

The rhomboids, together with the middle and lower trapezius, function concentrically to retract the scapula. They are also important stabilizers for elevation of the shoulder, at which time they will contract eccentrically to control scapular positioning.

Serratus Anterior

The serratus anterior is a thin, irregularly shaped muscle that originates from the anterior lateral ribs of the thoracic cage (see figure 10.23b). The serratus anterior has three functional components, with nine tendinous slips coming from the first nine ribs. The upper slip originates at ribs one and two and runs upward and posteriorly to insert onto the ante-

rior superior angle of the scapula. The next portion originates from ribs two through four and runs posteriorly to attach to the anterior medial border of the scapula. The lower portion originates on ribs five through nine and attaches to the anterior inferior angle of the scapula. The serratus anterior is supplied by the long thoracic nerve, formed by branches from the fifth through seventh spinal nerves.

The serratus anterior's primary function is scapular protraction. The serratus also works as a force couple with the other scapular muscles to perform scapular upward rotation.

Pectoralis Minor

The pectoralis minor is a small, thin, flat triangular muscle located deep in the anterior shoulder region (see figure 10.23b). The minor lies deep to the pectoralis major. Muscular slips coming from the third through fifth costal cartilages run obliquely upward and outward to form a flattened tendon that attaches to the inner border and upper surface of the coracoid process. The pectoralis minor is innervated by the medial and lateral pectoral nerve with the pectoralis major.

Pectoralis Major

The pectoralis major is a large, broad muscle that has three portions. An upper portion originates off of the medial half to two-thirds of the clavicle and then inserts onto the lateral portion of the bicepital groove. The middle portion attaches to the manubrium and the upper two-thirds of the body of the sternum and inserts directly behind the clavicular portion of the pectoralis major on the humerus. The inferior (lower) portion attaches from the distal body of the sternum and the fifth and sixth ribs, while it inserts onto the humerus at the same location of the other two heads. The exception is that the lower portion spirals and rotates 180° so that the inferior fibers attach to the superior portion of the humerus. As a whole, the pectoralis major adducts the humerus but can also function as a humeral elevator and depressor, dependent on its starting position. If the humerus is depressed, the clavicular head can participate in shoulder flexion, and if elevated, the lower sternal portion can depress the humerus. Innervation is supplied by two sources, the lateral pectoral nerve (C5, C6, and C7) innervates the clavicular portion, while the medial pectoral nerve (C8, T1) innervates the remaining portion. The blood supply comes from the deltoid branch of the thoracoacromial artery and the pectoral artery.

Subclavius

The subclavius is a small muscle that has a tendon only 1 to 1.5 in (2.5 to 4 cm) long (Cave & Brown 1952). The origin of the subclavius is the first rib and cartilage, and it inserts onto the inferior surface of the medial third of the clavicle. The subclavius is innervated by the fifth cervical spinal nerve.

Teres Major

The thicker of the two teres muscles, this teres is flattened and arises from the posterior aspect of the inferior angle of the scapula just below the infraspinatus origin (see figure 10.23a). Its fibers run in an upward and outward direction and attach as a flat tendon onto the medial lip of the bicipital groove of the humerus, which is an extension of the lesser tubercle. The teres major runs in a much more horizontal fashion than the group of rotator cuff muscles. The function of the teres major is internal rotation, humerus adduction, and extension. When the distal portion of the extremity is fixed, as in doing push-ups or the rings in gymnastics, the teres will function as an upward rotator of the scapula.

Muscles of the Rotator Cuff

The rotator cuff muscles (figure 10.24) are a special set of dynamic stabilizers for the glenohumeral joint. These muscles collectively provide compression and dynamic stabilization for the glenohumeral joint as it moves through the largest range of motion of any joint in the human body.

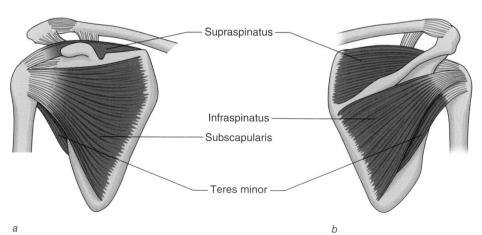

a   b

▶ **FIGURE 10.24**   *(a)* Anterior aspect and *(b)* posterior aspect of the right rotator cuff.

Supraspinatus

The supraspinatus is the most superior of all the rotator cuff muscles (see figure 10.24). It lines the supraspinatus fossa of the scapula. This muscle originates on the medial floor of the supraspinatus fossa and runs distally underneath the supraspinatus outlet across the top of the humeral head to attach on the highest portion of the greater tuberosity. The supraspinatus not only forms the floor of the subdeltoid bursa but also is very intimate with the capsule as it runs its course to the greater tuberosity. The tendon of the supraspinatus blends into a confluence with the infraspinatus and the coracohumeral ligaments just before they attach to the tuberosity. Additionally, all the cuff tendons have a transitional zone in which the tendon goes from fibrous connective tissue to fibrocartilage to cartilage and eventually to bone tissue at the insertion site (Codman 1934).

The action of the supraspinatus is to initiate and aid the deltoid during elevation of the shoulder (Howell et al. 1986). Both muscles fire synchronously during elevation. The fibers of the supraspinatus run in an orientation directly from the scapula to encircle the humeral head. The supraspinatus may be more important as a compressor of the humeral head during these elevation motions, its primary function as one of the rotator cuff muscles. The supraspinatus moment arm is constant through the range of motion and is actually larger than that of the deltoid through the first 60° of abduction (Poppen & Walker 1978). If these muscles do not work in concert with one another because of fatigue or injury, the compressive effect may be lost, creating a decrease in space between the supraspinatus and the coracoacromial arch overhead. This can create the problems described in Clinical Correlation 10.10. The supraspinatus's maximal moment arm and therefore muscle effort are produced around 30° of elevation; after that point the greater tubercle lever arm is decreased (Rievtveld et al. 1988; Atwater 1979). The supraspinatus is innervated by the suprascapular nerve.

Infraspinatus

The infraspinatus is a large, thick, and triangular-shaped muscle that inserts on the posterior surface of the medial scapula (see figure 10.24). Muscle fibers of the infraspinatus converge to form the posterior rotator cuff tendons that run behind the posterior portion of the glenohumeral joint, attaching to the middle facet of the greater tuberosity with the teres minor tendon. The infraspinatus is innervated by the suprascapular nerve. Infraspinatus muscle activity has been shown to linearly progress through a full range of motion in coronal and scapular plane abduction (Inman et al. 1944; Ito 1980; Saha 1971). If firing in isolation, the infraspinatus will cause external rotation of the humerus.

Teres Minor

The teres minor is a small muscle that lies below the infraspinatus (see figure 10.24). It arises from the posterior axillary border of the scapula and is separated by a lamina

CLINICAL CORRELATION 10.10

The area inside the supraspinatus outlet can be increased or decreased by various shoulder movements. This space is decreased with shoulder internal rotation and increased with the shoulder in a position of external rotation (Neer 1972). This may have implications in rotator cuff disease, as a smaller outlet may create compressive situations for the supraspinatus tendon. If a smaller outlet is created by a curved or hooked acromion, a procedure known as a subacromial decompression may be required. In this procedure, a surgeon uses a motorized shaver to create a larger subacromial area by removing offending bony tissue that has encroached this valuable space.

from the teres major below. The fibers of the narrow muscle run slightly upward and outward to attach below the infraspinatus on the lower facet as a flat conjoined tendon with the infraspinatus. Passing through the quadrilateral space, the posterior division of the axillary nerve innervates the teres minor. Occasionally this can be a source of problems if the axillary nerve becomes compressed in the quadrilateral space (see Clinical Correlation 10.11). The teres minor functions very similarly to the infraspinatus, but because it has a much smaller cross-sectional area, it is unable to create as much torque with an isolated muscle contraction. It does, however, contribute to elevation activities of the rotator cuff and will be discussed later as part of this muscle group.

Subscapularis

The rotator cuff muscle with the largest cross-sectional area is the subscapularis (see figure 10.24). The subscapularis is the anterior counterpart of the infraspinatus. It is a large, thick, triangular-shaped muscle that attaches broadly to the anterior surface of the scapula, on the medial two-thirds of the subscapularis fossa. The entire subscapularis runs obliquely in an upward and outward direction to insert on the lesser tuberosity of the humerus and by itself forms the entire anterior portion of the rotator cuff. The subscapularis is supplied by the upper and lower subscapular nerves. Subscapularis muscle activity is greatest with isolated internal rotation of the shoulder. Furthermore, it is very active during active elevation. Peak activity during elevation occurs when lifting the arm in the coronal plane from 90 to 120°, with a sharp decline after 120° (Inman et al. 1944).

Other Muscles

Many other muscles in the shoulder either produce primary movements or work as stabilizers. These muscles are listed secondarily simply because they are not as commonly involved in shoulder conditions as the muscles listed previously. However, these secondary muscles can create dysfunction and cause problems when injured or pathological.

CLINICAL CORRELATION 10.11

Compression of the axillary nerve or artery can occur in an area known as the quadrilateral space (figure 10.25)—bordered by the teres major inferiorly, the teres minor posteriorly, the humeral shaft laterally, and the subscapularis medially. Compression of the axillary the posterior shoulder that at times can radiate into the forearm (Manske et al. 2009). A high index of suspicion is required to make this diagnosis, as it is commonly missed during examination. A thorough examination is needed for any patient with a vague, generalized complaint of discomfort in the posterior shoulder.

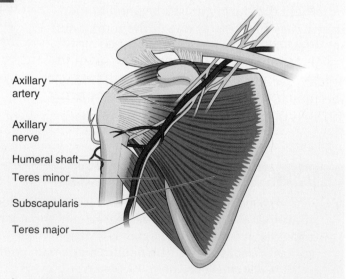

Axillary artery
Axillary nerve
Humeral shaft
Teres minor
Subscapularis
Teres major

▶ **FIGURE 10.25** The subscapularis and area known as the quadrilateral space.

Deltoid

The multipennate deltoid has a relatively large cross-sectional area and is composed of three different heads: anterior, middle, and posterior (see figure 10.23, *a* and *b*). It is the composite three heads that give the shoulder its characteristic rounded appearance. Although the middle deltoid is active in most shoulder motions, the anterior generates more force with flexion, while the posterior more with extension movements. Because of a small lever arm, the posterior deltoid is generally not as powerful as the other components and usually functions as a joint compressor, except when the shoulder is moved into horizontal abduction (DeLuca & Forrest 1973; Reinold et al. 2007). When the middle and anterior deltoid contract together, the humerus tends to elevate and medially rotate. When the posterior and middle deltoid contract together, the resultant movement is extension and lateral rotation. When all three portions work in concert, the movement is elevation in the scapular plane, and the anterior and posterior portions tend to nullify each other's action. The middle deltoid has consistently shown great activity in studies that assess elevation of the shoulder in the scapular plane (Happee & van der Helm 1995; Inman et al. 1944; Kronberg et al. 1990; Yoshizaki et al. 2009).

Part of the deltoid's ability to produce optimal torque is related to its length–tension relationship. With the arm at the side, the deltoid is at its optimal resting-length tension. The deltoid functions synergistically during shoulder elevation during a dynamic glenohumeral caudal glide caused by contractions of the infraspinatus, teres minor, and subscapularis muscles of the rotator cuff. EMG studies of the deltoid demonstrate a consistent increase in muscle activity during elevation in the frontal plane that peaks around 90° of abduction, where the moment arm of the upper extremity is the largest. Even though the moment arm decreases with further elevation past 90°, the deltoid still remains active as it moves into a position toward active insufficiency. Greater recruitment is required in the later stages of elevation to maintain an appropriate force output to move into the end ranges of elevation.

A study by Rodriques and colleagues (2008) found that after a 30-second isometric contraction, the middle and posterior deltoid were fatigued while the clavicular, or anterior, portion was unaffected. This finding occurred during fatiguing motions in both frontal and scapular plane elevation.

The deltoid originates from the lateral third of the clavicle, the acromion, and the crest of the spine of the scapula; it inserts via a conjoined tendon from the scapular and clavicular portions onto the deltoid tuberosity of the upper humerus. The deltoid is innervated by the axillary nerve.

Biceps

The biceps muscle is primarily an elbow flexor and forearm supinator, although it is also thought to assist in shoulder flexion. The biceps brachii (figure 10.26) arises from two heads—one from the coracoid process, the other from the supraglenoid tubercle—and travels distally to insert into the radial tuberosity. Landin and colleagues (2008) determined that the biceps brachii does have a role as a dynamic stabilizer of the shoulder and has a role in shoulder elevation. Their results indicate that both shoulder and elbow joint angles influence shoulder joint elevation produced by the biceps brachii. Sakurai et al. (1998) used surface EMG to assess biceps activity during elevation of the shoulder. The long head of the biceps was active in all motions tested, suggesting that the biceps muscle acts as a flexor and an abductor of the shoulder. The long head was even active with internal shoulder rotation. The role of the biceps brachii as an elbow flexor and forearm supinator is discussed in chapter 11.

As usual, these findings are not without controversy. Levy and colleagues (2001) tested biceps function as a shoulder elevator with the elbow locked in a brace and reported no electrical activity in the long head in response to shoulder motion. They report that any role

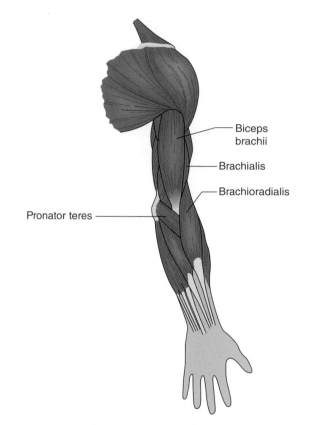

▶ **FIGURE 10.26**   Muscles of the biceps.

of the biceps in elevation must be based either on a passive role of the tendon or on tension in association with elbow and forearm activity. Yamaguchi et al. (1997) also found that using a brace to lock the elbow into neutral forearm rotation and 100° of flexion minimized biceps activity during active shoulder motion. These findings led the researchers to state that any function of the long head of the biceps in shoulder motion does not involve active contractions.

The biceps does appear to provide some joint stabilization. Pagnani et al. (1996) simulated contraction of the long head of the biceps brachii and examined its effects on glenohumeral translations. Application of a force to the biceps long head significantly reduced the amount of translation of the humeral head.

Shoulder Muscle Force Couples

In the shoulder, several instances of force couples occur. A force couple is two or more muscles or groups of muscles on opposite sides of a joint working in concert to stabilize the joint or move the associated segment.

▶ **KEY POINT**

A force couple occurs when two or more muscles or groups of muscles on opposite sides of a joint work in concert to stabilize the joint or move the associated segment.

The deltoid–rotator cuff force couple (figure 10.27) produces a large amount of force to rotate the humerus. The larger, more powerful deltoid has a moment arm that produces elevation of the humerus. If this elevation were to occur unopposed, the resultant force would cause migration of the humeral head into the acromion, impacting the greater tuberosity. A force couple exists in the frontal plane between the deltoid and the rotator cuff muscles; the superiorly directed force from the deltoid is counteracted by the inferior translatory force created by the rotator cuff muscles because they act in the opposite direction on either side of the center of rotation, creating the force couple. Pressure is increased on the coracoacromial arch if the rotator cuff is not working properly as the result of a decreased dynamic caudal glide provided by the cuff. The cuff muscles also provide a compressive force of the humeral head into the glenoid fossa, thus reducing shear forces. Compression occurs more commonly at middle ranges of humeral elevation but has also been seen at the end range of motion as joint forces are increased (Labriola et al. 2005; Wuelker et al. 1998).

The trapezius–serratus anterior force couple is able to upwardly rotate the scapula during motions that elevate the shoulder (figure 10.28). This synergistic force couple occurs when the lower portion of the serratus anterior and the lower portion of the trapezius contract in unison with the upper trapezius and levator scapulae to provide scapular rotation. Upward rotation of the scapula (1) maintains the glenoid surface in an optimal position for the humeral head during elevation; (2) maintains an efficient length–tension relationship for the deltoid muscle; (3) helps

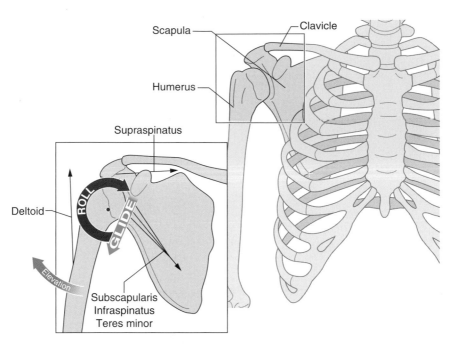

FIGURE 10.27 Deltoid–rotator cuff force couple. During active elevation of the shoulder, as the deltoid elevates the humerus, the rotator cuff muscles contract, exerting a compressive and downward force on the humeral head to counteract the superior translation.

prevent subacromial impingement from the overlying acromion; and (4) provides for a stable scapular base, allowing more efficient and appropriate recruitment of the scapulo-humeral muscles (Kelley 1995; Abboud & Soslowsky 2002). Faulty scapulohumeral rhythm occurs when there is an overdominance of the upper trapezius and late firing of the serratus and lower trapezius.

The rotator cuff group itself—the anterior and posterior cuff force couple—is very valuable at the glenohumeral joint. The anterior-based subscapularis and the posterior-based infraspinatus and teres minor provide dynamic stability because of their inferiorly directed force couple. The subscapularis and infraspinatus have about a 45° inferior line of pull, while the teres minor's is about 55° (Perry 1988). Additionally they, along with the supraspinatus, provide concavity compression in which the humeral head is compressed into the glenoid fossa, resisting unwanted shear forces across the glenohumeral joint (figure 10.29).

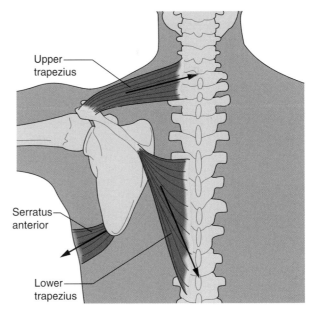

FIGURE 10.28 The trapezius–serratus anterior force couple. Shoulder elevation requires upward scapular rotation to allow full unrestricted motion. The upper, middle, and lower trapezius contract in succession with the serratus anterior to move the scapula into a position of upward rotation.

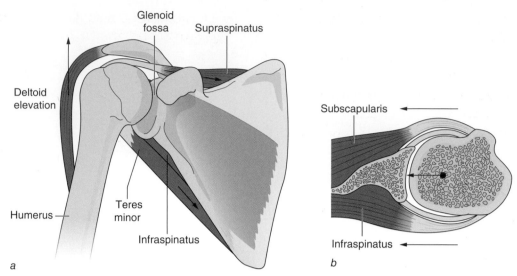

▶ **FIGURE 10.29**   Rotator cuff force couple. *(a)* The supraspinatus compresses the humeral head, providing compression during elevation by the deltoid muscle. *(b)* Further compression is provided by the subscapularis and the infraspinatus muscles anterior and posterior to the glenohumeral joint.

Conclusion

The shoulder complex is a highly organized set of joints designed for mobility and range of motion. The complexity of understanding shoulder injury comes from its many joints and articulations. Knowledge of the anatomy and biomechanics of this joint can help provide a pathway for developing appropriate treatment interventions for those who exhibit pathokinesiology.

REVIEW QUESTIONS

1. The shoulder joint is considered a simple joint.
   a. true
   b. false

2. The only true articulation of the upper extremity to the axial skeleton occurs at which joint?
   a. acromioclavicular
   b. glenohumeral
   c. scapulothoracic
   d. sternoclavicular

3. How many joints are in a unilateral shoulder complex?
   a. three
   b. four
   c. six
   d. eight

4. Common motions at the scapula during shoulder movement include all of the following except
   a. protraction
   b. winging
   c. tipping
   d. lateral flexion

5. Which ligament or ligaments at the sternoclavicular joint check elevation of the clavicle?
   a. anterior and posterior sternoclavicular
   b. medial and lateral sternoclavicular
   c. costoclavicular
   d. interclavicular

6. Which statement is true in regard to the humeral head?

a. The normal angle of inclination is 110 to 120°.

b. Posterior torsion is known as the angle of inclination.

c. The angle of torsion is normally retroverted 30°.

d. The articular surface is about three-quarters of a sphere.

7. Which of the following structures does not attach at the coracoid process?

    a. coracobrachialis

    b. pectoralis minor

    c. long head of the biceps

    d. short head of the biceps

8. The muscles collectively known as the rotator cuff exclude which muscle?

    a. supraspinatus

    b. infraspinatus

    c. subscapularis

    d. teres major

9. During normal shoulder motion, the clavicle rotates posteriorly how many degrees?

    a. 20 to 30

    b. 30 to 50

    c. 50 to 80

    d. 90 to 110

10. Which is a true statement regarding the clavicle?

    a. The ROM for elevation and depression is equal.

    b. The ROM for protraction and retraction is equal.

    c. Arthrokinematic motion is described by movement of the proximal end.

    d. The proximal end is a double condyloid synovial joint.

Elbow and Forearm

The elbow and forearm are important articulations that allow for functional activities such as eating, grooming, and dressing. At first glance, the elbow may seem to be a single simple joint. In reality, the elbow consists of three separate synovial joints: the humeroulnar, humeroradial, and superior (proximal) radioulnar joints. Primary motions that occur here are elbow flexion and extension and forearm **supination** and **pronation**.

Osteology

This section provides detailed information on the bones that make up the elbow joint complex. These bones are the humerus, ulna, and radius. Several clinical correlations are included to provide clinical relevance.

Mid-to-Distal Humerus

The humerus is the largest bone of the upper extremity. It widens distally into the medial and lateral supracondylar ridges (figure 11.1). One of the most common fractures in children is a supracondylar fracture. The mechanism

OBJECTIVES

After reading this chapter, you should be able to do the following:

> List the bones associated with the elbow complex.

> Describe the three joints that make up the elbow complex.

> Explain the ligamentous structures that support the elbow complex.

> Describe the major functions of the muscles that act on the elbow and forearm.

> Discuss a variety of injuries associated with the elbow complex.

of injury for this fracture is a fall on an outstretched hand. This mechanism commonly results in hyperextension of the elbow, causing the distal segments to be displaced posteriorly.

Continuing down from the supracondylar ridges are the medial and lateral **epicondyles**. The ovoid medial epicondyle is larger and more prominent than the lateral epicondyle. The epicondyles serve as attachment sites for the collateral ligaments. Pain along the medial epicondyle can be a sign of medial epicondylitis, or golfer's elbow. Medial epicondylitis is a **tendinosis** of the common wrist flexors at the medial epicondyle origin. Overuse of these muscles can occur in sports such as tennis, golf, throwing, and weightlifting.

The **trochlea** is just lateral to the medial epicondyle and occupies the anterior, lower, and posterior parts of the humerus. The hyperboloid trochlea is made up of the medial and lateral lips separated by the trochlear groove. The medial lip is prominent and extends farther distally than the lateral lip. The trochlea looks like a rounded empty spool, and the articulating surface of the trochlea is covered by hyaline cartilage in an arc of 330°. Superior to the trochlea on the anterior side is the **coronoid** fossa, which approximates with the coronoid process of the ulna during flexion of the forearm. Lateral to the trochlea on the anterior side is the smooth, convex capitulum, which articulates with the concave surface on the head of the radius. The capitulum is covered by hyaline cartilage, forming an arc close to 180°.

Superior to the capitulum on the anterior side is the radial fossa, which receives the anterior border of the head of the radius during flexion of the forearm. Superior to the trochlea on the posterior side is the **olecranon** fossa, which articulates with the olecranon process of the ulna during extension of the forearm. A thin sheet of bone or membrane separates the olecranon fossa from the coronoid. On the medial side is the ulnar groove for the ulnar nerve.

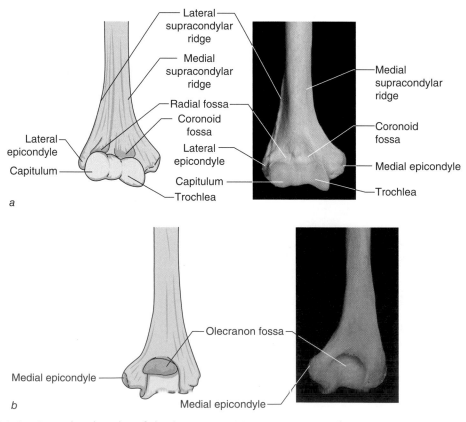

▶ **FIGURE 11.1** Bony landmarks of the humerus. (a) Anterior view, (b) posterior view.

Proximal Ulna

The ulna is the medial bone of the forearm (figure 11.2). The shaft of the ulna is relatively smooth with the exception of the lower two-thirds of the lateral aspect, which has a sharp ridge termed the **interosseous** border. The large, blunt proximal posterior end of the ulna is the olecranon process. The concave trochlear fossa lies between the olecranon process and the coronoid. The coronoid process projects from the anterior body of the proximal ulna. On the lateral side is the radial notch, a narrow articular concavity that receives the head of the radius. The trochlear notch (semilunar notch) is located between the anterior tips of the olecranon and coronoid processes. The trochlear notch is divided down its midline by the longitudinal crest. Inferior to the coronoid process is the ulnar tuberosity. The supinator crest is the distal attachment for part of the lateral collateral ligament and supinator muscle. Fracture to the proximal ulna with radial involvement is covered in Clinical Correlation 11.1.

Proximal Radius

The radius is the lateral bone of the forearm (see figure 11.2). Similar to the ulnar shaft, the radial shaft is smooth with the exception of the medial aspect, where there is a sharp ridge called the interosseous border of the radius. The proximal end of the radial shaft is the location of the radial head. The majority of this cylindrical structure is covered by hyaline cartilage, with the exception of the anterolateral portion. This is a common area for radial fractures (see Clinical Correlation 11.2). The superior surface of the radial head consists of the fovea, a shallow concavity, which articulates with the capitulum of the humerus. The head is supported on a smooth and narrowed portion called the neck. Inferior to the neck at the anteromedial edge of the proximal radius is the radial tuberosity (bicipital tuberosity).

▶ **KEY POINT**

The shafts of the radius and ulna are relatively smooth with the exception of the inner aspect, where there is a sharp ridge on each shaft for the attachment of the interosseous membrane.

Carrying Angle

In full extension, the elbow exhibits a valgus position. The angle between the axes of the ulna and humerus is measured in the anteroposterior plane. This orientation is also called the carrying angle because the angle tends to

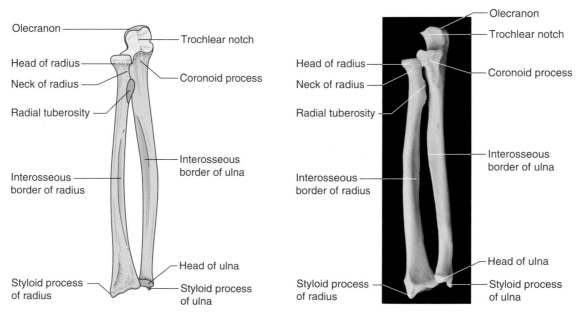

▶ **FIGURE 11.2**   Anterior aspect of the right ulna and radius.

A fracture to the ulna that involves dislocation of the radial epiphysis is termed a Monteggia fracture. There are four classifications of Monteggia fractures (figure 11.3). Type I is the most common and is described as an anterior dislocation of the radial head and fracture of the ulnar diaphysis, with anterior angulation. This type of fracture occurs from forced pronation. A type II fracture involves a posterior or posterolateral dislocation of the radial head and fracture of the ulnar diaphysis, with posterior angulation. The type II fracture occurs secondary to a posterior elbow dislocation. Lateral or anterolateral dislocation of the radial head and fracture of the ulnar metaphysis is a type III Monteggia fracture and is caused by forced forearm abduction. Finally, a type IV fracture is anterior dislocation of the radial head and fracture of the proximal third of the radius and ulna. Rehabilitation after a healed Monteggia fracture involves improving range of motion of the elbow and forearm. Patients may never achieve full pronation or supination after this type of fracture.

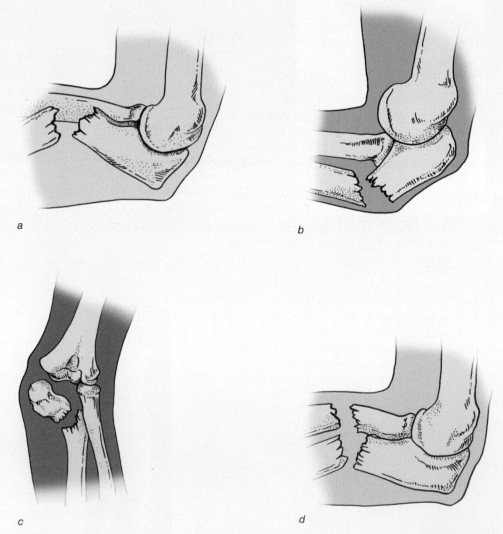

▶ **FIGURE 11.3** Monteggia fractures: *(a)* type I, *(b)* type II, *(c)* type III, and *(d)* type IV.

CLINICAL CORRELATION 11.2

A fracture to the radial head (figure 11.4) may occur when a person falls on an outstretched hand with the forearm supinated and the elbow flexed. Signs and symptoms of a radial head fracture include stiffness and extreme tenderness over the radiocapitellar joint. The person will also complain of pain with supination and pronation. A radial fracture requires immobilization for a few weeks, but patients should return to normal functioning within 4 to 6 weeks.

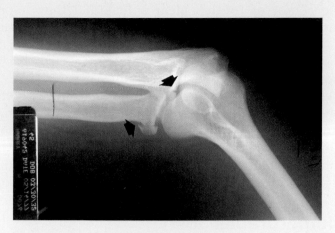

▶ **FIGURE 11.4**   Radial head fracture.

Photo courtesy of University of Virginia, Department of Radiology and Medical Imaging.

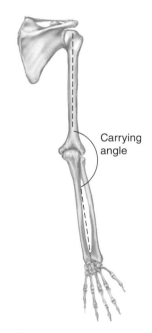

Carrying angle

▶ **FIGURE 11.5**   Carrying angle.

keep carried objects away from the side of the thigh while walking (figure 11.5). This angulation is present because the trochlea extends more distally than the capitulum. A normal valgus angle is 10 to 13°. This angle is smaller in children and greater in women than in men. An excessive angle (greater than 20°) is termed *cubitus valgus*. A diminished angle (less than 10°) is termed *cubitus varus*.

▶ **KEY POINT**

The carrying angle is the angle between the axis of the ulna and humerus. The normal carrying angle is 10 to 13°. This angle is smaller in children and greater in women than in men.

Joint Articulations

The elbow joint complex consists of the humeroulnar, humeroradial, and superior radioulnar joints. This section details the specifics of the bone articulations that make up the joint.

▶ **KEY POINT**

The elbow joint complex consists of three joints: humeroulnar, humeroradial, and superior radioulnar.

Humeroulnar Joint

The humeroulnar articulation occurs between the convex trochlea of the distal humerus and the reciprocally shaped concave trochlear notch of the proximal ulna. This provides the elbow with most of its stability. This is a diarthrodial modified hinge joint because the ulna also has slight axial rotation and side-to-side motion during flexion and extension. The axis of flexion and extension passes through the center of the trochlear sulcus and the capitulum (London 1981).

In pediatric athletes, separation of the medial epicondyle growth plate can occur from repetitive throwing or serving activities (see Clinical Correlation 11.3). The cocking phase of throwing places a forceful valgus stress on the elbow. This occurs in conjunction with the tensile loading produced by the muscular contractions of the wrist flexors and forearm pronator muscles.

Humeroradial Joint

The articulation occurs between the convex capitulum of the humerus and the concave fovea of the radial head. The articulation between the capitulum and radial head of the humeroradial joint provides minimal stability of the elbow. This is a diarthrodial sellar joint. When the arm is at rest in extension, there is little physical contact at the humeroradial joint. During elbow flexion, muscle contraction pulls on the radial head, compressing it against the capitulum.

During the valgus stress at the human elbow during the acceleration phase of both throwing and serving motions, lateral compressive forces occur in the lateral aspect of the elbow, specifically at the radiocapitellar joint. Of great concern in the immature pediatric throwing athlete is osteochondritis dissecans (Ellenbecker & Mattalino 1997; Joyce et al. 1995). Osteochondritis dissecans is a fragmentation and potential displacement of the articular cartilage within the joint due to lack of blood flow. In the older adult elbow, the radiocapitellar joint can be the site of joint degeneration and **osteochondral** injury from the compressive loading (Indelicato et al. 1979). This lateral compressive loading is increased in the elbow with medial ulnar collateral ligament laxity or ligament injury (Ellenbecker & Mattalino 1997).

Superior (Proximal) Radioulnar Joint

The articulation of the superior (proximal) radioulnar joint is formed by the ring of the concave radial notch and the annular ligament with the convex head of the radius. This is a diarthrodial pivot (**trochoid**) joint. In the anatomical position, the radius and ulna are situated so that their long axes are parallel to

CLINICAL CORRELATION 11.3

Little League elbow (figure 11.6) is sometimes considered a variant of medial epicondylitis, but this condition is technically a traction apophysitis of the medial epicondyle. As the name implies, Little League elbow commonly occurs in youth baseball overhand pitchers. The repetitive valgus stress associated with pitches such as curveballs is more stress than the growing bone can tolerate. Limiting pitch count and limiting the age when pitchers can throw breaking balls has helped decrease the occurrence of Little League elbow. Rehabilitation involves managing the pain and inflammation at the elbow and addressing muscle weakness and limited joint range of motion.

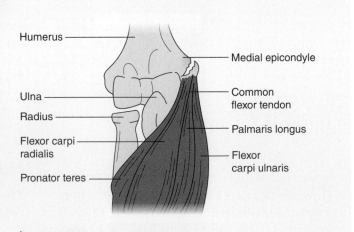

▶ **FIGURE 11.6**　Little League elbow

each other resulting in a supinated forearm position with the palm facing anteriorly. The superior (proximal) and distal radioulnar joints allow the forearm to rotate into pronation. The radius rotates about an axis passing through the head of the radius. Pronation of the radius about the ulna causes the radius to cross over the ulna (figure 11.7).

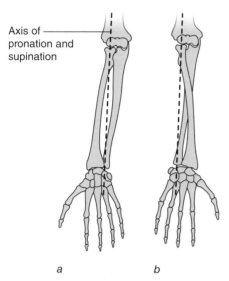

Axis of pronation and supination

a   *b*

▶ **FIGURE 11.7** Radius and ulna orientation during supination and pronation: *(a)* forearm supinated; *(b)* forearm pronated.

Distal Radioulnar Joint

The articulation of the distal radioulnar joint is formed by the convex head of the ulna with the concave ulnar notch of the radius. This joint helps stabilize the distal forearm during pronation and supination. More detail on this joint is found in chapter 12 which covers the wrist and hand.

Joint Anatomy

This section provides specific information on the joint capsule, ligaments, soft tissues, and nerve and blood supply of the elbow joint complex.

Elbow Joint Capsule

The elbow joint capsule encloses the humeroulnar, humeroradial, and proximal radioulnar joints. The capsule is thin and strengthened anteriorly by oblique bands of fibrous tissue and collateral ligaments. A synovial membrane lines the internal surface of the capsule. The joint capsule attaches anteriorly to the coronoid and radial fossas and in front of the medial and lateral epicondyles. Inferiorly, it attaches to the coronoid process of the ulna and the annular ligament. Posteriorly, the capsule attaches to the olecranon fossa of the humerus.

Ligaments

The elbow joint complex consists of strong ligaments that support the joint both medially and laterally. The primary ligaments are depicted in figure 11.8. The ligaments of the elbow have mechanoreceptors important for proprioception and detecting safe limits of passive tension in surrounding structures.

Radial Collateral Ligament

The radial collateral ligament (see figure 11.8a) originates on the lateral epicondyle and then splits into two fiber bundles (fanlike). One set of fibers inserts onto the annular ligament. A second fiber bundle, the true radial collateral ligament, inserts distally on the supinator crest of the ulna. The radial collateral ligament stabilizes the elbow against varus forces. The fibers become taut in full flexion. In elbow extension, the joint articulation offers the most resistance to varus force instead of the radial collateral ligament.

Posterolateral rotatory instability of the elbow may occur if there is injury to the radial collateral ligament, lateral portion of the UCL, and joint capsule (O'Driscoll et al. 1990). In this case, the ulna supinates on the humerus and the radial head dislocates in a posterolateral direction.

Ulnar Collateral Ligament

The ulnar collateral ligament (UCL) is made up of three bands (see figure 11.8b). The *anterior band* is the strongest and provides the most resistance to a valgus force at the elbow. The anterior fibers arise from the anterior aspect of the medial epicondyle and insert on the medial

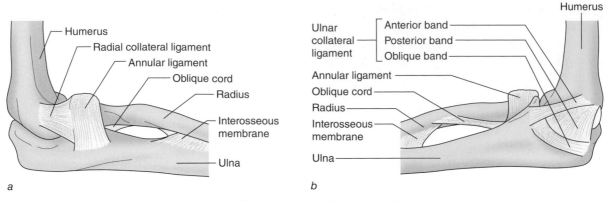

▶ **FIGURE 11.8** Major ligaments of the elbow: *(a)* lateral view and *(b)* medial view.

part of the coronoid process. The majority of anterior fibers become taut at full extension, with a few becoming taut at full flexion. This provides articular stability throughout the entire range of motion. The *posterior band* attaches on the posterior part of the medial epicondyle and inserts on the medial side of the olecranon process. These fibers become taut in extreme elbow flexion. The *oblique (transverse) band* attaches to the anterior and posterior bands. This band provides little stability, but it helps the ulna move on the humerus. In full elbow extension, resistance to valgus stress is shared between the UCL, capsule, and humeroradial joint articulation. In flexion, the primary restraint is the UCL (Morrey & An 1983).

▶ **KEY POINT**

The ulnar collateral ligament (UCL) is made up of three bands: the anterior, posterior, and oblique (transverse) bands. The anterior band is the strongest and provides the most resistance to a valgus force at the elbow.

Attenuation of the ulnar collateral ligament can produce valgus instability of the elbow, which can lead to medial joint pain; ulnar nerve compromise; and lateral radiocapitellar and posterolateral osseous dysfunction, which is a severely restricting injury for the throwing or racket sport athlete. Repetitive valgus elbow loading during the acceleration phase of the throwing or serving motion can attenuate the ulnar collateral ligament. Sprains and

partial-thickness tears of the UCL can occur and progress to complete tears and avulsions (Conway et al. 1992). Figure 11.9 pictures the gapping of the medial joint due to a tear of the ulnar collateral ligament.

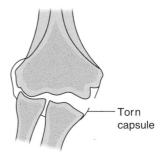

▶ **FIGURE 11.9** Gapping at the medial joint due to a torn UCL.

Annular Ligament

The annular ligament attaches from the anterior aspect of the radial notch of the ulna, wraps around the radial head, and attaches to the posterior aspect of the radial notch (figure 11.10). This ligament runs with the capsule and the lateral collateral ligament. It helps keep the radial head in the radial notch of the ulna, allowing it to rotate. This ligament is considered part of the superior radioulnar joint. Injury to the annular ligament can occur in young children when an adult pulls on a child's forearm, forcing it into full extension. In this instance, the radius can be subluxed from the annular ligament. If a child guards her elbow and has limited elbow extension, the clinician should suspect a pulled elbow,

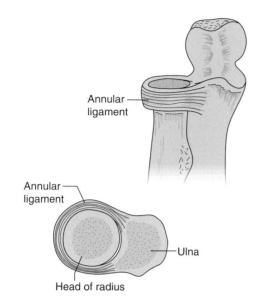

FIGURE 11.10 Annular ligament.

Another term for a pulled elbow is nursemaid's elbow.

Quadrate Ligament

The quadrate ligament attaches from the inferior aspect of the radial notch to the neck of the radius. This ligament reinforces the inferior joint capsule and limits forearm supination. It is considered part of the superior radioulnar joint.

Oblique Cord

The oblique (transverse) cord attaches from the inferior radial notch to the inferior aspect of the bicipital tuberosity of the radius (see figure 11.8b). This structure resists distal disarticulation of the radius during pulling movements. It is considered part of the superior radioulnar joint.

▶ **KEY POINT**

The oblique cord attaches from the inferior radial notch to the inferior aspect of the bicipital tuberosity of the radius.

Soft-Tissue Structures

The soft-tissue structures discussed in this section are the interosseous membrane and bursae.

Interosseous Membrane

The interosseous membrane is a sheet of tissue that connects and transmits forces between the radius and ulna. The location of the interosseous membrane is illustrated in figure 11.11. This membrane also provides longitudinal stability of the forearm.

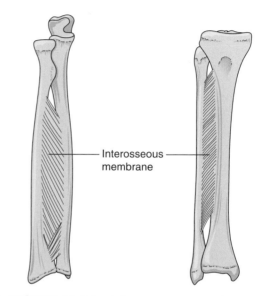

FIGURE 11.11 Interosseous membrane.

Bursae

Several bursae are present in the elbow joint complex. One is the subcutaneous bursa, which lies between the olecranon process and the subcutaneous tissues (see Clinical Correlation 11.4). Second is the intratendinous bursa, a deep bursa situated within the substance of the triceps muscle just superior to the olecranon process. A third is the subtendinous bursa, also between the triceps and olecranon process. The bicipitoradial bursa is a fourth bursa, located at the distal biceps tendon insertion onto the radial tuberosity.

Nerve Supply

Most of the articulating nerves are derived from the musculocutaneous and radial nerves. The ulnar, median, anterior, and posterior interosseous nerves do play a smaller role in the workings of the elbow joint, which is

CLINICAL CORRELATION 11.4

Inflammation of the olecranon bursa (figure 11.12) can result in bursitis. This form of bursitis is obvious because of the substantial swelling that occurs. The usual mechanism of injury is a fall on the olecranon or sustained compression. Treatment of an inflamed bursa may include a corticosteroid injection to help manage the bursal swelling. The area will also need to be protected with a foam donut pad to alleviate pressure on the bursa.

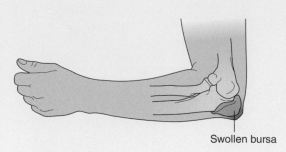

Swollen bursa

▶ **FIGURE 11.12**   Swollen olecranon bursa.

a common location for nerve entrapments. Clinical syndromes include cubital tunnel syndrome, radial tunnel syndrome, and pronator syndrome. Ulnar nerve compression within the cubital tunnel can occur from direct trauma or entrapment through repetitive movement. Signs and symptoms include paresthesia radiating to the dorsal portion of the fourth and fifth digits, a weak pinch or grasp, and in extreme cases claw hand. The radial nerve can become compressed in a tunnel situated along the lateral elbow. Signs and symptoms of radial tunnel syndrome include pain over the lateral epicondyle or radial head and numbness along the radial nerve distribution. Pronator teres syndrome is described in Clinical Correlation 11.5.

▶ **KEY POINT**

The primary nerves that innervate the joint, joint structures, and muscles about the elbow joint are the musculocutaneous, radial, ulnar, median, anterior, and posterior interosseous nerves.

Blood Supply

The elbow derives its blood supply from the articulations arising from the surrounding blood vessels that originate with the collaterals of the brachial artery and the recurrent of both the radial and ulnar arteries. This vascular supply can be compromised in cases of trauma to the elbow, such as a dislocation or fracture. If a person complains of pain, cold

CLINICAL CORRELATION 11.5

Compression of the median nerve can occur between the two heads of the pronator teres, resulting in pronator teres syndrome (figure 11.13). Because the symptoms are similar to those found in carpal tunnel syndrome, pronator teres syndrome is also referred to as pseudo carpal tunnel syndrome. Median nerve entrapment at the elbow is initially treated conservatively with median nerve glides and by addressing the causative factor. Occasionally, corticosteroid injections are used to combat inflammation of the tissue surrounding the entrapment site.

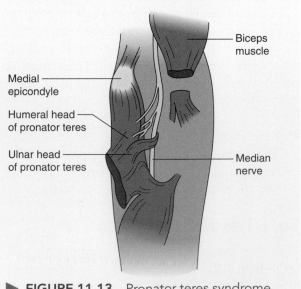

Biceps muscle

Medial epicondyle

Humeral head of pronator teres

Ulnar head of pronator teres

Median nerve

▶ **FIGURE 11.13**   Pronator teres syndrome

sensitivity, or a loss of color of the distal arm (cyanosis), and distal pulses are diminished, then the clinician should suspect a vascular compromise.

Joint Function

This section describes the axes of motion, arthrokinematics, range of motion, closed and loose packed positions, end feel, and capsular patterns of the three joints that make up the elbow joint complex. These articulations are the humeroulnar joint, humeroradial joint, and superior radioulnar joint. These three joints function very closely, and so injury or pathology at one joint will affect the function of the whole elbow complex.

Humeroulnar Joint

The humeroulnar (HU) joint is the main articulation of the elbow joint complex. This joint has the greatest surface area, and the majority of movement occurs here.

Axes of Motion

The humeroulnar joint has two degrees of freedom: (1) flexion and extension in a sagittal plane around a coronal axis and (2) abduction and adduction in a frontal plane around

a sagittal axis. The axis of rotation for flexion and extension passes through the center of the trochlea and capitulum in the lateral view, but it has been shown to change with varying degrees of flexion (Ishizuki 1979).

Arthrokinematics

Surface motion during flexion and extension is primarily gliding (figure 11.14). Rolling takes place near the end range of flexion and extension (London 1981). The articulating surfaces at the humeroulnar joint are the convex trochlea of the humerus and the concave trochlear notch on the ulna. In flexion, the trochlear ridge of the ulna glides anteriorly, superiorly, and laterally on the trochlear groove until the convex coronoid process reaches the concave coronoid fossa. With extension, the concave ulna glides posteriorly, superiorly, and medially on the convex trochlea. The olecranon rolls within the olecranon fossa.

Range of Motion

The normal range of flexion to extension is 0 to 150°, with a functional range of 30 to 130° (AAOS 1965).

Closed and Loose Packed Positions

The closed packed position of the humeroulnar joint is full extension with forearm

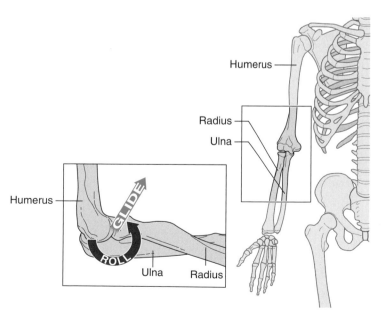

▶ **FIGURE 11.14** Humeroulnar joint flexion. The trochlear ridge of the ulna glides anteriorly, superiorly, and laterally on the trochlear groove.

supination. The loose packed position occurs at close to 70 to 90° of flexion and 10° of supination.

End Feel

The end feel for elbow flexion can be soft, hard, or firm. If it is soft, the restraining tissue is the muscle bulk between the arm and forearm. A firm end feel is due to tension in the posterior joint capsule and the triceps. A hard end feel is not common but can occur when the coronoid process contacts the coronoid fossa in the absence of muscle or in a thin person.

The end feel for elbow extension is commonly classified as hard because of the olecranon process contacting the olecranon fossa. In addition, there may be a firm end feel, and this is due to tension in the anterior joint capsule, collateral ligaments, and biceps muscle.

Capsular Pattern

The capsular restriction of movement for the humeroulnar joint is greater limitation in flexion than extension.

Humeroradial Joint

The humeroradial (HR) joint is the lateral joint of the elbow joint complex. It contributes motion (flexion and extension; pronation and supination) to the other two joints of the elbow complex.

Axes of Motion

The humeroradial joint has three degrees of freedom: (1) flexion and extension in the sagittal plane around a coronal axis, (2) abduction and adduction in a frontal plane around a sagittal axis, and (3) supination and pronation in a transverse plane around a vertical axis.

Arthrokinematics

The articulating surfaces at the humeroradial joint are the convex capitulum of the humerus and the concave fovea on the radial head. Flexion requires the rim of the radial head to glide anteriorly in the capitulotrochlear groove to enter the radial fossa. The concave

radial head glides posteriorly on the convex capitulum during elbow extension.

Range of Motion

The normal range of flexion to extension is 0 to 150°, with a functional range of 30 to 130° (AAOS 1965).

Closed and Loose Packed Positions

The closed packed position at the humeroradial joint is 90° flexion and 5° supination. There is not one loose packed position for this joint, rather, every range beyond the closed packed position is considered loose packed.

End Feel

The end feel for flexion and extension was previously described under the humeroulnar joint.

Capsular Pattern

The capsular restriction of movement for the humeroradial joint is greater limitation in flexion than extension.

Superior Radioulnar Joint

The superior (proximal) radioulnar joint is the primary forearm joint within the elbow joint complex. This joint is key for many functional movements of the wrist and hand such as opening a door and maneuvering a steering wheel.

Axes of Motion

The superior radioulnar joint has one degree of freedom: supination and pronation in the transverse plane around a vertical axis. The axis for pronation and supination is oblique and passes through the center of the capitulum, radial head, and distal ulnar articular surface.

Arthrokinematics

The articulating surfaces at the superior radioulnar joint are the concave radial notch and the convex head of the radius. During pronation and supination, the radial head rotates within the annular ligament, and the

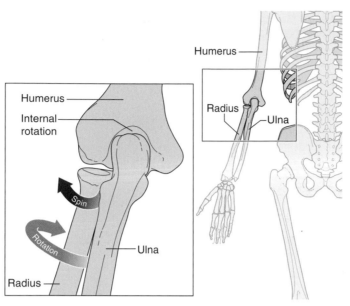

▶ **FIGURE 11.15** Superior radioulnar joint pronation. The convex rim of the head of the radius spins posteriorly in the concave radial notch.

distal radius rotates around the distal ulna. During pronation (figure 11.15), the convex rim of the head of the radius spins posteriorly in the concave radial notch. The ulnar head moves distally and dorsally (internal rotation). There is a relative superior migration of the radius with forearm pronation (Palmer et al. 1982). Supination involves the convex rim of the radial head spinning anteriorly in the concave radial notch. The ulnar head moves proximally and ventrally (external rotation) (O'Driscoll et al. 1990).

Range of Motion

The range of pronation and supination according to the AAOS is 80° in both directions. Most activities occur in a functional range of 50° of pronation and 50° of supination (AAOS 1965).

Closed and Loose Packed Positions

The closed packed position for the superior radioulnar joint is full pronation and full supination, while the loose packed position is 70° flexion and 35° supination.

▶ **KEY POINT**

The closed pack position of the superior radioulnar joint is full pronation and full supination.

End Feel

Pronation at the superior radioulnar joint can be hard because of the ulna contacting the radius. More commonly, pronation has a firm end feel due to tension in the radioulnar ligament, supinator muscle, and interosseous membrane. Supination has a firm end feel due to tension in the palmar radioulnar ligament (distal radioulnar joint), interosseous membrane, pronator quadratus muscle, and quadrate ligament.

Capsular Pattern

The capsular pattern for the superior radioulnar joint is equal limitation of supination and pronation.

Muscles

The muscles that function about the elbow and forearm are listed in table 11.1 along with their origins, insertions, actions, and innervations. Figure 11.16 illustrates these muscles.

Elbow Flexors

The elbow flexors include the biceps brachii, brachialis, brachioradialis, and pronator teres. The biceps brachii, with its two heads, has

TABLE 11.1 Muscles of the Elbow

Muscle	Origin	Insertion	Action	Innervation
Biceps brachii	Short head: apex of coracoid process of scapula Long head: supraglenoid tubercle	Radial tuberosity and aponeurosis of biceps brachii	Elbow flexion; forearm supination	Musculocutaneous nerve (C5-C6)
Brachialis	Anterior surface of lower humerus	Coronoid process of ulna	Elbow flexion	Musculocutaneous nerve (C5-C6)
Brachioradialis	Lateral supracondylar 41 ridge of humerus	Styloid process of radius	Elbow flexion	Radial nerve (C5-C6)
Extensor carpi radialis brevis	Lateral epicondyle of humerus	Base of 3rd metacarpal	Wrist extension	Radial nerve (C7-C8)
Extensor carpi radialis longus	Lateral supracondylar ridge of humerus	Base of 2nd metacarpal	Wrist extension	Radial nerve (C6-C7)
Extensor carpi ulnaris	Lateral epicondyle	Base of 5th metacarpal	Wrist extension	Radial nerve (C7-C8)
Flexor carpi ulnaris	Medial epicondyle	Pisiform, hamate, base of 5th metacarpal	Elbow flexion; wrist flexion	Ulnar nerve (C7-C8)
Flexor carpi radialis	Medial epicondyle	Base of 2nd and 3rd metacarpals	Forearm pronation; wrist flexion	Median nerve (C6-C7)
Pronator quadratus	Distal anterior surface of ulna	Distal anterior surface of radius	Forearm pronation	Median nerve (C5-C6)
Pronator teres	Medial epicondyle and coronoid process of ulna	Midway down lateral surface of radius	Elbow flexion; forearm pronation	Median nerve (C6-C7)
Supinator	Lateral epicondyle	Upper, lateral side of radius	Forearm supination	Radial nerve (C5-C6)
Triceps brachii	Long head: infraglenoid tubercle of scapula Lateral head: lateral and posterior surfaces of proximal 1/2 of body of humerus; lateral intermuscular septum Medial head: distal 2/3 of medial and posterior surfaces of humerus below the radial groove; medial intermuscular septum	Olecranon process and antebrachial fascia	Elbow extension	Radial nerve (C6-C8)
Anconeus	Lateral epicondyle of humerus	Olecranon process of ulna	Elbow extension	Radial nerve (C7-C8)

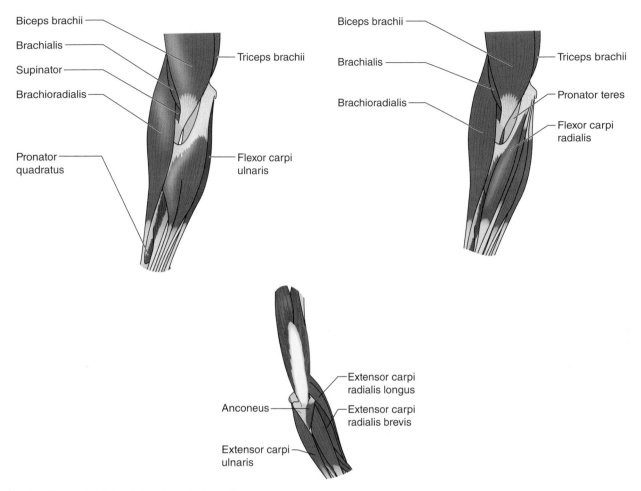

FIGURE 11.16 Muscles of the elbow.

two origins: the long head on the supraglenoid fossa of the scapula and the short head on the coracoid process. The biceps inserts onto the radius via the bicipital tuberosity. The biceps produces maximum force when performing flexion and supination simultaneously. It is able to produce its greatest torque between 80 and 100° of elbow flexion because of the length–tension relationship and lever system. The biceps is minimally active in elbow flexion when the forearm is pronated (Morrey 1993).

The brachialis is a single-joint muscle that originates on the distal aspect of the anterior surface of the humerus and inserts on the coronoid process and tuberosity of the proximal ulna. The brachialis is considered the workhorse for elbow flexion. It is able to generate the largest force of any muscle crossing the elbow. It produces its greatest torque at 100° of elbow flexion.

▶ **KEY POINT**
The brachialis is a single-joint muscle and is considered the workhorse of the elbow for the movement of flexion. It is able to generate the largest force of any muscle crossing the elbow.

The brachioradialis is the longest elbow muscle, spanning from the lateral two-thirds of the distal humerus to the distal aspect of the radius by the styloid. It is the primary elbow flexor during rapid movements against high resistance or when a weight is lifted during a slow flexion movement. The pronator teres is a weak elbow flexor.

Elbow Extensors

The elbow extensors include the triceps brachii and the anconeus. These muscles also provide dynamic stability of the elbow joint. The triceps has three heads; the long head originates from the infraglenoid tubercle and the medial and lateral heads from the posterior humerus. The three heads join together to form a single tendon that inserts onto the olecranon process. The triceps produces the majority of the total extensor torque at the elbow (An et al. 1981). The medial head of the triceps is considered the workhorse of the extensors and remains active for most elbow extension movements. The lateral head is recruited second, followed by the long head. The anconeus is a short single-joint muscle that spans from the posterolateral distal humerus to the posterolateral proximal ulna. The anconeus is the first muscle to initiate and maintain low levels of elbow extension force. It also acts as a dynamic joint stabilizer during pronation and supination. Maximal torque of the elbow extensors occurs when the elbow is flexed to 90°. The internal moment arm is greatest near full elbow extension.

Forearm Supinators

The supinators of the forearm are the supinator and biceps brachii. The supinator originates on the lateral epicondyle of the humerus and the proximal lateral aspect of the ulna, and it inserts on the anterior aspect of the proximal radius. The supinator is recruited for low-power tasks that require a supination motion only. The biceps is an effective supinator when the elbow if flexed to 90°.

Forearm Pronators

The pronators of the forearm are the pronator quadratus and pronator teres. The pronator teres originates on the medial epicondyle of the humerus and inserts onto the lateral aspect of the radius, near the midshaft. The pronator quadratus is much more distal, originating on the volar aspect of the ulna and inserting onto the lateral aspect of the radius. The pronator quadratus is involved in all pronation movements and produces a compression force to stabilize the distal radioulnar joint. The pronator teres helps stabilize the superior radioulnar joint. The wrist extensors and flexors originate above the elbow, but their primary function is at the wrist; more description of these muscles is found in chapter 12. A common injury that affects the wrist extensors is lateral epicondylitis (see Clinical Correlation 11.6).

CLINICAL CORRELATION 11.6

Tennis elbow is a term used by the layperson to describe lateral epicondylitis (figure 11.17). Lateral epicondylitis is a tendinosis of the wrist extensor tendons, primarily the extensor carpi radialis brevis. It is prevalent in tennis players (40 to 50% experience symptoms) and other athletes or laborers who perform repeated pinching or grasping. The injury is most common in adults between the ages of 30 and 60 years. In tennis the backhand stroke seems to be the primary stroke associated with lateral epicondylitis. Many treatments have been advocated for the client with tennis elbow. Eccentric exercise, use of a counterforce brace, and radial head mobilization are the treatments that seem the most effective for this condition (Loudon 2000).

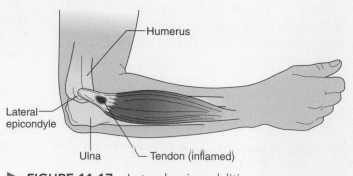

▶ **FIGURE 11.17** Lateral epicondylitis.

Conclusion

At first glance, the elbow joint complex appears to be a simple hinge joint. Further investigation reveals this structure is much more complicated and that the elbow is instrumental in the function of the upper extremity.

This chapter examines the bones, soft tissues, and muscles associated with the humeroulnar, humeroradial, and superior radioulnar joints. Clinical correlations discussing common injuries to the elbow complex are also presented.

REVIEW QUESTIONS

1. What motions occur at the elbow joint complex?

2. What is the functional range of motion for elbow flexion–extension?

3. Describe the arthrokinematics at the humeroulnar joint with elbow flexion.

4. What is the carrying angle?

5. What muscle is the primary flexor of the elbow?

6. What is the function of the anconeus muscle?

Wrist and Hand

Neena Sharma

The wrist and hand are a complex set of joints that are capable of numerous grips and grasps, allowing the hand to perform countless functions. Stability of the wrist is essential for optimal function of the finger flexors and extensors and mobility of the hand. The hand is the distal link to the upper extremity kinetic chain and is dependent on strength and mobility of the shoulder, elbow, forearm, and wrist.

Osteology

The wrist and hand joints are made up of 29 bones. The bones are classified as carpals, metacarpals, and phalanges (figure 12.1) according to their structure and functions. These bony elements form multiple articulations of the wrist and hand skeleton.

Distal Radius

The radius is the lateral bone of the forearm that connects the hand to the humerus. The radius widens distally and forms a large concave surface to articulate with the distal ulna

OBJECTIVES

After reading this chapter, you should be able to do the following:

> List the bones associated with the wrist and hand joint complex.

> Describe the multiple joints that make up the wrist and hand joint complex.

> Explain the ligamentous structure that supports the wrist and hand complex.

> Differentiate the function of the extrinsic and intrinsic wrist and hand muscles.

> Discuss a variety of injuries associated with the wrist and hand complex.

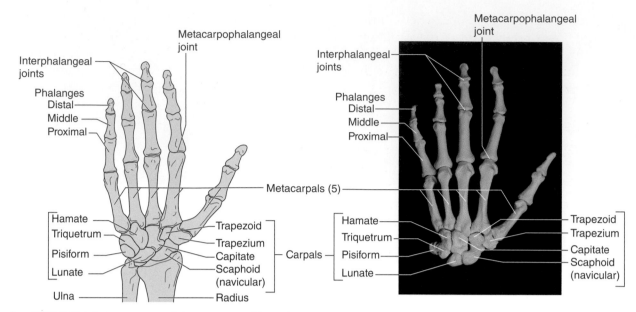

FIGURE 12.1 Bones of the wrist and hand.

and the scaphoid and lunate carpals. The distal surface extends farther on the radial side and slightly on the posterior side of the bone, giving the articular surface an ulnar inclination (see figure 12.1) and a slight volar angle. On the dorsal aspect, the radius contains a protuberance, **Lister's tubercle** (tubercle of the radius), which is located about one-third of the way across from the radial styloid process. Lister's tubercle is in alignment with the third metacarpal. The tubercle forms a pulley around which the extensor pollicis longus (EPL) tendon passes.

The medial aspect of the distal radius contains an ulnar notch, which articulates with the ulnar head during forearm motions. In addition, the medial edge of the ulnar notch serves as an attachment point for the triangular fibrocartilage complex (described in the section on joint anatomy). The lateral side of the radius contains a bony projection called the radial styloid process, which is a prominent anatomical landmark. The distal surface of the radius forms concave facets to articulate with the carpals. A **Colles fracture** is a fracture of the distal radius, with or without ulnar displacement. The radius is displaced in a dorsal direction. The mechanism of injury is commonly a fall on an outstretched arm. A

dinner-fork deformity may result from this fracture.

▶ **KEY POINT**
Lister's tubercle is in alignment with the third metacarpal. The tubercle forms a pulley around which the extensor pollicis longus (EPL) tendon passes.

Distal Ulna

The ulna is the medial bone of the forearm that connects the hand to the humerus. The ulna becomes more cylindrical and smaller in size distally. The distal end is considered the head of the ulna. The ulnar styloid process projects from the medial aspect of the distal ulna, whereas the ulnar head projects from the lateral aspect of the distal ulna. The ulnar styloid process is more prominent than its radial counterpart, but it is half an inch (1 cm) shorter than the radial styloid process. There is no direct contact of the ulna to the carpals; this connection occurs through the articular disc. The lateral aspect of the base of the ulnar styloid process serves as the medial attachment for the apex of the articular disc. These styloid processes (the radial and ulnar styloid processes) reinforce the concave surface of the wrist joint for proximal carpal articulation and form the basic reference points of the wrist. The articulation between the distal ulna and radius and the articular disc forms the distal radioulnar joint.

Carpals

The wrist is made up of 8 carpal bones arranged in two rows, proximal and distal. The proximal row contains the scaphoid, lunate, triquetrum, and pisiform carpals. The distal row contains the trapezium, trapezoid, capitate, and hamate carpals. Collectively, the carpals of the proximal row are shaped convex proximally and concave distally. See figure 12.1 for an illustration of these structures.

▶ **KEY POINT**

The proximal row contains the scaphoid, lunate, triquetrum, and pisiform carpals. The distal row contains the trapezium, trapezoid, capitate, and hamate carpals.

Scaphoid

The scaphoid (navicular) is the largest of the proximal carpal bones. The shape of the scaphoid is considered to resemble a boat or banana. The bone is divided into three anatomical divisions. The proximal and distal poles are separated by the midline of the waist. The scaphoid articulates with four carpal bones (capitate, lunate, trapezium, and trapezoid)

and the radius. The proximal convex surface articulates with the radius. The medial articulation is with the lunate; the lateral articulation is with the trapezium and trapezoid; and the distal articulation is with the capitate. The scaphoid tubercle on the palmar surface serves as the attachment of the flexor retinaculum and the abductor pollicis brevis. It represents the floor of the **anatomical snuffbox** (figure 12.2). The scaphoid is the most commonly fractured carpal bone (see Clinical Correlation 12.1). The incidence of fracture is most common at the waist, at approximately 60 to 70%, followed by the proximal and distal poles at approximately 10 to 20% (Berger 2001).

Lunate

The word *lunate* comes from the Latin word *luna*, which means moon. The lunate is the central bone of the proximal carpal row; it articulates laterally with the scaphoid and medially with the triquetrum. The lunate's proximal convex surface articulates with the radius and triangular disc, and the distal concave surface articulates with the capitate and the edge of the hamate. The lunate is the most frequently dislocated and the second most fractured bone of the wrist (Hoppenfeld 1976). A fall on an outstretched hand is the most common mechanism for dislocation. The direction of the dislocation is usually volar. Avascular necrosis of the lunate may occur after trauma because of the limited blood

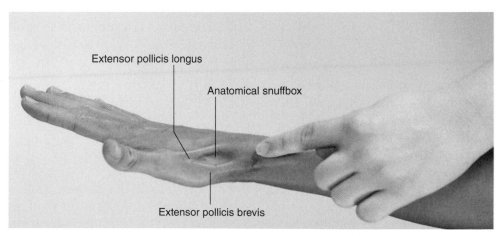

Extensor pollicis longus

Anatomical snuffbox

Extensor pollicis brevis

▶ **FIGURE 12.2** Anatomical snuffbox.

CLINICAL CORRELATION 12.1

The most commonly fractured carpal is the scaphoid. The mechanism of injury often associated with scaphoid fracture is falling on an outstretch hand in order to brace a fall. With the wrist in full extension and radial deviation, the scaphoid absorbs an unusually large load, often leading to fracture. Patients are typically young males or females who are active in sports. Patients typically present with painful wrist movements, snuffbox tenderness, and loss of mobility and strength. Early detection, while essential, can often be difficult as early radiographs may not identify fractures. Fracture of the scaphoid presents additional concerns not applicable at most other fracture sites because of its poor blood supply. Nonunion fractures and osteonecrosis of proximal fractures are common. Nondisplaced scaphoid fractures can be treated conservatively with casting or thumb spica splinting. Fractures that have minimal displacement can be approximated through closed reduction and transcutaneous pinning. For more severe displacement, open reduction with internal fixation using K-wires, plates, or screws is indicated. Surgical fixation of the scaphoid has become the trend in treating scaphoid fractures and has been associated with better and faster outcomes.

supply to the carpal. Avascular necrosis of the lunate is termed **Kienböck's disease**.

▶ **KEY POINT**
The lunate is the most frequently dislocated and the second most fractured bone of the wrist.

Triquetrum

The triquetrum is a pyramid-shaped bone that lies distal to the ulnar styloid process. Distally it articulates with the pisiform and hamate; proximally it articulates with the triangular fibrocartilage complex; and laterally it articulates with the lunate. The medial and dorsal surfaces of the triquetrum are the attachment sites for the ulnar collateral ligament.

Pisiform

The pisiform is a small sesamoid bone that is shaped like a pea. It is formed within the flexor carpi ulnaris tendon (FCU) (Hoppenfeld 1976) to increase the moment arm of the FCU. The pisiform articulates loosely with the palmar surface of the triquetrum. It is an attachment site for the flexor retinaculum, the abductor digiti minimi, and several ligaments of the carpal bones (ulnar collateral ligament, pisohamate ligament, and pisometacarpal ligament). It forms the medial border of Guyon's canal, through which the ulnar nerve and artery pass (see Clinical Correlation 12.2).

Trapezium

The trapezium has a unique shape and articulation related to the thumb. Its proximal surface is slightly concave for articulation with the scaphoid. Its saddle-shaped distal surface articulates with the base of the first metacarpal. Medially, its concave surface articulates with the trapezoid, and its distal convex surface articulates with the base of the 2nd metacarpal. The lateral aspect of the trapezium is an attachment point for the radial collateral ligament and the capsular ligament of the first carpometacarpal (CMC) joint. Other margins provide attachment sites for the flexor retinaculum.

Trapezoid

The trapezoid is a small and irregularly shaped bone (Dutton 2008) that is wedged between the capitate and the trapezium. The trapezoid articulates proximally with the scaphoid, laterally with the trapezium, medially with the capitate, and distally with the second metacarpal.

Capitate

The capitate is the largest and most prominent carpal bone. It is an oval shaped bone with a convex articulating surface. With its central location, it articulates with multiple neighboring carpals: proximally with the triquetrum, lunate, and scaphoid; medially with

CLINICAL CORRELATION 12.2

Compression of the ulnar nerve can occur in **Guyon's canal**. This type of compression neuropathy is common due to prolonged gripping of a bicycle handlebar or crutch walking (Ginanneschi et al. 2009). Another possible cause is a ganglion cyst formation in this region or fracture of the hamate. The clinician should assess motor and sensory functions of the nerve to determine the extent of injury. Signs and symptoms include numbness and tingling into the ulnar distribution of the hand. The symptoms usually start with a feeling of pins and needles in the fourth and fifth digits and progress to burning pain or numbness as well as weakness of intrinsic muscles of the hands and atrophy of the hypothenar muscles. The patient should be advised to avoid repetitive hand motions, heavy gripping, and pressure over the carpals against hard surfaces. Maintaining the wrist in the resting position with a brace may decrease compression of the ulnar nerve and prevent curling of the hand during the night. In later stages, a surgical incision of a ligament is performed to eliminate intrinsic pressure on the ulnar nerve.

the hamate; laterally with the trapezoid; and distally with the third metacarpal. In a neutral wrist position, the capitate presents with a concave dorsal surface, which is palpable as a depression (Hoppenfeld 1976) and used as a reference point.

Hamate

The hamate is shaped like a pyramid. It articulates proximally with the lunate and triquetrum, medially with the capitate, and distally with the bases of the fourth and fifth metacarpal bones. The hamate has a bony projection on the palmar surface, called the hook of the hamate, that forms the lateral border of Guyon's canal. The hamate also contributes to the wall of the carpal tunnel (Dutton 2008). The hook of the hamate and pisiform provide medial attachment sites for the flexor retinaculum.

Metacarpals

The metacarpals are the small bones of the palm that resemble long bones of the body, with a shaft, base, and head. The hand is made up of five metacarpals (see figure 12.1). The middle elongated portion (the shaft) provides attachment sites for intrinsic muscles. The base is the proximal end of the bone that articulates with the distal row of the carpal bones. The head is the distal end of each metacarpal, the knuckle that articulates with its corresponding phalanx. The head has a

biconvex surface that is broader anteriorly and narrower posteriorly. The posterior tubercles are the attachment sites for the collateral ligaments at the metacarpophalangeal (MCP) joints. The first metacarpal is shorter and broader compared with the other metacarpals.

▶ **KEY POINT**

In regard to the metacarpals, the base is the proximal end of the bone that articulates with the distal row of carpal bones. The head is the distal end (knuckle) that articulates with its corresponding phalanx.

Phalanges

The **phalanges** are the bones of the fingers and thumb (see figure 12.1). There are 14 phalanges in the hand; each consists of the base, shaft, and head, similar to the metacarpals. The proximal aspect (base) of each digit has a concave base with two separate depressions. The distal end (head) of each digit has a convex surface with two separate condyles. The base and head both have a pulley-shaped configuration. The tuberosity of each phalanx at the distal end anchors the fleshy pulp of soft tissue to the terminus of each digit.

Joint Articulations

The joint articulations of the wrist consist of the distal radioulnar, radiocarpal, midcarpal, and carpometacarpal joints. The hand consists of the metacarpophalangeal, proximal interphalangeal, and distal interphalangeal joints.

Distal Radioulnar Joint

The articulation occurs between the convex ulnar head and the concave ulnar notch of the radius (Loudon et al. 2008). This joint is also united by an articulating disc, a part of the triangular fibrocartilage complex (TFCC). The disc binds the distal radius to the ulna and is the primary stabilizer of the distal radioulnar joint. This is a diarthrodial double pivot joint, allowing the forearm rotation movements of supination and pronation. The function of the distal radioulnar joint is to transmit the load from the hand to the forearm. Fractures to this area are somewhat common. The most common is the Colles fracture, which was previously described. Another type of fracture is a Smith fracture, which is similar to a Colles fracture except the radius is displaced in a volar direction.

Radiocarpal Joint

This joint is formed by the articulation of the concave surface formed by the radial facets and the TFCC disc with the convex surface of the scaphoid and lunate carpals and the triquetrum. The radius articulates with the scaphoid and lunate. The TFCC disc articulates with the lunate and triquetrum. The ulna does not participate in the radiocarpal joint articulation apart from being an attachment site for the TFCC. This is a diarthrodial ellipsoid joint because its joint surfaces are ovoid and vary in length and curvature. The radiocarpal joints allow for wrist flexion and extension as well as radial and ulnar deviations.

▶ **KEY POINT**

The radiocarpal joint is the articulation of the radius and TFCC disc with the scaphoid, lunate, and triquetrum. The radius articulates with the scaphoid and lunate. The TFCC disc articulates with the lunate and triquetrum. The ulna does not participate in the radiocarpal joint articulation.

Midcarpal Joint

A complex articulation occurs between the proximal and distal rows of the carpal bones during wrist movements. It is considered complex because each row has both a concave and a convex segment. The midcarpal joint is the S-shaped joint space separating the proximal and distal rows of carpal bones (figure 12.3). The articulation between the scaphoid, lunate, and triquetrum with the capitate and hamate forms a saddle-shaped joint on the medial side, and the articulation of the scaphoid with the trapezoid and trapezium forms another saddle-shaped joint on the lateral side (Magee 1997). Classified as a synovial condyloid joint, the midcarpal joint aids in wrist mobility (flexion, extension, radial deviation, and ulnar deviation). Some references classify this joint as a plane synovial joint.

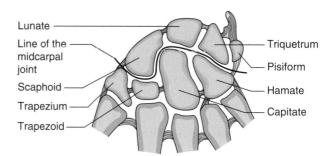

▶ **FIGURE 12.3**   Midcarpal joint.

First Carpometacarpal Joint (Thumb Joint)

The articulation occurs between the base of the first metacarpal and the distal surface of the trapezium. This is a diarthrodial saddle-shape joint with reciprocal concavo-convex joint surfaces. The trapezium is concave in an anteroposterior direction and convex in a mediolateral direction. The base of the first metacarpal is concave in a mediolateral direction and convex in an anteroposterior direction. The most common location for osteoarthritis (OA) of the hand is the CMC joint of the thumb. Females are more prone to the development of OA in their fourth and fifth decades. OA in this joint is quite debilitating because of its frequent use during activities of daily living and fine motor tasks.

Second Through Fifth Carpometacarpal Joints

The articulation occurs between the distal carpals and the base of the metacarpals: the second metacarpal with the trapezium, trapezoid, and capitate; the third metacarpal with the capitate; the fourth metacarpal with the capitate and hamate; and the fifth metacarpal with the hamate. The CMC joint is a diarthrodial condyloid joint. Some references classify it as a plane synovial joint (Magee 1997) that allows for gliding motion on each other. The mobility of the CMC joints is increased from the 2nd to the 5th joint. The stability of these joints is provided by the palmar and dorsal carpometacarpal and intermetacarpal ligaments.

Metacarpophalangeal Joint of the Thumb

The articulation occurs between the head of the first metacarpal and the first phalanx. The head of the metacarpal has a convex surface, and the base of the first phalanx has a concave surface. It is a hinge joint.

Metacarpophalangeal Joint of Digits Two Through Five

The articulation occurs between the head of each metacarpal and its respective proximal phalanx. It is a biaxial synovial joint. The head of the metacarpal has a biconvex shape that is broader anteriorly and narrower posteriorly.

Proximal Interphalangeal (PIP) Joint of Digits Two Through Five

All four fingers have a PIP joint. The thumb has one interphalangeal joint. The PIP articulation occurs between the head of each proximal phalanx and its respective middle phalanx. The head of the proximal phalanx of each digit has two separate convex-shaped condyles (pulley-shaped heads). The base of the middle phalanx of each digit has a concave base with two separate depressions. This is a hinged joint that allows flexion and extension motions.

Distal Interphalangeal (DIP) Joint of Digits Two Through Five

Only the fingers have DIP joints. Similar to the proximal interphalangeal (IP) joints, the head of the middle phalanx of each digit has a convex surface (with two separate condyles). The distal phalanx (base) of each digit has a concave base (with two separate depressions). This is also a hinge joint. Although the structures of the distal IP joints are similar to the proximal IP joints, the joints are less stable and allow for greater hyperextension.

Joint Anatomy

The anatomical structures of the wrist and hand are complex, and the coordinated actions of the wrist and hand are supported by many soft tissues. Some of the most common structures, particularly related to stability of the joints and soft-tissue injuries treated by physical therapists, are described in this section.

Wrist Joint Capsule

The wrist joint capsule encloses the distal radioulnar and the radiocarpal joints (the radius, ulna, triangular disc, and proximal row of carpals). The fibrous capsule is strong and loose and further reinforced by ligaments. It attaches proximally to the articular margins of the radius and ulna (just distal to the inferior epiphyseal line) and the triangular disc and distally to the proximal edge of the carpals of the proximal row on dorsal and volar aspects. Two synovial membranes line the internal surface of the fibrous layer. The synovial membrane of the wrist joint originates at the distal radius on both palmar and dorsal aspects and attaches to the proximal carpals. A separate synovial membrane encloses the distal radioulnar joint. Windischi and colleagues (2001) found no communication between the joint cavities of the radiocarpal and radioulnar joints, as both joints are separated by the TFCC. Forearm movement puts tension on the joint capsule. The anterior capsule (palmar) tightens with supination, and the posterior capsule tightens with pronation.

Joint Ligaments

The function of the ligaments is to limit joint motion and provide joint integrity. However, the ligaments of the wrist and hands are also capable of producing movement and transmitting load. The major ligaments of the wrist included here are presented for proximal to distal joints.

Ligaments of the Distal Radioulnar Joint

The stability of the distal radioulnar joint is reinforced by the palmar and radial ligaments and the TFCC. The palmar radioulnar ligament runs from the anterior margin of the ulnar notch of the radius to the front of the head of the ulna. The dorsal radioulnar ligament runs from the posterior margin of the

ulnar notch of the radius to the posterior head of the ulna.

Ligaments of the Wrist Joint

The medial aspect of the wrist joint is stabilized by the ulnar collateral ligament (UCL). The ulnar collateral ligament originates at the ulnar **styloid process** and TFFC (figure 12.4a) and is made of two bands. One band attaches to the pisiform and to the transverse carpal ligament. The second band attaches to the triquetrum. The UCL stabilizes the TFCC and is taut during radial deviation. The lateral aspect of the wrist joint is stabilized by the radial collateral ligament. The radial collateral ligament originates from the radial styloid process and attaches to the scaphoid. It is taut during ulnar deviation. The ulnar (medial) and radial (lateral) collateral ligaments are the largest ligaments of the wrist.

▶ **KEY POINT**
The ulnar collateral ligament originates at the ulnar styloid process and TFFC and has two bands. The ulnar collateral ligament stabilizes the TFCC and is taut during radial deviation.

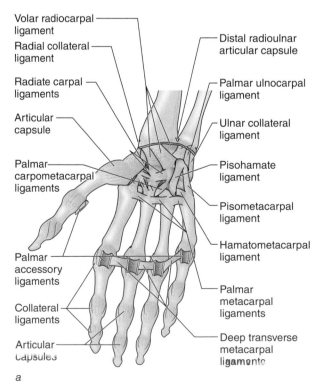

Volar radiocarpal ligament
Radial collateral ligament
Radiate carpal ligaments
Articular capsule
Palmar carpometacarpal ligaments
Palmar accessory ligaments
Collateral ligaments
Articular capsules

Distal radioulnar articular capsule
Palmar ulnocarpal ligament
Ulnar collateral ligament
Pisohamate ligament
Pisometacarpal ligament
Hamatometacarpal ligament
Palmar metacarpal ligaments
Deep transverse metacarpal ligaments

a

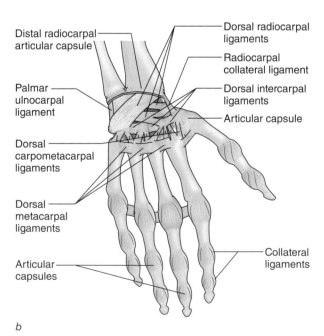

Distal radiocarpal articular capsule
Palmar ulnocarpal ligament
Dorsal carpometacarpal ligaments
Dorsal metacarpal ligaments
Articular capsules

Dorsal radiocarpal ligaments
Radiocarpal collateral ligament
Dorsal intercarpal ligaments
Articular capsule
Collateral ligaments

b

▶ **FIGURE 12.4**   Ligaments of the wrist and hand: (a) volar (palmar) view and (b) dorsal view.

Besides the collateral ligaments, the wrist joint is stabilized by many ligaments and the TFCC. The major ligaments of the wrist joint can be divided into the extrinsic and intrinsic ligaments of the dorsal and ventral aspects of the wrist (table 12.1). The extrinsic ligaments connect the radius and ulna to proximal and distal carpals. The extrinsic palmar ligaments provide the majority of wrist stability. The intrinsic ligaments connect the proximal and distal carpal rows (midcarpal and interosseous ligaments in table 12.1). They provide the rotational stability to carpals. This suggests that the stability of the carpals is primarily dependent on ligaments and joint capsules. These ligaments are also called intercarpal or interosseous ligaments. These ligaments lie internal to the synovial membrane (Levangie & Norkin 2011). The palmar ligaments are highly developed; they originate laterally from the radial styloid process and are directed in a distal ulnar direction.

Ligaments of the MCP Joints

The medial and lateral collateral ligaments run obliquely from the lateral aspect of the metacarpal head to the corresponding base of the phalanx on the side, facing palmarly. These ligaments are taut during abduction and adduction of fingers with the MCP joint flexed. The palmar aspect of the joint is covered by a dense fibrocartilaginous pad, the palmar ligament (**volar plate**), which prevents hyperextension of the joints. The dorsal aspect of the joint is covered by the extensor aponeurosis (extensor expansion). The heads of the metacarpals are connected by deep transverse metacarpal ligaments (figure 12.5).

Ligaments of the Interphalangeal Joints (PIP and DIP)

Similar to the MCP joints, the interphalangeal joints are protected by collateral ligaments with oblique fibers. The palmar aspect of the joint contains a volar plate (palmar ligament) that is fixed proximally to the neck of its proximal phalanx and distally to the base of its distal phalanx (see figure 12.5). The volar plate also attaches to the flexor sheath and collateral ligament and prevents hyperextension of the joint.

Soft-Tissue Structures

The remaining soft tissues that provide attachment surfaces for ligaments and tendons and

TABLE 12.1 Ligaments of the Wrist

		Dorsal aspect	Palmar aspect
Extrinsic		Dorsal radiocarpal	Palmar radiocarpal and palmar ulnocarpal ligaments: Radioscaphocapitate Long and short radiolunate Radioscapholunate Ulnolunate Ulnotriquetral Ulnocapitate
Intrinsic	Midcarpal	Scaphotriquetral Dorsal intercarpal	Scaphotrapeziotrapezoid Scaphocapitate Triquetrocapitate Triquetrohamate
	Interosseous		Trapeziotrapezoid Trapeziocapitate Capitohamate Scapholunate Lunotriquetral

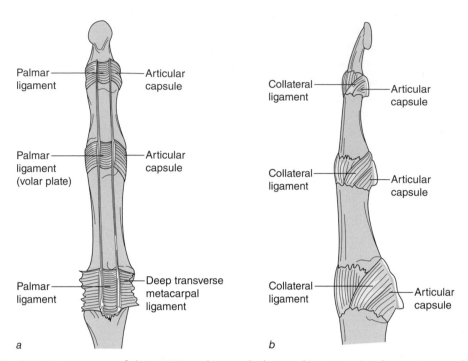

a   b

▶ **FIGURE 12.5**   Ligaments of the MCP and interphalangeal joints: *(a)* palmar view and *(b)* lateral view.

further reinforce the integrity of the joint surfaces are presented here and include the forearm fascia, TFCC, and retinacula.

Antebrachial Fascia

The antebrachial fascia is the deep fascia of the forearm. It is a dense connective tissue sheath that covers the tendons across the wrist. The fascia is firmly attached to the sub-cutaneous border of the ulna and radius and divides the forearm into anterior and posterior compartments. As the fascia approaches the wrist flexor tendons, it thickens and continues with the volar carpal ligaments and transverse carpal ligament.

Triangular Fibrocartilage Complex

The **triangular fibrocartilage complex (TFCC)** consists of the triangular radioulnar disc and several ligaments (the ulnarcarpal ligaments and the radioulnar ligaments). The radioulnar disc is made of fibrocartilaginous tissues, and its margins are surrounded by thick volar and dorsal ligamentous structures. The disc is located at the distal end of the ulna on the medial aspect of the hand (figure 12.6), extending from the medial aspect of the radius (ulnar notch) to the radial side of

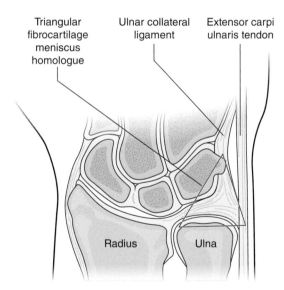

▶ **FIGURE 12.6**   Articular disc as part of the triangular fibrocartilage complex (TFCC).

the ulnar styloid process, connecting medial to the ulna via dense fibrous tissues. It is triangular shaped, with a broader base located medially and a pointed apex located laterally. The disc transmits the axial load from the hand to the forearm, cushions weight-bearing forces, and is an attachment site for several ligaments. Finally, it provides stability to the distal radioulnar joint and allows gliding motion between the ulna and carpals during forearm pronation and supination. As with any other disc complex in the body, the disc at the wrist is at risk for injury from falls and repetitive movements. Differential diagnosis of pain located on the ulnar side of the wrist should include injury to the TFCC.

▶ **KEY POINT**

The triangular fibrocartilage complex (TFCC) consists of the triangular radio-ulnar disc, the radioulnar ligament, the ulnarcarpal ligaments, and a meniscus homologue (ulnocarpal meniscus). The complex transmits the axial load from the hand to the forearm, cushions weight-bearing forces, is an attachment site for ligaments, and provides stability to the distal radioulnar joint.

Extensor Retinaculum

The extensor retinaculum is a ligamentous structure that keeps the extensor tendons from bowstringing at the wrist joint. The retinaculum runs dorsally from the lateral border of the distal radius to the posterior aspect of the distal ulna and ulnar styloid process; some of its fibers also attach to the triquetrum and pisiform. The retinaculum and the underlying bones form a fibro-osseous compartment or tunnels (details in the section on the extensor hood).

Flexor Retinaculum

The flexor retinaculum is called the transverse carpal ligament, which constructs the carpal arch into the carpal tunnel (figure 12.7). It attaches laterally to the pisiform and hamate and medially to the scaphoid and trapezium, forming a roof of the carpal tunnel through which the median nerve and some flexor tendons pass. The transverse ligament protects the median nerve, prevents extrinsic flexor tendons from bowstringing, and is an attachment point for the thenar and hypothenar muscles.

Carpal Tunnel

The carpal bones and ligaments form a tunnel through which nine flexor tendons and the median nerve pass. The floor of the **carpal tunnel** is covered by the palmar radiocarpal ligaments and palmar ligaments; the roof is formed by the flexor retinaculum (transverse carpal) ligament; the ulnar and radial borders are formed by the trapezium and hook of the hamate, respectively. The median nerve is subject to compression within this tunnel (see Clinical Correlation 12.3).

Palmar Aponeurosis

The palmar aponeurosis is a dense fibrous structure that continues with the palmaris longus tendon. It is the fascia that covers the thenar and hypothenar muscles. This fascia is

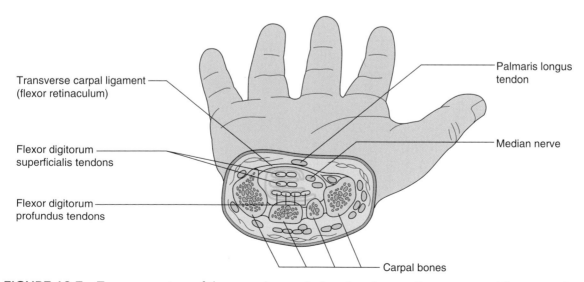

Transverse carpal ligament (flexor retinaculum)

Palmaris longus tendon

Flexor digitorum superficialis tendons

Median nerve

Flexor digitorum profundus tendons

Carpal bones

▶ **FIGURE 12.7**   Transverse view of the carpal tunnel, showing the median nerve and flexor tendons.

CLINICAL CORRELATION 12.3

Increased intracompartmental pressure within the carpal tunnel is associated with ergonomic risk factors such as repetitive loading and awkward wrist positions. Signs and symptoms of carpal tunnel syndrome (CTS) are burning pain and tingling along the dermatome of the median nerve (see figure 12.11); possible numbness in the hand and fingers innervated by the median nerve; nocturnal pain caused by a flexed wrist position during sleep; painful and decreased grip strength; and possible atrophy of the hand muscles innervated by the median nerve. A common clinical test for assessing CTS is the Phalen's test (figure 12.8). The patient places the dorsum of both hands together and points the fingers down, attempting to maintain this position for 60 seconds. The test is considered positive if it elicits the patient's symptoms or increases the patient's symptoms.

▶ **FIGURE 12.8**   Phalen's test.

subject to pathological scarring, which results in Dupuytren's contracture. This condition causes abnormal thickening of the palmar fascia and contractures of the MCP and PIP joints. The contracture primarily affects the fourth and fifth digits in males over the age of 50 years.

Extensor Hood

The extensor hood is a complex connection to the extensor tendons that creates a cable system to provide a mechanism for extension of MCP and IP joints. It also allows the intrinsic muscles (lumbrical and interosseous) to assist in flexion of the MCP joints (figure 12.9). The tendon of the extensor digitorum fans out like a hood to cover the MCP joint. Other extensor tendons (extensor indicis and extensor digiti minimi) also join the extensor digitorum tendon (see Clinical Correlation 12.4). In addition, the tendons of the lumbricales and interossei join the extensor digitorum proximal to the PIP joints, forming a complex extensor system. At this point, the complex extensor digitorum tendon splits into three parts: one central and two lateral bands. The central band inserts proximal to the DIP joint, whereas the lateral bands rejoin at the middle phalanx, form the terminal tendon, and insert to the distal phalanx. Rupture of the terminal tendon is called mallet finger (see Clinical Correlation 12.4).

Flexor Pulleys

On the palmar aspect of the hand and thumb, a number of flexor pulleys (annular and cruciate pulleys) restrain the flexor tendons to the metacarpals and phalanges and form fibroosseous tunnels through which the tendons travel. These pulleys arise from various parts of the digits; there are five annular (A1-A5) and three cruciate (C1-C3) pulleys. The thumb includes A1 (at the MCP joint) and A2 (at the IP joint).

Synovial Sheath

The tendons of the wrist and hand muscles are covered with a synovial sheath. The synovial

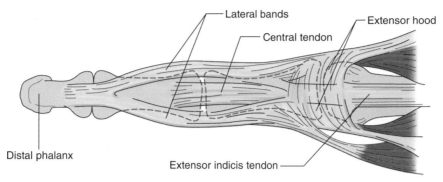

Lateral bands

Central tendon

Extensor hood

Distal phalanx

Extensor indicis tendon

▶ **FIGURE 12.9** Extensor hood and tendons.

CLINICAL CORRELATION 12.4

A direct force to the tip of the finger during a sporting activity or traumatic event can cause rupture of the extensor tendon or avulsion of the distal phalanx at the DIP joint (figure 12.10). The patient may or may not experience pain over the joint. The classic sign of a mallet finger is inability to straighten the DIP joint. The third and fourth fingers are the most commonly injured. A volar splint to hold the DIP joint in extension or surgical correction is recommended.

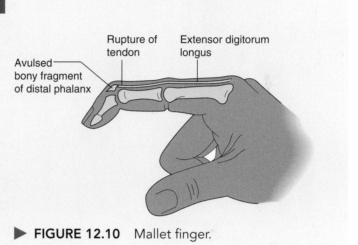

Rupture of tendon

Extensor digitorum longus

Avulsed bony fragment of distal phalanx

▶ **FIGURE 12.10** Mallet finger.

sheaths are filled with synovial fluid and wrap around tendons, including the flexor digitorum superficialis and flexor digitorum profundus. The function of the synovial sheath is to provide smooth motion and reduce friction during wrist motion.

Bursae

The bursae of the wrist communicate with the synovial sheath. There are two primary bursae of the wrist joint: the ulnar and radial. The ulnar bursa covers the tendons of the index, middle, and ring fingers, whereas the radial bursa covers the tendons of the thumb.

Nerve Supply

The peripheral nerves that supply the skin and muscles of the wrist and hand are the median, ulnar, and radial nerves and their branches

(figure 12.11). The TFCC is innervated by the branches of the posterior interosseous, ulnar, and dorsal sensory ulnar nerves (Unglaub et al. 2010).

Blood Supply

The wrist and the hand derive their blood supply from the radial artery, the ulnar artery, and their recurrent branches. The brachial artery bifurcates at the elbow into radial and ulnar branches. The ulnar and radial arteries merge within the palm of the hand, creating vascular arches of the hand (dorsal arches and palmar arches). The dorsum of the hand is supplied by the dorsal arches, and the palm of the hand is supplied by the palmar vascular arches. Their collateral branches supply the dorsal and palmar aspect of the fingers. The peripheral surface of the TFCC disc is well

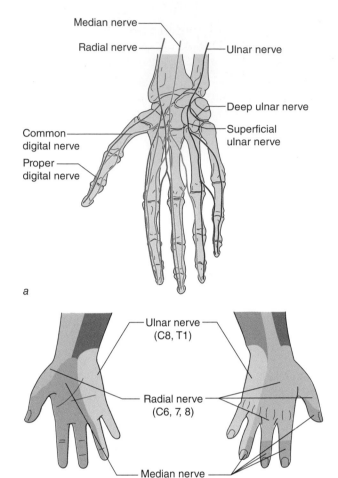

Median nerve

Radial nerve

Ulnar nerve

Deep ulnar nerve

Superficial ulnar nerve

Common digital nerve

Proper digital nerve

a

Ulnar nerve (C8, T1)

Radial nerve (C6, 7, 8)

Median nerve (C6, 7, 8)

b

▶ **FIGURE 12.11** (a) Nerves of the palmar aspect of the wrist and hand. (b) Innervations of the palmar and dorsal aspects of the hands.

vascularized, whereas the center of the disc is avascular.

Functional Arches of the Hand

The three arches of the hand (proximal, distal transverse, and longitudinal) allow the hand to conform to a variety of objects to improve **prehensile** capability. These arches also maximize the contact surface area, providing grip stability and enhancing sensory input. Loss of the arches results in severe functional impairment of the hand. The proximal transverse arch is at the level of the carpometacarpal joint with reference to the capitate. The distal transverse arch is at the level of the metacarpophalangeal joints with reference to the second and third metacarpals. The longitudinal arch is observed in reference to the third digit.

Joint Function

The wrist and hand perform complex movements. The function (the degree and direction of motions) of a joint is determined by its configuration (i.e., how the bones articulate). This section describes the multiple joints of the wrist and hand from proximal to distal, their articulations, and the associated soft tissues that limit their physiological motion. Included are the joints' axes of motion, arthrokinematics, range of motion, closed and loose packed positions, end feel, and capsular patterns.

Forearm (Distal Radioulnar Joint)

The motions of the forearm are supination and pronation, or axial rotation. Although supination and pronation are not wrist joint motions, they are important for positioning the hand in space and for intricate wrist function.

Axis of Motion

The distal radioulnar joint has one degree of freedom. The movements of supination and pronation occur simultaneously, with movement at the proximal radioulnar joint in the transverse plane around a vertical axis. This vertical axis extends from the radial head to the ulnar head (Kapandji 1970).

Arthrokinematics

The distal radioulnar joint consists of the convex ulnar head and the concave ulnar notch on the radius. With supination, the concave ulnar notch of the radius glides dorsally (posteriorly) on the ulnar head. The ulnar head moves proximally and medially in supination. Pronation is opposite, with the concave ulnar notch gliding volarly (anteriorly) on the ulnar head. The ulnar head moves distally and dorsally.

Range of Motion

Supination and pronation have an equal range of motion of 80° according to the American Academy of Orthopaedic Surgeons (AAOS 1965; Norkin & White 2009).

Closed and Loose Packed Positions

The closed packed position of the radioulnar joint is 5° supination. The loose packed position is 10° supination.

End Feel

The end feel for supination is firm because of ligamentous restraints surrounding the joint and the interosseous membrane. For pronation, the end feel is firm because of musculature structures.

Capsular Pattern

The capsular pattern of the forearm is equal limitation of pronation and supination, which mainly occurs with significant limitation of the elbow joint rather than the wrist joint.

Wrist (Radiocarpal and Midcarpal Joints)

The wrist joint proper consists of the distal radius and ulna, the carpal bones, and the bases of the metacarpals. Simultaneous and synchronized motion occurs in these joints for wrist motions.

Axes of Motion

The radiocarpal and midcarpal joints have two degrees of freedom: flexion–extension and radial and ulnar deviation. Flexion and extension occur in the sagittal plane around a mediolateral axis that goes through the head of the capitate. Radial and ulnar deviations occur in the frontal plane around an anteroposterior axis that goes through the head of the capitate.

Arthrokinematics

At the radiocarpal joint, the relatively convex proximal carpal row articulates with the concave radial facets and the radioulnar disc. During flexion, the proximal carpal row (scaphoid and lunate) glides dorsally on the radius, with the triquetrum gliding dorsally on the TFCC. About 35° of flexion motion of the wrist occurs at the radiocarpal joint, with the remaining motion occurring at the midcarpal joint (figure 12.12). Flexion is also accompanied by slight ulnar deviation and supination. Wrist extension involves the proximal carpal row (scaphoid and lunate) gliding volarly (anteriorly) on the concave surface of the radius, whereas the triquetrum glides volarly on the radioulnar disc. Most of the extension motion occurs at the radiocarpal joint (45°), with the remainder occurring at the midcarpal joint. Extension is accompanied by slight radial deviation and pronation of the forearm.

Ulnar deviation of the wrist is characterized by convex on concave movement at both the radiocarpal and midcarpal joints. The proximal carpal row rolls ulnarly and glides radially at the radiocarpal joint. At the midcarpal joint, the capitate and hamate roll ulnarly and glide radially, and the trapezium and trapezoid glide palmarly. Radial deviation of the wrist is the opposite of ulnar deviation. The proximal carpal row rolls radially and glides ulnarly at the radiocarpal joint. At the midcarpal joint, the capitate and hamate roll radially and glide dorsally, and the trapezium and trapezoid glide dorsally.

Range of Motion

Wrist motion equals 85° for wrist flexion and 70° for wrist extension. Radial deviation is 20°, and ulnar deviation is 30° (AAOS 1965; Norkin & White 2009).

Closed and Loose Packed Positions

The closed packed position of the wrist is extension with radial deviation. The loose packed position is 10° of flexion and slight ulnar deviation.

End Feel

The end feel for wrist flexion is firm because of tension in the dorsal radiocarpal ligament, dorsal joint capsule, and extensor muscles. The extension end feel is firm because of

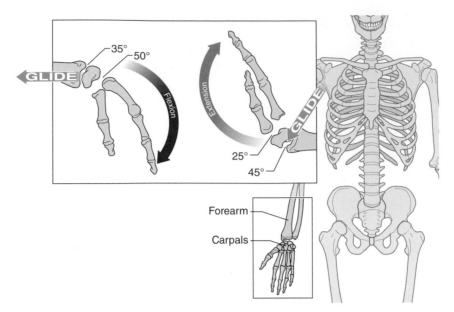

▶ **FIGURE 12.12**   Radiocarpal and midcarpal flexion and extension.

tension in the palmar radiocarpal ligament, ulnarcarpal ligament, palmar joint capsule, and flexor muscles. Radial deviation has a hard end feel due to contact between the radial styloid process and the scaphoid. The end feel for ulnar deviation is firm due to tension in the radial collateral ligament, radial portion of the joint capsule, extensor pollicis brevis, and abductor pollicis longus.

Capsular Pattern

The capsular pattern of the radiocarpal joint is equal limitation of flexion and extension, with slight limitations of radial and ulnar deviation.

Second Through Fifth Carpometacarpal (CMC) Joints

The irregular shapes of the CMC joint surfaces don't allow for standard arthrokinematic description. The second and third digits have very little movement due to the interlocking articular surfaces. Movement at the fourth and fifth digits allows the ulnar border of the hand to fold toward the center of the hand, deepening the palmar concavity. Movement at the fourth and fifth carpometacarpal (CMC) joint is referred to as ulnar mobility. There is 10° of flexion at the fourth MC joint and 20 to 25° of flexion at the fifth CMC joint that allow cupping motions of the hand.

Second Through Fifth Metacarpophalangeal (MCP) Joints

The metacarpal phalanges connect carpal bones to fingers, allowing various hand motions, including prehension activities. Prehension activities of the hand involve the grasping of an object between any two surfaces in the hand, with or without the thumb. Prehension can be categorized as either power grip or **precision handling**. Power grips include hook, spherical, cylindrical, and fist grasps. Precision handling includes pad-to-pad prehension, tip-to-tip prehension, and pad-to-side prehension (figure 12.13).

Axes of Motion

The metacarpophalangeal (MCP) joints have two degrees of freedom: flexion–extension and abduction–adduction. Flexion and extension occur in the sagittal plane around a mediolateral axis through the head of the metacarpal. Abduction and adduction occur in the frontal plane around an anteroposterior axis through the head of the metacarpal.

Arthrokinematics

The MCP articulation consists of the concave base of the phalanx and the convex metacarpal

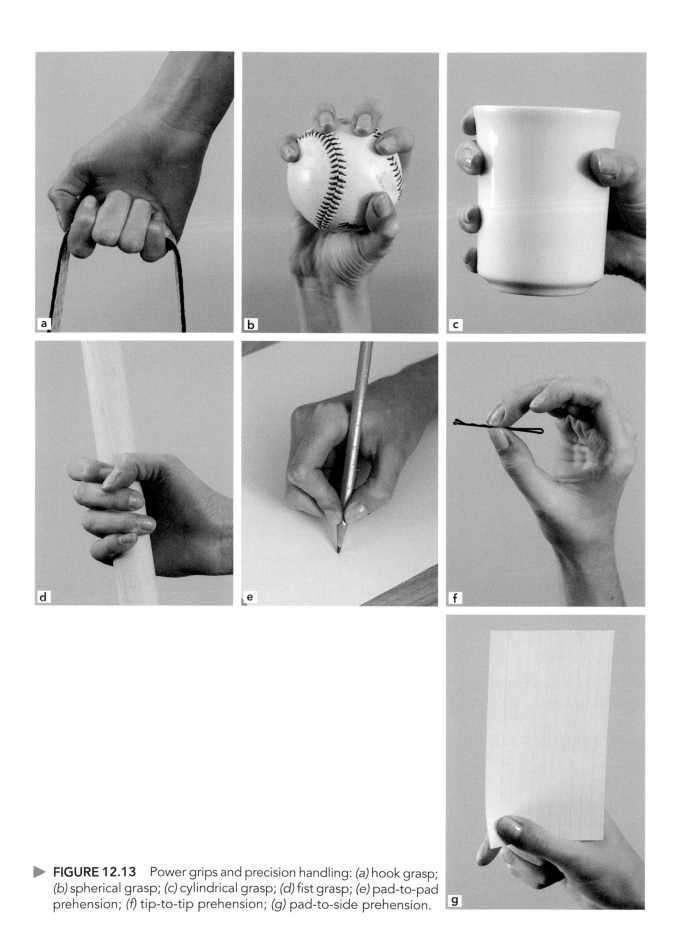

▶ **FIGURE 12.13** Power grips and precision handling: *(a)* hook grasp; *(b)* spherical grasp; *(c)* cylindrical grasp; *(d)* fist grasp; *(e)* pad-to-pad prehension; *(f)* tip-to-tip prehension; *(g)* pad-to-side prehension.

head. During flexion, the phalanx rolls and glides palmarly on the convex metacarpal. Extension involves the phalanx rolling and gliding dorsally on the convex metacarpal. During abduction and adduction, the proximal phalanx rolls and glides in the same direction of movement as the fingers (radially or ulnarly).

Range of Motion

Flexion of the MCP can reach a position of 90°. Extension can move up to 45° beyond neutral (AAOS 1965; Norkin & White 2009).

Closed and Loose Packed Positions

The closed packed position for the MCP is full flexion. The loose packed position is slight flexion.

End Feel

The end feel for MCP flexion is hard due to contact between the proximal phalanx and the metacarpal. This end feel can also be firm because of tension in the dorsal joint capsule and collateral ligaments. Extension end feel is firm because of tension in the palmar joint capsule and the palmar plate. The end feel for MCP abduction is firm from tension in the collateral ligaments, the fascia of the web space between the fingers, and the palmar interossei. The end feel for MCP adduction is soft due to contact with the next finger.

Capsular Pattern

The capsular pattern of the MCP joint is equal limitation of flexion and extension.

Proximal Interphalangeal (PIP) Joint

The joints of the hands are the proximal and distal interphalangeal, allowing prehensile motion of the hand.

Axis of Motion

The PIP joint has one degree of freedom: flexion–extension. Flexion and extension occur in the sagittal plane around a mediolateral axis.

Arthrokinematics

The PIP joint consists of the concave end of the middle phalanx and the convex end of the proximal phalanx. During flexion, the concave middle phalanx glides palmarly. Extension is achieved with the middle phalanx gliding dorsally.

Range of Motion

PIP joint flexion averages 100° for the second through fifth digits. Extension is a return to 0°. Hyperextension should not occur at this joint (AAOS 1965; Norkin & White 2009).

Closed and Loose Packed Positions

The closed packed position of the PIP joint is full extension, and the loose packed position is slight flexion.

End Feel

The end feel for PIP joint flexion is hard due to contact between the palmar aspect of the middle phalanx and proximal phalanx. PIP joint extension has a firm end feel due to tension in the palmar joint capsule and the palmar plate.

Capsular Pattern

The capsular pattern of the PIP joint is equal restriction of flexion and extension.

Distal Interphalangeal (DIP) Joint

The distal interphalangeal joints allow precision grips and fine motions of the hands.

Axis of Motion

The DIP joint has one degree of freedom: flexion–extension. Flexion and extension occur in the sagittal plane around a mediolateral axis.

Arthrokinematics

The DIP joint consists of the concave end of the distal phalanx and the convex end of the middle phalanx. During flexion, the concave distal phalanx glides palmarly. Extension is achieved with the distal phalanx gliding dorsally.

Range of Motion

DIP joint flexion averages 90° for the second through fifth digits. Extension is a return to 0°. Hyperextension should not occur at this joint (AAOS 1965; Norkin & White 2009).

Closed and Loose Packed Positions

The closed packed position of the DIP joint is full extension, and the loose packed position is slight flexion.

End Feel

The end feel for DIP joint flexion is firm due to tension in the dorsal joint capsule and collateral ligament. DIP joint extension has a firm end feel due to tension in the palmar joint capsule and the palmar plate.

Capsular Pattern

The capsular pattern of the DIP joint is equal restriction of flexion and extension.

First Carpometacarpal (CMC) Joint

The carpometacarpal joint of the thumb has different types of joint surfaces and articulations than the other four CMC joints of the hand and therefore is described separately. The independency of the thumb from the rest of the hand is important for grasping objects of different sizes and shapes and for performing both power and precision grips (figure 12.13). Degeneration of the CMC joint can severely affect hand function (see Clinical Correlation 12.5).

Axes of Motion

The CMC joint of the thumb has two degrees of freedom: flexion–extension and abduction–adduction. Flexion and extension occur in the frontal plane around an anteroposterior axis through the trapezium (figure 12.14).

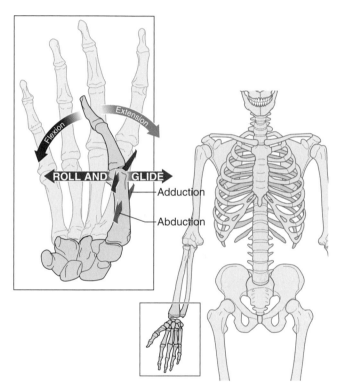

▶ **FIGURE 12.14** First CMC joint planes and axes of motion.

CLINICAL CORRELATION 12.5

The functional position of the hand and wrist is (a) wrist in slight extension (20°); (b) wrist in slight ulnar deviation (10°); (c) finger MCP joints moderately flexed (45°); (d) PIP joints slightly flexed (30°); and (e) DIP joints slightly flexed (10 to 20°). This position gives the finger flexors optimal power, and the wrist muscles are under equal tension. Inability to achieve the functional position of the hand can affect grip ability and dynamic stability of the wrist joint complex during fine movements. Traumatic injuries and conditions such as osteoarthritis, which is the most common disease of the hand joints, lead to pain, restricted motions in capsular pattern, joint crepitus, and eventually functional deficits. The clinician should evaluate joint play, joint motion, and soft-tissue flexibility of the involved joints; perform muscle testing; and assess functional limitations to guide conservative interventions.

Abduction and adduction occur in the sagittal plane around a mediolateral axis through the metacarpal. The motion of **opposition** is a combination of movements (including varying amounts of flexion, internal rotation, and abduction).

Arthrokinematics

The CMC joint of the thumb consists of the concavo-convex base of the first metacarpal and the concavo-convex trapezium. The metacarpal is concave in the mediolateral direction, and it is convex in the anteroposterior direction. Thumb flexion involves the concave surface of the metacarpal rolling and gliding in an ulnar (medial) direction. Extension is the reverse and consists of the metacarpal rolling and gliding in a radial (lateral) direction across the transverse diameter of the joint.

The arthrokinematics of thumb abduction involve the convex surface of the metacarpal rolling palmarly and gliding dorsally on the concave surface of the trapezium. During adduction, the metacarpal rolls dorsally and glides palmarly on the concave surface of the trapezium.

Range of Motion

Thumb flexion is measured as 0 to 15°. Extension equals 20 to 80° since extension may start with the thumb in some flexion. Abduction can reach 70°. Opposition is measured by the distance between the thumb and the little finger; these two surfaces touch with full opposition (AAOS 1965; Norkin & White 2009).

Closed and Loose Packed Positions

The closed packed position of the thumb CMC joint is full opposition, and the loose packed position is midrange.

End Feel

The end feel for thumb CMC joint flexion is soft due to contact between the thenar eminence and the palm. The end feel can also be firm because of tension in the dorsal joint capsule, the extensor pollicis brevis, and the abductor pollicis brevis. Extension end feel is firm because of tension in the anterior joint capsule, flexor pollicis brevis, adductor pollicis, opponens pollicis, and first dorsal interosseous. The end feel for abduction is firm from tension in the fascia and the skin of the web space between the thumb and the index finger. Adduction has a soft end feel due to contact of the thenar eminence and the metacarpal. Finally, opposition has a soft end feel due to contact between the thenar eminence and the palm.

Capsular Pattern

The capsular pattern of the thumb CMC joint is a limitation of abduction and slight limitation of extension.

First Metacarpophalangeal (MCP) Joint

The kinetic function of the thumb MCP joint is similar to the other four MCP joints of the hand.

Axes of Motion

The metacarpal joint of the thumb has two degrees of freedom: flexion–extension and abduction–adduction. The orientation of these movements differs from digits two through five. Flexion and extension occur in the frontal plane around an anteroposterior axis. Abduction and adduction occur in the sagittal plane around a mediolateral axis through the metacarpal. It also allows a slight degree of axial rotation.

Arthrokinematics

During flexion, the concave base of the proximal phalanx glides toward the palmar surface of the thumb. The opposite occurs during extension—the concave base of the proximal phalanx glides toward the dorsal surface. The arthrokinematics for thumb MCP joint abduction and adduction involve the proximal phalanx rolling and gliding in the same direction as movement of the thumb.

Range of Motion

Flexion of the thumb MCP joint can reach a position of 50°. Extension can move to neutral (AAOS 1965; Norkin & White 2009).

Closed and Loose Packed Positions

The closed packed position for the thumb MCP joint is full flexion. Slight flexion is the loose packed position of the thumb MCP joint.

End Feel

Similar to the other four MCP joints of the hand, the end feel for MCP joint flexion is hard due to contact between the proximal phalanx and the metacarpal. This end feel can also be firm because of tension in the dorsal joint capsule, collateral ligaments, and extensor pollicis brevis muscle. Extension end feel is firm because of tension in the palmar joint capsule, palmar plate, and flexor pollicis brevis muscle. The end feel for MCP abduction is firm from tension in the collateral ligaments and the fascia of the web space between the thumb and index finger. The end feel for MCP adduction is soft due to contact with the index finger and hypothenar muscles.

Capsular Pattern

The capsular pattern of the first MCP joint is greater restriction of motion in flexion than in extension.

Interphalangeal (IP) Joint

Unlike the fingers, the thumb has only one interphalangeal joint, which moves with the fingers to allow finger–thumb prehension such as pad-to-pad, tip-to-tip, and pad-to-side movements for the purpose of writing, picking up a coin, or manipulating fine objects.

Axis of Motion

The interphalangeal joint has one degree of freedom: flexion–extension. These motions occur in the sagittal plane about a mediolateral axis.

Arthrokinematics

The IP articulation consists of the concave base of the distal phalanx and the convex head of the proximal phalanx. Flexion involves the concave base of the distal phalanx gliding in a palmar direction. During finger extension, the base of the distal phalanx glides toward the distal surface of thumb.

Range of Motion

Interphalangeal flexion reaches 80°, and extension past neutral can be 20°, although this is usually a passive motion (AAOS 1965; Norkin & White 2009).

Closed and Loose Packed Positions

The closed packed position of the IP joint is full extension, and the loose packed position is slight flexion.

End Feel

End feel for IP joint flexion is firm because of tension in the collateral ligaments and dorsal joint capsule. Extension also has a firm end feel due to tension in the palmar joint capsule and the palmar plate.

Capsular Pattern

The capsular pattern for the IP joint is equal restriction in flexion and extension.

Muscles

The muscles of the forearm and wrist can be divided into extrinsic muscles (24 muscles) and intrinsic muscles (19 muscles). The wrist and hand muscles that originate in the forearm and insert within the hand are considered extrinsic muscles. The muscles that are located entirely within the hand (originate and insert within the hand) are considered intrinsic muscles. This anatomical design allows for a large number of muscles to control hand function without creating excessive muscle bulk. In addition, the tendons of the extrinsic muscles of the wrist flexors and extensors provide a balancing force to stabilize the wrist and compress the carpals. The muscles of the wrist and hand (figure 12.15 and table 12.2) are designed for control and for producing fine movements rather than torque.

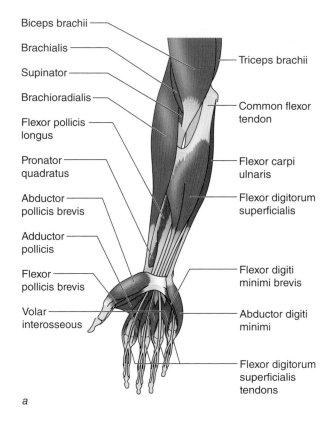

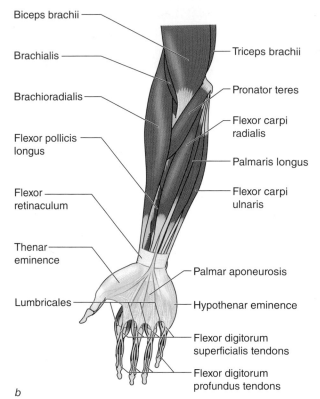

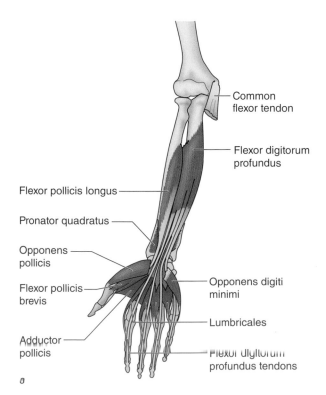

▶ **FIGURE 12.15**   Muscles of the forearm, wrist, and hand: (a) intermediate, (b) superficial, and (c) deep.

TABLE 12.2 Muscles of the Wrist and Hand

Muscle	Origin	Insertion	Action	Innervation
Extrinsic wrist and hand				
Extensor carpi radialis longus	Lateral supracondylar ridge and lateral epicondyle of humerus	Base of metacarpal 2	Wrist extension, radial deviation	Radial nerve (C6-C7)
Extensor carpi radialis brevis	Lateral epicondyle	Base of metacarpal 3	Wrist extension, radial deviation	Radial nerve (C6-C7)
Extensor carpi ulnaris	Lateral epicondyle	Base of metacarpal 5	Wrist extension, ulnar deviation	Posterior interosseous branch of radial nerve (C6-C7)
Extensor digitorum communis	Lateral epicondyle	Bases of middle and distal phalanges of digits 2-5 via extensor aponeurosis	Wrist extension, extension of MP joints of digits 2-5	Posterior interosseous branch of radial nerve (C7-C8)
Extensor indicis	Lower posterior ulna and interosseous membrane	Extensor expansion of index finger	Extension of index finger	Posterior interosseous branch of radial nerve (C7-C8)
Extensor digiti minimi	Lateral epicondyle	Extensor aponeurosis of little finger	Extension of digit 5	Posterior interosseous branch of radial nerve (C7-C8)
Extensor pollicis longus	Middle posterior ulna and interosseous membrane	Base of distal phalanx of thumb	Extension of IP joint of thumb, radial deviation	Posterior interosseous branch of radial nerve (C7-C8)
Extensor pollicis brevis	Distal posterior radius and interosseous membrane	Base of proximal phalanx of thumb	Extension of MP joint of thumb, radial deviation	Posterior interosseous branch of radial nerve (C7-C8)
Flexor carpi radialis	Medial epicondyle distal to origin of pronator teres	Base of metacarpals 2 & 3	Wrist flexion, radial deviation	Median nerve (C6-C7)
Flexor carpi ulnaris	Humeral head from medial humeral epicondyle; ulnar head from medial border of olecranon and proximal 2/3 of posterior border of ulna	Pisiform, hamate, metacarpal 5	Wrist flexion, ulnar deviation	Ulnar nerve (C7-T1)
Palmaris longus	Medial epicondyle	Apex of palmar aponeurosis	Wrist flexion	Median nerve (C8-T1)
Flexor digitorum superficialis	Humeroulnar head: medial epicondyle, ulnar collateral ligament, and ulnar coronoid process Radial head: upper 1/2 of anterior border of radius	Tendons split and insert on sides of middle phalanx of each digit	Flexion of PIP joints of digits 2-5, wrist flexion	Median nerve (C8-T1)

> *continued*

TABLE 12.2 > *continued*

Muscle	Origin	Insertion	Action	Innervation
Extrinsic wrist and hand				
Flexor digitorum profundus	Upper 3/4 of anterior ulna and interosseous membrane	Bases of distal phalanges of digits 2-5	Flexion of DIP joints of digits 2-5, wrist flexion	Median nerve (to middle and index fingers) (C8-T1) Ulnar nerve (to ring and little fingers) (C8-T1)
Flexor pollicis longus	Anterior radius and interosseous membrane	Base of distal phalanx of thumb	Flexion of thumb (IP flexion)	Anterior interosseous branch of median nerve (C8-T1)
Abductor pollicis longus	Upper posterior radius and ulna, interosseous membrane	Base of 1st metacarpal and trapezium	Extension of thumb, abduction of CMC joint of thumb, radial deviation	Posterior interosseous branch of radial nerve (C7-C8)
Intrinsic hand				
Lumbricales (4 muscles)	Tendons of flexor digitorum profundus	Extensor digitorum expansion; each muscle attaches distally to the radial side of extensor digitorum	MP flexion of the fingers (2-5), extension of IP joints with MCP joint held in extension	Median nerve (1st and 2nd) (C8-T1) Ulnar nerve (3rd and 4th) (C8-T1)
Dorsal interossei (4 muscles)	Base of metacarpals; each muscle has 2 heads from adjacent sides of metacarpals	Dorsal extensor expansion (dorsal hood) at the base of the proximal phalanges; 1st and 2nd on the radial side and 3rd and 4th on the ulnar side	Finger abduction	Ulnar nerve (C8-T1)
Abductor digiti minimi	Pisiform, tendon of flexor carpi ulnaris, and pisohamate ligaments	Ulnar side of the base of the 5th proximal phalanx and its dorsal expansion of extensor digiti minimi	Abduction of 5th finger	Ulnar nerve (C8-T1)
Palmar interossei (3 muscles)	Base of 2nd, 4th and 5th metacarpals	Dorsal extensor expansion of corresponding proximal phalanx	Finger adduction	Ulnar nerve (C8-T1)
Flexor pollicis brevis	Across 3 distal carpals (capitate, trapezoid, and trapezium) and their palmar ligaments and flexor retinaculum	Thumb; base of 1st proximal phalanx	Flexion of thumb (MP flexion), thumb opposition	Median nerve: superficial head (C8-T1) Ulnar nerve: deep head (C8-T1)
Abductor pollicis brevis	Tubercle of scaphoid and trapezium, flexor retinaculum, and tendon of abductor pollicis longus	Base of 1st proximal phalanx	Thumb abduction, thumb opposition	Median nerve (C8-T1)

Muscle	Origin	Insertion	Action	Innervation
Adductor pollicis	Oblique head: base of 2nd and 3rd metacarpals, capitate, and palmar ligaments of carpals Transverse head: shaft of 3rd metacarpal on palmar aspect	1st proximal phalanx on ulnar side	Thumb abduction	Ulnar nerve (C8-T1)
Opponens pollicis	Tubercle of trapezium and flexor retinaculum	Length of 1st metacarpal; lateral aspect and palmar surface	Thumb opposition	Median nerve (C8-T1)
Opponens digiti minimi	Hook of hamate and flexor retinaculum	Length of 5th metacarpal; ulnar aspect and palmar surface	Opposition	Ulnar nerve (C8-T1)

▶ **KEY POINT**

The muscles of the wrist and hand are designed for control and for producing fine movements rather than torque.

Wrist Flexors

The wrist flexors are located on the palmar aspect of the forearm and hand. Collectively they flex the wrist and fingers. A number of superficial muscles, including pronator teres, flexor carpi radialis, palmaris longus, flexor digitorum superficialis, and flexor carpi ulnaris, share a common tendon, called the common flexor tendon. The common flexor tendon serves as the origin for these muscles and attaches to the medial epicondyle of the humerus. The tendons of the flexor digitorum superficialis muscle can produce relatively independent action of PIP joint flexion of each finger when other fingers are held in extension. The tendons of the flexor digitorum profundus muscle work simultaneously at DIP joints. The flexor carpi ulnaris is the most powerful of the wrist flexors (Nordin & Frankel 2001).

Most actions of the fingers involve actions of both extrinsic and intrinsic muscles. A coordinated sequence of flexion motion occurs during the closing of the hand. The active force is produced primarily by the flexor digitorum profundus at all joints of the fingers during low-power hand closing. In contrast, the flexor digitorum profundus, flexor digitorum superficialis, and interossei muscles are activated during high-power hand closing. The extensor digitorum stabilizes flexion torque at the MCP joints (Neumann 2010).

Wrist Extensors

The wrist extensors are located on the dorsal aspect of the forearm and hand. Collectively they extend the wrist and fingers. A number of superficial muscles, including extensor carpi radialis brevis, extensor digitorum, extensor digiti minimi, and extensor carpi ulnaris, share a common tendon called the common extensor tendon. The common extensor tendon serves as the origin for these muscles and attaches to the lateral epicondyle of the humerus. Of all wrist extensors, extensor carpi radialis brevis has the greatest tension and moment arm and therefore is the most effective wrist extensor. The extensor pollicis longus serves as the ulnar border, while the extensor pollicis brevis and abductor pollicis longer serve as the radial border of the anatomical snuffbox. Overuse of abductor pollicis longus and extensor pollicis brevis cause tenosynovitis (see Clinical Correlation 12.6).

The primary muscle responsible for exerting force during opening of the hand is the extensor digitorum communis. This muscle extends the MCP joints. The intrinsic muscles (the lumbricales more so than the interossei) assist with extension of the IP joints by pulling on bands of the extensor mechanism and producing a flexion torque at the MCP joints to prevent hyperextension (Neumann 2010).

CLINICAL CORRELATION 12.6

DeQuervain's syndrome is a tenosynovitis of the abductor pollicis longus and extensor pollicis brevis tendons. This condition develops when repetitive radial deviation causes irritation of the synovial sheath around the two tendons. Signs and symptoms include localized pain and swelling over the involved tendons that is exacerbated by thumb movement. Occasionally, crepitus is present in this area. The clinician can perform the Finkelstein test (flexion of the thumb into the palm and passive deviation of the wrist toward the ulnar side) to confirm the diagnosis. Splint mobilization with a spica brace to rest the thumb and wrist along with modalities and patient education to avoid thumb movement are recommended in the acute stage. Stretching and strengthening exercises to improve hand grip and finger skills are recommended after the acute stage.

Conclusion

The wrist and hand consist of several joints that collectively allow for specialized functions such as gripping and writing. The soft tissues (e.g., TFCC, joint capsule) and numerous ligaments provide joint stability to carry out these unique functions. The muscles contribute to fine movements of the wrist and hand and allow for controlled functional movements. Because of their complexity and their being the most active part of the upper extremity, the wrist and hand are susceptible to injury.

REVIEW QUESTIONS

1. Name the carpal bones located in the distal row.

2. What is the function of the triangular fibrocartilage complex?

3. What muscle is the most powerful wrist flexor?

4. Describe the movement of thumb opposition.

5. What is the name of the grip that is used to hold a pencil?

Hip

The hip joint is a large synovial joint that attaches the lower limb to the trunk. Because of its architectural arrangement and ligament support, the hip is inherently a very stable joint. It also has a moderate amount of mobility that allows for walking and activities of daily living. Large forces are transmitted through the hip, and articular damage is not uncommon in the older adult. The numerous muscles that attach about the hip are important for control and balance of the lower extremity and spine.

Osteology

This section provides detailed information on the bones that make up the hip joint complex. These bones are the pelvis (ilium, ischium, pubis) and proximal femur. Although the hip may seem like a simple joint, it is the main stabilizing structure for the lower extremity. The hip is important for balance attainment and steady ambulation.

Os Coxae

The **os coxae**, or pelvis, is formed from three bones: the ilium, ischium, and pubis.

OBJECTIVES

After reading this chapter, you should be able to do the following:

> List the bones associated with the hip joint complex.

> Describe the primary joint that makes up the hip joint.

> Explain the ligamentous structures that support the hip joint.

> Describe the major functions of the muscles that affect the hip.

> Discuss a variety of injuries associated with the hip joint.

This continuous ring connects anteriorly at the pubic symphysis and posteriorly at the sacrum. The three bones come together inferiorly to make up the acetabulum (figure 13.1). The concavity formed by the three bones is crucial for weight bearing. Unfortunately, this area can break down with excessive loading.

▶ **KEY POINT**

The pelvis is formed by three bones: the ilium, ischium, and pubis. The concavity formed by the three bones is crucial for weight bearing. Unfortunately, this area can break down with excessive loading.

Ilium

The distal lateral portion of the pelvis is the ilium[2], which forms the upper portion of the acetabulum. On the external surface are posterior, anterior, and inferior gluteal lines. These lines help identify attachment sites of the gluteal muscles to the pelvis. Anteriorly are two bony projections that are muscular insertion sites and commonly used landmarks for pelvic alignment: the anterior superior iliac spine (ASIS) and the anterior inferior iliac spine (AIIS). The most superior part of the ilium (the rim) is the iliac crest. Posteriorly are two more bony projections: the posterior superior iliac spine (PSIS) and posterior inferior iliac spine (PIIS). Also found posteriorly is the greater sciatic notch where the sciatic nerve travels. This notch is formed by the sacrospinous ligament. The internal aspect houses the iliac fossa, the attachment of the iliacus muscle. The auricular surface of the ilium articulates with the sacrum at the sacroiliac joint.

Pubis

The lateral portion of the superior pubic ramus forms the frontal portion of the acetabulum. The superior pubic ramus extends anteriorly from the anterior wall of the acetabulum to the large flattened body of the pubis[2]. The pectineal line on the pubis is the marking for the attachment of the pectineus muscle. The pubic tubercle projects anteriorly from the superior pubic ramus and is an attachment for the inguinal ligament. The inferior pubic ramus extends from the body of the pubis posteriorly to the junction of the ischium. The pubic symphysis joint is the articulation of the pubic bones at the midline. This is a fibrocartilaginous joint and is classified as an amphiarthrosis.

Ischium

The third part of the pelvis is the ischium[2]. The body of the ischium forms the posterior portion of the acetabulum. The ischial spine projects from the posterior side of the ischium, just inferior to the greater sciatic notch. The lesser sciatic notch is located just inferior to the spine. The ischial ramus extends anteriorly from the ischial tuberosity, ending at the junction with the inferior pubic ramus. The ischial tuberosity projects posteriorly and inferiorly from the acetabulum. It is large and stout. This serves as the proximal attachments for many muscles of the lower extremity, including the hamstrings.

Acetabulum

The articulating surface of the hip is called the **acetabulum** and is the junction of the ilium, ischium, and pubis (figure 13.2). The ilium and ischium contribute 80%, and the pubis contributes the remaining 20%. The acetabulum is located just above the obturator foramen and is the socket (fossa) on the lateral aspect of the pelvic bone. It faces anteriorly, laterally, and inferiorly. A plane that would parallel the opening of the acetabulum would face 40° posterior to the sagittal plane (Nordin & Frankel 2001).

The concave fossa secures the head of the femur. Within the acetabulum is a smooth, horseshoe-shaped lunate surrounded by the acetabular rim. This area is covered by hyaline cartilage that thickens peripherally and laterally. An acetabular labrum (also known as the glenoid lip) deepens the fossa for added stability (figure 13.2). A labrum is a ring (or lip) of fibrocartilage. In addition, a fat pad is located within the acetabular fossa for added shock absorption. The weight-bearing surface of the acetabulum is superior, anterior, and posterior. Cartilage degeneration in this region

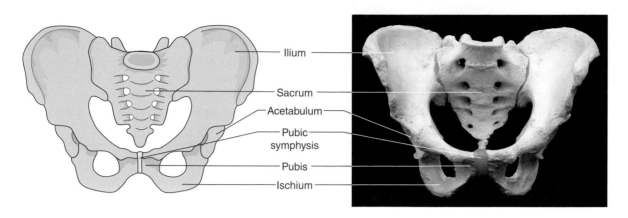

a

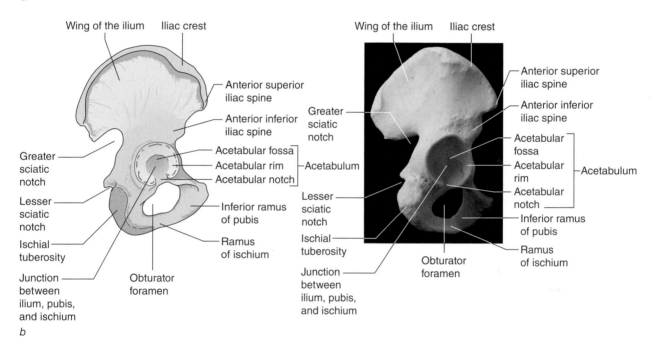

b

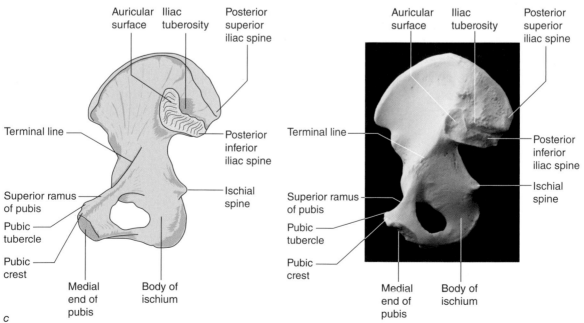

c

▶ **FIGURE 13.1** *(a)* Anterior aspect of the pelvis. *(b)* Lateral aspect of right innominate bone. *(c)* Medial aspect of right innominate bone.

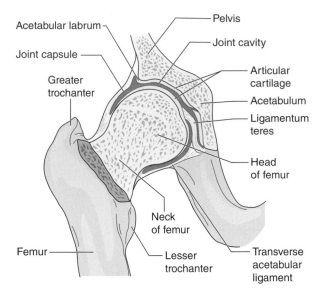

FIGURE 13.2 Acetabular labrum.

is common after the age of 65 years (see Clinical Correlation 13.1).

The gap between the two ends of the acetabular rim is continuous with the fossa and is called the acetabular notch. The acetabular notch forms a foramen with the transverse acetabular ligament to allow the artery in the ligamentum teres femoris to pass though.

Femur

The femur (figure 13.3) is the longest and strongest bone of the human body. The proximal part of the femur is the femoral head, which projects superiorly, medially, and anteriorly to articulate with the acetabulum. The head forms two-thirds of a sphere. The articular cartilage that covers the femoral head is thickest on the mediocentral surface and becomes progressively thinner as it reaches the periphery. It is thought that most of the

stresses on the femoral head during daily activities are anterior and medial (Bergmann et al. 1993).

▶ **KEY POINT**
The acetabulum faces anteriorly, laterally, and inferiorly. The proximal part of the femur is the femoral head, which projects superiorly, medially, and anteriorly to articulate with the acetabulum.

On the tip of the head is the fovea, which is the site of the ligament of the head of the femur. Moving from the tip of the head toward the femoral shaft is the femoral neck, which connects the femoral head to the shaft. The interior of the neck is composed of cancellous bone with a medial and lateral **trabecular** system (figure 13.4). It is likely that the medial system supports joint reaction forces from the femoral head and that the lateral system resists compressive forces from the abductor muscles (Nordin & Frankel 2001). The neck allows for a great amount of movement of the femur, but it is also a common site for fractures in the aged adult (see Clinical Correlation 13.2).

Other important landmarks on the femur include the greater trochanter, lesser trochanter, trochanteric fossa, linea aspera, intertrochanteric line and crest, and adductor tubercle. The greater trochanter extends laterally and posteriorly from the junction of the femoral neck and shaft. It is easily palpable and serves as the distal attachment for many muscles, primarily the hip external rotators. It is also a common landmark used for measuring hip and knee range of motion. The lesser trochanter projects sharply from the inferior

CLINICAL CORRELATION 13.1

Total hip replacement surgery (also known as total hip arthroplasty, THA) is a surgical procedure where the surgeon replaces the acetabulum and femoral head and neck with a prosthesis (usually titanium alloy). Approximately 120,000 THA surgeries are performed in the United States annually, and the procedure is more common in women. The average age of the person who undergoes a THA is 66. Rehabilitation after a hip replacement includes education on precautions for motion limitations (e.g., extreme hip flexion), strength training, and functional training (gait).

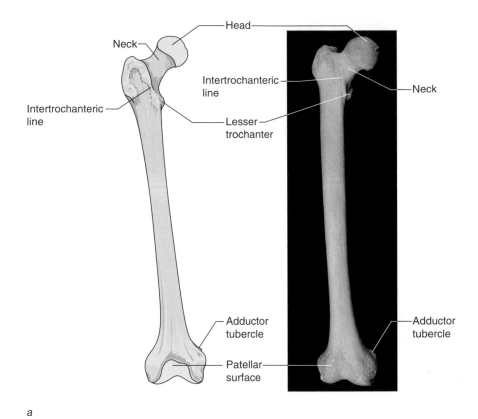

Head

Neck

Intertrochanteric line

Intertrochanteric line

Lesser trochanter

Neck

Adductor tubercle

Patellar surface

Adductor tubercle

a

Head

Greater trochanter

Head

Neck

Neck

Intertrochanteric crest

Lesser trochanter

Spiral line

Gluteal ridge

Linea aspera

Medial supracondylar ridge

Adductor tubercle

Lateral supracondylar ridge

Lateral epicondyle

Lateral condyle

Medial epicondyle

Lateral epicondyle

Medial epicondyle

Medial condyle

Lateral condyle

Medial condyle

Intercondylar notch

b

▶ **FIGURE 13.3** The right femur: *(a)* anterior and *(b)* posterior aspect.

end of the crest in a posteromedial direction. This trochanter is an attachment site for the psoas major and iliacus. The **linea aspera** is a vertical ridge located on the posterior side of the femoral shaft. Proximally, the linea aspera splits into the pectineal (spiral) line medially and the gluteal line laterally. Distally, the linea aspera splits into the lateral supracondylar line and the medial supracondylar line. The intertrochanteric line lies on the anterior surface of the femur between the greater and lesser trochanters. It serves as an attachment site for capsular ligaments. The intertrochanteric crest is located on the posterior femur and separates the greater and lesser trochanters. The posterior joint capsule attaches to this crest. Finally, distally the adductor tubercle located on the extreme medial supracondylar line is the attachment site for the adductor magnus.

Angle of Inclination

The femur is not perpendicular to the ground but inclined medially at the distal end. A frontal plane angle occurs between a line bisecting the neck of the femur and a line bisecting the shaft. This is called the **angle of inclination**[2] (figure 13.5a). The purpose of this angle is to keep the lower leg parallel when standing, and it changes with age because of the continual stress of weight bearing. The average value for the angle is 150° for infants, 125° for adults, and 120° for older adults.

A pathological increase (coxa valga; figure 13.5b) or decrease (coxa vara; figure 13.5c) in the angle of inclination can occur from trauma or congenital hip issues. These abnormal angles can alter the mechanical loading of the femur. Coxa vara will create a tensile stress on the superior portion of the femoral neck and a compression stress on the inferior surface. A congenital hip issue that may predispose a person to coxa varum is slipped capital femoral epiphysis (SCFE). SCFE is a transphyseal Salter I fracture of the femoral head resulting in slippage of the femoral head (usually posteriorly and inferiorly). This condition is most common in 11- to 13-year-old females and 13-

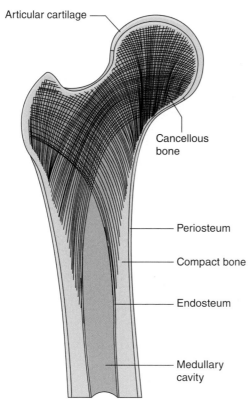

Articular cartilage

Cancellous bone

Periosteum

Compact bone

Endosteum

Medullary cavity

▶ **FIGURE 13.4**   Trabecular pattern of the femoral head.

CLINICAL CORRELATION 13.2

With aging, the femoral neck gradually undergoes degenerative changes. The longitudinal trabeculae of the cortical bone becomes thinner, and some of the transverse trabeculae of the cancellous bone disappear (Siffert & Levy 1981). The neck of the femur is the most common fracture site in the elderly person and often occurs secondary to a fall. Ninety percent of cases involve people over 65 years, three-quarters of whom are female. In the United States, there is an estimated 340,000 fractured hips per year. The fracture may involve the anatomical or surgical neck. After a hip fracture, therapy consists of regaining hip range of motion and working on balance and gait activities.

to 16-year-old males. SCFE is more common in males by a 3:1 ratio. It occurs bilaterally in 30% of cases.

▶ KEY POINT

The angle of inclination is the angle between the neck of the femur and its shaft. A value of 120 to 125° is a normal angle for adults. An angle less than 120° is termed coxa vara, and an angle greater than 135° is termed coxa valga.

Angle of Torsion

In the transverse plane, an angle is formed between a line bisecting the neck of the femur and a line parallel to the plane of the shaft of the femur. This is called the **angle of torsion** and averages 8 to 12° (figure 13.6). This twist in the femur is necessary for the femoral condyles to present in a frontal plane facing anteriorly. An excessive angle of torsion (>12°) is termed **anteversion**, whereas a decrease in this angle (<8°) is termed **retroversion**. In infants the femur presents more anteverted, and the angle of torsion is commonly around 20°. As with the angle of inclination, the angle of torsion declines as the person grows. People with an anteverted hip will tend to stand and walk with their toes pointed inward (internal rotation) as a compensation. People with a retroverted hip will tend to stand and walk with their toes pointed outward (external rotation). In addition, the acetabulum can be retroverted, leading to a femoroacetabular pincer-type impingement. Clinical Correlation 13.3 illustrates this concept.

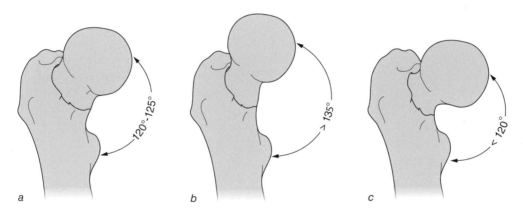

▶ **FIGURE 13.5** Angle of inclination: (a) normal, (b) coxa valga, and (c) coxa vara.

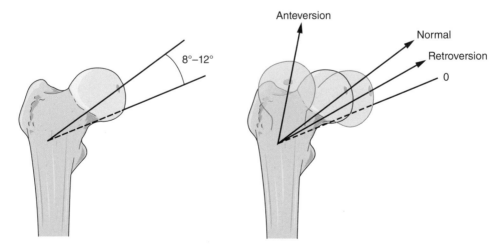

▶ **FIGURE 13.6** Angle of torsion: normal, anteversion, and retroversion.

CLINICAL CORRELATION 13.3

Femoroacetabular impingement (FAI) occurs when the femur impinges on the acetabulum because of faulty bony approximation. Two types of FAI have been identified: cam and pincer. Both types can result in damage to the acetabular cartilage and labrum. Cam impingement, where the femoral head has an abnormally large radius, causes shear forces on the acetabular rim. This type of impingement is more common in young athletic men and is caused by the jamming of an abnormal femoral head or head–neck junction against the acetabulum, especially with flexion and internal rotation (Ito et al. 2001). Cam impingement often occurs in patients with post-traumatic deformities such as slipped capital femoral epiphysis or coxa vara.

The pincer type, where there is an abnormal acetabulum with increased overcoverage, causes impingement from linear contact between the acetabular rim and the femoral head–neck junction. This type is often seen in patients with coxa profunda, coxa protrusio, or acetabular retroversion. Many of these patients are middle-aged females engaging in activities that require extreme ranges of motion such as yoga and ballet. Cam and pincer impingement rarely occur in isolation, and the combination has been termed mixed cam–pincer impingement (Beck et al. 2005). In an epidemiological study of 149 hips with impingement, 17.4% had isolated cam impingement, 10.7% had isolated pincer impingement, and 71.8% had combined cam–pincer impingement (Beck et al. 2005).

The patient with FAI typically describes a gradual onset of sharp groin pain that is worse with athletic activities requiring an excessive demand on hip flexion. The patient may report mechanical symptoms (locking, catching, and giving way) indicative of a labral tear or injury of the articular cartilage (Beall et al. 2005). Ninety-one percent of FAI patients had activity-related pain and 71% had night pain. Surgical intervention for FAI focuses on improving the clearance of hip motion and alleviating femoral abutment against the acetabular rim, relieving the pathological changes in the labrum and articular cartilage. The presence of other related pathology should be treated accordingly. This is especially the case with acetabular labral tears, as FAI was determined to be an underlying cause in 55% of a series of 300 consecutive labral tear cases (Philippon 2001).

▶ **KEY POINT**

The range of external and internal rotation at the hip is largely determined by the angle of torsion. The normal angle of torsion ranges between 8 and 12°. People with an angle of torsion greater than 12° have an anteverted hip and will compensate in their stance with an inwardly rotated lower limb.

The center-edge angle of Wiberg is formed by a line drawn perpendicular to a baseline that passes through the center of the femoral heads and a line connecting the center of the femoral head and the superior border of the acetabulum (figure 13.7). This angle is used in radiographic evaluation of the hip joint, and

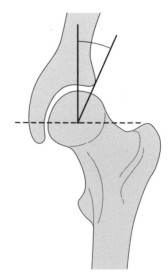

▶ **FIGURE 13.7**   Center-edge angle of Wiberg.

an angle of 25° is considered normal. An angle less than 20° is prevalent in developmental dysplasia of the hip.

Joint Articulation

The hip joint consists of one single articulation between the femoral head and the acetabulum, referred to as the femoroacetabular joint. This joint is a diarthrodial spheroidal joint, more commonly known as a synovial ball-and-socket joint. The spherical (convex) head of the femur articulates with the concave acetabulum. This articulation is primarily on the lunate surface of the acetabulum. Motion at this joint occurs in all three planes. The axis of motion in all three planes is in the center of the femoral head, indicating almost pure spin in these motions.

Joint Anatomy

This section provides specific information on the joint capsule, ligaments, soft tissues, and nerve and blood supply of the hip joint complex.

Joint Capsule and Labrum

The joint capsule that encompasses the external portion of the hip joint is a loose, fibrous capsule. The fibers of the capsule spiral from the joint to the intertrochanteric line, forming a fibrous collar around the femoral neck. Longitudinal fibers are found within the anterior capsule. A band of circular fibers, called the zona orbicularis, encircles the neck of the femur. The internal portion of the joint capsule is the synovial membrane. This membrane covers the intracapsular neck of the femur up to the articular cartilage of the head of the femur (see figure 13.2). Along the outer surface of the acetabulum is an acetabular labrum. The labrum is similar to the labrum of the shoulder. It is made of fibrocartilage and increases the articular surface of the acetabulum by 10%. The labrum allows the acetabulum to hold more than half the head of the femur and thus provides stability to the joint (see Clinical Correlation 13.4). The labrum contains free nerve endings and sensory end organs in its superficial layer that may participate in nociceptive and proprioceptive function (Kim & Azuma 1995). There is a gap in the labrum near the acetabular notch. This area is bridged by the transverse acetabular ligament. The ligament creates a foramen for blood vessels. The transverse acetabular ligament is thought to provide extra support to the articulation.

▶ **KEY POINT**
The labrum of the hip is made of fibrocartilage and increases the articular surface of the acetabulum by 10%, allowing the acetabulum to hold more than half the head of the femur, thus providing stability to the joint.

Ligaments

Three ligaments arise from thickenings of the joint capsule: the iliofemoral, pubofemoral, and ischiofemoral ligaments (figure 13.8). The position of hip extension winds the ligaments tighter, adding to the stability of the joint in this position. The reverse, flexion, unwinds these three ligaments, allowing for more mobility to take place.

CLINICAL CORRELATION 13.4

Tears of the fibrocartilage labrum can be detected using advanced diagnostic procedures. Labral tears can be caused by direct trauma or repetitive stress, especially external rotation. People with a history of hip dysplasia may be susceptible to labral tears. Clinical tests such as FABER (Flexion, Abduction, External Rotation), leg log rolling, impingement tests, the quadrant test, and compressive rotation tests are usually positive. Magnetic resonance angiography (MRA) is helpful for diagnosis.

The iliofemoral ligament is also called the ligament of Bigelow or Y ligament. This ligament covers the anterior and superior portion of the hip. Its two bands attach from the anterior inferior iliac spine and acetabular rim and insert on the intertrochanteric line. The iliofemoral ligament is the strongest, thickest ligament and is able to hold the trunk upright with little muscular support when a person is in hip extension. This ligament prevents hyperextension. The superior band of the ligament checks adduction, and the inferior band checks abduction.

The pubofemoral ligament sits anterior and inferior to the hip joint. Coursing from the obturator crest of the pubis, it passes laterally and inferiorly to join with the medial part of the iliofemoral ligament. The function of the pubofemoral ligament is preventing overabduction of the hip. It also assists in preventing hip hyperextension.

The ischiofemoral ligament is the primary posterior ligament. Anatomically, it runs from the ischial part of the acetabular rim and then spirals superolaterally near the apex of the greater trochanter. The ischiofemoral ligament is the weakest of the intrinsic ligaments. It tightens during internal rotation and extension.

The ligament of the head of the femur (ligamentum teres) is an intracapsular ligament that conducts a small artery to the head of the femur for nutrition (see figure 13.2). This ligament travels from the margin of the acetabular notch to the fovea. It is surrounded by a fat pad within the acetabular fossa. The ligamentum teres can be injured by trauma to the hip such as a hip dislocation (see Clinical Correlation 13.5). This type of injury can lead to avascular necrosis of the femoral head.

▶ **KEY POINT**
The ligament of the head of the femur (ligamentum teres) can be injured by trauma to the hip such as a hip dislocation. This type of injury can lead to avascular necrosis of the femoral head.

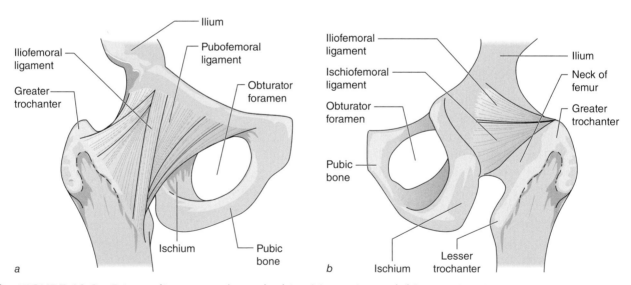

▶ **FIGURE 13.8** Primary ligaments about the hip: (a) anterior and (b) posterior views.

CLINICAL CORRELATION 13.5

Ligamentum teres lesions have been increasingly reported in the literature and are the third most common finding during hip arthroscopy in athletes. Typically, these patients present with deep anterior groin pain, with or without mechanical symptoms. Activities reported to be associated with these injuries include motor vehicle accidents; falls from a height; and football, skiing, and hockey. The diagnosis of ligamentum teres lesions continues to be made based on history, clinical examination, and arthroscopic evaluation.

Soft-Tissue Structures

The primary soft-tissue structures discussed in this section are the hip bursae, the iliotibial band, and the nerve and blood supply.

Bursae

Several bursae are found around the hip. Laterally, there are two situated around the greater trochanter, superficial and deep (figure 13.9). Posteriorly, the ischial bursa is situated between the ischial tuberosity and the hamstrings tendon. Anteriorly, the iliopsoas bursa is located anterior to the iliopsoas muscle. Bursitis is somewhat common about the hip. The three most commonly affected bursae are the trochanteric, ischial, and iliofemoral. This condition is most frequent in sedentary females aged 40 to 60 years. Symptoms include hip pain that is tender to palpation.

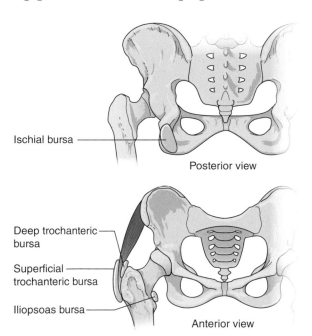

▶ **FIGURE 13.9** Hip bursae.

Iliotibial Band

The iliotibial band (ITB) is a thickened piece of fascia that originates along the lateral hip and inserts distally below the knee at Gerdy's tubercle, with slips to the patella (figure 13.10). Proximally, the gluteus maximus and the tensor fasciae latae insert into the band. The ITB may become irritated proximally (external snapping hip syndrome) with repeti-

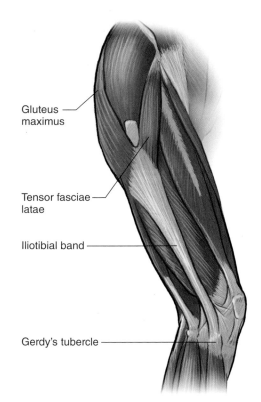

▶ **FIGURE 13.10** Iliotibial band.

tive activity such as hiking, cycling, or running. MRI or diagnostic ultrasound can help identify irritated tissue.

Nerve Supply

Five nerves supply the hip. Posteriorly, the sciatic nerve is the primary nerve for the posterior musculature beyond the gluteal muscles. The superior and inferior gluteal nerves innervate the gluteal muscles and tensor fasciae latae. The femoral nerve innervates the anterior hip joint. The obturator nerve innervates the inner thigh. Hilton's law states that nerves supplying the muscles extending directly across and acting at a given joint also innervate the joint.

Blood Supply

The hip joint is supplied with blood by the medial and lateral circumflex femoral arteries. There is also a small contribution from an artery in the ligament of the head of the femur, which is a branch of the posterior division of the obturator artery. The trochanteric anastomoses supplies most of the blood to the head of the femur.

Joint Function

The hip joint is very stable, with motion occurring in three planes. This section covers the axis of motion, arthrokinematics, range of motion, closed and loose packed positions, end feel, and capsular pattern of the single articulation that makes up the hip joint, the femoroacetabular joint.

Axis of Motion

The axis of motion in all three planes is in the center of the femoral head, indicating almost pure spin in these motions. Motions at the hip are flexion–extension, abduction–adduction, and internal–external rotation.

Arthrokinematics

Motion at the hip joint occurs between the convex femoral head and the concave acetabulum. Three degrees of motion occur at the hip: flexion–extension; abduction–adduction; and internal–external rotation. Arthrokinematically, the motion of the hip is primarily a spin of the femoral head within the acetabulum, accompanied by slight gliding (figure 13.11). The tight fit and steep walls of the acetabulum limit significant translation between the joint surfaces.

Flexion

To bring the knee to the chest the hip must flex, which requires the femoral head to spin within the acetabulum and slightly glide posteriorly and inferiorly. In standing, to bend forward as if to touch your toes, the pelvis glides anteriorly on the fixed femur.

Extension

Non-weight-bearing (NWB) hip extension is defined as the femur moving on the pelvis; it requires the femoral head to spin and slightly glide anteriorly. With the foot on the ground, relative hip extension requires the pelvis to glide posteriorly on the fixed femur.

Abduction

For the femur to abduct from the pelvis, the femoral head needs to inferiorly glide within

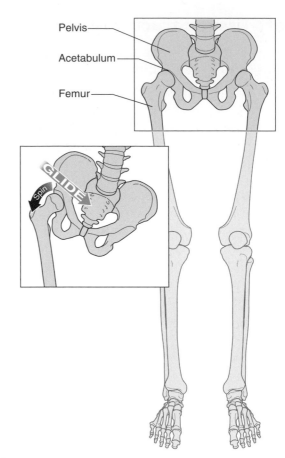

▶ **FIGURE 13.11** Frontal plane arthrokinematics of the hip in weight bearing.

the concave acetabulum. Because of the geometry of the femoral head, if the hip is flexed close to 90°, the arthrokinematics change. If the hip is flexed to 90°, the femoral head will glide anteriorly with abduction. In a weight-bearing state, the concave acetabulum glides toward the opposite pelvis, with the contralateral side of the pelvis hiking.

▶ **KEY POINT**
Because of the geometry of the femoral head and neck, the arthrokinematics of the hip change depending on the amount of hip flexion.

Adduction

The femur glides superiorly in non-weight-bearing hip adduction, which is the opposite of hip abduction. With the hip flexed to 90°, the femoral head glides posteriorly to achieve adduction. In a weight-bearing state, the concave acetabulum glides toward the ipsilateral femur, with the contralateral side of the pelvis dropping.

Internal Rotation

In a non-weight-bearing state, internal rotation is accomplished by posterior glide of the femoral head. With the hip flexed to 90°, internal rotation requires the femoral head to glide inferiorly. In stance, internal rotation is achieved by the acetabulum spinning about the femoral head toward the side of rotation. For right lower extremity internal rotation, the pelvis will rotate to the right.

External Rotation

External rotation is accomplished by anterior glide of the femoral head. With the hip flexed to 90°, the femoral head glides superiorly with external rotation. With the foot fixed, external rotation is achieved by the acetabulum spinning about the femoral head opposite the side of rotation. For right lower extremity external rotation, the pelvis will rotate to the left.

Range of Motion

Sagittal plane motion includes flexion and extension. The greatest hip motion occurs with flexion and can reach up to 140° with the knee flexed. If the knee is extended, hip flexion is 80 to 90° because of passive tension of the hamstring and gracilis muscles. Range of motion of hip extension is 20° beyond the neutral position. When the knee is fully flexed, hip extension is limited by the passive tension of the rectus femoris. Frontal plane motion includes abduction and adduction. The ROM of hip abduction is 40°. The pubofemoral ligament and adductor and hamstring muscles limit this motion. The ROM of hip adduction is 25° beyond the neutral position. The passive tension of the hip abductor muscles, iliotibial band, and superior fibers of the ischiofemoral ligament limit full adduction.

Transverse plane motion includes internal (medial) and external (lateral) rotation. The average ROM of internal rotation is 45° from a neutral position. In healthy young adults, there is no significant change in internal rotation between a sitting and prone position. The average ROM of external rotation is 45°. Tension in the tensor fasciae latae, iliotibial band, and iliofemoral ligament may limit full external rotation. Reduced hip range of motion is an early indicator of hip disease and imposes functional limitations in activities of daily living (e.g., walking, tying shoes). Clinical Correlation 13.6 discusses the functional range of motion at the hip. Loss of hip range of motion may be associated with pain and muscle weakness.

Closed and Loose Packed Positions

The closed packed position of the hip is approximately 20° of hip extension (maximal extension), with slight internal rotation and abduction. The hip capsule and ligaments are maximally stretched. This position closely resembles the orientation of the hip during the preswing of gait. A position of congruence has been described as 90° of flexion, 5° of abduction, and 10° of external rotation, which simulates the quadruped position. The loose packed position of the hip is 30° of flexion, 30° of abduction, and slight external rotation. This position is commonly used as the starting point for joint mobilization (e.g., inferior glide).

CLINICAL CORRELATION 13.6

The three planes of motion needed during common daily activities have been reported in the literature. For daily activities, the hip needs to achieve up to 120° flexion, abduction of at least 20°, and external rotation of at least 20°. For example, to tie your shoes with your foot on the floor, you need 124° of flexion, 19° of abduction, and 15° of external rotation. To ascend stairs, you need 67° of flexion, 16° of abduction, and 18° of external rotation (Johnston & Smidt 1970). Therefore, a primary component of any hip rehabilitation program is addressing the lack of hip range of motion.

End Feel

The end feel for hip flexion is soft due to contact between the muscles of the anterior thigh and lower abdomen and stretch of the hip extensors, primarily the gluteus maximus. The extension end feel is firm because of tissue stretch of the iliofemoral, ischiofemoral, and pubofemoral ligaments; the hip flexor muscles; and the anterior hip joint capsule. At end range, femoral abduction normally stops because of a firm end feel due to tissue stretch of the pubofemoral and ischiofemoral ligaments, the inferior capsule, the inferior band of the iliofemoral ligament, and the adductor muscles. With the foot fixed, pelvic-on-femoral hip abduction has a hard end feel due to the femoral neck snugging against the acetabulum. The end feel for hip adduction is soft because of the tension in the hip adductor muscles. Pelvic-on-femoral hip adduction has a firm end feel because the abductor muscles become taut. The motion of internal rotation has a firm end feel because a tissue stretch occurs in the external rotator muscles of the hip, the ischiofemoral ligament, and the posterior joint capsule. External rotation end feel is firm because of a tissue stretch of the internal rotator muscles, the iliofemoral and pubofemoral ligaments, and the anterior joint capsule.

Capsular Pattern

The capsular pattern of the hip as described by Cyriax (1982) is equal limitation in flexion, abduction, and internal rotation, with a small loss in extension of the hip and little to no loss of external rotation.

Muscles

The hip joint is dependent on a complex set of muscles that create three-dimensional movement and dynamic stability. Figure 13.12 displays the primary muscles about the hip. Table 13.1 lists the muscles about the hip along with their origins, insertions, actions, and innervations. With the massive muscle system that supports the hip, muscle impair-

ments and imbalances can render the hip joint susceptible to dysfunction (see Clinical Correlation 13.7).

Hip Flexors

The hip flexors bring the femur forward, or anteriorly. With the femur fixed, the hip flexors will flex the trunk forward. The hip flexors are found anterior to the hip joint axis and include the iliopsoas (psoas major, iliacus), sartorius, rectus femoris, tensor fasciae latae, and pectineus. The iliopsoas is partially formed by the psoas major, which originates on the transverse processes of all five lumbar vertebrae and the vertebral bodies and anterior discs of the 12th thoracic vertebra and all five lumbar vertebrae and inserts on the lesser trochanter. The psoas major is commonly short in athletes and will cause the hip to remain in a position of flexion and the lumbar spine to extend. The other component of the iliopsoas is the iliacus, which originates from the iliac fossa and also inserts on the lesser trochanter.

A second hip flexor, the sartorius, is a long strappy muscle that originates on the anterior superior iliac spine and inserts on the proximal medial tibia as a common tendon with the semitendinosus and gracilis, into the pes anserine. Besides flexing the hip, the sartorius also abducts and externally rotates the femur. The rectus femoris is the only muscle of the quadriceps that crosses the hip joint and serves as a hip joint flexor. It originates on the anterior inferior iliac spine and inserts on the tibial tubercle. Another strong hip flexor is the tensor fasciae latae (TFL), which originates on the external rim of the iliac crest and is the anterior muscle that attaches to the iliotibial band (see figure 13.10). At the hip, the TFL can be tight, causing not only hip flexion but also femoral internal rotation. The final hip flexor, the pectineus, is a short, broad muscle that attaches from the superior ramus of the pubic bone to the pectineal line of the femur. Besides hip flexion, the pectineus also adducts the femur. Other muscles that assist with hip flexion are the adductor longus, adductor brevis, adductor magnus, gluteus minimus, and gluteus medius (anterior fibers).

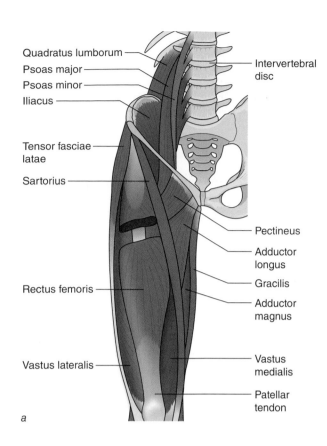

a

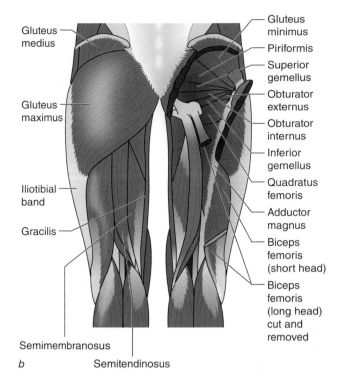

b

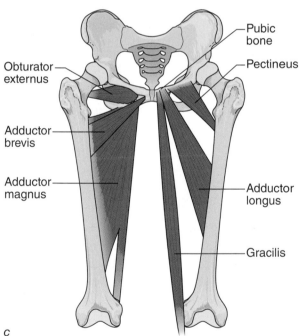

c

▶ **FIGURE 13.12** Hip musculature: *(a)* anterior, *(b)* posterior, and *(c)* adductors of the hip.

One consideration for designing a hip rehabilitation program is the amount of load that occurs at the hip joint during various activities. Much of these forces are a result of gravity and muscle contraction. The joint reaction force while unilateral standing is 2 to 5 times body weight (BW), while ascending stairs is 3 times BW, and running is 5 times BW. Although these values seem to be very high, the normal healthy hip can tolerate 12 to 15 times BW. If pathology exists, a cane can alter the magnitude of the joint reaction force significantly if carried on the opposite side of the involved extremity.

TABLE 13.1 Muscles of the Hip

Muscle	Origin	Insertion	Action	Innervation
Anterior thigh				
Psoas	Transverse processes and body of L1-L5 and T12	Lesser trochanter of femur	NWB: hip flexion WB: decelerates hip internal rotation and controls hip extension	L1-L3
Iliacus	Iliac fossa and sacrum	Lesser trochanter of femur	NWB: hip flexion WB: controls hip extension	Femoral nerve (L2-L3)
Rectus femoris	Anterior inferior iliac spine (AIIS)	Tibial tuberosity via patellar ligament	NWB: hip flexion WB: controls hip extension	Femoral nerve (L2-L4)
Sartorius	Anterior superior iliac spine (ASIS)	Upper part of medial surface of tibia	NWB: hip flexion, abduction, external rotation WB: controls hip extension, adduction, internal rotation	Femoral nerve (L2-L3)
Pectineus	Pectineal line of pubis	Upper half of pectineal line of femur	NWB: hip flexion, adduction, internal rotation WB: controls hip extension, abduction, external rotation	Femoral nerve (L2-L3)
Medial thigh				
Gracilis	Inferior ramus of pubis	Upper part of medial surface of tibia	NWB: hip flexion, adduction, internal rotation WB: controls hip abduction, extension, external rotation	Obturator nerve (L2-L3)
Adductor longus	Inferior ramus of pubis	Middle 1/3 of posterior femur	NWB: hip flexion, adduction, internal rotation WB: controls hip extension, abduction, and external rotation	Obturator nerve (L2-L4)
Adductor brevis	Inferior ramus of pubis	Upper 1/2 of posterior femur	NWB: hip flexion, adduction, internal rotation WB: controls hip extension, abduction, external rotation	Obturator nerve (L2-L4)
Adductor magnus	Anterior pubis and ischial tuberosity	Linea aspera and adductor tubercle	NWB: hip extension, adduction WB: controls hip flexion, abduction	Obturator nerve and sciatic nerve (L2-L5, S1)
Posterior thigh				
Semimembranosus	Ischial tuberosity	Medial condyle of tibia	NWB: hip extension WB: controls hip flexion	Sciatic nerve (L5, S1-S2)
Semitendinosus	Ischial tuberosity	Upper part of medial surface of tibia	NWB: hip extension WB: controls hip flexion	Sciatic nerve (L5, S1-S2)

Muscle	Origin	Insertion	Action	Innervation
Biceps femoris (long head)	Ischial tuberosity	Lateral condyle of tibia and head of fibula	NWB: hip extension, helps with hip external rotation WB: controls hip flexion	Sciatic nerve (L5, S1-S2)
Gluteal muscles				
Gluteus minimus	Ilium between anterior and inferior gluteal lines	Anterior border of greater trochanter	NWB: hip abduction, internal rotation WB: controls hip adduction, external rotation	Superior gluteal nerve (L5, S1)
Gluteus medius	Anterior, lateral ilium	Lateral surface of greater trochanter	NWB: hip extension, abduction, internal rotation (anterior fibers), external rotation (posterior fibers) WB: controls hip flexion, adduction, external rotation (anterior fibers), internal rotation (posterior fibers)	Superior gluteal nerve (L5, S1)
Gluteus maximus	Posterior ilium, sacrum, coccyx	Gluteal tuberosity and iliotibial tract	NWB: hip extension, abduction, external rotation WB: controls hip flexion, adduction, internal rotation	Inferior gluteal nerve (L5, S1-S2)
Tensor fasciae latae	Anterior superior iliac spine (ASIS)	Iliotibial tract	NWB: hip abduction from hip flexed position, internal rotation WB: controls hip adduction, external rotation	Superior gluteal nerve (L4-L5)
Lateral rotators				
Piriformis	Anterior, lateral sacrum	Superior greater trochanter	NWB: hip external rotation WB: controls hip internal rotation	L5, S1-S2
Superior gemellus	Ischial spine	Greater trochanter via obturator internus tendon	NWB: hip external rotation WB: controls hip internal rotation	Sacral plexus (L5, S1)
Obturator internus	Sciatic notch and margin of obturator foramen	Greater trochanter	NWB: hip external rotation WB: controls hip internal rotation	Sacral plexus (L5, S1)
Inferior gemellus	Ischial tuberosity	Greater trochanter via obturator internus tendon	NWB: hip external rotation WB: controls hip internal rotation	Sacral plexus (L4-L5, S1)
Obturator externus	Pubis, ischium, and margin of obturator foramen	Upper, posterior femur	NWB: hip external rotation WB: controls hip internal rotation	Obturator nerve (L3-L4)
Quadratus femoris	Ischial tuberosity	Greater trochanter	NWB: hip external rotation WB: controls hip internal rotation	Sacral plexus (L4-L5, S1)

NWB = non–weight bearing; WB = weight bearing

Hip Extensors

The primary hip extensors are the gluteus maximus, hamstrings, posterior gluteus medius, and piriformis. The gluteus maximus is the largest muscle of the hip, accounting for 16% of the total cross-sectional area (Winter 2005). It has several anatomical landmarks including the ilium, sacrum and coccyx, and sacrotuberous ligament as an origin (Kendall et al. 1993). Eighty percent of the gluteus maximus inserts into the iliotibial band; the remainder inserts on the distal portion of the femur's gluteal tuberosity. The hamstrings (semimembranosus, semitendinosus, biceps femoris long head) have a common origin at the ischial tuberosity, although the origin of the semimembranosus is slightly more proximal and lateral to the other two components. The short head of the biceps femoris originates on the lateral lip of the linea aspera, proximal two-thirds of the supracondylar notch, and lateral intermuscular septum. The insertion site of the semimembranosus is the posteromedial aspect of the medial condyle of the tibia. The semitendinosus inserts just below the medial condyle of the tibia, along with the gracilis at the pes anserine. The biceps femoris has a lateral attachment on the fibular head. The gluteus medius is described in detail under hip abduction. The piriformis origin and insertion are found in table 13.1.

The hip extensors and abdominal muscles act opposite to the hip flexors and back extensors. For example, the hip extensors produce posterior pelvic tilt and decrease lumbar lordosis, whereas the hip flexors assist with anterior pelvic tilt and increase lumbar lordosis. The hamstrings control forward lean of the upper body because of their longer lever arms as two-joint muscles. There is also passive tension from the connective tissue. In weight bearing, the gluteus maximus is a powerful hip extensor and is often used to move the body forward and upward from a position of hip flexion ranging from 45 to 60° (e.g., sprinting, squatting, climbing a steep hill). Additionally, the gluteus maximus is active during a plant-and-cut maneuver to the opposite side (Neumann 2010).

Hip Adductors

The primary hip adductors are the adductor longus, adductor brevis, adductor magnus, pectineus, and gracilis. Muscles that assist with adduction are the obturator externus and the lower fibers of the gluteus maximus (Kendall et al. 1993). The adductor longus originates from the anterior surface of the pubis and inserts on the middle third of the medial lip of the linea aspera of the femur. The adductor brevis extends from the outer surface of the inferior ramus of the pubis to the proximal third of the linea aspera. The adductor magnus is a substantial muscle, making up the majority of the muscle mass of the adductor group. It originates on the inferior pubic ramus, with the anterior fibers attaching to the ramus of the ischium and the posterior fibers attaching to the ischial tuberosity. The magnus inserts along the entire length of the linea aspera and the adductor tubercle. Finally, the gracilis originates on the anterior aspect of the pubic symphysis and inserts into the pes anserine (proximal to the semitendinosus and lateral to the sartorius tendons).

The adductor muscles insert on the femur from several different directions, so they create motion in all three planes. In the frontal plane, adduction can occur with the femur moving on the pelvis, such as with kicking. With the pelvis moving on the femur, the hip will tilt laterally. In the sagittal plane, the posterior fibers of the adductor magnus perform hip extension. The adductor longus can flex the hip when the hip is in less than 60° of hip flexion. It can perform hip extension when the hip is in more than 60° of flexion. This action of the adductors assists the primary hip flexors and extensors to produce a more forceful stride in running and cycling. In the transverse plane, many of the adductors assist in internal rotation.

Hip Abductors

The hip abductors are the tensor fasciae latae, gluteus medius, gluteus minimus, and gluteus maximus (upper fibers). The gluteus medius is a broad, fan-shaped muscle that originates from the middle of the external aspect of the ilium and inserts on the oblique ridge on the lateral aspect of the greater trochanter. It stabilizes the femur and pelvis during weight bearing.

The gluteus minimus has a similar origin to the gluteus medius and inserts anteriorly on the greater trochanter. The origins and insertions of the tensor fasciae latae and gluteus maximus were previously given. Other muscles that help with hip abduction are the iliopsoas, sartorius, inferior gemellus, obturator internus, and piriformis (Kendall et al. 1993).

In non-weight-bearing activities, the gluteus medius is the largest of the hip abductors and accounts for 60% of the total abductor muscle cross-sectional area (Clark & Haynor 1987). The anterior and middle portions of the gluteus medius help initiate hip abduction, while the posterior portion extends, abducts, and externally rotates the hip (Delp et al. 1999).

In weight-bearing activities, the main action of the abductor muscles is to stabilize the pelvis and hips while walking (Gottschalk et al. 1989; Lyons et al. 1983). These muscles produce a torque that is opposite that of body weight while on one leg in the walking cycle. The force of the abductors must be approximately twice the person's body weight (Ward et al. 2010). This creates an equilibrium that causes the hips to remain fairly level during gait. The combination of the abductors with the body weight creates a large joint reaction force while walking (2.5 to 3 times body weight) and running (5.5 times body weight). It has been proposed that weak hip abductors are a contributing factor to medial lower extremity collapse, commonly seen in patellofemoral joint pain syndrome (DiStefano et al. 2009; Powers 2003).

Hip Internal Rotators

Internal rotators of the hip include the adductor longus, adductor brevis, TFL, gluteus minimus, gluteus medius (anterior fibers), and medial hamstrings (Kendall et al. 1993). The origins and insertions of each of these muscles have been previously described. The internal rotators are commonly short and biased with functional movements.

While walking, the internal rotators rotate the pelvis over the femur during the stance phase, causing the opposite leg to move anteriorly. The amount of activation of the muscles depends on the length and speed of your stride. The greatest amount of internal rotation torque occurs when the hip is flexed to 90° (Delp et al. 1999).

Hip External Rotators

The primary hip external rotators are the deep six: obturator externus, obturator internus, inferior gemellus, superior gemellus, quadratus femoris, and piriformis. The origins and insertions of these muscles can be found in table 13.1. In addition, external rotation is performed by the gluteus maximus, iliopsoas, sartorius, posterior fibers of the gluteus medius, and long head of the biceps femoris muscle (Kendall et al. 1993).

External rotation is the most common action produced during pelvic-on-femoral motion, such as during an activity when a change in direction is required. With the leg planted, the external rotators rotate the body away from the planted leg. The gluteus maximus is the most powerful external rotator of the hip (Delp et al. 1999). The external rotators are commonly weak in lower extremity injuries such as patellofemoral pain syndrome and iliotibial band syndrome (Ferber et al. 2010).

Conclusion

The hip joint is composed of the head of the femur and the acetabulum of the pelvis. The geometrically aligned joint allows for moderate mobility for normal daily activities such as walking, sitting, and squatting. Ligaments and strong muscles allow for a stable joint.

REVIEW QUESTIONS

1. List the bones that make up the os coxae.

2. What is the normal angle of inclination between the neck of the femur and its shaft?

3. Describe slipped capital femoral epiphysis.

4. Describe the closed packed position of the hip joint.

5. What is the function of the gluteus medius in gait?

6. Describe the arthrokinematics for hip internal rotation with the hip at 90° of flexion.

7. What two muscles attach to the iliotibial band?

8. Name and describe the two types of femoral acetabular impingement.

14

Knee

At first glance the knee seems to be a fairly simple synovial structure. In reality, it is one of the more complex structures in the human musculoskeletal system. The knee is the center point between two of the longest bones in the human body, the tibia and the femur. Its location dictates that incredible forces are placed on the articulating components. If the knee predominantly functioned in an open kinetic chain (as does its upper limb counterpart, the elbow, a unilateral hinge joint), this may not be as problematic. The knee, however, functions in a closed kinetic chain system; thus weight-bearing forces are borne through the knee as ground reaction forces are transferred superiorly to the hip and low back. In addition, there are no direct muscular attachments at the joint itself; origins and insertions are all either proximal or distal to the actual joint proper. Tremendous forces are placed on the knee during running, jumping, and cutting. In closed-chain situations, movements of the foot on the ground can cause obligate translations proximally at the knee.

OBJECTIVES

After reading this chapter, you should be able to do the following:

> Identify the knee articulations involved in active movement patterns.

> Detect abnormal anatomical alignment that may create abnormal forces or stress on the knee.

> Recognize the importance of intra-articular structures such as the meniscus in dampening force through the knee during weight-bearing activities.

> Determine what high- and low-level joint reaction forces occur in the patellofemoral joint during functional activities.

> Interpret which muscular actions create various movement patterns in and around the knee.

> Differentiate passive from active insufficiency.

Osteology

The articulation of the knee joint includes three bones: the femur, the tibia, and the patella. Because only small portions of the femur (distal portion) and tibia (proximal portion) are part of the knee, only the relevant portions are described in this chapter. The proximal femur and distal tibia are described in their respective sections covering the hip, foot, and ankle.

Distal Femur

The most proximal portion of the knee is the distal femur (figure 14.1), which is covered with articular cartilage to allow weight bearing, shearing, and stress to the femoral condyles and the patella. The most distal portion of the femur is formed by the **femoral condyles**. These two large rounded protuberances are centrally divided posteriorly and inferiorly by the intercondylar fossa, or notch, which is the location for the attachments of the cruci-

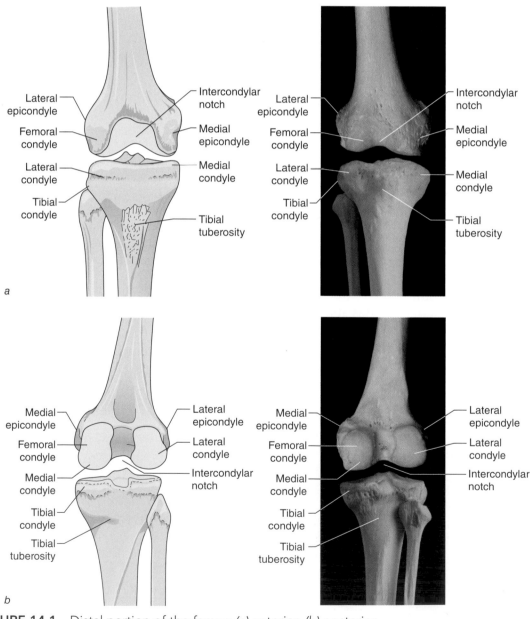

▶ **FIGURE 14.1** Distal portion of the femur: *(a)* anterior, *(b)* posterior.

ate ligaments, while this division turns into the trochlear groove anteriorly. There has been tremendous debate concerning notch type and size and its relationship to anterior cruciate ligament (ACL) injury, especially in female athletes. Notch types are described as either U-shaped or A-shaped, with an A-shape possibly predisposing an athlete to higher rates of ACL disruption because of its space-occupying effect.

▶ **KEY POINT**

The knee is the confluence of the two longest bones in the human body—the femur and the tibia.

The surfaces of the two condyles are each unique, which contributes to the complex motions described later in this chapter. Anatomically, when viewed from a lateral direction, the condyles are a cam shape, with curves both anteriorly and posteriorly, while having a flatter distal end. The posterior curve is much tighter (smaller) than that of the anterior condyle (see figure 14.1). The lateral femoral condyle is wider both antero-posteriorly and mediolaterally than that of the medial. Although the medial condyle extends distally farther than the lateral, because the shaft of the femur angles medially, the two condyles lie in the same horizontal plane. Additionally, the medial femoral condyle projects farther anteriorly. The lateral condyle projects farther posteriorly than the medial. Although both condyles are slightly convex, the lateral is slightly less so. Proximal to each condyle in the frontal plane is a bony projection known as the epicondyle. Each epicondyle is an attachment for various muscle tendons and ligaments. The epicondyles are important bony landmarks for palpation and for use as a standardized location for the axis of rotation during goniometry. As you will soon see, though, the axis of motion at the knee is not quite as simple as a single static bony landmark. The variances in shape of the condyles facilitate the screw-home mechanism described later in this chapter.

The most superior portion of the condyle turns into the trochlear portion of the patel-

lofemoral joint. A large sulcus runs down the middle of the trochlea and corresponds well with the ridge running down the middle of the posterior patellar surface. This central ridge divides the surface into medial and lateral facets. The lateral facet of the trochlea normally sits higher than that of the medial and is thought to provide a lateral bony restraint (figure 14.2). Some people have trochlear dysplasia in which the facets are not properly formed. Those who have a smaller lateral trochlea than normal tend to be more prone to lateral dislocations or subluxations of the patellofemoral joint.

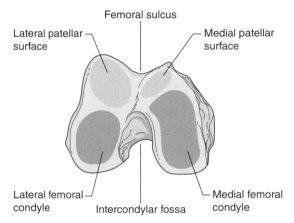

▶ **FIGURE 14.2** The patellar trochlea of the distal femur.

▶ **KEY POINT**

Patients with trochlear alterations in which the facets are not properly formed may be more prone to patellar instabilities.

Proximal Tibia

The proximal portion of the tibia articulates with the distal femur through the femoral condyles and forms the tibiofemoral joint (see figures 14.1 and 14.3). The tibiofemoral joint is the largest in the human body as it is the convergence of the two longest bones in the body, the femur and the tibia. The proximal portion of the tibia is much more expansive than the shaft because it forms the medial and lateral condyles. The medial **tibial plateau** exhibits a slight concavity that reciprocally accepts the more convex femoral condyle, while the lateral tibial plateau is flat or even slightly convex (Ateshian et al. 1991; Freeman & Pinskerova 2005). The contact area for the medial plateau is larger than that of the lateral, which may also account for the fact that the articular cartilage is thicker on the lateral plateau (Kettelkamp & Jacobs 1972). This may assist with the greater forces that are placed on the medial joint during functional weight-bearing activities (Ateshian et al. 1991). Using magnetic resonance imaging and dual fluoroscopy during gait, Liu and colleagues (2010) determined that contact areas throughout the stance phase were greater in the medial joint as compared with the lateral compartment. This might lead one to suspect that more damage due to excessive joint loading would always be in the lateral compartment. However, Winby and colleagues (2009) caution that although most of the force is placed on the medial compartment, muscles provide sufficient stability to counter the tendency of the external adduction moment to allow unloading that compartment. Therefore, dynamic muscular stability will help unload the medial side and place a greater demand onto the lateral compartment. This external moment unloading may help account for the high incidence of osteoarthritis that is typically seen after meniscus injury on the lateral side of the joint (Chatain et al. 2003) (see Clinical Correlation 14.1). The medial and lateral sides of the tibial plateau clearly display different loading characteristics that ultimately affect joint geometry. This disparity in shape may be a moot point because of the interposed meniscus, at least when it is healthy and functional.

Situated between the two plateaus, running from the anterior to posterior tibia, is a ridge known as the intercondylar eminence. Proximal to the eminence are the medial and lateral intercondylar tubercles, which are the attachments of the ACL and posterior cruciate ligament (PCL). Anterior and midline to the tibia is an easily palpable landmark called the tibial tuberosity. Covered only by skin and a small bursa, the tibial tubercle is the distal attachment site for the patellar tendon (ligament).

Two important locations on the lateral side of the tibia are the lateral facet, which accepts the fibula, and Gerdy's tubercle. The lateral

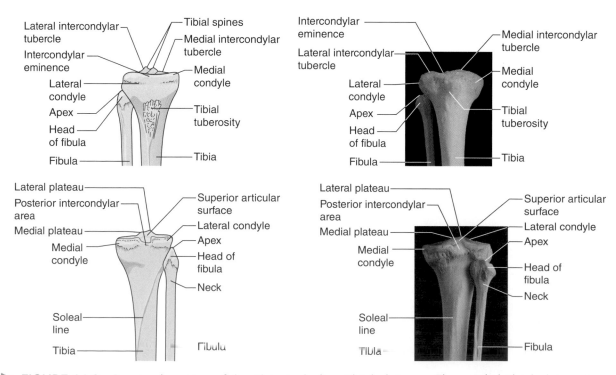

▶ **FIGURE 14.3**   Proximal portion of the tibia, including tibial plateaus. The medial tibial plateau is concave, whereas the lateral is flat or convex.

facet is situated inferior to the lateral plateau and is a flattened area that faces inferiorly, posteriorly, and laterally. Gerdy's tubercle is located between the lateral condyle and the tibial tuberosity and is the attachment site of the iliotibial band.

Patella

The largest sesamoid bone in the human body, the **patella** is an inverted triangular-shaped bone embedded within the extensor mechanism (figure 14.4). Both the base (superior)

CLINICAL CORRELATION 14.1

Anatomically, the flat or convex lateral tibial plateau may be in part why patients who have a lateral meniscectomy do not fare as well as those who have injury to the medial meniscus. Because of the convexity of the lateral tibial plateau, any loss of meniscal tissue will substantially increase the compressive forces on the lateral side. Historically, those with lateral meniscectomy have less favorable results compared with their medial counterparts.

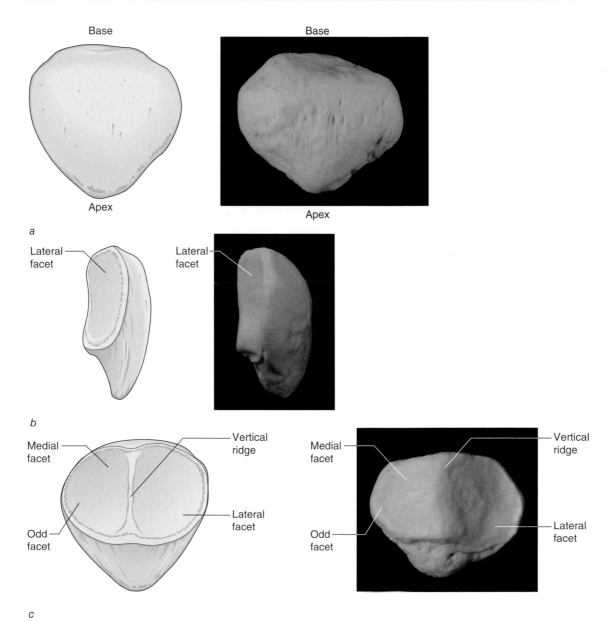

▶ **FIGURE 14.4** *(a)* Anterior, *(b)* right lateral, and *(c)* posterior views of the patella.

and the apex (inferior) are rounded, although the inferior apex has a smaller radius. With the patient in long sitting and the quadriceps fully relaxed, the apex of the patella rests just at or slightly proximal to the joint line. The anterior surface of the patella is convex in both the anteroposterior and mediolateral planes. It is easily palpable because it is covered only by skin and the prepatellar bursa. The posterior surface is much more complex. The articular cartilage on the posterior patella ranges from 1 mm on the periphery to between 5 and 7 mm centrally (Fulkerson 1997; Heegaard et al. 1995; Grelsamer & Weinstein 2001). This thick cartilage is thought to dissipate large joint reaction forces created during forceful contractions of the quadriceps muscle. A vertical ridge down the center of the posterior patella longitudinally divides the patella equally into medial and lateral facets. The lateral facet is slightly concave, while the medial facet is variable. A second smaller vertical ridge separates the medial facet proper from the odd facet on the far medial side of the medial facet. The posterior surface of the patella is articular except for the inferior pole (apex), which is the attachment of the patellar tendon. The superior and inferior portions of the patella are slightly roughened to allow attachment of the quadriceps and patellar tendons.

▶ **KEY POINT**

The patella is a large sesamoid bone that is engulfed by the patellar tendon and patellar ligament. The patella is tethered by the quadriceps tendon superiorly and the patellar ligament inferiorly.

Joint Articulations

To the untrained eye, the knee looks like a single joint. It actually consists of two separate joints—the tibiofemoral and the patellofemoral. Controversy exists in classifying the tibiofemoral joint. Because the largest amount of tibiofemoral motion is in the sagittal plane, it appears that this joint is a pure hinge joint. A small amount of rotation and varus and valgus motion occurs with normal joint mobility; therefore, it is more appropriately described

as a double condyloid joint (three degrees of freedom—flexion and extension, medial and lateral rotation, varus and valgus) or a modified synovial hinge joint (two degrees of freedom—flexion and extension, medial and lateral rotation). The patellofemoral joint is more of a gliding joint, while the superior tibiofibular joint is a synovial joint with motion in three degrees of freedom. Although quite small, there are gliding motions anteroposteriorly and superoinferiorly and limited amounts of rotation.

Joint Anatomy

The overall joint anatomy of the knee involves precise form meeting function. Very high stresses and strains are placed on both the tibiofemoral and patellofemoral joints during activities of daily living, and even higher forces with recreation or sporting activities. This section describes the general alignment and anatomy of the various joints of the knee.

Tibiofemoral Joint

When one thinks of the knee joint, it is usually the tibiofemoral joint that comes to mind. It is the intersection of two of the longest bones in the human body. The joint consists of the proximal portion of the tibia and the distal portion of the femur. Interposed between these two surfaces is the meniscus of the knee.

Overall Alignment

The proximal portion of the femur at the hip determines its distal position at the knee. A normal 125° angle of inclination at the hip directs the shaft of the femur in a medial and inferior oblique direction toward the knee. This anatomical axis runs longitudinally through the shaft of the femur. This does not alter the position of the femoral condyles at the knee; the medial condyle extends slightly longer than the lateral so that they actually rest evenly on the nearly horizontal proximal tibia. The medially deviated femur on a vertical tibia creates an angle on the medial knee of about 185 to 190°. Figure 14.5a illustrates the normal slight valgus angulation that occurs in the frontal plane. This normal 5 to 10° valgus angulation is called **genu valgum**. The lay term *knock-knee* occurs when the medial frontal plane angulation is greater than 190°; clinically this is known as excessive genu valgum (figure 14.5*b*). When the medial angle is less than 175° in the frontal plane, the person is considered to have bowlegs, or **genu varum** (figure 14.5*c*). Both these anatomical positions can create altered joint forces on the tibiofemoral joint. With excessive genu valgum, the lateral side of the joint incurs excessive forces, while with genu varum, the medial side

of the joint incurs excessive forces (see Clinical Correlation 14.2). Although these frontal plane alterations can be a normal genetic knee position, they can also be seen after trauma or degenerative knee conditions.

▶ **KEY POINT**
Excessive genu valgum is known as knock-knees, while excessive genu varum is known as bowlegs.

When viewing the knee from a lateral position, excessive knee extension range of motion is termed **genu recurvatum**. In most people,

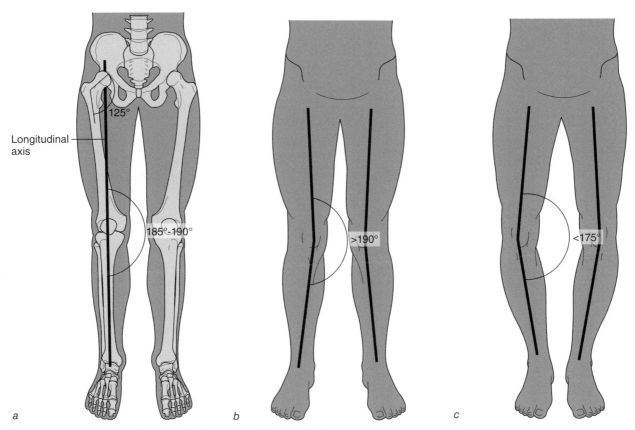

▶ **FIGURE 14.5**   *(a)* Normal valgus angulation. *(b)* Excessive genu valgum. *(c)* Excessive genu varum.

CLINICAL CORRELATION 14.2

Frontal plane abnormalities can be genetic or the result of degeneration. A valgus knee position can be caused by osteoarthritis or a degenerative joint disease of the lateral compartment of the knee joint, while a genu varum position can be the result of an arthritic condition of the medial joint. Either of these deviations can be seen in people with knee pathology. It must be stressed, though, that not everyone with either a valgus or varus knee will have pathology; it can be to some degree congenital.

5 to 10° of hyperextension is acceptable, but any more is labeled **recurvatum**.

Ligaments and Capsule

The tibiofemoral joint is stabilized statically by four major ligaments and an extensive joint capsule. Two of the four ligamentous structures that give support to the tibiofemoral joint are the anterior and posterior cruciate ligaments, named because they cross within the **intercondylar notch** of the distal femur. Both cruciates supply anterior, posterior, and rotary stability to the knee. The other two ligaments are the medial and lateral collateral ligaments, which supply frontal plane varus–valgus stability.

The arcuate ligament complex is a functional tendoligamentous unit made up of the following: lateral collateral ligament, biceps femoris tendon, popliteus muscle and tendon, popliteal meniscal and popliteal fibular ligaments, oblique popliteal ligament, arcuate and fabellofibular ligaments, and lateral gastrocnemius muscle tendon (Recondo et al. 2000).

Anterior and Posterior Cruciate Ligaments   The cruciates are both intracapsular ligaments located between layers of fibrous and synovial linings. The fibrous lining extends as two folds anteriorly, forming a pocket of tissue that wraps around the cruciates; therefore, they are also extrasynovial. Their names are derived from their attachment location on the tibia.

The function of the **anterior cruciate ligament** (ACL) has been described as far back as 3,000 BC, where it is mentioned on Egyptian papyrus scrolls (Zantop et al. 2006). The ACL attaches to the anterior tibia in front of and lateral to the intercondylar eminence (figure 14.6). The tibial attachment is wider and stronger than the femoral attachment (Gigris et al. 1975), found posteriorly on the medial surface of the lateral femoral condyle. The posterior portion of the attachment site is convex, while the anterior is straight. This femoral site is at the back of the intercondylar notch. The ACL courses anteriorly, medially, and distally from the femur to the tibia, and because of these attachment sites, the anterior cruciate actually

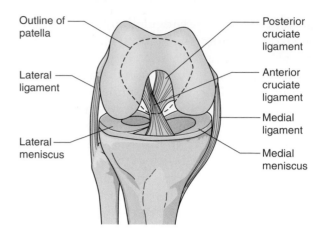

▶ **FIGURE 14.6**   Ligaments of the tibiofemoral joint.

turns on itself laterally in a spiral fashion. The dimensions of the anterior cruciate describe a mean length of 3.5 cm, with a mean width of 1.1 cm at its midportion (Gigris et al. 1975). The ACL is the primary restraint to anterior tibial translation relative to the femur at all angles of knee flexion, providing 80 to 85% of total restraint (Takai et al. 1993; Butler et al. 1980). The ACL provides a secondary restraint to tibial rotation, especially near full extension (Markolf et al. 1976).

▶ **KEY POINT**

The anterior cruciate courses anteriorly, medially, and distally from the femur to the tibia.

Although Amis and Dawkins (1991) described the fiber bundle arrangement of the ACL as a three-band system (anteromedial, intermediate, and posterolateral), most utilize a two-bundle description (anteromedial and posterolateral) (Gigris et al. 1975; Bach et al. 1997; Brantigan & Voshell 1941; Edwards et al. 2007, 2008). As mentioned previously, the ACL spirals on itself within these fiber bundles. It is this spiraling of fibers that allows a portion of the cruciate to remain taut throughout full range of motion of knee flexion to extension. When the knee is in extension, the posterolateral band is tightened and the anteromedial band is slack. When the knee is in flexion, the posterolateral band is slack while the anteromedial band is taut (Gigris et al. 1975; Petersen

& Tillmann 2002; Bach et al. 1997; Beynnon et al. 1995; Beynnon et al. 1996; Jordan et al. 2007). Recent evidence has demonstrated that the posterolateral band specifically has a significant effect in controlling tibiofemoral rotation (Lorbach et al. 2010).

The **posterior cruciate ligament** (PCL) attaches to the proximal tibia in a small recess on the posterior surface of the tibia and intercondylar space and to the posterior lateral portion of the medial femoral condyle (see figure 14.6). The lower boundary of the femoral attachment is convex, while the upper boundary is more horizontal (Gigris et al. 1975). The dimensions of the PCL describe a mean length of 3.8 cm and a mean width of 1.3 cm at its midportion (Gigris et al. 1975). The PCL runs posteriorly, laterally, and distally from the femur to the tibia and is narrowest at its midportion. Its attachment sites at both the tibia and the femur fan out, making it wider at the ligament–bone interface compared with its midsubstance. Although not injured nearly as often as its counterpart the ACL, the PCL can be injured as discussed in Clinical Correlation 14.3.

▶ **KEY POINT**

> The posterior cruciate ligament courses posteriorly, laterally, and distally from the femur to the tibia.

Similar to the ACL, the PCL is thought to be composed of either two (Amis et al. 2003; Markolf et al. 2006) (anterolateral and posteromedial) or four (Covey et al. 2008; Markolf et al. 2006) bundles. Regardless, the twisting fibers allow some part of the ligament to maintain tension throughout full motion. The anteromedial band is significantly larger than the posterolateral. The posterior cruciate is on more slack during knee extension, and tension in the anteromedial band gradually increases as knee flexion increases.

Medial and Lateral Collateral Ligaments   The medial collateral ligament (MCL) is reported by some to be the most commonly injured knee ligament in competitive and recreational sports (McMahon & Skinner 2003). A reported 90% of all knee ligament injuries in young active patients occur to either the MCL, the ACL, or an MCL–ACL combination (Miyasaka et al. 1991). The primary functions of the MCL and lateral collateral ligament (LCL) are to provide a restraint from valgus and varus stress about the knee and to control rotation (Warren et al. 1974; Grood et al. 1981). A valgus stress at the knee would cause the lower leg (tibia) to move into abduction relative to the femur, while a varus stress at the knee would cause an adduction of the tibia relative to the femur.

Their location along the medial and lateral sides of the tibiofemoral joint makes these two ligaments perfect for restraining and stabilizing the knee from varus and valgus stress. The MCL is a much broader and flatter ligament than the LCL (figure 14.7). It covers a large area of the medial side of the knee joint and is divided into both superficial and deep portions. The anterior is more superficial, while the posterior is the deep portion of the MCL. A unique feature of the MCL is that its deeper portion also has attachments to the capsule and the medial meniscus. Both superficial and deep portions attach proximally to the medial femoral epicondyle, while the distal portions attach at separate locations. The superficial portion attaches to the shaft of the tibia, with slips of fibrous tissue attaching to the retinaculum, while the deep attaches slightly posterior to the tibial condyle. Injuries to the proximal

CLINICAL CORRELATION 14.3

Posterior cruciate tears are the result of a sudden twisting that applies a passive external force to the tibiofemoral joint. This can occur when the anterior tibia is hit on the dashboard during a motor vehicle accident or when a person falls on a flexed knee and lands directly on the tibia with the foot in plantar flexed position. Finally, it is possible that a hyperextension injury can rupture the PCL.

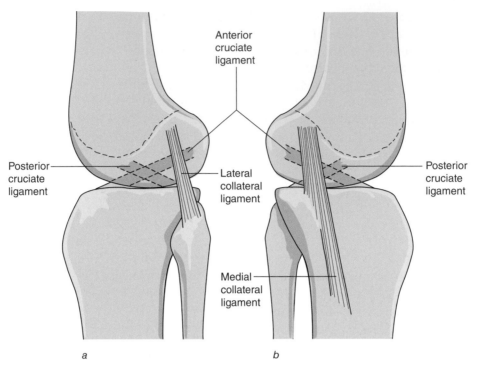

Anterior
cruciate
ligament

Posterior
cruciate
ligament

Lateral
collateral
ligament

Posterior
cruciate
ligament

Medial
collateral
ligament

a   b

▶ **FIGURE 14.7** The lateral and medial collateral ligaments: *(a)* lateral and *(b)* medial aspects.

CLINICAL CORRELATION 14.4

The Pellegrini-Stieda sign is usually seen as a calcium deposit at the femoral attachment of the medial collateral ligament. This is thought to occur after trauma to the MCL as the proximal portion of the ligament goes through acute hemorrhage. An area of tenderness will be noted at the MCL's proximal attachment at the distal medial femur.

portion of the MCL can cause calcific deposition as described in Clinical Correlation 14.4. The LCL is more tubular in shape and runs almost vertically from the lateral epicondyle of the femur to the head of the fibula (Sanchez et al. 2006; Mendes & Vieira da Silva 2006). Unlike its medial counterpart, the LCL does not contain attachments to the lateral meniscus. Historically, the LCL is rarely injured in isolation, but it can be disrupted by a straight varus stress to the knee or by complete knee dislocations causing multiple ligament injuries (Romeyn et al. 2006). Deep to the LCL are several other structures that provide restraint to varus stress, including the fabellofibular ligament and the arcuate complex (Seebacher et al. 1982).

▶ **KEY POINT**
The medial collateral ligament has deep attachments to the capsule and the medial meniscus.

Joint Capsule The external fibrous knee joint capsule is large and encompasses both compartments of the tibiofemoral joint and the patellofemoral joint. This expansive capsule has an outer more fibrous portion and an inner synovial portion. The capsule is attached directly to bone around both the femur and the tibia. The superior anterior portion of the capsule extends about a hand's breadth superiorly above the joint line, much farther than the posterior portion. The capsule invaginates the intercondylar notch posteriorly, and anteriorly it has expansive tissue attachments to the vasti muscles. Because of the great degree of motion at the knee joint, the capsule is reinforced by numerous soft-tissue structures that can be seen in table 14.1.

The synovial lining is the portion that more closely follows the femoral condyles

TABLE 14.1 Knee Capsule Soft-Tissue Reinforcements

Portion of capsule	Tissue reinforcements
Anterior	Quadriceps
	Patellar tendon
	Retinaculum
Lateral	Biceps femoris
	Gastrocnemius (lateral head)
	Iliotibial band
	Lateral collateral ligament
	Lateral retinaculum
	Popliteus
Posterior	Gastrocnemius
	Hamstrings
	Arcuate complex
	Oblique popliteal ligament
Posterolateral	Arcuate complex
	Lateral collateral ligament
	Popliteus
Medial	Medial collateral ligament
	Pes anserine
	Posteromedial capsule
	Posterior oblique ligament
	Retinaculum

and invaginates into the intercondylar notch. A proximal expansion superiorly forms the superior capsule of the knee joint. The superior portion of the capsule is on slack with the knee in extension and is placed on tension during knee flexion. Joint fluid can accumulate in this area after an injury, often creating limitations of motion. As long as the amount of fluid is not excessive, it can move around the joint with knee movements. When the knee is extended, the posterior portion of the capsule becomes taut and forces fluid anteriorly. The opposite is true with knee flexion, in which the fluid is moved posteriorly. The position of maximal volume in the tibiofemoral joint is 30° of knee flexion. Therefore, with a significant amount of knee intra-articular effusion, the joint will be held in 30° of flexion as a position of comfort.

Embryonic remnants of chambers formed during fetal development are called synovial **plicae**. These are soft-tissue pleats or folds along the inner surface of the synovial membrane. With trauma or repetitive minor insult, these linings can become thickened and fibrotic. During knee flexion and extension, these irritated folds can rub across the femoral condyle and patella. The folds can become pathologically inflamed, creating what is commonly called plicae syndrome, as described in Clinical Correlation 14.5.

Meniscus

In the earlier days of modern medicine, the menisci were thought to be of little value. In fact, Sutton (1897) described the cartilage structures as "functionless remains of leg muscles." Even Dr. McMurray (1942) claimed that "a far too common error is shown in the incomplete removal of the injured meniscus." Meniscus injuries are the most common injury

CLINICAL CORRELATION 14.5

Synovial plicae can be a tremendous source of knee pain. Most commonly, the anterior medial synovial fold becomes irritated and bowstrung over the medial femoral condyle. Pain is perpetuated with flexion of the knee. A common differential diagnosis for plicae syndrome is a meniscus tear, as their clinical symptoms are very similar.

treated by orthopedic surgeons (Garrett et al. 2006). These two structures are fibrocartilaginous discs located directly between the femoral condyles and the tibial plateau (figure 14.8). Each **meniscus** is wedge (pie) shaped when viewed in the frontal plane. This allows for a thicker periphery and a thinner central portion. This fits reciprocally with the femoral condyles by making a concavity atop the tibial plateau. Small coronary ligaments attach each meniscus to the tibial plateau and the joint capsule.

Although very similar, each meniscus has some unique characteristics. The medial meniscus is more C-shaped than the lateral. It is approximately 3.5 cm in length. The horns, or attachments of the meniscus to the tibia, are wider posteriorly than anteriorly. These horns are farther apart on the medial side versus the lateral. Because the medial tibial plateau is larger than the lateral, the medial meniscus has a larger diameter. The medial meniscus has a firm attachment to the MCL and the medial capsule. An additional attachment to the medial meniscus occurs from the semimembranosus, which is attached to the posterior horn. This muscular attachment may allow some excursion during flexion activities at the knee.

The lateral meniscus is much more circular and O-shaped. It is actually about four-fifths of a ring. The anterior and posterior horns are of similar width. There is no attachment to the LCL, but there is a muscular attachment to the popliteus tendon to the posterior horn (Brantigan & Voshell 1941; Messner & Gao 1998). The lateral meniscus covers a larger surface area of the tibia than does the medial meniscus (Ateshian et al. 1991; Fukubayashi et al. 1982; Johnson et al. 1995; Kettlekamp & Jacobs 1972). Seedhom has shown that the meniscus in the medial side of the joint takes only 50% of the load, while the lateral meniscus takes 70% of the load (Seedhom et al. 1974). Therefore, even though the medial meniscus covers a larger area than the lateral, injury to the lateral may be ultimately more problematic because of its enhanced joint protective capabilities.

▶ **KEY POINT**
The medial meniscus is C-shaped. The lateral meniscus is more circular, or O-shaped.

Bursae

Bursae are fluid-filled synovial pockets of tissue that surround or communicate with other joint cavities in an attempt to reduce friction between muscles, tendons, ligaments, and bones. Multiple bursae exist in the knee, as outlined in table 14.2.

Fat Pads

The **infrapatellar fat pad** between the patellar tendon and the tibia is sometimes known as Hoffa's pad. This extensive fat pad extends superiorly under the patella and separates it from the tibia. According to Dye and colleagues (1998), this pad is one of the most sensitive structures in the knee.

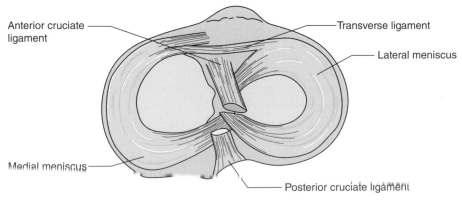

▶ **FIGURE 14.8** Superior surface of the tibial plateau

TABLE 14.2 Knee Bursae

Bursa	Structures separated	
	Bone	Soft tissue
Suprapatellar	Femur	Quadriceps
Prepatellar	Patella	Skin
Deep infrapatellar	Tibia	Patellar tendon
Pretibial	Tibia	Skin
Lateral gastrocnemius	None	Gastroc and joint capsule
Fibular	None	LCL and biceps femoris
Fibulopopliteal	None	LCL and popliteus
Subpopliteal	Femur	Popliteus
Medial gastrocnemius	None	Medial gastroc and joint capsule
Pes anserine	None	MCL and pes anserine
Bursa semimembranosa	None	MCL and semimembranosus

Gastroc = gastrocnemius; MCL = medial collateral ligament; LCL = lateral collateral ligament

Patellofemoral Joint

The patellofemoral joint contains the posterior surface of the patella and the trochlear surface of the anterior femur. This is a very shallow and incongruent joint, as the patella does not fit deeply into the trochlea. The anatomy of the patellofemoral joint is described earlier in this chapter. Dynamic stability of this joint is provided by the quadriceps muscles and the iliotibial band. Static passive stability is provided by the retinaculum of the anterior knee, the knee capsule, and the medial and lateral patellofemoral ligaments.

Specific alignment at the patellofemoral joint is termed the **quadriceps angle (Q angle)**. This angle is a description of the pull of the quadriceps on the patellar tendon. It is thought to be an indicator for patellofemoral joint dysfunction. A larger Q angle is thought to create a larger lateral vector and potentially a greater predisposition for lateral patellar tracking when compared with a smaller Q angle (Schulthies et al. 1995; Powers 2003). Excessive lateral tracking of the patella could increase pressure on the lateral knee structures during contact between the lateral patellar facet and the lateral femoral trochlea, leading to degeneration of the articular cartilage, or chondromalacia (Boden et al. 1997; Grelsamer & Klein 1998; Guerra et al. 1994; Livingstone 1998; Mizuno et al. 2001; Reider et al. 1981).

The Q angle is formed between a line connecting the anterior superior iliac spine to the midpoint of the patella and the bisecting line that runs from the tibial tuberosity to the midline of the patella (figure 14.9). Normal Q angles for healthy people range from 10 to 15° (Horton & Hall 1989; Livingstone & Mandigo 1997). It is generally accepted that women have higher Q angles than do men (Aglietti et al. 1983; Nguyen & Shultz 2007; Guerra et al. 1994; Horton & Hall 1989; Hsu et al. 1990; Woodland & Francis 1992). The reason for this variance, once thought to be that females tend to have wider pelvises, has been disputed by several studies (Horton & Hall 1989; Kernozek & Greer 1993). The exact reason for females having larger Q angles remains unclear.

One of the main problems with using the Q angle as a clinical measurement is the lack of agreement on how to measure it. The Q angle was originally described by Brattstrom (1964) as a static measurement taken with the knee in an "extended, end-rotated position." Traditionally, the Q angle was measured in supine with either the quadriceps relaxed or contracted (Insall et al. 1976; Fairbank et al. 1984; Guerra et al. 1994), while others have assessed it in

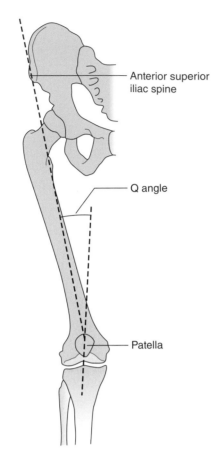

Anterior superior
iliac spine

Q angle

Patella

▶ **FIGURE 14.9** Quadriceps angle.

standing (Caylor et al. 1993; Cowan et al. 1996; Roy & Irvin 1983; Woodland & Francis 1992) with various foot positions (Cowan et al. 1996; Guerra et al. 1994; Reider et al. 1981). It is even possible that what is being measured as a static position can be altered dynamically. Therefore, the best way to assess the Q angle may be during dynamic active function (Kernozek & Greer 1993). Additional concerns arise because the relationship between an excessive Q angle and signs and symptoms of patellofemoral pain are not always consistent (Aglietti et al. 1983; Haim et al. 2006; Messier et al. 1991; Thomee et al. 1995; Witvrouw et al. 2000; Caylor et al. 1993).

Superior Tibiofibular Joint

The superior tibiofibular joint is just distal to the knee and considered to be completely separate from the knee, but because of its close proximity to the tibiofemoral joint, it is dis-

cussed in this section. This joint is composed of the proximal portions of both the tibia and the fibula. The superior tibiofibular joint has the following three functions: (1) dissipation of torsional stresses applied at the ankle joint, (2) dissipation of lateral tibial bending movements, and (3) tensile rather than compressive weight bearing (Ogden 1974). The superior tibiofibular joint is surrounded by a joint capsule reinforced by anterior and posterior tibiofibular ligaments (Levangie & Norkin 2005). Both these surfaces are covered with articular cartilage (Stranding 2009). Movement of this joint occurs during any movement involving the ankle (Magee 2008).

Nerve Supply

Sensory innervation of the knee and surrounding ligaments is supplied by the L3 to L5 spinal nerve roots. These nerve roots travel to the cord via posterior tibial, obturator, and femoral nerves (Neumann 2010). Motor innervation is listed in table 14.3 later in this chapter. The quadriceps is innervated primarily by the femoral nerve, while the flexors and rotators are innervated by the lumbar and sacral plexus, primarily by the tibial portion of the sciatic nerve (Neumann 2010).

Blood Supply

The two major arteries that supply blood to the area surrounding the knee are the femoral and popliteal arteries, which divide into six further branches. The six branches are the superior medial and superior lateral genicular arteries, the inferior medial and inferior lateral genicular arteries, the descending genicular artery, and the recurrent branch of the anterior tibial artery.

Joint Function

This section discusses knee joint function, including the axes of motion, arthrokinematics, range of motion, closed and loose packed positions, end feel, and capsular patterns of the region. This foundation information will be valuable in the future for discussing pathological conditions that have caused either excessive motion or restricted motion at the knee joint.

Tibiofemoral Joint

The tibiofemoral joint is one of the largest joints in the body and is located between the femur and the tibia. This joint is commonly known as the knee joint unless specified otherwise.

Axes of Motion

The tibiofemoral joint is a double condyloid synovial joint, displaying up to three degrees of freedom of movement. Movements of flexion and extension occur around a mediolateral axis in the sagittal plane. Internal and external rotation motions occur around a vertical axis in the transverse plane, while abduction and adduction, which are limited in comparison with the other motions, occur around an anteroposterior axis in the frontal plane.

The majority of movement that occurs at the tibiofemoral joint is flexion and extension in the sagittal plane. Smaller amounts occur in the other two planes if the knee is in some degree of flexion. Full extension is the closed packed position of the knee; the joint has minimal accessory movements or physiological movement capabilities in that position. Once some amount of flexion is induced, both rotations and adduction and abduction movements can occur.

Arthrokinematics

Specific arthrokinematics of the tibiofemoral joint include accessory motions of rolls and glides of either the femur or the tibia depending on whether the motion is in open-chain or closed-chain patterns.

Flexion and Extension Flexion and extension involve rolling and gliding of either component of the joint (tibia or femur) and depend on whether the motion occurs in a weight-bearing position or not. During weight-bearing activities (a closed kinetic chain), the femur moves on a relatively fixed tibia, or a femur on tibia movement pattern (figure 14.10a). During a non-weight-bearing movement, (an open kinetic chain), the tibia moves on a relatively fixed femur, or a tibia on femur movement pattern (figure 14.10b).

During a closed kinetic chain movement such as a squat, the convex condyles of the femur roll posteriorly on the concave tibial plateau. Because the condyles are larger than the tibial plateau, this could create a problem—a pure rolling motion would cause the condyles to roll completely off the back of the tibia if it were not for an anterior gliding motion. This anterior gliding motion occurs in part because of an intact ACL. As the femur rolls posteriorly (a relative anterior translation of the tibia), the ACL becomes taut and limits further motion, creating the gliding that occurs to create almost a pure spinning of the condyle on the tibia. When moving from a position of flexion to extension such as standing from a seated position, the femur rolls anteriorly with a simultaneous posterior glide.

In open kinetic chain movements, the tibia moves on a relatively fixed femur. During the motion of knee extension, the concave tibia rolls and glides anteriorly across the convex condyles of the femur. With knee flexion in an open chain, the tibia rolls and glides posteriorly in a similar manner.

This text gives the classic descriptions of tibia on femur and femur on tibia motion. Recent evidence using three-dimensional studies has questioned these theories as being inaccurate, more than likely because of the classic use of two-dimensional imaging techniques to describe these motions (DeFrate et al. 2004; Hill et al. 2000; Patel et al. 2004). Future studies will hopefully more clearly demonstrate the complexity of these motions.

The motions of lateral and medial rotation are normally present during knee flexion and extension at the tibiofemoral joint. The tibia can laterally rotate from 0 to 20°, and it can medially rotate from 0 to 15°.

As long as the knee is not in full extension, movements of lateral and medial rotation can occur. These movements occur as tibia on femur motion in open kinetic movements and femur on tibia motion in a closed kinetic chain (figure 14.11). The motions occur in

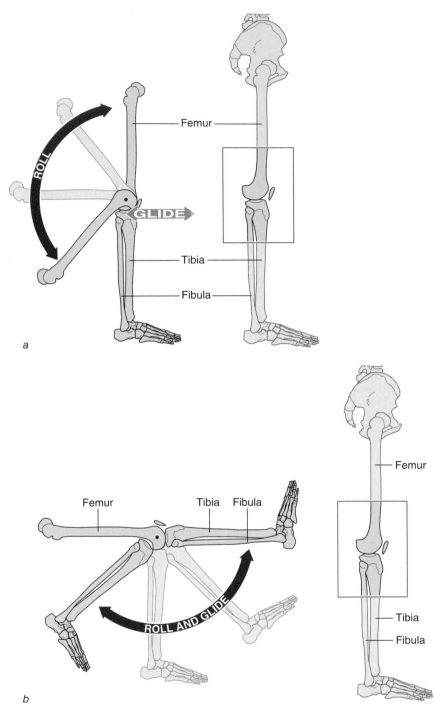

▶ **FIGURE 14.10** *(a)* Femur on tibia knee flexion and extension; *(b)* tibia on femur knee flexion and extension.

the transverse plane around a superoinferior (vertical) axis. The amount of available axial rotation through the long axis of the tibia increases with knee flexion. In full terminal extension, no rotation exists because of the tight capsular and ligamentous structures and the constraint of full joint congruency.

Also as the knee moves from a position of 90° of flexion to full knee flexion, the amount of axial rotation decreases once again. The amount of rotation sequentially increases to a maximum of about 45° of total rotation once the knee reaches the position of 90° of flexion (Almquist et al. 2002; Mossber & Smith 1983).

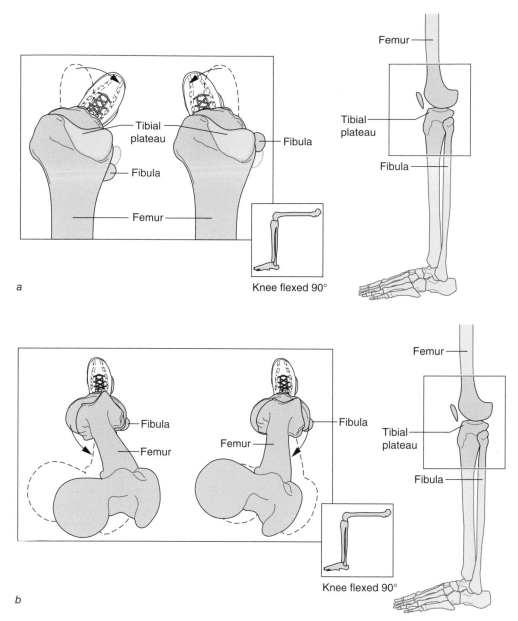

▶ **FIGURE 14.11**   (*a*) Tibia on femur internal and external rotation; (*b*) femur on tibia internal and external rotation.

Screw-Home Mechanism   The **screw-home mechanism** is an important concept for functional activities and knee stability. During the final degrees of knee extension range of motion, an obligatory lateral rotation of the tibia occurs. Obligatory implies that this coupled pattern (knee extension and rotation) is not voluntary. It occurs without thought or conscious effort. This has been called the locking or screw-home mechanism. This automatic obligatory motion occurs in part because of the sizes of the femoral condyles

and the tibial plateau. As mentioned earlier in this chapter, the medial femoral condyle and medial tibial plateau are larger and longer than those of the lateral compartment. When the knee goes through its arthrokinematic motions, the femoral condyles roll and glide. Because the medial femoral condyle is longer than the lateral, it continues to roll and glide (closed kinetic chain). In a weight-bearing position, this may occur as femoral medial rotation at the end range (a relative tibial lateral rotation) (figure 14.12). When the tibia

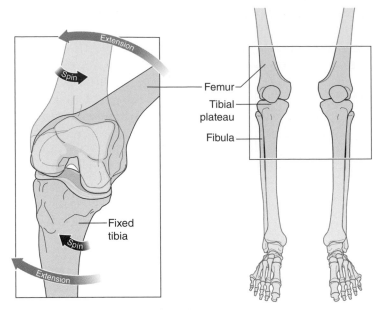

▶ **FIGURE 14.12**   Screw-home mechanism of the knee.

rolls and glides over the femoral condyles (open kinetic chain), the medial plateau has a longer distance to travel because of the longer medial femoral condyle. This continuation of medial-side movement while the lateral side has stopped creates lateral rotation of the tibia, which is most evident in the last 5 to 10° of knee extension. This is a very stable position (closed packed) for the knee, as the meniscus is wedged tightly in the joint and the capsule and ligaments in and surrounding the joint are pulled taut.

▶ **KEY POINT**

The screw-home mechanism is an obligatory external tibial rotation that occurs in the later degrees of open kinetic knee extension.

Unlocking of the knee occurs via a medial rotation of the tibia (open kinetic chain). This is also an obligatory motion found in normal arthrokinematic movement patterns. It is thought to occur by initiation of a flexion moment by the popliteus muscle. In a weight-bearing position, this rotation may occur through a lateral rotation of the femur.

Varus and Valgus   Additionally, there is a slight amount of movement in the frontal plane along an anteroposterior axis. This is the varus and valgus motion at the knee in which the foot moves laterally to the midline of the lower leg (valgus) or medially to the lower leg (varus). It should not be confused with the angulation of the knee, which is also described as valgus and varus. This motion is minimal to none at full extension because of the tightness of the capsule and surrounding ligaments, but it does increase slightly in the loose packed position of about 30°.

Range of Motion

The largest motions at the tibiofemoral joint include flexion and extension of the knee. Knee flexion is from 0 to 150°, while extension usually achieves 0°, or full extension (AAOS 1965).

Flexion and extension motion varies depending on body morphology, age, and health status. A person who is healthy and thin may have up to 150° or more of movement. A patient with osteoarthritis may have limitations of 90° or less. Generally, around 135° is considered the norm for flexion (Hemmerich et al. 2006). Knee extension also varies and is generally around 5 to 10° of hyperextension, allowing most to bend past the neutral fully extended knee position. The mediolateral axis of motion does not remain fixed but rather follows a curved path dictated by the shape of the femoral condyles. This instantaneous axis of rotation begins at a more proximal and anterior position on the femoral condyle in extension and moves to a slightly inferior and posterior position upon full knee flexion. This has clinical implications for measuring joint motion with a goniometer. Normal goniometric technique requires finding a fixed axis to allow the movable arm to follow the tibia while still maintaining a stationary axis. Because the true axis of rotation occurs, there probably exists some slight degree of error in recording range of motion at the knee. Therefore, the lateral epicondyle of the knee—a position close to the instantaneous axis of rotation—is recommended as the goniometric stationary axis.

Closed and Loose Packed Positions

The closed packed position is a position of stability for the tibiofemoral joint and is full extension with lateral rotation of the tibia. The loose packed position is where the largest amount of capsular volume occurs and is in a position of 25 to 30° of knee flexion.

End Feel

The tibiofemoral joint has several normal end feels. During knee extension, the end feel is firm because of tension in the posterior knee joint capsule, the oblique and arcuate popliteal ligaments, the collateral ligaments, and the cruciate ligaments. The end feel for knee flexion is soft-tissue approximation due to contact between the muscle bulk of the posterior thigh and calf between the heel and the buttocks.

Capsular Pattern

If there is a true capsular pattern of the tibiofemoral joint, knee flexion range of motion will be limited more than knee extension motion. This will occur with capsular shortening due to immobilization or trauma.

Patellofemoral Joint

Although a role of the patella may be to act as a bony anterior shield to protect structures that lie dorsal to it, the main function of the patellofemoral joint is to increase the moment arm for the quadriceps muscle, helping to provide greater force production during functional activities. The ability to increase torque is greatly enhanced by increasing the lever arm of the mechanical system in use. The patella essentially increases the lever arm of the quadriceps muscles, augmenting the force provided by the quadriceps to the tibia.

The moment arm created by the patella changes as the knee goes through a full arc of motion from full extension to full flexion. In general, the moment arm starts out small and increases up to about 20 to 60° of knee flexion, at which point it then decreases again to full flexion range (Schmidt 1973; Krevolin et al. 2004).

▶ **KEY POINT**
The moment arm afforded by the patella changes as the knee goes through a range of flexion and extension. In general, it starts small in full extension, then increases as the knee flexes.

Patellar Motions

Movements of the patellofemoral joint include gliding motions as the knee flexes and extends. These motions occur between the dorsal surface of the patella and the ventral surface of the femur. Because of the distal attachment of the patella to the tibial tubercle of the tibia, open kinetic chain movements of the patella follow the path of the tibia (figure 14.13). In closed kinetic chain movements, the patella is relatively tethered at the tibia while the femur moves underneath the patella. In this instance, it is the femoral surface that glides behind the patella.

Motions of the patellofemoral joint are commonly described as movements of the patella on the femur as if in an open kinematic chain pattern. These motions include flexion and extension, medial and lateral glides, medial and lateral tilts, medial and lateral rotation, and anterior and posterior tilts. These are much easier to visualize than movements of the femur on the patella. When the knee moves from a position of full extension into

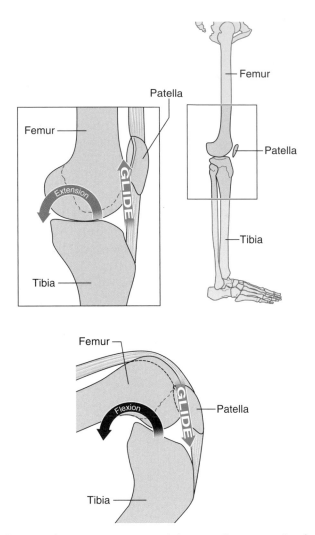

▶ **FIGURE 14.13**   Open kinetic chain movements of the patella across the femur during knee extension and flexion.

flexion, the associated movement of the patella is described as patellar flexion, or inferior glide. Movements that occur from a flexed position to that of extension are patellar extension, or superior glide. These two motions occur as the patella enters the femoral trochlea (extension to flexion) or as the patella exits the trochlea (flexion to extension).

Patellar flexion and extension occur as gliding motions in the sagittal plane. The patella also goes through medial and lateral gliding motions (figure 14.14). During lateral glide, the lateral edge of the patella moves closer to the lateral side of the knee, and during medial glide the medial side moves toward the medial edge of the knee. Medial and lateral glide of the patella occur as translations in the frontal plane.

Rotation of the patella around a vertical axis is termed tilting. Tilts are described by which direction the reference facet is moving. In a medial tilt, the medial posterior facet moves closer to the medial femoral condyle, while a lateral tilt is movement of the lateral posterior patellar facet closer toward the lateral femoral condyle (figure 14.14). Rotation of the patella can occur around an anteroposterior axis and is described by the direction of the inferior pole of the patella. A lateral rotation occurs when the inferior pole is directed toward the lateral side of the knee, while a medial rotation occurs when the inferior pole is directed medially (figure 14.14). A laterally rotated tibia is easily seen in a person with excessive lateral tibial torsion, known as a bayonet sign. Because the tibial tubercle is lateralized com-

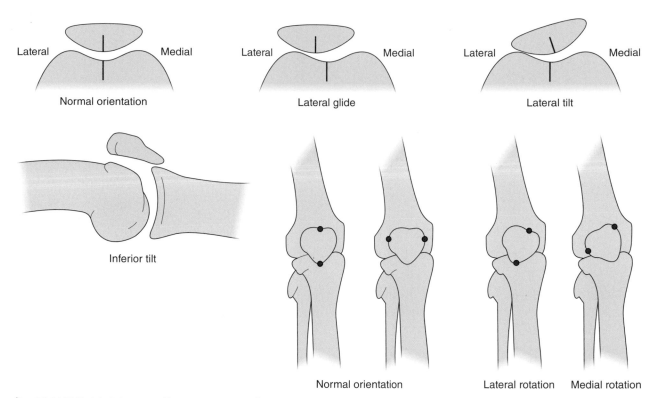

Lateral   Medial

Normal orientation

Lateral   Medial

Lateral glide

Lateral   Medial

Lateral tilt

Inferior tilt

Normal orientation   Lateral rotation   Medial rotation

▶ **FIGURE 14.14** Patellar motions and orientations.

pared to normal, the patella's inferior pole will be in a laterally rotated position, even when the quadriceps are relaxed. Lastly, the motions of anterior and posterior tilts can occur at the patellofemoral joint. These motions are described by the location of the inferior pole of the patella in either a depressed (posterior tilt) or elevated (anterior tilt) position. An inferiorly tilted patella can be problematic, as it may pinch or irritate the patellar fat pad that lies underneath the patellar tendon.

Patellar Articular Congruence

At full knee extension, the patella is usually superior to the trochlear surface of the femur. There is no joint contact and thus no to minimal joint congruence. In the range of motion near full extension, the patella relies on soft-tissue structures and dynamic support to remain in its normal position. Indeed, it is here in the earliest ranges of knee flexion that the potential for subluxation and dislocation are the greatest. Stability can be enhanced by patellar position superiorly and inferiorly. Someone who has a longer patellar tendon will

typically also have a patella that rests higher on the femur, described as a high-riding **patella alta** (figure 14.15). In this elevated position, it takes longer for the patella to reach the bony constraint of the femoral trochlea, and thus the patella is at a greater risk for subluxation (see Clinical Correlation 14.6). The knee will have to go through a greater range of flexion motion before the patella becomes stable in the trochlea. A low-riding patella, or **patella baja**, has a smaller patellar tendon length. The patella enters the trochlea sooner than normal, possibly placing excessive compression forces on the patellofemoral joint.

▶ **KEY POINT**

A high-riding patella is known as patella alta. A low-riding patella is known as patella baja.

During normal knee motion as flexion occurs, the patella enters the femoral trochlea around 20 to 30° of flexion (Goodfellow et al. 1976). Because the extended lateral wall of the trochlea provides a lateral buttress to patellar

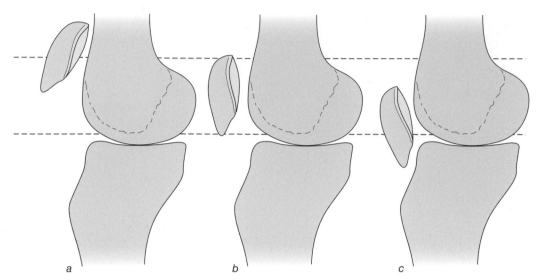

▶ **FIGURE 14.15**   Patella reference positions: *(a)* high-riding patella, or patella alta; *(b)* normal patella; and *(c)* low-riding patella, or patella baja.

CLINICAL CORRELATION 14.6

Acute patellar dislocations account for approximately 2 to 3% of all knee injuries. These injuries occur as a result of noncontact lower extremity internal rotation as well as knee valgus stress on a fixed distal extremity. Subluxations and dislocations more commonly occur in the motion closer to knee extension as the patella nears leaving the confines of the femoral trochlea. Repair of the medial patellofemoral ligament has become one of the more recent surgical procedures for recurrent dislocations.

motion, in the normal knee little excessive medial or lateral motion occurs during flexion because the patella remains relatively centered on the trochlea. In the pathological knee, various amounts of tilting and medial or lateral gliding can occur, disturbing knee symmetry and resulting in pathological conditions. During normal knee motion, the point of contact when the patella first enters the trochlea is at the distal pole region, and as flexion motion increases, that contact area moves proximally. This is just the opposite for the trochlear surface, in which the contact during early flexion is proximal and moves distally during increased flexion motion. The entire area of contact is altered as the knee goes through full range of motion. The area of greatest contact between the patella and the trochlea occurs at around 45 to 60° of knee flexion, when joint reaction forces are great-

est (figure 14.16). Both the contact area and the contact stress increase until up to 90° of knee flexion (Huberti & Hayes 1984; Luyckx et al. 2009; Salsich & Perman 2007). This is by nature a protective function. When joint reaction force is increased, so is the amount of contact area. By 90° of knee flexion, almost the entire posterior surface of the patella has had some contact except for the far medial odd facet, which usually does not generate contact until full flexion has been attained (Goodfellow et al. 1976; Komistek et al. 2000; Nakagawa et al. 2003). In those with normally aligned patellofemoral joints, this distribution of force allows the knee to resist the deleterious effects that could occur from routine high-compressive forces. In deep flexion, the patella actually bridges the intercondylar notch, as there is contact only on the far medial and lateral edges (Feller et al. 2007). It is worth

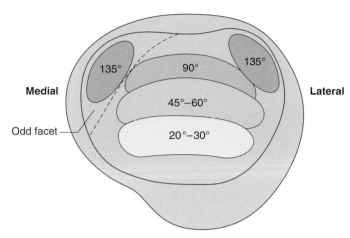

FIGURE 14.16 Patellar contact forces during knee motion. Only the inferior pole makes contact early in knee flexion. Contact moves proximally as the knee moves into further flexion.

noting that although absolute contact areas are greater in males than females, this disparity disappears when measurements are normalized to body weight (Salsich et al. 2003; Besier et al. 2005; Gold et al. 2004).

Patellofemoral joint reaction forces and joint stress can be tremendous during even the simplest activities of daily living, not to mention with sports and recreational activities. Studies have demonstrated forces of 1.3 times body weight (BW) during level ambulation, 3.3 times BW during stair ambulation, and up to 7.8 times BW during a deep knee bend or squat (Feller et al. 2007; Reilly & Martens 1972). The actual amount of joint stress (per unit of area) on the posterior patella or femoral trochlea varies depending on the amount of surface area to distribute the compressive forces. When larger portions of the surface are in contact with each other, the actual amount of joint stress (per unit of area) is small since the force is distributed over a larger surface area. When smaller portions of the joint surface are in contact, the amount of joint stress (per unit of area) is greater since the force is distributed over a smaller area. Compare a neutrally positioned patella and the amount of contact area (figure 14.14, normal orientation) with one that is in a position of excessive lateral tilt (figure 14.14, lateral tilt). This clearly demonstrates the effect a malpositioned patella may have on articular cartilage.

Therefore, the interplay between force and area is of extreme importance because a high force across a large area may be less injurious than smaller forces across an even smaller contact area.

The two main factors that contribute to contact forces are (1) the amount of resistance of the quadriceps muscles and (2) the knee angle. Because the patella is tethered superiorly by the quadriceps tendon and inferiorly by the patellar tendon, a contraction of the vasti muscles superiorly results in a posterior compressive force upon the patella. The posterior compression force is accentuated the farther the knee is flexed as the knee angle is decreased (knee flexion is increased). When the knee is in or near full extension, there is only a small posterior force vector to the pull of the quadriceps, as most of the force is directed superiorly. As the angle becomes more acute as the knee is flexed, the posterior compressive force is accentuated (figure 14.17). As the knee becomes flexed, even passive tension in the muscles and tendons create an increased force. Superimposition of a quadriceps muscle contraction greatly increases this compressive force.

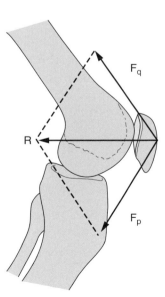

FIGURE 14.17 Patellofemoral compressive forces are increased as the force of the quadriceps increases and as the knee moves into deeper knee flexion.

► **KEY POINT**

Patellofemoral contact forces are increased as the force of resistance of the quadriceps is increased and the knee is increasingly flexed.

Superior Tibiofibular Joint

Although not commonly thought of as a knee joint, the superior tibiofibular joint is very near the tibiofemoral joint and thus is considered one of the articulations of the knee joint proper.

Axes of Motion

The superior tibiofibular joint is a plane synovial joint formed by the head of the fibula and the proximal portion of the tibia. Despite the fact that the joint surfaces are relatively flat, some describe the proximal fibular facet as slightly convex, while the tibial facet is slightly concave (Eichenblat & Nathan 1983). The superior tibiofibular joint has three axes of movement and three degrees of freedom. Motions at this joint are typically gliding motions that occur as an anterior and posterior glide of the fibula on the tibia, a superior and inferior glide of the fibula on the tibia, and rotation of the fibula (Radakovich & Malone 1982).

The majority of movement that occurs at the superior tibiofibular joint is gliding of the fibula on the tibia.

Arthrokinematics

Because of the shape of the joint surfaces of the fibula on the tibia, the joint accessory motions are gliding of the fibula on the tibia in the directions just described.

Closed and Loose Packed Positions

The closed packed position of the tibiofibular joint is full ankle dorsiflexion, while the loose packed position is 0° of ankle plantar flexion.

End Feel

The normal end feel for the tibiofibular joint is firm or ligamentous because of its support via ligamentous attachments, including the thick interosseous membrane.

Capsular Pattern

The capsular pattern of the tibiofibular joint is pain with contraction of the biceps femoris muscle (Kaltenborn 2011).

Muscles

Since the knee moves predominantly in a single plane, the muscles surrounding the knee can be divided easily into either knee extensors or knee flexors. The knee extensor group includes the quadriceps muscle group, while the flexors include the hamstrings, gracilis, and sartorius. All the knee muscles and their origins, insertions, actions, and innervations can be seen in table 14.3.

Multiple muscles form the quadriceps femoris group: the rectus femoris, the vastus lateralis, the vastus medialis, and the vastus intermedius (figure 14.18). All these muscles form a common tendon that attaches distally to the patella. The rectus femoris attaches proximally at the hip, while the other three of the group arise from the femur. Because the rectus femoris crosses both the hip and the knee, it is the only knee extensor that is biarticular. The quadriceps muscles are the largest and most powerful muscles in the body.

The rectus femoris has a proximal attachment at the anterior inferior iliac spine of the hip. It travels distally to blend with the remaining quadriceps muscles to insert into the patella. A portion of the rectus femoris continues past the patella to become the patellar tendon (ligament) and attaches to the tibial tubercle. Because this portion of the quadriceps tissue is between the patella and the tibia (two bones), it probably should be called the patellar ligament, although patellar tendon is used interchangeably.

The vastus lateralis is a large muscle of the quadriceps femoris that is located on the lateral side of the joint. This thick muscle has several attachments to the femur including the intertrochanteric line, the greater trochanter, the gluteal tuberosity, the lateral linea aspera, and the lateral intermuscular septum. It runs distally to attach with the rectus femoris onto the lateral portion of the patella.

TABLE 14.3   Muscles of the Knee

Muscle	Origin	Insertion	Action	Innervation
Anterior knee muscles (extensors)				
Quadriceps				
Rectus femoris	Anterior inferior iliac spine (AIIS)	Quadriceps tendon and patellar tendon	Knee extension; hip flexion	Femoral nerve (L2, L3, L4)
Vastus lateralis	Greater trochanter, lateral lip of linea aspera, and lateral intermuscular septum	Quadriceps tendon and patellar tendon	Knee extension	Femoral nerve (L2, L3, L4)
Vastus medialis	Intertrochanteric line of femur and medial linea aspera	Quadriceps tendon and patellar tendon	Knee extension	Femoral nerve (L2, L3, L4)
Vastus intermedius	Anterior lateral femur	Quadriceps tendon and patellar tendon	Knee extension	Femoral nerve (L2, L3, L4)
Posterior knee muscles (flexors)				
Hamstrings				
Biceps femoris, long head	Ischial tuberosity	Fibula and lateral tibia	Flexes and laterally rotates knee; extends thigh	Tibial nerve (L5, S1, S2)
Biceps femoris, short head	Linea aspera	Fibula and lateral tibia	Flexes and laterally rotates knee; extends thigh	Common peroneal nerve (L5, S1)
Semimembranosus	Ischial tuberosity	Posterior medial tibia	Flexes and medially rotates knee; extends thigh	Tibial nerve (L5, S1, S2)
Semitendinosus		Medial tibia		
Sartorius	Anterior inferior iliac spine (AIIS)	Superior medial surface of tibia at pes anserine	Flexes hip and knee Laterally rotates thigh	Femoral nerve (L2, L3)

The vastus medialis is another thick and powerful muscle that covers the medial portion of the knee. Its proximal attachment is to the femur at the medial intermuscular septum, the intertrochanteric line, and the spiral line. It blends distally with the adductor magnus, adductor longus, and rectus femoris to attach on the medial side of the patella.

The vastus intermedius lies deep to the rectus femoris and arises from the anterior and lateral surface of the upper two-thirds of the shaft of the femur. Other bony attachments come from the distal portion of the lateral intermuscular septum and from the lateral lip of the linea aspera. It attaches deep to the rectus femoris and in the middle of the vastus lateralis and vastus medialis.

▶ **KEY POINT**
The quadriceps muscle group consists of the rectus femoris, the vastus lateralis, the vastus intermedius, and the vastus medialis.

The gracilis muscle (figure 14.19) arises from the anterior aspect of the inferior pubic rami and courses distally to attach at the medial surface of the tibia. The sartorius

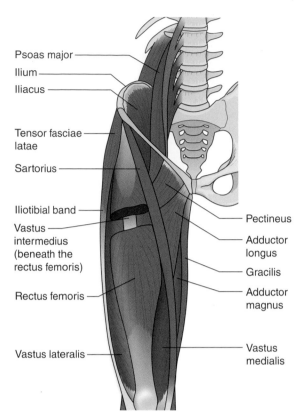

▶ **FIGURE 14.18** Anterior view of the knee muscles.

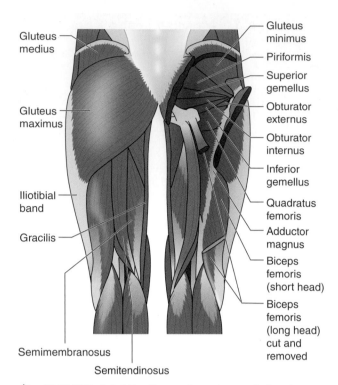

▶ **FIGURE 14.19** Posterior view of the knee muscles.

muscle (figure 14.18) arises from the anterior superior iliac spine. It is a long, strap-like muscle that wraps around the anterior to medial thigh. Both muscles travel together across the medial side of the knee to insert onto the upper medial surface of the tibia alongside the semitendinosus, where together they form the pes anserine. The hamstring muscle group occupies most of the posterior portion of the upper thigh (figure 14.19). Collectively known as the hamstrings, individually they are the biceps femoris, semitendinosus, and semimembranosus.

The biceps femoris is the most lateral hamstring muscle. It is so named because it has two heads, a long and a short. The short head arises from the lateral lip of the linea aspera and the lateral intermuscular septum; the long head, along with the remaining hamstrings, arises from the ischial tuberosity. The biceps femoris is innervated by the tibial part of the sciatic nerve. At about the middle of the thigh, the long head joins the short head, where they

both attach to the fibular head, the fascia of the lower leg, the lateral collateral ligament, and the lateral capsule of the knee.

The semimembranosus and the semitendinosus arise with the biceps femoris long head at the ischial tuberosity. As their names imply, the semimembranosus remains muscular for a longer distance distally than the semitendinosus. The semitendinosus proximally is muscular but becomes a long and slender tendon as it courses distally. For much of its length it runs behind the semimembranosus, but toward the distal end it courses anterior to its partner. Each of these muscles attaches distally to the fascia of the lower leg and the proximal medial portion of the tibia. The semimembranosus attaches deeper than the semitendinosus. The semitendinosus is commonly harvested for surgical use, as described in Clinical Correlation 14.7.

▶ **KEY POINT**

The hamstring muscle group consists of the biceps femoris, the semimembranosus, and the semitendinosus.

CLINICAL CORRELATION 14.7

The semitendinosus tendon is commonly harvested for use as a replacement tissue substitute for surgical repairs of the ulnar collateral ligament and anterior cruciate ligament. Despite sacrificing this hamstring muscle, patients rarely loose knee flexion strength after its removal. After harvest of the semitendinosus, it may take up to 6 weeks before it is safe to perform a hamstring contraction without fear of irritating these muscles. A 4- to 6-week wait for hamstring contractions allows adequate soft-tissue healing time for reattachment or scar tissue formation to tether the remaining portion of the tendon to adjacent structures.

To some extent, all the flexor muscles flex the knee. All of these muscles are biarticular in that they cross two joints, the hip and the knee. Therefore they can produce action at not only the knee but also the hip. During open kinematic chain movements, the three hamstrings can extend the hip, while in closed kinetic chain movements they can provide restraint of the pelvis and trunk.

The hamstrings not only flex the knee but also rotate it. This rotation (tibia on femur) can occur either medially or laterally in the transverse plane around a longitudinal superior to inferior axis. Because of the screw-home position being closed packed, any rotation of the knee must be done with some degree of flexion. The closed packed position implies a tightly locked and restrained knee. With the knee flexed at 90°, the tibia will rotate internally with a contraction of the medial hamstrings—the semitendinosus and semimembranosus. As a group the pes anserine also flexes and internally rotates the tibia. External tibia on femur rotation occurs with an active contraction of the biceps femoris.

Active and passive insufficiency can occur in both knee extensors and knee flexors. The hamstring group and the rectus femoris are two-joint muscles whose function can be impaired when stretched or shortened over two joints. When the hip is placed in a position of extension and the knee is concomitantly flexed, **active insufficiency** of the active hamstrings' contraction can occur because they are in an extremely shortened range, enough that overlapping myofilaments may not be able to recruit adequate tension. The same can occur if the hip is flexed and the knee is extended, although usually **passive insufficiency** of the hamstrings will not allow this to occur. Passive insufficiency of the knee occurs when the hip is flexed and the knee is extended, as the hamstrings (if tight) will not allow full knee extension. Alternatively, the rectus femoris can become passively insufficient if the hip and knee are flexed at the same time because the rectus femoris (if tight) will not have adequate length to allow full knee flexion (figure 14.20).

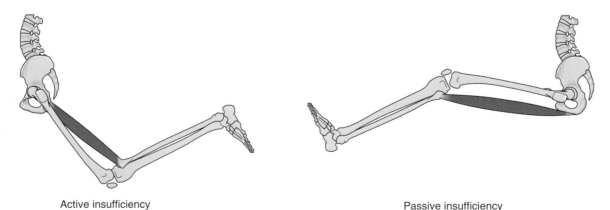

Active insufficiency   Passive insufficiency

▶ **FIGURE 14.20**   Active and passive insufficiency of the hamstrings.

Conclusion

Although the knee is a simple double condyloid synovial joint, tremendous stresses are placed on the knee during functional activities. Understanding the kinesiology and movement patterns of the knee is integral to fully understanding pathology in and around the knee. The ability to distinguish between normal and abnormal movement patterns and alignment is needed to determine proper treatment parameters. To fully appreciate and recognize anatomical variances, one needs to fully comprehend the important structures of the knee.

REVIEW QUESTIONS

1. Contractions of which muscles would place strain on the PCL?
   a. quadriceps
   b. hamstrings
   c. soleus
   d. sartorius

2. A decrease in the normal medial angle at the knee is called?
   a. genu recurvatum
   b. tibial torsion
   c. genu varum
   d. genu valgum

3. What happens during the screw-home mechanism of the knee in open kinetic chain knee extension?
   a. Internal rotation of the tibia occurs in the last few degrees of extension from flexion.
   b. From 15 to 17° of external rotation occurs in the last 30° of extension from knee flexion.

   c. External rotation of the knee partly occurs because the lateral femoral condyle is round and shorter while the medial femoral condyle is longer and oblique.
   d. The tibia is "locked" into external rotation by help of the popliteus muscle.

4. Which of the following is false about the patella?
   a. displaces quadriceps tendon anteriorly
   b. increases internal moment arm
   c. augments quadriceps force production
   d. not a true sesamoid bone

5. Axial rotation at the knee reaches its maximum limit in which position?
   a. 0-10° of knee flexion
   b. 20-40° of knee flexion
   c. 90° of knee flexion
   d. 135° of knee flexion

Foot and Ankle

The structure of the foot and ankle is complex, allowing multiple functions including shock absorption, forming a rigid lever for push-off, and the ability to adapt to uneven terrain. These functions occur with every gait cycle. The foot and ankle contain 26 bones (14 phalanges, 5 metatarsals, and 7 tarsal bones) plus 2 sesamoid bones. This results in 33 joints and more than 100 muscles and ligaments. This chapter discusses the osteology, soft-tissue structures, motions, and muscles of the foot and ankle.

Osteology

This section provides detailed information on the bones that make up the foot and ankle joint complex (figure 15.1). The foot is functionally divided into the **rearfoot**, **midfoot**, and **forefoot**. The rearfoot, also termed the hindfoot, is made up of the subtalar joint. The midfoot consists of five tarsal bones: the navicular, the cuboid, and the three cuneiforms. The forefoot consists of the five metatarsals and their phalanges. The first ray is the first metatarsal and the first cuneiform. The fifth ray is the fifth metatarsal.

OBJECTIVES

After reading this chapter, you should be able to do the following:

> List the bones associated with the foot and ankle complex.

> Describe the multiple joints that make up the foot and ankle complex.

> Explain the ligamentous structure that supports the foot and ankle complex.

> Differentiate the function of the extrinsic and intrinsic foot and ankle muscles in non-weight-bearing and weight-bearing activities.

> Discuss a variety of injuries associated with the foot and ankle complex.

Tarsus

The seven tarsal bones are the calcaneus, talus, cuboid, navicular, and first, second, and third cuneiforms. The **tarsals** make up the primary joints of the rearfoot and midfoot. The intricate movements that occur at these joints are key for the function of the foot.

Calcaneus

The largest of the tarsal bones, the calcaneus forms the heel bone (see figure 15.1). The superior surface of the calcaneus has three articulating facets (posterior, middle, and anterior) that articulate with the body, neck, and head of the talus. The posterior facet is located on the posterolateral part of the calcaneus, with a convexity anterior to posterior. The anterior facet extends along the superomedial aspect of the calcaneus and is biconcave. The middle facet is concave and articulates with the middle calcaneal articular surface of the talus.

The plantar surface is rough, with the back a little wider and more bulbous than the front. The sides are convex from side to side. The calcaneal tuberosity, found posteriorly, is depressed in the middle and prolonged at either end into a process.

The lateral surface of the calcaneus is broad behind and narrow in front to form a trumpet-like shape. Located on this surface are two grooves, separated by the peroneal tubercle, that allow passage of the tendons of the peroneus longus and brevis muscles. Near its center is a tubercle for the attachment of the calcaneofibular ligament. The upper anterior surface provides attachment for the lateral talocalcaneal ligament.

The medial surface is relatively concave; this area serves as a passageway for the plantar vessels and nerves and is the origin of the quadratus plantae. At the most superior surface is a shelf, the **sustentaculum tali**, which is the attachment site to a slip of the tendon of the tibialis posterior. Other attachments on the medial side are the flexor hallucis longus muscle, plantar calcaneonavicular ligament, and part of the deltoid ligament of the talo crural joint.

The posterior surface of the calcaneus is prominent, flat superiorly, and convex inferiorly. The midportion of this surface is rough and is the point of insertion of the Achilles tendon. Proximal to this is a smooth surface covered by a bursa that intervenes between the calcaneus and the Achilles tendon.

The anterior surface of the calcaneus is the articulating surface for the cuboid and is shaped like a triangle. It is concave from superior to inferior and convex at right angles to where its medial border gives attachment to the plantar calcaneonavicular ligament.

Talus

The talus is the second largest tarsal bone, shaped like a truncated cone. It is surrounded by the tibia and fibula superiorly, the calcaneus inferiorly, and the malleoli laterally and medially. The talus consists of a body, neck, and head.

The superior surface of the talus is the trochlea. The trochlea is wedge shaped and 4.2 mm broader in the front than behind, which tightens the joint mortise during dorsiflexion. The head is positioned forward and medialward; its anterior articular surface is large, oval, and convex for articulation with the navicular. The neck of the talus attaches the head to the body.

Three primary facets on the talus (medial, anterior, and posterior) articulate with the calcaneus. The posterior facet is the largest and is relatively concave in the anteroposterior direction, corresponding with the facet on the upper surface of the calcaneus. The middle calcaneal articular surface is small, oval in form, and slightly convex; it articulates with the upper surface of the sustentaculum tali of the calcaneus. The anterior facet is the smallest and most distal.

A depression found on the inferior lateral side of the talus is called the **sinus tarsi**. The cervical talocalcaneal ligament lies within the sinus tarsi. This location can be the site of pain after trauma (see Clinical Correlation 15.1). The medial surface is fairly flat; the upper part has a pear-shaped articular facet for the medial **malleolus**. This surface is roughened

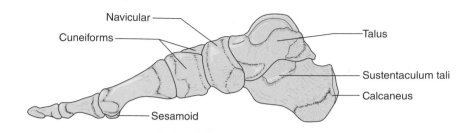

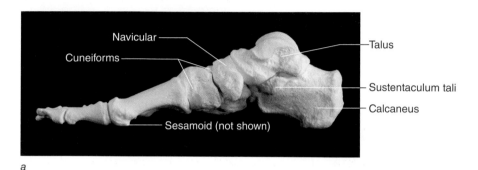

a

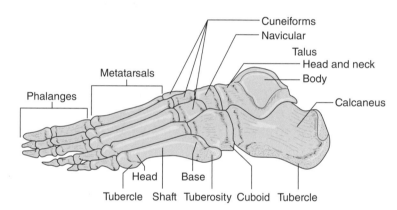

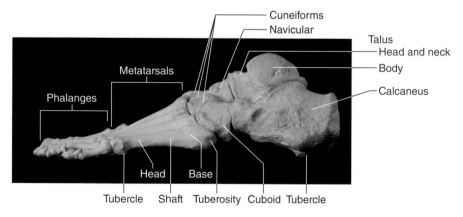

b

▶ **FIGURE 15.1** Bones of the foot and ankle: *(a)* medial aspect and *(b)* lateral aspect. *(continued)*

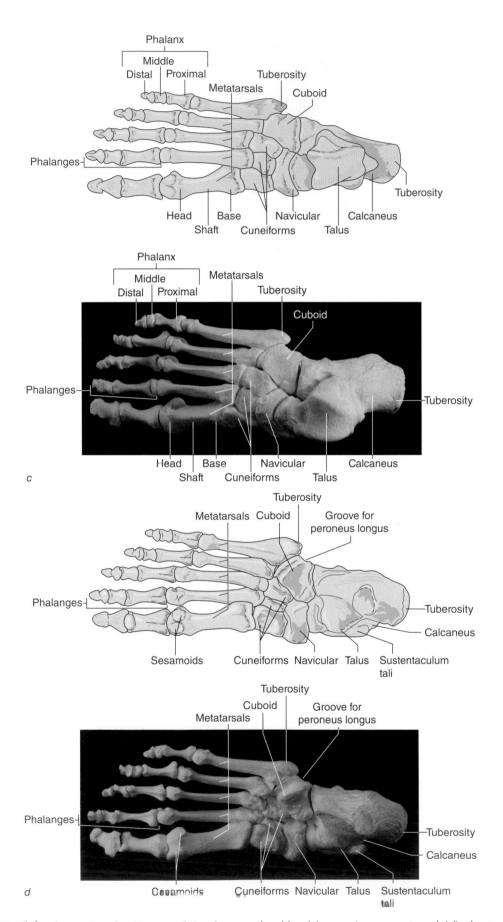

► **FIGURE 15.1** *(continued)* Bones of the foot and ankle: *(c)* superior aspect and *(d)* plantar aspect.

Sinus tarsi syndrome is an inflammatory reaction secondary to trauma, commonly a lateral ankle sprain. Another mechanism for injury is chronic overuse of the subtalar joint. Although forced inversion is the most common mechanism, forced eversion can also be a culprit. This type of stress can be found in sports such as softball pitching, bowling, and ballet. The diagnosis of sinus tarsi syndrome is made by direct palpation of the sinus tarsi. Additionally, the individual will complain of pain during inversion and eversion range of motion. X-rays may be taken to rule out fracture and arthritis. MRIs can identify inflamed tissue within the sinus tarsi. Treatment includes a period of rest and then a gradual return to function.

for attachment of the deltoid ligament (tibiofibular, talofibular, and calcaneofibular ligaments). A unique feature of the talus is that no tendons attach to this tarsal.

▶ **KEY POINT**
The talus is the only tarsal bone without muscle attachments.

Cuboid

The cuboid is part of the midfoot and the key to the lateral arch. This tarsal is located in front of the calcaneus and behind the fourth and fifth metatarsal bones. It is shaped like a pyramid (see figure 15.1). The peroneal sulcus for the tendon of the peroneal longus is located on the plantar surface. The cuboid can sublux during an inversion ankle sprain (see Clinical Correlation 15.2). The anterior surface is triangular shaped and houses two facets, one medial that articulates with the fourth metatarsal and one lateral that articulates with the fifth metatarsal. The articulating surface with the calcaneus is saddle shaped. An oval facet is found on the medial side for articulation with the third cuneiform. A smaller facet is located posteriorly for articulation with the navicular.

Navicular

On the medial side of the midfoot is the navicular, which sits between the talus and the cuneiforms (see figure 15.1). Sometimes termed the scaphoid bone, the navicular is the center of the medial longitudinal arch (figure 15.2a). The anterior surface of the navicular is convex from medial to lateral. Situated on this surface are three facets for the three cuneiforms. The posterior surface is shaped like an oval, with a concave facet for the rounded head of the talus. The dorsal surface is convex from side to side and rough for the attachment of ligaments. The plantar surface is irregular and also rough for the attachment of ligaments. The lateral surface occasionally presents a small facet for articulation with the cuboid bone. The medial surface presents a rounded tuberosity, the lower part of which gives attachment to part of the tendon of the tibialis posterior muscle.

▶ **KEY POINT**
The keystone to the medial longitudinal arch is the navicular bone. This bone is commonly used in the clinic as a landmark for measurement of arch height.

Cuboid syndrome is a subluxation of the cuboid in a plantar medial direction. The mechanism of injury can be an acute event, such as a lateral ankle sprain, or repetitive stress to the supporting ligaments of the cuboid. Pain is directly over the cuboid bone and may refer to the plantar surface toward the medial longitudinal arch. Pain is worse in weight bearing. The diagnosis of cuboid syndrome is made by clinical symptoms, palpation, and joint accessory mobility of the cuboid. Treatment consists of manipulation of the cuboid, followed by padding or strapping.

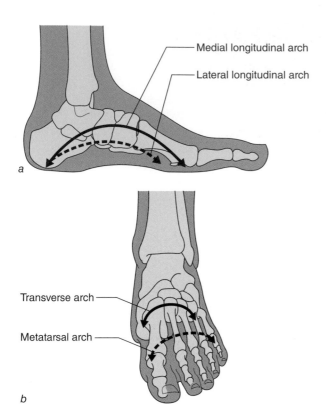

FIGURE 15.2 Arches of the foot: *(a)* lateral and *(b)* anterior.

Cuneiforms

The distal tarsals consist of the first, second, and third cuneiforms (see figure 15.1). The largest of the three is the first cuneiform, which articulates with the navicular, first and second metatarsals, and second cuneiform. The first cuneiform and the first metatarsal together make up the first ray. The first cuneiform houses partial insertions of the tendons of the tibialis anterior, tibialis posterior, and peroneus longus. The middle cuneiform has a wedge-like shape and is the smallest of the three cuneiforms. It articulates with the navicular, first and third cuneiforms, and

second metatarsal. The third cuneiform also has a wedge-like shape and lies between the second cuneiform medially, the cuboid laterally, and the navicular and the third metatarsal anteriorly. The articulation between the cuneiforms and the metatarsals, clinically, is referred to as the Lisfranc joint. This can be the site of a fracture or dislocation called a Lisfranc fracture. These injuries can cause extreme pain and long-term issues.

Metatarsals

The metatarsals are five bones numbered from the medial side (see figure 15.1). All five bones have a prismoid body that is curved longitudinally, concave inferiorly, and slightly convex superiorly. The first metatarsal is the shortest. Its body is thick and strong. The plantar surface of its head has two grooved facets for the sesamoid bones. The base of the first metatarsal articulates with the medial cuneiform. When the first metatarsal is excessively short, it is called a Morton's toe. Some people will develop excessive callusing under the second metatarsal head because of Morton's toe. In addition, entrapment of the digital nerve can cause Morton's neuroma (see Clinical Correlation 15.3).

▶ **KEY POINT**
Morton's toe is a condition where the first metatarsal is excessively short. On appearance, the second toe is longer than the first.

The second metatarsal is the longest and articulates with all cuneiforms and the base of the third metatarsal. The third articulates with the third cuneiform and the second and

CLINICAL CORRELATION 15.3

Morton's neuroma is an entrapment of the common digital nerve between the metatarsal heads, resulting in a fibrous thickening of the nerve. It is most common in the area of the third and fourth toes. This injury is due to repetitive stress and is somewhat common in runners. Pain can be reproduced in the forefoot by compressing together the metatarsal heads of the second and third or the third and fourth toes. Contributing factors include tight shoes, especially slip-on shoes, and biomechanical dysfunction such as overpronation.

fourth metatarsals. The fourth articulates with the cuboid, third cuneiform, and third and fifth metatarsals. The fifth articulates with the cuboid and fourth metatarsal. The fifth metatarsal has several soft-tissue attachments. The peroneus tertius (medial) and peroneus brevis (lateral) both attach to the base of the fifth metatarsal.

Phalanges

The phalanges and the metatarsals together form the forefoot. There are five phalanges that articulate with each of the corresponding metatarsals (see figure 15.1). The base is concave, while the head is convex. The middle row of phalanges is smaller and shorter but broader than the proximal row of phalanges. Middle phalanges are present in the four lateral toes. These phalanges are smaller and shorter than the proximal row. The distal phalanges are small and flattened from above downward. They have a broad base that extends distally for support of the toenail.

Sesamoids

The sesamoids (see figure 15.1) are small rounded masses embedded in the tendon of the flexor hallucis brevis. The function of the sesamoids is to give the flexor hallucis brevis more of a mechanical advantage with push-off of the foot. Injury to these bones is called sesamoiditis and can occur secondary to faulty foot biomechanics.

Joint Articulations

This section covers all articulations of the ankle and foot, starting most proximally and working distally.

Distal Tibiofibular Joint

The distal tibiofibular joint (figure 15.3) is classified as a synarthrodial syndesmosis. This articulation consists of the distal tibia and fibula. The surfaces of the bones present flat, oval facets covered with cartilage. The bones are connected by an articular capsule and by the anterior and posterior tibiofibular ligaments, which provide stability for the mortise of the ankle. The amount of distal tibiofibular joint separation during maximum dorsiflexion has been estimated at 3 to 5 mm.

Talocrural Joint

The talocrural joint is the true "ankle" joint consisting of the distal tibia (**tibial plafond**) and fibula (medial and lateral malleoli) and the trochlea of the talus (see figure 15.3). This joint is classified as a uniaxial hinge joint, and the motions that occur at this articulation are plantar flexion and dorsiflexion. The axis of the talocrural joint runs just distal to both the medial and lateral malleoli. The axis deviates approximately 10° above a mediolateral axis. In the horizontal plane, it is 6° anterior to the mediolateral axis. Because of the cone-shaped trochlea, during dorsiflexion and plantar

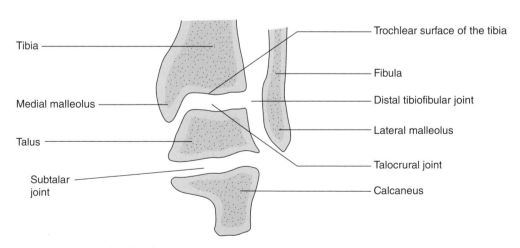

Tibia

Medial malleolus

Talus

Subtalar joint

Trochlear surface of the tibia

Fibula

Distal tibiofibular joint

Lateral malleolus

Talocrural joint

Calcaneus

▶ **FIGURE 15.3** Distal tibiofibular, talocrural, and subtalar joints.

flexion, the amount of trochlear rotation occurring laterally is greater than medially. The stability of the joint depends on the joint congruency and the ligamentous support on the medial and lateral syndesmosis.

Subtalar Joint

The **subtalar** joint consists of the calcaneus and talus (see figure 15.3). This joint is a diarthrodial joint. The anterior and posterior portions of the joint are separated by the sinus tarsi. The posterior articulation is largest and is saddle shaped, formed by a concave (anteroposterior) talus facet and convex (anteroposterior) calcaneus facet. The two anterior articulations are flatter, allowing gliding motion between the convex facets on the body and neck of the talus and the concave facets on the calcaneus.

The subtalar joint is a uniaxial joint that produces triplane motions: supination and pronation (figure 15.4). The axis of rotation can vary among individuals but averages around 42° above the horizontal plane axis from the horizontal plane and 23° medial to the AP axis from the sagittal plane (Inman 1976; Sangeorzan 1991) (figure 15.5). Non-weight-bearing pronation consists of calcaneal eversion (frontal plane), abduction (transverse plane), and dorsiflexion (sagittal plane). Non-weight-bearing supination consists of calcaneal inversion (frontal plane), adduction (transverse plane), and plantar flexion (sagittal plane). When the subtalar joint is in a neutral position, the talus is equidistant over the calcaneus, without medial or lateral tilt. This position is used by clinicians when identifying rearfoot and forefoot positions and can be used to describe pathomechanics of the foot.

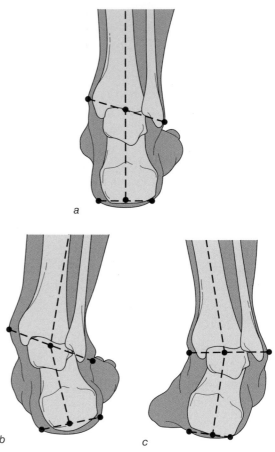

FIGURE 15.4   Motions of the subtalar joint: (a) neutral, (b) pronation, and (c) supination.

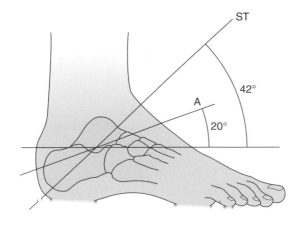

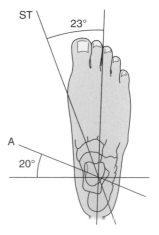

FIGURE 15.5   Orientation of the axes of rotation of the ankle (A) and subtalar (ST) joints.

Non-weight-bearing pronation consists of calcaneal eversion, abduction, and dorsiflexion. When describing pronation in weight bearing, the motion of the talus is included since the calcaneus is relatively fixed. Weight-bearing pronation is described as calcaneal eversion, talar adduction, and talar plantar flexion. Supination is the opposite motion of pronation.

In weight bearing, this triplane motion is described as the talus moving on the calcaneus. Pronation is described as talar plantar flexion, adduction, and calcaneal eversion. Supination is described as talar dorsiflexion, abduction, and calcaneal inversion. Grossly, the motions that identify pronation and supination are inversion and eversion. A pronated foot will have the lateral sole showing, whereas a supinated foot has the medial sole showing.

Transverse Tarsal Joint

The transverse tarsal joint (also known as the midtarsal joint or Chopart's joint) consists of the calcaneocuboid and talonavicular joints (figure 15.6). The calcaneocuboid joint is a synovial saddle joint. The posterior cuboid is convex in a medial to lateral direction and concave in a superior to inferior direction. The articulating calcaneus is just opposite. The talonavicular joint is a condyloid synovial joint. The articulation consists of the anterior talus head and the concave posterior navicular.

Two axes of motion exist at the transverse tarsal joint, the longitudinal and oblique. The longitudinal axis runs 15° upward from the transverse plane and 9° medially from the longitudinal reference. Inversion and eversion occur about the longitudinal axis. The oblique axis runs 52° upward from the transverse plane and 57° medially from the frontal plane. Flexion and extension occur about the oblique axis (Manter 1941). These joints allow for slight gliding motion in three planes and closely follow the motion of the subtalar joint. When the transverse tarsal axes are parallel, as with foot pronation, the entire foot becomes mobile or flexible. When the transverse tarsal axes are nonparallel, as with foot supination, the entire foot becomes rigid (figure 15.7).

Tarsometatarsal Joint

The tarsometatarsal joint (also known as the Lisfranc joint) consists of the articulations between the first, second, and third cuneiforms and the cuboid with the bases of the metatarsal bones. These joints are plane synovial joints limited to gliding movements. The first articulation is between the concave base of the first metatarsal and the convex surface of the first cuneiform. Hypomobility of this articulation may lead to hallux valgus deformity (see Clinical Correlation 15.4) (Klaue et al. 1994). Next is the articulation between the second metatarsal and the second cuneiform, which forms a key-like configuration. The third cuneiform and third metatarsal form the next articulation, followed by the articulation of the cuboid with the fourth and fifth metatarsals. Injury to the tarsometatarsal joint is commonly referred to as a Lisfranc injury, named after a French surgeon. These injuries occur from a variety of mechanisms but usually involve a forceful blow or twist that results in a fracture or fracture-dislocation.

The tarsometatarsal joint is also referred to as the Lisfranc joint. Injury to this joint is also termed a Lisfranc fracture or dislocation.

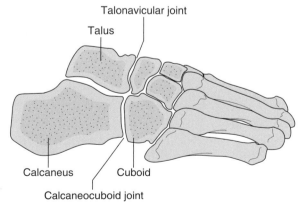

▶ **FIGURE 15.6** Transverse tarsal joint, consisting of the calcaneocuboid and talonavicular joints.

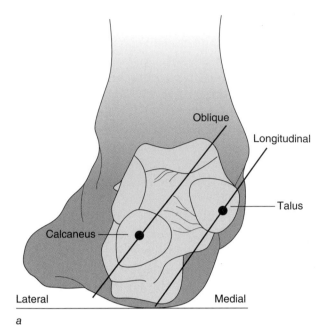

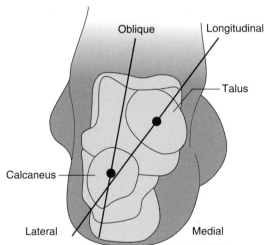

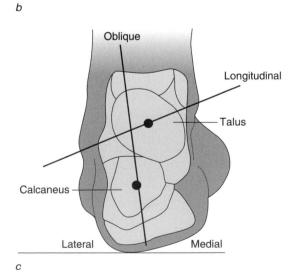

▶ **FIGURE 15.7** The oblique and longitudinal axes of the transverse tarsal joint in (a) pronation, (b) neutral, and (c) supination.

Intertarsal Joint

The intertarsal joints consist of the cuneonavicular, cuneocuboid, and cuboideonavicular joints. These joints are plane synovial joints with little movement because of the congruency of the articulations.

Metatarsophalangeal Joint

The metatarsophalangeal (MTP) joints are condyloid synovial joints. The articulation exists between the rounded heads of the metatarsal bones in shallow cavities on the ends of the first phalanges. These joints are biaxial with two degrees of freedom: flexion–extension and abduction–adduction. Flexion and extension occur about a longitudinal axis in the sagittal plane. Abduction and adduction occur about a vertical axis in the horizontal plane.

Interphalangeal Joints

The interphalangeal joints are synovial hinge joints with one degree of freedom about a mediolateral axis. The movement at this joint is flexion and extension.

Joint Anatomy

This section provides specific information on the joint capsule, ligaments, soft tissue, and nerve and blood supply of the foot and ankle joint complex. Figure 15.9 is included to depict the primary ligaments of the rearfoot and midfoot.

Distal Tibiofibular Joint

The joint capsule of the tibiofibular joint is often continuous with the synovial membrane lining the talocrural joint. Two syndesmotic ligaments are present that stabilize the distal tibiofibular joint, the anterior and posterior tibiofibular ligaments (see figure 15.9). The anterior tibiofibular ligament is a flattened band that extends obliquely downward and laterally between the adjacent margins of the anterior tibia and fibula on the front of the syndesmosis. The posterior tibiofibular ligament is stronger than the anterior and

Hallux abducto valgus (HAV) (figure 15.8) is an excessive adduction of the first metatarsal about the tarsometatarsal joint. The adducted position eventually collapses the proximal **phalanx** into excessive abduction, thereby exposing the metatarsal head as a bunion. This abnormal position of the proximal phalanx decreases its ability to depress the metatarsal head during toe-off; as a result, a person with HAV tends to push off the medial side of the great toe. Contributing factors are thought to be genetics, improper footwear, and pronated feet that cause strain at the hallux. Conservative treatment for HAV includes great toe mobilization, correction of subtalar joint pronation, and shoewear with a wide toebox.

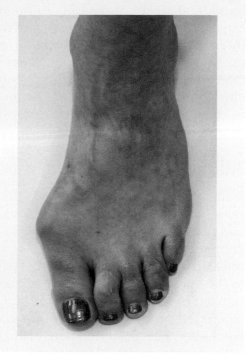

▶ **FIGURE 15.8** Hallux abducto valgus.

Reprinted from S.J. Shultz and P.A. Houglum and D.H. Perrin, 2009, *Examination of musculoskeletal injuries*, 3rd ed. (Champaign, IL: Human Kinetics), 406. By permission of P.A. Houglum.

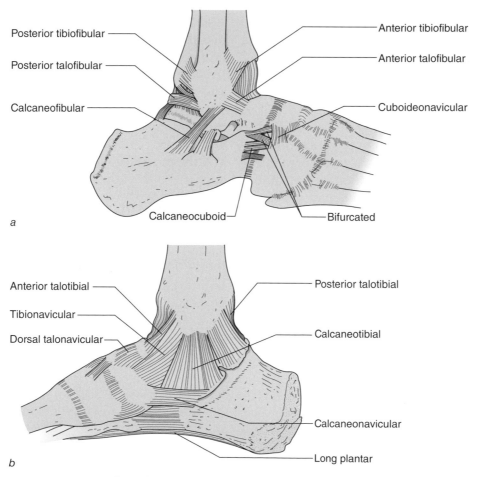

▶ **FIGURE 15.9** Ligaments of the foot and ankle: *(a)* lateral and *(b)* medial.

runs downward and laterally from the posterolateral aspect of the tibia to the posterior aspect of the lateral malleolus. The crural interosseous tibiofibular ligament connects the tibia and fibula and is continuous with the interosseous membrane; this constitutes the chief bond of union between the lower ends of the bones. Injury to this joint is discussed in Clinical Correlation 15.5.

Talocrural Joint

A fibrous joint capsule surrounds the talocrural joint. It is thin and membranous in front and behind and is attached above to the borders of the articular surfaces of the tibia and malleoli; below it is attached to the talus, close to the margins of the trochlear surface except in front of the superior articular surface. It is supported on each side by strong collateral ligaments. The posterior part of the capsule consists principally of transverse fibers. The capsule blends with the inferior transverse ligament and is somewhat thickened laterally where it reaches as far as the malleolar fossa of the fibula.

Located medially is a triangular-shaped deltoid ligament (see figure 15.9) that is covered by the tendons of the tibialis posterior and the flexor digitorum longus. The deltoid ligament has two layers, superficial and deep. The superficial layer consists of, anteriorly, the tibionavicular ligament, which attaches from the medial malleolus to the navicular bone and spring ligament; the middle tibiocalcaneal ligament, which attaches from the medial malleolus to the sustentaculum tali and posterior fibers; and the posterior tibiotalar ligament, which attaches from the medial malleolus to the posterior talus process. The deep layer is

the anterior tibiotalar ligament, which travels from the medial malleolus to the navicular bone. The strong deltoid ligament stabilizes the medial ankle and prevents excessive eversion, external rotation, and plantar flexion.

▶ **KEY POINT**
The deltoid ligament consists of the tibionavicular, tibiocalcaneal, posterior tibiotalar, and anterior tibiotalar ligaments.

The lateral ligaments are not as strong as the medial ligaments. These ligaments include the anterior talofibular ligament, which is the shortest and weakest of three primary ligaments. The anterior talofibular ligament is a flat band that passes from the anterior margin of the lateral malleolus, forward and medially, to the neck of the talus. The thick posterior talofibular ligament is the strongest. It runs horizontally from the lateral malleolus to the lateral tubercle of the talus. It functions to stabilize the posterolateral ankle. The calcaneofibular ligament, the longest of the three, is a narrow, rounded cord that passes from the apex of the lateral malleolus downward and slightly backward to the lateral surface of the calcaneus. It is covered superficially by the tendons of the fibularis (peroneus) longus and brevis. The anterior talofibular ligament and the posterior talofibular ligament provide support to the talocrural joint. In the sagittal plane, the anterior ligament is under greater tension in plantar flexion. The calcaneofibular ligament plays an important role in providing stability to both the talocrural and the subtalar joints. This ligament is under greater tension in ankle dorsiflexion (Bulucu et al. 1991) (see Clinical Correlation 15.6).

CLINICAL CORRELATION 15.5

A **high ankle sprain** is not as common as a lateral ankle sprain and occurs secondary to twisting of the tibia inward while the foot is dorsiflexed and everted. Structures that are torn include the anterior tibiofibular ligament and the interosseous membrane (syndesmosis) between the tibia and fibula. This type of sprain is more common in sports such as football and soccer. Healing of this type of sprain can be tricky. Weight bearing may need to be limited because of the stress placed on the injured structures with walking.

CLINICAL CORRELATION 15.6

A lateral ankle sprain is a partial or complete rupture of the lateral ligaments of the ankle. Most commonly, the anterior talofibular ligament is torn, followed by the calcaneofibular ligament and finally the posterior talofibular ligament. Ninety-five percent of all ankle sprains are lateral. The mechanism of injury is forced inversion and plantar flexion. Examples include stepping in a hole or a basketball player coming down from a jump and landing on an opponent's foot. Recurrence of lateral ankle sprains is high. Rehabilitation after a lateral ankle sprain should include regaining proper joint mobility, functional strength, and balance. There is some evidence that bracing or taping is helpful in preventing further lateral ankle sprains (Quinn et al. 2000).

▶ **KEY POINT**

The lateral ligaments can be injured from a combination of inversion and plantar- or dorsiflexion. The anterior talofibular ligament is most commonly injured from forced plantar flexion and inversion. The calcaneofibular ligament is most commonly injured from forced dorsiflexion and inversion. The posterior tibiofibular ligament is injured with forced dorsiflexion and a posterior displacement of the talus.

Subtalar Joint

These two bones are connected by two fibrous capsules that envelop the anterior and posterior joint. The posterior joint capsule is lined with a synovial membrane, and the joint cavity does not communicate with any of the other tarsal joints. The anterior joint capsule surrounds the articulation between the talus and navicular. The primary ligaments that support the subtalar joint are the cervical talocalcaneal ligament and the interosseous talocalcaneal ligament. The cervical talocalcaneal ligament runs from the inferolateral aspect of the talar neck, downward and lateral to the dorsum of the calcaneus, near the origin of the extensor digitorum brevis. This ligament restricts inversion. The interosseous talocalcaneal ligament is a thick quadrilateral ligament that travels from the underside of the talus at the sustentaculi tali, downward and lateral to the dorsum of the calcaneus. It helps limit eversion. There are three slender ligaments named by their location that provide secondary stability,

primarily posteriorly; these are the medial, posterior, and lateral talocalcaneal ligaments.

The subtalar joint is unique in that it has several functions. When the joint is pronated, the joint is somewhat flexible to allow for shock absorption of the ground reaction forces. When the joint is supinated, it becomes very stable, allowing for a rigid lever for push-off. During the terminal stance of gait, the subtalar joint supinates (locks) for this very purpose. Lack of motion can lead to lateral ankle sprains.

Transverse Tarsal Joint

The joint capsule of the calcaneocuboid joint is strong and reinforced by ligamentous support. The bifurcated ligament (see figure 15.9) is a Y-shaped band with its stem attached to the calcaneus. The medial fibers reinforce the dorsolateral side of the talonavicular joint; the lateral fibers cross dorsal to the calcaneocuboid joint and form the primary bond between the two bones. The long and short **plantar** ligaments reinforce the plantar side of the joint. The long plantar ligament arises from the plantar surface of the calcaneus and attaches to the cuboid bone, continuing distally to form a tunnel for the peroneus longus and inserting on the plantar surface of the bases of the lateral three or four metatarsal bones. The short plantar ligament runs from the anterior tubercle of the calcaneus to the plantar surface of the cuboid.

The talonavicular joint is enclosed by a thin, irregularly shaped capsule and is totally independent of the posterior joint capsule.

Posteriorly, the capsule is thickened by the interosseous ligament of the subtalar joint. Three primary ligaments support the talo-calcaneonavicular joint: the plantar calcaneonavicular ligament (spring ligament), the dorsal talonavicular ligament, and laterally the calcaneonavicular fibers of the bifurcated ligament. The spring ligament helps support the medial longitudinal arch by forming a sling for the talar head, preventing excessive medial and plantar movement of the talus. The talonavicular ligament is a broad, thin band connecting the neck of the talus to the dorsal surface of the navicular; it is covered with the extensor tendons.

▶ **KEY POINT**
The plantar fascia, short and long plantar ligaments, and spring ligament are the passive structures that support the medial longitudinal ligament.

Tarsometatarsal Joint

The ligaments that reinforce the tarsometatarsal joints are the dorsal and plantar tarsometatarsal and interosseous ligaments. The dorsal ligaments are strong, flat bands. The first metatarsal is joined to the medial cuneiform bone by an articular capsule; the second metatarsal receives three bands, one from each cuneiform; the third receives a band from the lateral cuneiform; the fourth receives one from the lateral cuneiform and another from the cuboid; the fifth receives one from the cuboid.

The plantar ligaments consist of longitudinal and oblique bands, disposed with less regularity than the dorsal ligaments. Those for the first and second metatarsal bones are the strongest. The first metatarsal ligament is the major restraint for dorsal angulation of the first metatarsal head. The second and third metatarsal bones are joined by oblique bands to the medial cuneiform; the fourth and fifth metatarsal bones are connected by a few fibers to the cuboid.

The interosseous ligaments consist of three separate bands. The first and strongest passes from the lateral surface of the medial

cuneiform to the adjacent angle of the second metatarsal bone; this ligament is also referred to as the Lisfranc ligament. The second band of the interosseous ligament connects the lateral cuneiform with the adjacent angle of the second metatarsal; the third connects the lateral angle of the lateral cuneiform with the adjacent side of the base of the fourth metatarsal bone.

Intertarsal Joint

The intertarsal joints are reinforced by dorsal and plantar ligaments. The cuneocuboid joint is a fibrous articulation between the lateral cuneiform and the cuboid. The cuboideonavicular joint is a fibrous joint between the lateral side of the navicular and the proximal one-fifth of the medial side of the cuboid. These bones are connected by the dorsal, plantar, and interosseous ligaments. The dorsal ligament extends obliquely forward and laterally between the cuboid and navicular bones. The plantar ligament passes nearly transversely from the cuboid to the navicular. The interosseous ligament connects the rough nonarticular portions of the two bones.

Metatarsal Joint

The primary ligaments of the metatarsal joint include the collateral ligaments, the plantar ligament, and the transverse metatarsal ligaments. The collateral ligaments are located on both the medial and lateral sides of the joint and blend with the joint capsule. These ligaments help support frontal plane motion. The plantar ligament is on the plantar side of the joint and is grooved for the passage of the flexor tendons. Fibers from the deep plantar fascia connect into the plantar plates and flexor tendon sheaths. The four transverse metatarsal ligaments blend with and join the adjacent plantar plates of all five joints. They correspond closely to the deep transverse metacarpal ligaments, but they also are connected to the plantar ligament of the first metatarsophalangeal joint. An injury (usually hyperextension) to the capsuloligamentous structures of the first metatarsophalangeal joint is termed *turf toe*. This term was created because of the high incidence of MTP injuries in athletes who play on artificial turf in sports such as soccer and football.

The extensor hood mechanism covers the dorsal side of each joint. The hood is a thin layer of connective tissue that is essentially inseparable from the dorsal capsule and extensor tendons. Both the extrinsic (long toe flexors and extensors) and intrinsic (lumbricales, interossei) muscles contribute to the extensor hood. The extensor hood is key in controlling motion of the metatarsophalangeal and interphalangeal joints.

Interphalangeal Joints

The articular capsule surrounds the entire joint, and it is reinforced by a medial and lateral collateral ligament. The plantar surface of the articular capsule is strengthened to form a fibrous plate, called the plantar ligament (see figure 15.9).

Metatarsophalangeal Joint

The fibrous joint capsule surrounds the joint and is attached to the margins of the articular surfaces. Dorsally, the capsule is thin and may be separated from the tendons of the long extensors by small bursae. The capsule is inseparable from the deep surfaces of the plantar and collateral ligaments.

Soft-Tissue Structures

Numerous soft-tissue structures are found in the foot and ankle complex. Specifically, the plantar fascia, heel pad, Achilles tendon, bursae, and tarsal tunnel are discussed in this section.

Plantar Fascia

The plantar fascia is a fibrous band that extends from the medial plantar tuberosity of the calcaneus to the metatarsophalangeal plantar plates, collateral ligaments, and hallucal **sesamoids**. The function of the plantar fascia is to support the medial longitudinal arch. When the toes are extended, the plantar fascia is placed under tension, which inverts the calcaneus and elevates the arch. This mechanism is called the *windlass effect* (Hicks 1954) (figure 15.10). This position helps supinate the subtalar joint, creating a rigid foot for push-off. The plantar fascia takes on high loads during activities such as running (1.3 to 2.9 times body weight) (Scott & Winter 1990), which when excessive can lead to medial heel pain (see Clinical Correlation 15.7).

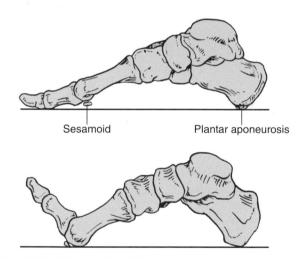

Sesamoid   Plantar aponeurosis

▶ **FIGURE 15.10** Windlass mechanism.

CLINICAL CORRELATION 15.7

Plantar fasciopathy (fasciitis) is a condition that occurs when the long, fibrous plantar fascia along the bottom of the foot develops tears in the tissue, resulting in pain. Excessive running, jumping, or other activities can easily place repetitive or excessive stress on the tissue and lead to tears, resulting in moderate to severe pain. Symptoms include burning, stabbing, or aching pain in the heel of the foot. Pain is localized to an area 1 to 2 cm distal to the medial calcaneal tuberosity. Most people who have this condition will be able to feel it in the morning because the fascia tightens up during sleep at night. One of the primary goals of rehabilitation for a person with plantar fasciopathy is to limit pain with weight bearing. Supportive shoes, arch taping, and foot orthoses may be used for this purpose.

Heel Pad

The heel pad is a unique anatomical structure designed to absorb shock. It consists of vertically arranged fat-filled columns. The heel pad can undergo atrophy with aging, leading to potential foot injuries (Jahss et al. 1992).

Achilles Tendon

The Achilles tendon is the tendinous extension of the triceps surae complex (gastrocnemius, soleus, and plantaris) that attaches to the posterior calcaneus (figure 15.11). The function of

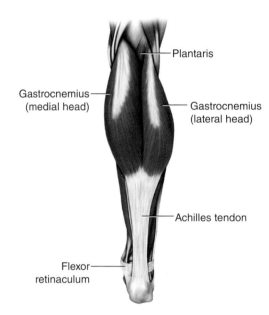

Plantaris

Gastrocnemius (medial head)

Gastrocnemius (lateral head)

Achilles tendon

Flexor retinaculum

▶ **FIGURE 15.11**   Achilles tendon.

the Achilles tendon is to transmit forces from these muscles to produce plantar flexion. The Achilles tendon is the thickest and strongest tendon in the human body. It has a relatively poor blood supply, especially at the insertion of the tendon to the calcaneus. During activities such as running and jumping, forces on the tendon can exceed 10 times body weight (Burdett 1982). A common injury to the Achilles tendon structure is tendinosis. This injury is further presented in Clinical Correlation 15.8.

Bursae

Two bursae are located just superior to the insertion of the Achilles (calcaneal) tendon. Anterior or deep to the tendon is the retrocalcaneal (subtendinous) bursa, which is located between the Achilles tendon and the calcaneus. Posterior or superficial to the Achilles tendon is the subcutaneous calcaneal bursa, also called the Achilles bursa. This bursa is located between the skin and the posterior aspect of the distal Achilles tendon. Inflammation of either or both of these bursae can cause pain at the posterior heel and ankle region.

Tarsal Tunnel

The tarsal tunnel is found along the medial side of the foot, posterior and inferior to the medial malleolus. The tunnel is formed by

CLINICAL CORRELATION 15.8

Achilles tendinosis is a chronic degenerative change of the Achilles tendon. An acute inflammation would be termed *Achilles tendinitis*. The cause of Achilles tendinosis is repetitive stress to the Achilles tendon, common in jumping sports and running. Signs and symptoms include mild pain during or after exercise. The localized pain is felt along the tendon and may continue a few hours after a workout. The tendon itself is very tender in the mornings, with noticeable swelling. A complete tear of the Achilles tendon can also occur and is most common in middle-aged (fourth decade) males. Four primary mechanisms for tendon rupture have been identified: (1) sudden dorsiflexion of a plantar flexed foot; (2) pushing off in plantar flexion with simultaneous knee extension; (3) sudden excess tension on an already taut tendon; and (4) a taut tendon struck by a blunt object (Mahan & Carter 1992). This rupture usually occurs 2 to 6 centimeters proximal to the calcaneal insertion. Surgery is required for an Achilles tendon rupture. Rehabilitation for Achilles tendinosis requires implementation of eccentric strengthening (Alfredson et al. 1998).

the flexor retinaculum that travels between the calcaneus and tibia (figure 15.11). The contents of the tunnel include the posterior tibial artery and vein, tibial nerve, tibialis posterior tendon, flexor digitorum longus muscle, and flexor hallucis longus muscle. The tarsal tunnel can be a site for entrapment of the tibial nerve, as further discussed in Clinical Correlation 15.9.

Arches of the Foot

Three primary arches exist in the foot (see figure 15.2). The longitudinal arch runs from the calcaneus to the distal ends of the metatarsals and consists of a medial and lateral component. The medial longitudinal arch (MLA) consists of the calcaneus, the talus, the navicular, the three cuneiforms, and the three medial metatarsals. The MLA is maintained by the plantar fascia, short and long plantar ligaments, spring ligament, and tibialis posterior muscle. The lateral longitudinal arch is flatter than the MLA and consists of the lateral border of the foot; the cuboid is the keystone of this arch. The third arch is the transverse arch, which extends medially to laterally across the plantar surface of the foot. This arch is formed by the cuboid, cuneiforms, and metatarsals.

Nerve Supply

The nerves that innervate the foot and ankle stem from the tibial and common peroneal nerves. The tibial nerve innervates the posterior muscle group. The terminal branches of the tibial nerve are the medial and lateral plantar nerves that innervate the muscles of the foot. The common peroneal nerve branches just below the knee joint to form the deep peroneal nerve and the superficial peroneal nerve. The deep peroneal nerve innervates primarily the extensors of the foot and digits. The superficial peroneal nerve innervates the lateral musculature.

A summary of the anatomy of each joint is found in table 15.1.

Blood Supply

The popliteal artery bifurcates distal to the popliteal muscle to form the anterior and posterior tibial arteries. Distal to this, the peroneal artery branches off the posterior tibial artery in the deep compartment of the leg. The distal branch of the anterior tibial artery is the dorsal pedis artery, a common place for palpation of the foot pulse. The location for palpation of the dorsalis pedal pulse is between the first and second metatarsal. The anterior tibial artery supplies the anterior leg and dorsum of the foot.

The posterior tibial artery runs through the deep compartment of the leg and is located medially to the tibial nerve. It is the main blood supply to the posterior leg and plantar surface of the foot. This artery runs through the tarsal tunnel. After exiting the tunnel it bifurcates into the medial and lateral plantar arteries.

The peroneal artery supplies blood flow to the flexor hallucis longus, peroneus tertius, peroneus brevis and longus, and soleus. Additionally, it supplies blood to the skin overlying the lateral calcaneus.

CLINICAL CORRELATION 15.9

Tarsal tunnel syndrome is an entrapment of the posterior tibial nerve as it travels through the tarsal tunnel. The mechanism of injury may be direct trauma to the medial foot, causing inflammation of the nerve, or repetitive stress. Contributing factors may include pes planus (flat feet) or ill-fitting shoes. Signs of tarsal tunnel syndrome are pain and tingling along the medial heel and into the medial longitudinal arch. A Tinel tapping test over the tarsal tunnel may reproduce symptoms. A nerve conduction velocity test may show tibial nerve irritation.

TABLE 15.1 Joint Anatomy Summary

Joint	Type	Articular surface	Articular capsule
Talotibial (talocrural)	Hinge type of synovial joint	Superior convex dome of talus fits into the concave surface formed by the medial malleolus, distal tibia, and lateral malleolus	Fibrous capsule is attached superiorly to tibia and malleoli and inferiorly to talus
Talocalcaneal (subtalar)	Plane type of synovial joint	Inferior surface of body of talus articulates with superior surface of calcaneus	Fibrous capsule is attached to margins of articular surfaces
Talocalcaneonavicular	Synovial joint; part is ball-and-socket type	Head of talus articulates with calcaneus and navicular bones	Fibrous capsule incompletely encloses joint
Calcaneocuboid	Plane type of synovial joint	Anterior end of calcaneus articulates with posterior surface of cuboid	Fibrous capsule encloses joint
Tarsometatarsal	Plane type of synovial joint	Anterior tarsal bones articulate with bases of metatarsal bones	Fibrous capsule encloses each joint
Intermetatarsal	Plane type of synovial joint	Bases of metatarsal bones articulate with each other	Fibrous capsule encloses each joint
Metatarsophalangeal	Condyloid type of synovial joint	Heads of metatarsal bones articulate with bases of proximal phalanges	Fibrous capsule encloses each joint
Interphalangeal	Hinge type of synovial joint	Head of one phalanx articulates with base of one distal to it	Fibrous capsule encloses each joint

Ligaments	Movement	Blood supply	Nerve supply
Lateral ligament: anterior and posterior talofibular and calcaneofibular support talocrural joint laterally Medial (deltoid) ligament: anterior and posterior tibiotalar, tibiocalcaneal, and tibionavicular support talocrural joint medially	Dorsiflexion and plantar flexion	Plantar flexion: tibial and superficial peroneal artery Dorsiflexion: deep peroneal artery	Plantar aspect: tibial nerve and its medial and lateral branches of deep peroneal nerve Dorsal aspect: tibial and medial sural cutaneous nerves
Medial, lateral, and posterior talocalcaneal ligaments support capsule; interosseous talocalcaneal ligament binds bones together	Inversion and eversion of foot	Posterior tibial and fibular arteries	Plantar aspect: medial or lateral plantar nerves Dorsal aspect: deep fibular nerve
Plantar calcaneonavicular (spring ligament) supports head of talus	Gliding and rotatory movements are possible	Anterior tibial artery via lateral tarsal artery	
Dorsal calcaneocuboid ligament, plantar calcaneocuboid ligament, and long plantar ligament support fibrous capsule	Inversion and eversion of foot	Anterior tibial artery via lateral tarsal artery	
Dorsal, plantar, and interosseous ligaments	Gliding	Lateral tarsal artery, a branch of dorsalis pedis artery	Deep fibular medial and lateral plantar and sural nerves
Dorsal, plantar, and interosseous ligaments bind bones together	Little individual movement of bones possible	Lateral metatarsal artery, a branch of dorsalis pedis artery	Digital nerves
Collateral ligaments support capsule on each side; plantar ligament supports plantar part of capsule	Flexion; extension; and some abduction, adduction, and circumduction	Lateral tarsal artery, a branch of dorsalis pedis artery	
Collateral and plantar ligaments support joints	Flexion and extension	Digital branches of plantar arch	

Joint Function

The ankle joint complex consists of the distal tibiofibular joint and the talocrural joint. Movement at the foot affects the ankle and vice versa. The foot consists of the subtalar, transtarsal, metatarsophalangeal, and interphalangeal joints. In this section, we describe the arthrokinematics, range of motion, closed and loose packed positions, end feel, and capsular patterns of the joints of the foot and ankle.

Axes of Motion

The axes of motion are previously discussed in joint anatomy. Please see those specific sections for axes of motion.

Arthrokinematics

The arthrokinematics for each primary motion at the foot and ankle are first described in non–weight bearing because for the most part, this is the position in which the clinician would mobilize the joint.

Non–Weight Bearing

Non-weight-bearing arthrokinematics are described as the distal segment moving on the proximal segment. The convex–concave rule is followed for these descriptions.

Dorsiflexion   Non-weight-bearing dorsiflexion consists of movement at the distal tibiofibular joint and at the talocrural joint. The talus is wider anteriorly than posteriorly, and therefore a small amount of mortise separation (1 to 2 mm) occurs with dorsiflexion. The fibula rotates laterally (2 to 3°) and glides proximally (superiorly) to allow for the movement of the talus into the mortise. The talus rolls anteriorly and glides posteriorly and abducts on the tibia as it moves into the mortise (figure 15.12).

During dorsiflexion, the calcaneofibular ligament and posterior capsule become taut. After a lateral ankle sprain, the talus may stay in an anterior position, causing an anterior impingement of the talus with the tibial plafond. People may describe anterior joint pain when trying to increase dorsiflexion (e.g., with gastroc–soleus stretching). During dorsiflexion, the subtalar (ST) joint is forced into pronation.

▶ **KEY POINT**
The arthrokinematics of non-weight-bearing dorsiflexion involve the talus rolling anteriorly, gliding posteriorly, and abducting on the tibia as it moves into the mortise.

Plantar Flexion   The motion of plantar flexion is opposite to dorsiflexion. The fibula rotates medially, which narrows the mortise, meaning the talus is moving out of the mortise. The talus rolls posteriorly and glides anteriorly on the tibia, and supination at the ST joint occurs (figure 15.13). The anterior talofibular ligament, tibionavicular ligament, and anterior capsule become taut with plantar flexion.

Inversion   The motion of inversion occurs primarily in the subtalar and forefoot joints. Inversion commonly occurs with plantar flexion. The calcaneus glides medially on a fixed talus, and the navicular glides medially and toward the plantar surface of the talus.

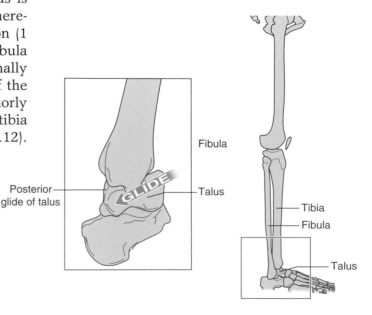

▶ **FIGURE 15.12**   Movement of the talus relative to the tibia during dorsiflexion.

Eversion   Eversion of the foot is accompanied by the calcaneus gliding laterally on the fixed talus. The navicular glides laterally and toward the plantar surface on the talus.

MTP Flexion and Extension   The MTP articulation consists of the convex metatarsal and a concave phalanx. Following the convex–concave rule, MTP extension involves a dorsal glide of the phalanx, and MTP flexion requires a plantar glide of the phalanx.

Weight Bearing

Weight-bearing motion at the foot produces a chain of events more proximally. The following motions occur together: subtalar joint pronation, talocrural joint dorsiflexion, tibial internal rotation, knee flexion, femoral internal rotation, and hip flexion.

Dorsiflexion   The weight-bearing dorsiflexion arthrokinematics are the same as the non-weight-bearing ones, although the motion is more dramatic and usually described as the proximal segment moving on the fixed distal segment. With weight-bearing dorsiflexion, the distal tibia and fibula glide anteriorly on a relatively fixed talus. This motion is more pronounced in the tibia on the talus, resulting in internal rotation of the tibia. There is slight posterior glide of the talus, and the ST joint pronates.

Plantar Flexion   Weight-bearing plantar flexion will result in a heel-raise maneuver. During weight-bearing plantar flexion, the fibula glides posteriorly more profoundly than the tibia on the talus, resulting in external rotation of the tibia, anterior and lateral glide of the talus, and ST joint supination. At the MTP joint, the metatarsal glides plantarly to allow toe extension.

Range of Motion

Table 15.2 gives active range of motion values for all the primary joints of the foot and ankle. Most are based on values from the American Academy of Orthopaedic Surgeons (1965).

Talocrural Joint

Normal dorsiflexion range of motion at the talocrural joint is 20°. Lack of dorsiflexion can result in compensation that may include increased subtalar joint pronation, hyperextension at the knee, or early heel rise during gait. Normal plantar flexion range of motion is 50°. Some people with hypermobility of the mid- and forefoot may display more range than this, so the clinician must be careful when measuring plantar flexion.

Subtalar Joint

Pure calcaneal inversion–eversion is a passive movement that occurs at the subtalar joint in the frontal plane. The amount of inversion is normally 20° (Donatelli 1996). Calcaneal eversion range of motion is one-third of the motion of total inversion/eversion and equals 10° (Donatelli 1996). Functional subtalar joint motion during gait is 10 to 15°. During the gait cycle, the calcaneus strikes the ground in slight inversion, followed by rapid pronation to midstance (Sarrafian 1983).

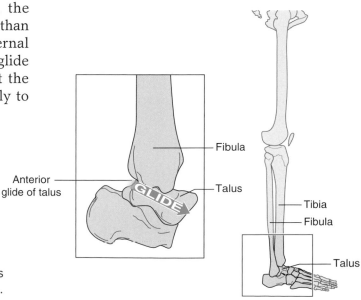

▶ **FIGURE 15.13** Movement of the talus relative to the tibia during plantar flexion.

TABLE 15.2   Range of Motion Values for the Foot and Ankle

Joint	Movement	Range of Motion
Talocrural	Dorsiflexion	0-20°
	Plantar flexion	0-50°
Subtalar	Inversion	0-20°
	Eversion	0-10°
Forefoot	Inversion	0-25°
	Eversion	0-25°
MTP, great toe	Flexion	0-30°
	Extension	0-90°
MTP, 2nd to 5th digits	Flexion	0-20°
	Extension	0-70°
ITP (PIP), great toe	Flexion	0-90°
	Extension	0°
ITP (PIP) 2nd to 5th digits	Flexion	0-35°
	Extension	0°
ITP (DIP) 2nd to 5th digits	Flexion	0-60°
	Extension	0°

ITP = interphalangeal; PIP = proximal interphalangeal; DIP = distal interphalangeal

Forefoot

Gross foot inversion and eversion are measured from the dorsum of the foot. Forefoot inversion is 25°, and forefoot eversion is 25°.

Metatarsophalangeal Joint

The great toe has the most range of motion of all the metatarsals to accommodate for many functional activities, such as the push-off phase of gait. The range is 30° of flexion to 90° of extension. The remaining MTP joints have a maximal amount of 20° of flexion and 70° of extension. There is a small amount of abduction at these joints that reaches close to 10°.

Interphalangeal Joints

Proximal interphalangeal (PIP) flexion of the great toes is 90°. PIP flexion for the remaining digits (two through five) is 35°. Extension for all digits at the PIP and DIP joints is 0°.

Distal interphalangeal (DIP) flexion for digits two through five is 60°.

Talocrural Joint

This section describes the closed and loose packed positions, end feel, and capsular pattern of the talocrural joint.

Closed and Loose Packed Positions

Maximal congruence of the talocrural joint occurs at maximum dorsiflexion. This occurs because of the wedge-shaped trochlear surface of the talus. When the joint is at 10° of plantar flexion midway between inversion and eversion, the joint is at its loosest position.

End Feel

End feel is the sensation felt by the clinician while passively moving a joint to the end of its range in a particular direction. The end feel for dorsiflexion is firm because of tension in the posterior capsule, Achilles tendon, posterior portion of the deltoid and calcaneofibular ligaments, and posterior talofibular ligament. The end feel for plantar flexion can be firm or hard. If it is firm, this is due to tension in the anterior capsule, anterior portion of the deltoid and anterior talofibular ligaments, anterior tibial muscle, and long extensors of the toes. If the end feel for plantar flexion is hard, this is due to the posterior tubercle of the talus contacting the posterior tibia.

Capsular Pattern

According to Cyriax, the capsular pattern of the talocrural joint is plantar flexion limited more than dorsiflexion (Cyriax 1982).

Subtalar Joint

This section describes the closed and loose packed positions, end feel, and capsular pattern of the subtalar joint.

Closed and Loose Packed Positions

Supination is the rigid position of the ST joint. The joint is relatively locked. Unfortunately, more direct injuries to the ankle occur with the ST joint in supination. The loose packed position of the ST joint is pronation, the position when the tarsals are less congruent and the foot is mobile.

End Feel

The end feel for subtalar joint inversion is firm because of the lateral joint capsule and lateral ligaments. When a person has had a moderate to severe lateral ankle sprain, the end feel for inversion can be empty (due to pain) or soft (due to lack of ligamentous integrity). The normal end feel for eversion is either firm because of the joint capsule, deltoid ligament, and tibialis posterior muscle or hard because of the abatement of the calcaneus and sinus tarsi.

Capsular Pattern

The capsular pattern of the subtalar joint is supination limited more than pronation,

and therefore inversion is limited more than eversion.

Transverse Tarsal Joint

The transverse tarsal joint does not have voluntary motion associated with it, so closed and loose packed positions and capsular pattern are not described for this joint.

Closed and Loose Packed Positions

The closed and loose packed positions are not applicable to the transverse tarsal joint.

End Feel

The end feel at the transverse tarsal joint is similar to the subtalar joint. Supination has a firm end feel because of the lateral joint capsule and lateral ligaments. The end feel for pronation may be firm due to the joint capsule, deltoid ligament, and tibialis posterior muscle or hard due to the approximation of the calcaneus and sinus tarsi.

Capsular Pattern

The capsular pattern is not described for the transverse tarsal joint.

Forefoot

This section describes the closed and loose packed positions, end feel, and capsular pattern of the forefoot.

Closed and Loose Packed Positions

The closed packed position of the forefoot is supination. The loose packed position is when the forefoot is pronated.

End Feel

The end feel for the forefoot is firm in all planes because of ligamentous constraints.

Capsular Pattern

The capsular pattern described at the forefoot is equal limitations in all directions.

Metatarsophalangeal Joint

This section describes the closed and loose packed positions, end feel, and capsular pattern of the metatarsophalangeal (MTP) joint.

Closed and Loose Packed Positions

The closed pack position of the MTP joint is full extension. The surface contact area of the metatarsal head moves dorsally with extension. Therefore in full extension, joint compression is at its maximum. The loose packed position is found at 10° of extension.

End Feel

Flexion at the MTP has a firm end feel because of tension in the dorsal joint capsule, collateral ligaments, and short toe extensors. The end feel for extension is firm due to tension in the plantar joint capsule, short toe flexors, and plantar fascia. As with the other two motions, abduction also has a firm end feel due to tension in the joint capsule, collateral ligaments, plantar interosseous, and adductor muscle fascia.

Capsular Pattern

The great toe has a capsular pattern of greater limitation in extension than flexion. The other four digits at the MTP joint have no specific capsular pattern.

Interphalangeal Joints

This section describes the closed and loose packed positions, end feel, and capsular pattern of the interphalangeal joints.

Closed and Loose Packed Positions

Different from the MTP joint, the closed packed position of the IP joints is full flexion. The loose packed position is slight flexion.

End Feel

Flexion of the PIP joints may be soft due to the approximation of soft tissues on the plantar surfaces or firm due to tension in the dorsal joint capsule and collateral ligaments. PIP joint extension has an end feel that is firm because of tension in the plantar joint capsule and plantar fascia. Flexion at the DIP joints is firm because of tension in the dorsal joint capsule, collateral ligaments, and oblique retinacular ligament; extension is also firm because of tension in the plantar joint capsule and plantar fascia.

Capsular Pattern

At the interphalangeal joints, the capsular pattern is flexion more limited than extension.

Muscles

The foot and ankle are an important mechanical part of the lower extremity. The movement of this complex is dependent on ground reaction forces along with the numerous muscles that control its movement. The muscles that surround the ankle–foot complex can be classified as intrinsic and extrinsic.

Intrinsic of Muscles

The intrinsic muscles of the foot are divided into four plantar layers (figure 15.14). The first layer, which is most superficial, consists of the abductor hallucis, flexor digitorum brevis, and abductor digiti minimi. The second layer consists of the quadratus plantae and the lumbricales. The third layer consists of the flexor hallucis brevis and flexor digiti minimi. The two sesamoids lie within the flexor hallucis brevis (FHB) muscle beneath the head of the first metatarsal. The sesamoids create a longer lever arm for the FHB muscle, resulting in a greater MTP flexion torque. The sesamoids also transfer loads from the ground to the first metatarsal head. The fourth layer consists of the dorsal and plantar interossei. Two intrinsic muscles on the dorsal side of the foot are the extensor digitorum brevis and the extensor hallucis brevis.

These intrinsic muscles act to flex the MTP joint and extend the IP joints. The long toe extensors extend the MTP joint through the action of the sagittal bands by lifting the proximal phalanges into extension. Table 15.3 details the intrinsic muscles of the foot.

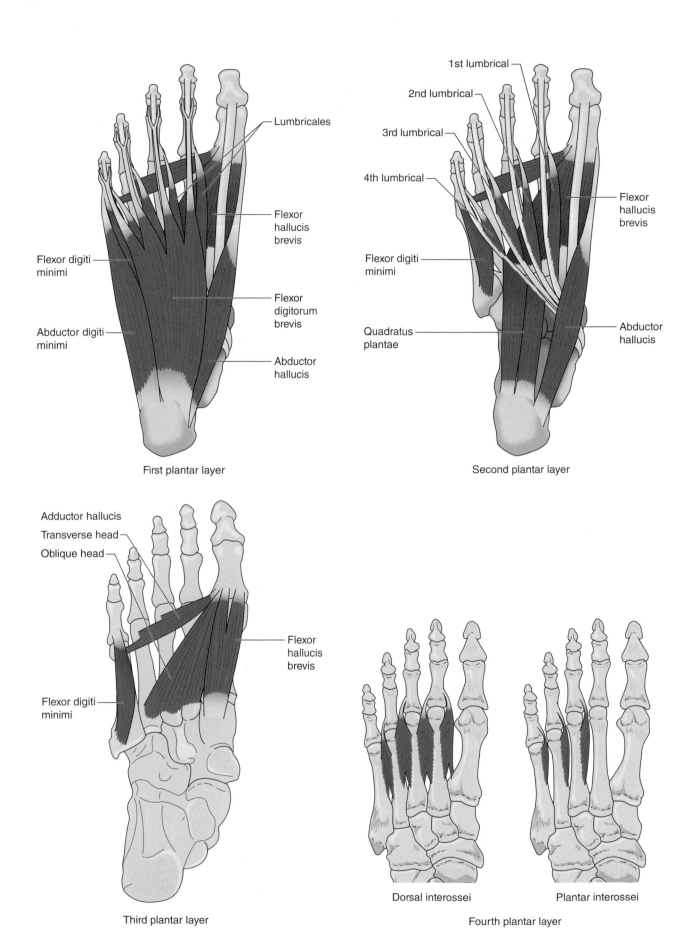

First plantar layer

- Lumbricales
- Flexor hallucis brevis
- Flexor digitorum brevis
- Abductor hallucis
- Flexor digiti minimi
- Abductor digiti minimi

Second plantar layer

- 1st lumbrical
- 2nd lumbrical
- 3rd lumbrical
- 4th lumbrical
- Flexor hallucis brevis
- Flexor digiti minimi
- Quadratus plantae
- Abductor hallucis

Third plantar layer

- Adductor hallucis
- Transverse head
- Oblique head
- Flexor digiti minimi
- Flexor hallucis brevis

Fourth plantar layer

- Dorsal interossei
- Plantar interossei

▶ **FIGURE 15.14** Intrinsic plantar muscles.

329

TABLE 15.3   Intrinsic Muscles of the Foot

Muscle	Origin	Insertion	Action	Innervation
First layer				
Abductor hallucis	Medial process of tuberosity of calcaneus; plantar aponeurosis	Medial side of base of proximal phalanx of great toe	Abducts great toe; helps with great toe flexion	Medial plantar nerve (S1- S2)
Flexor digitorum brevis	Medial process of tuberosity of calcaneus; plantar aponeurosis	Middle phalanx of digits 2-5	Flexes the PIP joints; assists in flexion of MTP joints 2-5	Medial plantar nerve (S1-S2)
Abductor digiti minimi	Medial and lateral process of tuberosity of calcaneus; plantar aponeurosis	Lateral side of base of the proximal phalanx of 5th toe	Abducts and flexes 5th toe	Lateral plantar nerve (S1, S2, S3)
Second layer				
Quadratus plantae	Medial: medial surface of calcaneus Lateral: lateral border of plantar surface of calcaneus	Lateral margin, dorsal and plantar surface of tendon of FDL	Modifies line of pull of FDL; assists in flexion of digits 2-5	Lateral plantar nerve (S1, S2, S3)
Lumbricales	1: medial side of first FDL tendon 2: adjacent sides of first and second FDL tendons 3: adjacent sides of second and third FDL tendons 4: adjacent sides of third and fourth FDL tendons	Medial side of proximal phalanx and dorsal expansion of EDL tendons 2-5	Flexes MTP joints 2-5; assists in extension of IP joints	Lumbricalis 1: medial plantar nerve (S1, S2) Lumbricales 2,3,4: lateral plantar nerve (S2, S3)
Third layer				
Flexor hallucis brevis	Medial plantar surface of cuboid; adjacent lateral cuneiform	Medial lateral surface of base of proximal phalanx; sesamoids	Flexion of great toe	Medial plantar nerve (S1, S2)
Adductor hallucis	Oblique head: base of metatarsals 2-4 Transverse head: plantar MTP ligaments of digits 3-5	Lateral side of base of proximal phalanx of great toe	Adducts great toe; helps with great toe flexion	Lateral plantar nerve (S2, S3)
Flexor digiti minimi	Medial plantar surface of base of 5th MT; sheath of peroneus longus	Lateral base of proximal phalanx of 5th digit	Flexes MTP of 5th toe	Lateral plantar nerve (S1, S2, S3)

Muscle	Origin	Insertion	Action	Innervation
Fourth layer				
Dorsal interossei (4)	Two heads attach to adjacent sides of MT bones	Sides of proximal phalanx; capsule of MTP joint. 1: medial side of digit 2; 2-4: lateral side of digits 2-4	Abducts digits 2-4; assists in MTP joint flexion (2-4)	Lateral plantar nerve (S2, S3)
Plantar interossei (3)	Medial bases of MT bones 3-5	Medial bases of phalanges 3-5	Adducts digits 3-5; assists in MTP flexion (2-5)	Lateral plantar nerve (S2, S3)
Dorsal surface				
Extensor digitorum brevis	Distal part of superior and lateral surfaces of calcaneus	Join lateral side of EDL tendon to digits 2-4	Extends MTP joints 1-4; assists in extension of IP joints (2-4)	Deep peroneal nerve (L5, S1)
Extensor hallucis brevis	Distal part of superior and lateral surfaces of calcaneus	Dorsal surface of base of proximal phalanx of great toe	Extends the MTP joint of great toe	Deep peroneal nerve (L5, S1)

FDL = flexor digitorum longus; EDL = extensor digitorum longus; MT = metatarsal; MTP = metatarsophalangeal

Extrinsic Muscles

The extrinsic muscles can be divided into three anatomical divisions: posterior, anterior, and lateral (figure 15.15). The superficial and intermediate posterior groups include the gastrocnemius, soleus, and plantaris. These three muscles make up the triceps surae and are responsible for plantar flexion of the ankle. The deeper posterior group consists of the tibialis posterior, flexor digitorum longus (FDL), and flexor hallucis longus (FHL). These muscles all pass posterior to the talocrural joint axis. These muscles help with plantar flexion and also help with inversion. The FDL is the primary flexor of the DIP joints and the FHL flexes the great toe.

The anterior muscles, sometimes referred to as the pretibial muscles, consist of the tibialis anterior, extensor hallucis longus (EHL), extensor digitorum longus (EDL), and peroneus tertius. The anterior muscles are primarily responsible for ankle dorsiflexion.

The lateral muscles are the peroneus longus and brevis. These muscles in the lateral group are the primary evertors of the foot and ankle. The strongest invertor of the foot and ankle is the tibialis posterior muscle. Table 15.4 details the extrinsic muscles of the ankle and foot.

Weight-Bearing Function

The previous paragraphs describe the main actions of the muscles in a non-weight-bearing position. However, lower extremity muscles work primarily in a weight-bearing position. Additionally, since the muscles are working to overcome gravity, they are working eccentrically. Using gait as an example, we now describe the weight-bearing muscle function. Gait terminology can be found in chapter 17. Starting with the posterior muscles, the soleus slows the momentum of the tibia over the fixed foot. This happens during the midstance of gait when the body is transferring weight over the fixed foot. The gastrocnemius decelerates internal rotation of the lower limb and via the Achilles tendon helps decelerate pronation. The gastrocnemius is the primary muscle for push-off during preswing. The tibialis posterior contributes the most to dynamic arch support and therefore eccentrically controls pronation. This muscle also inverts the subtalar joint during mid- and late stance

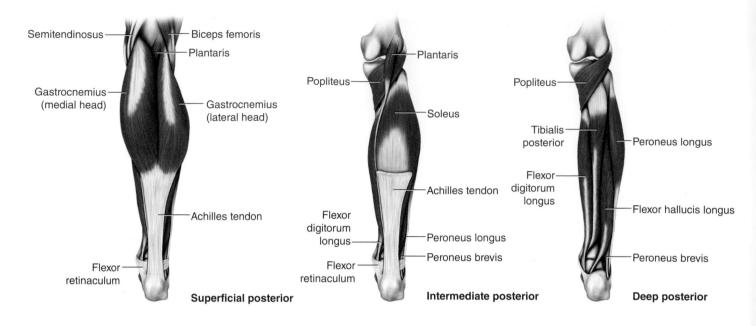

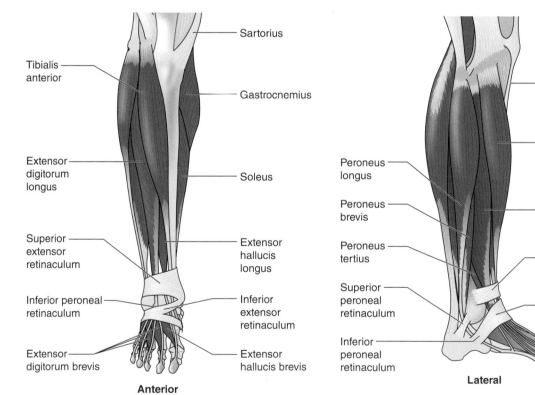

▶ **FIGURE 15.15**   Extrinsic muscles of the ankle.

and in the sagittal plane controls dorsiflexion. People lacking the function of the tibialis posterior present with the inability to lift the arch and a very flat foot. This condition is termed posterior tibial insufficiency and is commonly caused by a tibial nerve injury. Clinical Correlation 15.10 describes another

injury that is fairly common in long-distance runners. Similar to the soleus, the FDL and FHL decelerate the tibia. Because of their distal attachments, the FDL and FHL also control extension of the toes.

The tibialis anterior muscle dorsiflexes the ankle to provide clearance in the swing

TABLE 15.4 Extrinsic Muscles of the Ankle and Foot

Muscle	Origin	Insertion	Action	Innervation
Posterior				
Gastrocnemius	Medial epicondyle of femur; lateral epicondyle of femur	Calcaneus via Achilles tendon	NWB: plantar flexion of foot at ankle; flexion of knee WB: deceleration of internal rotation of the lower limb; helps decelerate pronation; plantar flexes the foot during push-off	Tibial nerve (S1, S2)
Soleus	Upper fibula; soleal line of tibia	Calcaneus via Achilles tendon	NWB: plantar flexes foot WB: decelerates the forward momentum of the tibia when foot is on the ground	Tibial nerve (S1, S2)
Plantaris	Above the lateral head of gastrocnemius on femur	Calcaneus via Achilles tendon	NWB: weak plantar flexion of the foot at ankle WB: controls weak dorsiflexion	Tibial nerve (S1, S2)
Tibialis posterior	Proximal posterior tibia; interosseous membrane; fibula	Navicular; cuneiform; cuboid bones and bases of 2nd to 4th metatarsals	NWB: Inverts foot; plantar flexes ankle (medial malleolus serves as a pulley) WB: Decelerates subtalar joint pronation. Strong stabilizer of the midtarsal joint in direction of supination.	Tibial nerve (L5, S1)
Flexor digitorum longus	Lower 2/3 of tibia	Bases of distal phalanges of lateral four digits	NWB: plantar flexes and inverts foot; flexes toes 2-5, supports longitudinal arch WB: controls dorsiflexion, eversion, and extension of toes	Tibial nerve (L5, S1)
Flexor hallucis longus	Inferior 2/3 of posterior fibula	Base of distal phalanx of big toe	NWB: flexes big toe; weak plantar flexion of ankle WB: controls extension of big toe; dorsiflexion	Tibial nerve (L5, S1, S2)
Anterior leg				
Tibialis anterior	Lateral tibia; proximal lateral surface of tibia; interosseous membrane	Medial cuneiform; first metatarsal	NWB: inversion of foot; dorsiflexion of ankle WB: eccentrically controls plantar flexion at heel strike; helps control pronation of forefoot	Deep peroneal nerve (L4, L5, S1)
Extensor digitorum longus	Lateral tibial condyle; fibula	Dorsal surface of phalanges 2-5	NWB: extension of toes 2-5; dorsiflexion of ankle; eversion of foot WB: controls flexion of toes, plantar flexion, and inversion	Deep peroneal nerve (L4, L5, S1)
Extensor hallucis longus	Medial aspect of fibula; interosseous membrane	Distal phalanx of big toe	NWB: extends big toe; dorsiflexion of the ankle; inversion of the foot WB: controls flexion of big toe, plantar flexion, and eversion	Deep peroneal nerve (L4, L5, S1)

> continued

TABLE 15.4 > *continued*

Muscle	Origin	Insertion	Action	Innervation
Peroneus tertius	Distal 1/3 of anterior fibula; interosseous membrane	Dorsal surface of base of 5th metatarsal	NWB: dorsiflexes ankle joint; everts foot WB: controls plantar flexion and inversion	Deep peroneal nerve (L4, L5, S1)
Lateral leg				
Peroneus longus	Upper lateral fibula	Medial cuneiform; base of 1st metatarsal	NWB: eversion and abduction of foot; weak plantar flexion of foot WB: stabilizes base of 1st ray; controls inversion, adduction, and dorsiflexion	Superficial peroneal nerve (L4, L5, S1)
Peroneus brevis	Lower lateral 2/3 of fibula	5th metatarsal	NWB: eversion and abduction of foot; weak plantar flexion of foot WB: stabilizes 5th ray; controls inversion, adduction, and dorsiflexion	Superficial peroneal nerve (L4, L5, S1)

NWB = non-weight-bearing; WB = weight bearing

CLINICAL CORRELATION 15.10

Medial tibial stress syndrome (MTSS) is a specific overuse injury that produces pain along the posteromedial aspect of the distal two-thirds of the tibia. The pathogenesis of MTSS is controversial, with some authors describing the condition as a periostitis (inflammation of the periosteum) due to strain of the medial tibial fascia (Eickhoff et al. 2000), while others describe it as tearing of the muscle–bone interface (Bouche & Johnson 2007; Tweed et al. 2008). Provocative tests to rule in MTSS include pain with passive ankle dorsiflexion, resisted plantar flexion, toe raises, and single-leg hops. Radiographs and compartment pressures will be normal.

phase of gait. It also works eccentrically to control foot descent during initial contact in the loading response of gait. People with neurological disorders that affect the nerve supply to the tibialis anterior present with a foot slap–type gait. With its distal attachment to the first cuneiform and first metatarsal, the tibialis anterior muscle can also decelerate pronation of the forefoot. The EHL and EDL control flexion of the toes and talocrural plantar flexion.

The lateral muscles are primary evertors but have slightly different functions in weight bearing. The attachment of the peroneus longus to the plantar base of the first metatarsal and medial cuneiforms allows this muscle to stabilize the first ray during weight bearing. The cuboid acts as a fulcrum for the peroneus

longus to allow plantar flexion of the first ray in the push-off phase of gait. Prolonged pronation of the foot results in a loss of this function, and weight distribution shifts to the second metatarsal, rendering it liable to stress fractures. The peroneus brevis stabilizes the fifth ray and controls foot inversion, adduction, and dorsiflexion.

The weight-bearing function of the intrinsic muscles is huge and often overlooked. This group of muscles is sometimes referred to as the rotator cuff of the foot. As a group, they control arch collapse during the midstance of gait and help lift the arch during the later phase of gait when the heel is off the ground. Specifically, the interosseous muscles are active during late stance to stabilize the forefoot during terminal stance.

Conclusion

The foot and ankle joint complex is an amazing system of bones, ligaments, and articulations that allows, in one instance, a flexible foot to adapt to uneven terrain, and in another, a rigid lever for push-off. Injury or pathomechanics to one component of the foot and ankle will have an effect on the whole complex.

REVIEW QUESTIONS

1. List the joints that are referred to as the transverse tarsal joint.

2. Describe the type of injury that is likely to result from extreme inversion and plantar flexion.

3. Describe the closed pack position of the talocrural joint.

4. What is the function of the soleus in gait?

5. Describe the arthrokinematics for plantar flexion at the talocrural joint in non–weight bearing.

6. Describe the origin and insertion of the tibialis posterior muscle.

7. What type of foot dysfunction will occur with injury to the deep peroneal nerve?

Basic Movements and Clinical Application

The previous 15 chapters have provided detailed information regarding human biomechanics, including joint mechanics and muscle action. This information will be brought together in the last 4 chapters of this text by analyzing full-body movements.

Chapter 16 describes the specifics of posture and posture examination, which is key information for the student learning to become a rehabilitation specialist. In detail, this chapter reviews normal and abnormal standing posture in the sagittal and frontal views. A discussion on sitting and lying posture is presented. Muscle imbalances that accompany faulty posture and specific postural deformities such as scoliosis and thoracic kyphosis are also included.

The next chapter of part IV, chapter 17, focuses on walking gait. The details of gait are introduced. Stance and swing phases are described, followed by the specifics of lower extremity joint kinematics and kinetics. The muscle activity that occurs at each phase of the gait cycle is presented. Once normal gait

mechanics are established, the chapter discussion shifts to abnormal gait. Both specific joint faults and gross gait deformities are reviewed.

The information from chapter 17 is expanded in the next chapter's topic of running gait. In chapter 18, similarities and contrasts between walking and running gait are listed. This information is followed by the introduction of running gait stance and swing phases. Kinematics and kinetics of running are then discussed, with interspersed dialogue related to injury. It is important for the student to understand the biomechanics of running to help design prevention and treatment programs for injuries associated with the running athlete.

The final chapter of the text, chapter 19, is dedicated to the mechanics of jumping and cutting. Jumping and cutting activities are commonplace in almost all competitive sporting activities and are a common source of injury mechanism. This chapter presents the ideal mechanics and neuromuscular recruitment patterns for optimal jumping, landing, and cutting.

Posture

Posture can be defined as the orientation of the body segments to each other in a relatively static position. Several factors can influence posture such as body type, habit, disease, and genetic anatomical predisposition. Habitual posture is common in people who maintain a particular posture for long periods. For example, an administrative assistant who sits at a desk for 8 hours may develop a forward head posture with rounded shoulders. Disease also influences posture; for example, a 70-year-old female with osteoporosis who has developed several thoracic compression fractures will present with increased thoracic **kyphosis**. There also seems to be a genetic predisposition to posture. It is not uncommon for certain postural traits such as increased **lordosis** to be passed down from generation to generation.

This chapter reviews normal and abnormal standing posture in the sagittal and frontal views. Normal posture is probably a misnomer in that most people do not possess "normal" posture. However, by identifying normal, or ideal, posture the clinician has a standard posture against which to make comparisons. The chapter also discusses sitting and lying postures. The last section describes faulty

OBJECTIVES

After reading this chapter, you should be able to do the following:

> Define normal standing posture for the sagittal and frontal views.

> Describe the muscle activity that occurs in quiet standing.

> Explain the differences in standing, sitting, and lying posture.

> Define scoliosis, and explain the biomechanical changes that occur in the spine because of scoliosis.

postures. (For more information on posture, including primary and secondary curves, see chapter 6.)

Body Types

Posture can be affected by body mass and a person's body type. **Somatotype** is the term used to describe body type. In the human race, there are variations in body types and sizes. Body type is a combination of bone structure, bone density, and musculature, which are genetically predetermined. Three primary classifications have been determined: endomorph, mesomorph, and ectomorph (figure 16.1). Endomorphs have a heavy or fat body build, and their bodies are round and soft. This person may have a large waist and short limbs, with the upper arms and legs being disproportionately larger than the distal part of the segment. Endomorphs may present with postural deviations directly related to their excessive body mass. For example, a male with an increase in abdominal mass may stand with excessive lordosis to counterbalance the weight that sits anterior to the line of gravity. Mesomorphs have a muscular, athletic build with apparent muscle definition. Mesomorphs generally have good posture and strong bones. An ectomorph has a very thin body build with little muscle mass. Some will even describe ectomorphs as being fragile because of their lean frames. People with this body type may be at risk for osteoporosis with aging. Body type cannot be changed in the sense that an endomorph will always have a greater tendency to put on weight compared with a mesomorph or ectomorph. Additionally, most people are a combination of two body types.

▶ **KEY POINT**
Classifications of body type are endomorph, mesomorph, and ectomorph.

Standing Posture

The majority of posture analyses are performed with the client standing. This section presents normal standing alignment in the anterior, posterior, and lateral views. Key landmarks are identified for the practitioner to observe when performing a postural analysis. Faulty postural deviations with associated muscle imbalances are also listed.

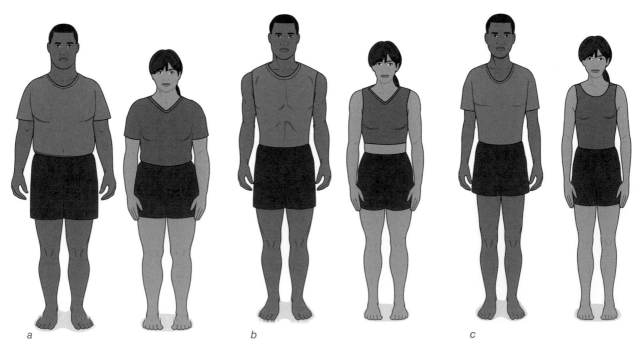

▶ **FIGURE 16.1** Somatotypes: (a) endomorph, (b) mesomorph, and (c) ectomorph.

In normal standing, the center of mass for the average adult is located at approximately the second sacral segment. In standing, there are two primary forces acting on the body: body weight and ground reaction force (figure 16.2). As a whole, the body weight exerts a force downward from the point of the center of mass. At each joint is an external moment (gravity) that is resisted by the internal moments (passive, active tissue). Table 16.1 lists the major joints, external moments, and active and passive tissues resisting these moments.

The point of application of the resultant force on the feet is the center of pressure. The center of pressure is used to measure the amount of sway of the person and is associated with neuromuscular control. In normal quiet standing posture, there is a slight sway both in the anteroposterior direction (7 mm)

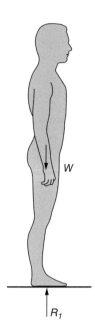

FIGURE 16.2 Forces during standing. W = weight and R_1 = ground reaction forces.

TABLE 16.1 Major Joints and Gravitational Moments

Joints	Line of gravity	Gravitational moment	Passive opposing forces	Active opposing forces
Atlantooccipital	Anterior	Flexion	Ligamentum nuchae, tectorial membrane	Posterior neck muscles
Cervical	Posterior	Extension	Anterior longitudinal ligament	Deep cervical flexors
Thoracic	Anterior	Flexion	Posterior longitudinal ligament, ligamentum flavum, supraspinous ligament	Thoracic extensors (paraspinal muscles)
Lumbar	Posterior	Extension	Anterior longitudinal ligament	Abdominals, hip flexors
Sacroiliac joint	Anterior	Flexion	Sacrotuberous ligament, sacrospinous ligament, long dorsal ligament, interosseous ligament	Hip extensors, lumbar extensors
Hip joint	Posterior	Extension	Iliofemoral ligament	Iliopsoas
Knee joint	Anterior	Extension	Posterior joint capsule, posterior cruciate ligament	Hamstrings
Ankle joint	Anterior	Dorsiflexion	Posterior tibiofibular ligament	Soleus

and the mediolateral direction (<7mm). This posture requires the least amount of muscle support as well as places the least stress on joints. Electromyographic (EMG) activity of the muscles in quiet standing is as follows:

- Muscles of the feet are quiet.
- Soleus is active to maintain upright position.
- Quadriceps and hamstrings for the most part are quiescent, although they may show slight activity from time to time.
- Iliopsoas remains constantly active.
- Gluteus maximus is quiescent.
- Gluteus medius and tensor fasciae latae are active to control lateral pelvic tilt.
- Erector spinae is active to counteract anterior moment.
- Abdominal muscles remain quiescent.
- Minimal activity is present in the upper trapezius, serratus anterior, supraspinatus, and posterior deltoid.

The pressure distribution of body weight has been described as follows: Approximately 45 to 65% of body weight should be carried over the heels, 30 to 47% of body weight should be carried on the forefoot, and 1 to 8% of body weight should be carried over the midfoot. In people with neuromuscular compromise, such as Parkinson's disease, the sway pattern will be quite different.

▶ **KEY POINT**
In standing, 45 to 65% of body weight is carried over the heels, 30 to 47% over the forefoot, and 1 to 8% over the midfoot.

Now that we have discussed the gravitational moments and pressure distribution that result in standing posture, we can look at the effects of these forces on each joint in the body using posture analysis. A systematic plan should be followed when assessing posture. Clinicians can begin their assessment by looking at the feet and working their way up to the head or start at the head and work down to the feet. (In this chapter, we follow a head-to-feet progression of assessment.) Either way, clinicians should keep the procedure consistent between clients. Foot position should be standardized. One way to do this is to have the client stand so that the distance between his feet is equivalent to the distance between the left and right anterior superior iliac spines. Shoes should be off.

One way to standardize the assessment of standing posture is to use a **plumb line**. Normally the plumb line is suspended from the ceiling. It is lined up with fixed points on the person, such as midway between the heels in the posterior view and slightly anterior to the lateral malleolus in the lateral view. The plumb line is used to determine if the person's posture falls within the points of reference for ideal posture. (Later sections outline these points of reference as well as provide a visual of plumb alignment.) Many reasons exist for why posture would fall outside of the ideal range. Table 16.2 lists postural faults and muscles that are short and weak, which can lead to postural defects.

The following sections outline an ideal posture from the anterior, posterior, and lateral views. In addition, text is provided that gives readers clues to abnormal posture. Many of the postural abnormalities mentioned are described in more detail under the appropriate anatomical chapters, and readers are encouraged to revisit those chapters.

Anterior View

When assessing anterior posture using a plumb line (figure 16.3), the plumb line should bisect the body into two equal halves. The alignment should be such that the line:

- Is equidistant between each eye orbit
- Bisects the nose into equal halves
- Bisects the chin
- Bisects the suprasternal notch
- Courses along the sternum
- Bisects the umbilicus
- Is equidistant between the knee joints
- Is equidistant between the medial malleoli

TABLE 16.2 Muscle Imbalances as a Result of Postural Defects

Postural defect	Muscles that are short	Muscles that are long (may be weak)
Forward head posture Atlantooccipital extended to allow person to keep eyes level from lower cervical flexion Lower cervical spine flexion TMJ clenched Scapulae abducted and elevated Humeral internal rotation	Levator scapulae Sternocleidomastoid Scalenes Suboccipital muscles Upper trapezius Pectoralis major and minor Serratus anterior Latissimus dorsi Subscapularis	Hyoid muscles Deep neck flexors (e.g., longus coli and capitis) Lower cervical and thoracic erector spinae Middle and lower trapezius Rhomboids
Anterior tilt of scapulae	Pectoralis minor Biceps	Lower trapezius Middle trapezius Serratus anterior
Scapular downward rotation	Rhomboids Levator scapulae Latissimus dorsi Pectoralis minor Supraspinatus	Upper trapezius Serratus anterior Lower trapezius
Scapular abduction	Serratus anterior Pectoralis major Pectoralis minor Shoulder external rotators	Middle trapezius Lower trapezius Rhomboids
Humeral medial rotation	Pectoralis major Latissimus dorsi Shoulder internal rotators	Shoulder external rotators
Humeral anterior glide	Shoulder external rotators Pectoralis major	Shoulder internal rotators
Thoracic kyphosis	Pectoralis major Pectoralis minor Internal obliques	Thoracic spine extensors Middle trapezius Lower trapezius Rhomboids
Flat thoracic spine	Thoracic erector spinae Scapular retractors	Scapular protractors Anterior intercostal muscles
Lordotic (flexible spine) Pelvis: anterior Lumbar: extension Hip: flexion	Lumbar extensors Hip flexors	Abdominals Hip extensors
Flat back (stiff spine) Pelvis: posterior Lumbar: flexion Hip: extension	Abdominals Hip extensors	Lumbar extensor muscles Hip flexor muscles

> continued

TABLE 16.2 > *continued*

Postural defect	Muscles that are short	Muscles that are long (may be weak)
Swayback Pelvis: posterior Lumbar: flexion Hip: extension Thoracic spine relatively posterior in relation to lumbar spine	Upper abdominals Hip extensors	Lower abdominals Hip flexors Lower lumbar extensors
Handiness pattern Shoulder girdle low on dominant side Iliac crest hide on dominant side	Ipsilateral trunk muscles	Contralateral trunk muscles Ipsilateral shoulder girdle muscles
Scoliosis	Muscles on concave side Hip adductors Foot supinators on short side	Muscles on convex side Hip abductors Foot pronators on long side
Femoral anteversion	Hip internal rotators	Hip external rotators
Femoral retroversion	Hip external rotators	Hip internal rotators
Genu valgum	Hip internal rotators Hip adductors	Gluteus medius
Genu varum	Hip adductors	Iliotibial band
Genu recurvatum	Quadriceps	Hamstrings

In the anterior view, the head should sit without tilt or rotation and be centered over the shoulders. Clues for deviation include the eyes not aligned with the horizontal plane or one ear that sits lower than the other. The cause of altered head position may include cervical spine trauma, muscle imbalance in the cervical region, or abnormal vision or hearing. The shoulders should sit level, with equal slope on both sides. In throwing athletes, the dominant side commonly sits low and is termed the handiness pattern (Kendall 2005). The clinician should note any protrusion or depression of the acromioclavicular joint, sternum, ribs, or costocartilage. People with previous AC joint separation will present with a "bump" at the joint. At the chest level, the pectoralis muscles should be symmetrical; if visible, the nipples should be level. Look for asymmetry of pectoralis muscle bulk and protrusion or hollowing of the chest wall.

The arms should hang symmetrically, with the carrying angle equal (5 to 15°). The palms should face in toward the side of the body. If the palms are facing posteriorly, this indicates tight internal rotators. Inequality of arm distance is linked to a lateral shift of the spine.

The waist line should be equal, and the arms should be equidistant from the waist, comparing right to left. The rib cage should be symmetrical, without protrusion on one side and not the other. The pelvis is identified by palpating the iliac crest, which should be level. The anterior iliac spine should be at the same level from right to left. Asymmetry of these landmarks may indicate an abnormal pelvic rotation.

Femoral position should be straight, without excessive transverse plane rotation (internal or external). The knee joint normal posture is slight valgus (7°), with the fibular heads level (Neumann 2010). If the person presents with the knees touching and the feet apart, this is indicative of genu **valgum** (knock-knees) (figure 16.4*a*). Excessive gapping between the knees with bowing inward indicates genu **varum** (bowlegs) (figure 16.4*b*).

The patellae should point straight ahead. If the patellae face inward, it is termed squinting; if the patellae face outward, it is termed frog

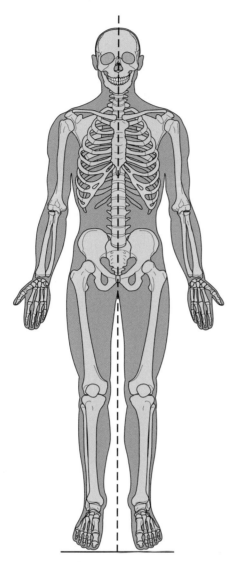

▶ **FIGURE 16.3** Ideal posture, anterior view. In this and subsequent similar figures, the dashed line represents a plumb line.

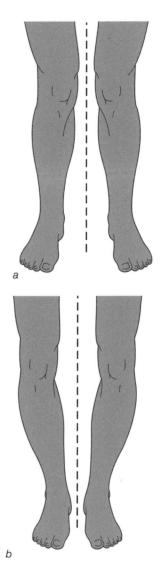

a

b

▶ **FIGURE 16.4** *(a)* Genu valgum and *(b)* genu varum.

eyes. Keep in mind that femoral or tibial position (**torsion**) can affect the patellar position. In the frontal plane, the tibia should have a slight amount of curvature with the convexity laterally. Excessive curvature is termed tibial varum. Abnormal rotation in the transverse plane is tibial torsion. Tibial torsion may be identified by noting that the patellae are facing forward, yet the feet point outward.

The feet angle out slightly (5 to 18°) but symmetrically. The medial longitudinal arch should be visible. If the arch is not visible, the clinician may suspect pes **planus** (flexible feet); if the arch appears high, the clinician might suspect pes **cavus** (rigid feet).

▶ **KEY POINT**

In standing, the feet are angled out 5 to 18° but symmetrically.

The five toes should point straight ahead. Abnormalities usually involve deviation of the toes from the midline. Deviation of the great toe away from the midline is termed hallux abducto valgus. The fifth toe may rotate outward; this is called a tailor's bunion and is associated with hypermobility of the fifth ray. The second through fifth toes may present with one of the following deformities: claw toes or hammer toes (figure 16.5). Claw toes occur when the proximal and distal

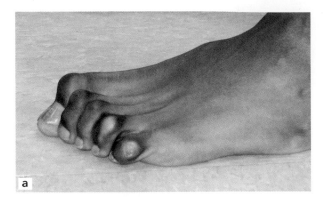

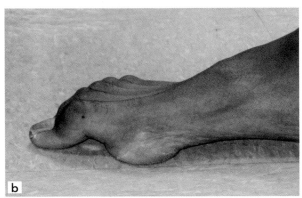

▶ **FIGURE 16.5** Toe deformities: *(a)* claw toes and *(b)* hammer toes.

Reprinted from S.J. Shultz and P.A. Houglum and D.H. Perrin, 2009, *Examination of musculoskeletal injuries*, 3rd ed. (Champaign, IL: Human Kinetics), 407. By permission of P.A. Houglum.

interphalangeal joints are hyperflexed and the metatarsophalangeal joint is subluxed dorsally. Hammer toes occur when the proximal interphalangeal joint is flexed and the MTP is extended. Possible causes of these deformities include foot intrinsic muscle imbalance or dysfunction, hereditary factors, and poor-fitting shoes.

Posterior View

When assessing posterior posture using a plumb line as in figure 16.6, the plumb line should bisect the body into two equal halves. The alignment should be such that the line does the following:

- Bisects the cranium into two equal halves
- Is equidistant between the scapulae
- Is equidistant between the posterior superior iliac spine

- Is equidistant between each knee
- Is equidistant between the medial malleoli

Some of the same landmarks and positions that were identified for the anterior view are also identified for the posterior view. The head should be centered over the shoulders without a tilt or rotation. The upper trapezius is level right to left, with equal slope on both sides. As previously mentioned, with the handiness pattern the dominant shoulder will be lower.

In normal posture, the scapulae sit between the second and seventh thoracic vertebrae, 2 to 3 in (5 to 8 cm) from the spine, and the scapular plane is 30° anterior to the frontal

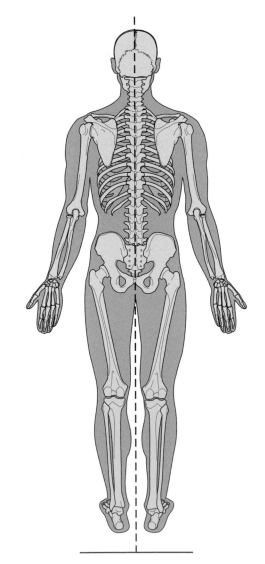

▶ **FIGURE 16.6** Ideal posture, posterior view.

plane. Numerous scapular dysfunctions can be present including abduction, winging, downward rotation, anterior tilt, upward rotation, elevation, and depression. More description, including illustrations, of each of these is given in chapter 10.

The spinal processes of the vertebrae should be straight and without any noticeable curvature. Excessive curvature is present in **scoliosis**, which is described in more detail at the end of this chapter. At the pelvis, the iliac crest and posterior iliac spine should be level. In the handiness pattern, the iliac crest on the dominant side will be high. The gluteal folds and the gluteal mass should be symmetrical. The popliteal crease at the knee should be level and relatively horizontal. Obliqueness of the crease indicates torsion of the tibia or femur, usually medially. From the posterior view, the calcaneal position should be vertical. The Achilles tendon should also be relatively vertical. Pes planus would present with flattening of the medial longitudinal arch, and a navicular "bulge" may be seen. Additionally, the Achilles tendon would display a curvature with a medial convexity, and the calcaneus would be everted. A rigid pes cavus would have the opposite effect: calcaneus inverted and Achilles tendon bowing outward. No more than the fourth and fifth digits should be seen from the rear. If more digits are seen, then the foot is considered abducted, and commonly excessive pronation is present.

Lateral View

When assessing lateral posture using a plumb line, the clinician should note the following alignments (figure 16.7) (Kendall 2005):

- Slightly anterior to the lateral malleolus, through the calcaneocuboid joint
- Just in front of the center of the knee joint
- Through the greater trochanter of the femur, slightly posterior to the center of the hip joint
- Midway through the trunk, through the bodies of the lumbar vertebrae
- Through the shoulder joint

- Through the bodies of the cervical vertebrae
- Through the odontoid process
- Through the lobe of the ear, through the external auditory meatus

The lateral view should be assessed from both the left and right. In assessing the lateral view starting at the top, the head should sit above the shoulders such that the earlobe is in line with the acromion process. Commonly, people present with a forward head, and the chin will protrude forward. The plumb line should bisect the acromion above the shoulder, and less than one-third of the humeral head should sit anteriorly to the acromion. The

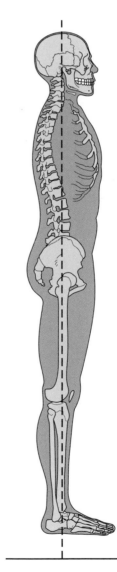

▶ **FIGURE 16.7** Ideal posture, lateral view.

palms should face in toward the body. Similar to the head, the shoulder position is commonly forward. Accompanying this posture is tightness of the anterior chest and shoulder muscles and the pectoralis minor and major.

There should be a slight posterior convexity in the thoracic spine. Abnormalities include an excessive posterior curve termed kyphosis, or a flattening (extension) of the spine. In young males, excessive thoracic kyphosis may be indicative of Scheuermann's kyphosis (see Clinical Correlation 16.1). Pay attention to the scapulae, as their position can give the illusion of thoracic kyphosis. In postmenopausal women, increased kyphosis can be associated with osteoporosis (Bradford 1995).

The lumbar spine should have a slight lordosis. Excessive lordosis may be present in people with hypermobility, which can be accompanied by tight hip flexors. Other faulty postures from the lateral view include kyphotic–lordotic, flat-back, and swayback (figure 16.9). The kyphotic–lordotic posture is characterized by increased thoracic kyphosis and subsequent increased lumbar lordosis. Some people will present with minimal to no lordosis in the lumbar spine, which is known as flat back. These people usually have a very stiff lumbar spine. The swayback posture is found in people with a flat lumbar spine; the thoracic spine is swayed posteriorly relative to the plumb line (Kendall 2005).

CLINICAL CORRELATION 16.1

Scheuermann's kyphosis is a structural deformity involving abnormal growth of the vertebral bodies in the thoracic spine due to abnormal ossification. This condition is characterized by wedging of the vertebral bodies in which the height of the bodies anteriorly is less than posteriorly (figure 16.8). Scheuermann's is idiopathic and more common in males between the ages of 11 and 17 years. The diagnosis is confirmed when wedging occurs in three or more vertebrae. Schmorl's nodes may also be present. These nodes are protrusions of the nucleus pulposus through the vertebral body end plate. In severe cases, surgery is performed to straighten the spine. Rehabilitation focuses on maintaining spine range of motion, especially in extension.

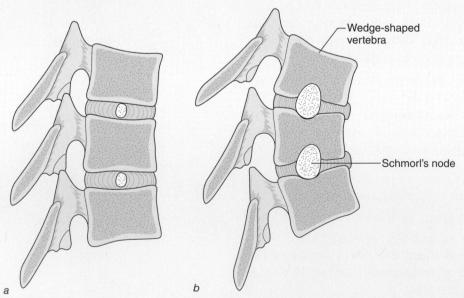

▶ **FIGURE 16.8**   Scheuermann's kyphosis. (a) Normal shape and orientation of thoracic vertebrae. (b) Scheuermann's kyphosis showing wedging of vertebrae and Schmorl's nodes.

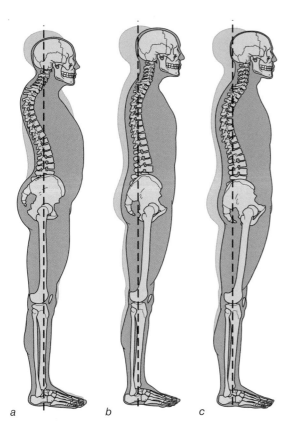

FIGURE 16.9 *(a)* Kyphotic–lordotic posture, *(b)* flat-back posture, and *(c)* swayback posture.

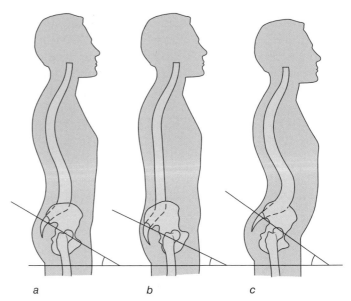

FIGURE 16.10 Lumbosacral angle. *(a)* Normal standing creates a lumbosacral angle of approximately 30°. *(b)* Tilting the pelvis backward decreases the lumbosacral angle (<30°) and flattens the lumbar spine. *(c)* Tilting the pelvis forward increases the lumbosacral angle (>30°) and exaggerates the lumbar lordosis.

The lumbosacral angle is the angle between the top of the sacrum and the horizontal plane, measured via radiograph. Normally this angle is approximately 30° but will change depending on the lumbar curvature (figure 16.10). A person with a flat lumbar spine will have a smaller angle, while a person with increased lumbar lordosis will have a greater angle. The pelvic tilt is associated with the curvature of the lumbar spine in that a lumbar lordosis is accompanied by an anterior tilt of the pelvis. A flat back is accompanied by a posterior tilt of the pelvis.

▶ **KEY POINT**

The normal lumbosacral angle is the angle between the top of the sacrum and the horizontal plane and is equal to 30 to 40°.

From the lateral view, the lower extremity should be relatively straight. The lateral hip should be in line with the plumb line. Tight-ness in the hip flexors will cause standing hip flexion. Excessive hip extension may indicate tightness in the hip extensors. The knees should be slightly bent, but not more than 5°. Abnormal knee flexion may be due to a knee flexion contracture or tightness in the hamstrings or gastrocnemius. Hyperextension of the knee (genu recurvatum) usually indicates hypermobility, which is accompanied by laxity in the posterior knee capsule. The feet should be aligned so that the plumb line is slightly anterior to the lateral malleolus, through the calcaneocuboid joint.

Sitting and Lying Postures

An orthopedic clinician must address postures beyond standing, such as sitting and lying. Various sitting positions will affect the load on the spine. Any deviation from an upright position will increase the load. Wilke et al. (1999) reported the intradiscal pressure in a single subject using an in vivo pressure transducer inserted in the L4-L5 disc. This group found the following results: lying prone, 0.1 MPa (megapascals); side lying, 0.12 MPa; relaxed standing, 0.5 MPa; standing flexed forward, 1.1 MPa; sitting unsupported, 0.46 MPa; sitting with maximum flexion, 0.83 MPa; nonchalant sitting, 0.3 MPa; and lifting a 20 kg (44 lb) weight with a round flexed back, 2.3 MPa; with flexed knees, 1.7 MPa; and close to the body, 1.1 MPa. Generalizing these data to a larger population must be done cautiously because of the single-subject design. According to these results, forward flexion and lifting of the trunk produce the highest stresses on the spine.

Sitting

Sitting posture is important to assess because so much of our time is spent in sitting. Some people spend more than 8 hours a day sitting because of work responsibilities. A good sitting posture is characterized by minimal muscle effort and is highly dependent on sitting support. Ideally, the spine is erect and supported, the shoulders are aligned over the hips, the hips are flexed to 90°, and the feet are supported (Sahrmann 2002). If a person sits without back support, the result is a significant increase in muscle activity to maintain an upright posture (Adams & Hutton 1985). If one has to sit for a long time with no support, then flexion in the spine is inevitable. Using support (e.g., a rolled towel) under the ischial tuberosities will help maintain the normal lumbar lordosis if sitting without support.

Sitting with or without support has different influences on sitting posture and on the spine. Further, if the chair improperly fits, other posture deviations occur (Nordin & Frankel 2001). An ill-fitting chair can lead to postural deviations:

- Absence of an arm rest causes muscle activity of the shoulder girdle in order to support the dangling arms.
- Sitting without a back rest leads to a flexed spine.
- The length of the seat can be too long for shorter people.

- A back support that is too high will result in lumbar flexion.
- A back support that is too low will not give stability to the trunk.
- A seat that is too high causes excessive pressure on the thighs, which might cause anoxia to the nerves of the thigh.

Assessment of sitting posture can be performed from the front, back, and side. Take note if there is a dramatic change in the spinal curvature. The legs should protrude the same distance, as measured by knee position.

▶ **KEY POINT**
Impairments in sitting posture commonly include excessive thoracic flexion, side bending, and extension.

Lying

Lying posture is also very important in our daily lives because we spend as much as 9 hours sleeping per day. Muscular activity is minimal in the recumbent posture because all the body segments are supported directly by the support surface. To examine this posture, have the client lie supine without support at the head, spine, or legs. The examiner should note the position of the head (tilted or rotated). If this position causes the head to tilt into cervical extension, then tightness in the cervical upper extensor muscles or joints is probable. The pelvis should be level as specified by the ASIS position. If the lumbar spine does not contact the table, then the person may have tightness in the hip flexors or lumbar extensors. Leg length can be measured in this position. A tape measure can be used to assess the distance from the ASIS to the lateral or medial malleolus. Another set of landmarks are the greater trochanter and the lateral malleolus. A slight difference up to 1.0 cm is considered normal (Magee 1997). If a person experiences low back pain with lying, a pillow under the hips or waist when sleeping prone or between the knees when side lying can improve alignment and relieve stress on the spine (Kendall 2005). A pillow under the knees will flatten the lumbar lordosis and relieve lumbar strain if the person is lying supine. If pain is experienced in the cervical spine and the person presents with a forward head and round shoulders, more than one pillow may be indicated. If no fixed deformity exists, then one pillow should suffice.

Postural Faults

The previous section set the stage for identification of normal posture. This section describes the possible postural faults that may be found throughout the body.

Movement System Impairment Syndromes

The movement system impairment syndromes are based on Sahrmann and colleagues' premise that proper alignment is the foundation for optimal movement (Sahrmann 2002). Faulty alignment leading to faulty movement patterns puts a person at risk for development of a painful movement syndrome. For example, a swimmer with tight internal shoulder rotators may be more prone to shoulder impingement because her swimming movement promotes further internal rotation of the humerus, placing the joint in a position that is closed packed. Identification of muscle imbalances is the key to categorizing the diagnostic movement syndrome in order to prescribe the appropriate treatment. An example of this system is found in Clinical Correlation 16.2. Treatment commonly focuses on strengthening the antagonist

of a short muscle. Further readings on this classification system are found in the reference list (Sahrmann 2002; Sahrmann 2011).

Crossed Syndromes

Vladimir Janda, a Czech physiatrist, introduced the concept of neuromuscular imbalances between groups of muscles resulting in poor posture, loss of mobility, and an increase in joint loads (Janda 1987). His theory states that certain muscle groups tend to be either tight (tonic muscles) or weak (phasic muscles). Muscles prone to tightness generally have a "lowered irritability threshold" and are readily activated with any movement, thus creating abnormal movement patterns. Abnormal movement patterns can lead to dysfunction and pain. Dr. Janda identified three specific muscle imbalance syndromes associated with chronic musculoskeletal pain: lower crossed syndrome, upper crossed syndrome, and layered syndrome.

The lower crossed syndrome (figure 16.11a) is characterized by tightness of the thoracolumbar extensors on the dorsal side that crosses with tightness of the iliopsoas and rectus femoris. Weakness of the deep abdominal muscles ventrally crosses with weakness of the gluteus maximus and medius. The upper crossed syndrome (figure 16.11b) is also referred to as proximal or shoulder girdle crossed syndrome (Janda 1988). In this syndrome, tightness of the upper trapezius and levator scapulae on the dorsal side crosses with tightness of the pectoralis major and minor. Weakness of the deep cervical flexors

CLINICAL CORRELATION 16.2

Scapular downward rotation syndrome is characterized by insufficient scapular upward rotation. In general, the scapula insufficiently upwardly rotates during humeral elevation, potentially creating tissue stress such as shoulder impingement or thoracic outlet syndrome. Muscle impairments include dominance, shortness, or stiffness of the rhomboids, levator scapulae, latissimus dorsi, and pectoralis minor and major. The serratus anterior and trapezius are insufficiently active. Laborers who work overhead commonly present with this muscle imbalance. Rehabilitation focuses on strengthening the serratus anterior and lower trapezius. In addition, the timing of the serratus–trapezius force couple will need to be addressed.

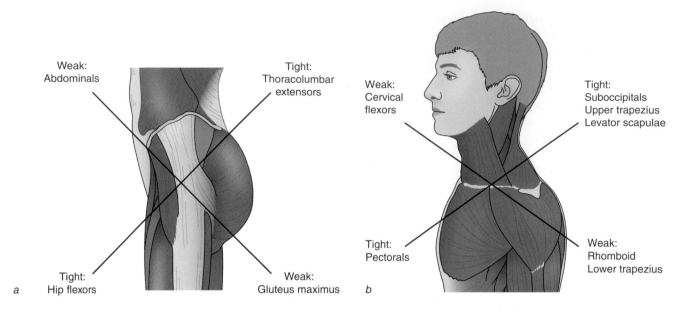

Weak: Abdominals

Tight: Thoracolumbar extensors

Tight: Hip flexors

Weak: Gluteus maximus

a

Weak: Cervical flexors

Tight: Suboccipitals
Upper trapezius
Levator scapulae

Tight: Pectorals

Weak: Rhomboid
Lower trapezius

b

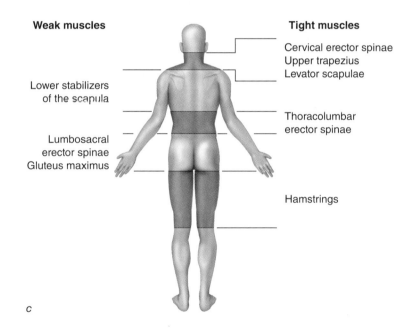

Weak muscles

Lower stabilizers of the scapula

Lumbosacral erector spinae
Gluteus maximus

Tight muscles

Cervical erector spinae
Upper trapezius
Levator scapulae

Thoracolumbar erector spinae

Hamstrings

c

▶ **FIGURE 16.11** *(a)* Lower crossed syndrome, *(b)* upper crossed syndrome, and *(c)* layered syndrome.

CLINICAL CORRELATION 16.3

Lower crossed syndrome is also referred to as distal or pelvic crossed syndrome. In lower crossed syndrome, tightness of the thoracolumbar extensors on the dorsal side is accompanied by (crosses with) tightness of the iliopsoas and rectus femoris. Weakness of the deep abdominal muscles crosses with weakness of the gluteus maximus and medius. Theoretically, this pattern of imbalance creates joint dysfunction, particularly at the L4-L5 and L5-S1 segments, SI joint, and hip joint. Specific postural changes seen include anterior pelvic tilt, increased lumbar lordosis, lateral lumbar shift, external femoral rotation, and knee hyperextension (Janda & VaVrova 1996). This posture type is not uncommon in young female athletes who participate in gymnastics or dance. Treatment to improve lower crossed syndrome includes strengthening the weak muscles (deep abdominals, gluteus maximus and medius) and stretching the hip flexors.

ventrally crosses with weakness of the middle and lower trapezius. This pattern of imbalance creates joint dysfunction, particularly at the atlantooccipital joint, C4-C5 segment, cervicothoracic joint, glenohumeral joint, and T4-T5 segment. (More information on lower crossed syndrome is found in Clinical Correlation 16.3.) The layered syndrome (figure 16.11c) is a combination of both upper and lower crossed syndromes. There is marked impairment of motor regulation that has increased over a period of time. This pattern is often seen in older adults and in patients who have undergone unsuccessful surgery for herniated nucleus pulposus (HNP) (Page et al. 2010).

Scoliosis

Scoliosis is a frontal and transverse plane deviation of the normal alignment of the spinal curve. A lateral curvature in the cervical spine is termed torticollis. Scoliosis is named by the shape (C-curve, S-curve), the spine location (thoracic, lumbar), and the side of convexity (right or left) (figure 16.12). For example a right thoracic C-curve describes a single curve in the thoracic spine, with the convexity on the right.

Scoliosis can be classified as functional or structural. Functional scoliosis is caused by postural habit, leg-length difference, or muscle imbalances. Structural scoliosis involves a fixed bony deformity. This type of deformity may be congenital or acquired, but the most common type is idiopathic scoliosis. Idiopathic scoliosis accounts for 75 to 85% of structural scoliosis and is most common in adolescent females. Structural scoliosis can be progressive. The Cobb angle is used to measure structural scoliosis. Clinical Correlation 16.4 describes the measurement of the Cobb angle.

▶ **KEY POINT**
Functional scoliosis is caused by postural habit, leg-length difference, or muscle imbalances. Structural scoliosis involves a fixed bony deformity.

A rib hump is always present in a person with structural scoliosis. It is located on the side of convexity. The rib hump is prominent when the patient bends forward from the waist. The hump is due to the transverse rotation of the vertebrae toward the side of convexity, causing the ribs to protrude posteriorly on the convex side. The disc space is compromised on the concave side. If the curve exceeds 60°, vital capacity and organ function may be hindered.

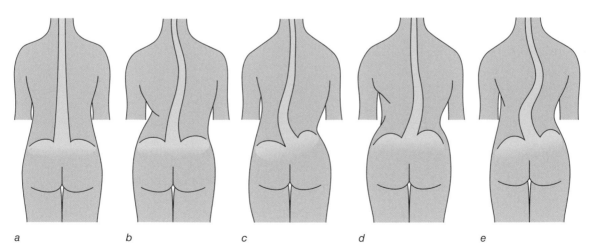

a   b   c   d   e

▶ **FIGURE 16.12** Forms of scoliosis: *(a)* normal orientation of the vertebral column; *(b)* right thoracic scoliosis; *(c)* left lumbar scoliosis; *(d)* right thoracolumbar scoliosis; *(e)* and bilateral scoliosis (right thoracic and left lumbar).

CLINICAL CORRELATION 16.4

The Cobb angle is a measurement used for evaluation of curves in scoliosis on an antero-posterior radiographic projection of the spine. When assessing a curve, the apical vertebra is first identified; this is the most likely displaced and rotated vertebra, with the least tilted end plate. The end, or transitional, vertebrae are then identified through the curve above and below. The end vertebrae are the most superior and inferior vertebrae that are least displaced and rotated and have the maximally tilted end plate. A line is drawn along the superior end plate of the superior end vertebra, and a second line is drawn along the inferior end plate of the inferior end vertebra. If the end plates are indistinct, the line may be drawn through the pedicles. The angle between these two lines (or lines drawn perpendicular to them) is measured as the Cobb angle (figure 16.13). In S-shaped scoliosis where there are two contiguous curves, the lower end vertebra of the upper curve represents the upper end vertebra of the lower curve. Because the Cobb angle reflects curvature only in a single plane and fails to account for vertebral rotation, it may not accurately demonstrate the severity of three-dimensional spinal deformities. As a general rule, a Cobb angle of 10° is regarded as a minimum angulation to define scoliosis.

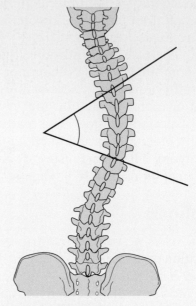

▶ **FIGURE 16.13**   Cobb angle.

Conclusion

A standard posture assessment helps identify asymmetries that may potentially lead to musculoskeletal dysfunction. This chapter describes normal and abnormal postures for each major joint associated with posture. Accompanying this description are the muscle imbalances associated with this faulty posture. Sitting and lying posture are also discussed. Rehabilitation of a postural dysfunction focuses on improving length of tight structures and strengthening weak ones.

REVIEW QUESTIONS

1. List the primary structures that should coincide with a plumb line used to assess posture from a lateral view.

2. Name the muscles that are active during quiet standing.

3. What is lower crossed syndrome?

4. Explain how to identify tibial varum.

5. Describe kyphosis.

Walking Gait

Human locomotion is bipedal and requires synchronized joint motion and muscle activity of the lower extremity. It is one of the most common functional activities, and therefore gait assessment should be a component of every musculoskeletal examination. This chapter uses the common terminology of gait examination, describes the gait sequence (Rancho Los Amigos terminology), and identifies joint range of motion and muscle activity associated with each phase of gait for all major joints. The chapter begins with a discussion of the determinants of gait and then continues with normal gait descriptors. Faulty gait characteristics for each major joint and where they occur in the gait cycle are also described. Gait examination is a technical skill that requires precision. The breakdown of each gait phase should help the clinician with identifying gait faults.

Determinants of Gait

In general, walking gait is a very mechanical and efficient way of locomotion. Much of this is due to the symmetry of joint angular motion and muscle activation patterns. If a

OBJECTIVES

After reading this chapter, you should be able to do the following:

> Define the various phases of walking gait.

> Describe the three-dimensional kinematics for the pelvis, hip, knee, ankle, and subtalar joint that occur during the different phases of walking gait.

> Analyze the muscle activity of the major muscles associated with walking gait.

> Explain the lower extremity kinetics during walking.

> Describe the most common gait abnormalities.

motion system marker is placed on the lateral trochanter and videotaped while a person walks several feet, the marker will display a sinusoidal curve indicating a small vertical displacement (3 cm) of the center of mass that occurs during walking. This displacement is due to six determinants:

- Pelvic rotation in the transverse plane: Swing-side pelvis moves anteriorly to advance the limb forward.

- Lateral pelvic tilt: On the swing side, the pelvis drops approximately 1 inch (2.5 cm).

- Lateral shift: The body shifts approximately 1 to 2 inches (2.5 to 5 cm) toward the stance limb.

- Knee flexion: The knee flexes up to 45° during preswing.

- Ankle dorsiflexion: The ankle dorsiflexes during the late phase of midstance as the body proceeds forward.

- Heel rise: During the last portion of stance, the heel rises for push-off.

Gait Sequence

A full **gait cycle** (figure 17.1) is the sequential completion of a single limb's **stance phase** and **swing phase**. Further, it can be described as the period between right initial contact and the sequential right initial contact (or left initial contact and the sequential left initial

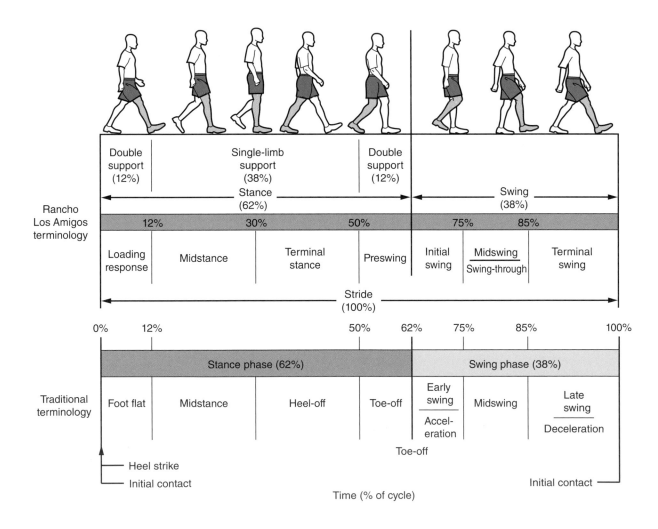

▶ **FIGURE 17.1** Rancho Los Amigos and traditional terminology for the walking gait cycle.

contact), or **stride length**. **Step length** is the distance between right foot contact and left foot contact. The stance (stride) phase is when the reference foot is on the ground; it is made up of five subphases: initial contact, loading response, midstance, terminal stance, and preswing. The swing phase is when the reference foot is off the ground; it includes the initial swing, midswing, and terminal swing. These phases are taken from the Rancho Los Amigos (RLA) terminology (Perry 1992). A comparison of traditional gait sequence terminology with Rancho Los Amigos terminology is included in figure 17.1. Traditional terminology refers to points in time, whereas the RLA terminology refers to lengths of time. In the next section, we look at the individual phases of gait.

▶ **KEY POINT**

Stride length is the distance between right initial contact and right initial contact (or left initial contact and left initial contact). Step length is the distance between right foot contact and left foot contact.

Stance Phase

The stance phase occurs when the foot is on the ground. With average walking speed, the stride phase makes up 60 to 62% of the whole gait cycle and takes on average 0.06 seconds to complete. **Single support** occurs when only one foot is on the ground (during midstance and terminal stance). **Double support** occurs when both feet are on the ground. This happens twice in one gait cycle: during initial contact and loading response and during preswing. The top part of figure 17.1 displays each phase of the gait sequence.

Initial contact is defined as the point when the foot contacts the ground. In normal gait, this contact is made by the heel. However, in gait sequences like those of people with cerebral palsy, initial contact may be with the mid- or forefoot. Initial contact is the beginning point of the stance phase.

Loading response is the phase from initial contact until the contralateral leg leaves the ground. During loading response, weight is rapidly transferred onto the outstretched limb, and the limb is decelerating. This along with initial contact is the first period of double-limb support, as the contralateral limb is on the ground. The initial contact phase and the loading response phase together make up 10% of the gait cycle.

Midstance is the period when the contralateral extremity lifts off the ground and continues to a position in which the body has progressed over and ahead of the supporting extremity. Midstance makes up 20% of the gait cycle and is a period of single-limb support.

Terminal stance occurs from midstance to a point just before initial contact of the contralateral extremity. In normal gait, the heel is leaving the ground as progression over the stance limb continues. The body moves ahead of the limb, and weight is transferred onto the forefoot. Terminal stance lasts for 20% of gait cycle and is the second period of single-limb support.

Preswing is the final phase of stride and takes place from just after heel-off to toe-off. A rapid unloading of the limb occurs as weight is transferred to the contralateral limb. Preswing is the second period of double-limb support and makes up 10% of the gait cycle. Preswing occurs during loading response on the opposite side.

Swing Phase

The swing phase—when the foot is not on the ground—consists of initial swing, midswing, and terminal (late) swing. The swing phase is 38 to 40% of the total gait cycle. These three phases are equally divided with regard to time.

Initial swing is the acceleration phase and is described as the time the foot leaves the ground until maximal knee flexion in swing. During this phase, the thigh begins to advance as the foot lifts off the floor. Midswing begins after initial swing and continues until the airborne tibia is in a vertical position. The thigh continues to advance as the knee begins to extend, and the foot clears the ground. Terminal (late) swing is the deceleration phase. The knee is extending as the limb prepares to contact the ground.

▶ **KEY POINT**
The swing phase consists of initial swing, midswing, and terminal swing.

Gait Kinematics

As defined in chapter 1, kinematics is the branch of mechanics that describes the motion of objects. With regard to gait, we will discuss the motion that occurs primarily in the lower extremity in all three planes of movement. Table 17.1 lists the primary joints, joint positions, external moments, and muscle movements for each phase of the gait cycle. Details of the major lower extremity joints in the sagittal, frontal, and transverse planes are presented in this section. The head, arms, and trunk (sometimes referred to as HAT) are usually considered in literature as a rigid unit and will not be discussed.

Sagittal Plane

The primary motion of the lower extremity occurs in the sagittal plane. This section discusses the sagittal plane motion at the pelvis, hip, knee, and ankle.

TABLE 17.1   Lower Extremity Kinematics During Gait Phase

	Anatomical position	Muscles	External moment
Subtalar joint			
IC	Inverted (2°)	Tibialis posterior (eccentric)	ND
LR	Pronated (5°)	Tibialis posterior (eccentric)	ND
MS	Supinating	Tibialis posterior (concentric)	ND
TS	Supinated (max)	Peroneals (eccentric)	ND
PS	MTP dorsiflexed (70°)	Peroneals (eccentric)	ND
S	Supinated	Peroneals (concentric)	
Ankle joint			
IC	Dorsiflexed	Pretibials (eccentric)	Plantar flexion
LR	Plantar flexed	Pretibials, soleus (eccentric)	Plantar flexion
MS	Early: neutral; late: dorsiflexed	Soleus (eccentric)	Dorsiflexion
TS	Dorsiflexed (10°)	Soleus (eccentric)	Dorsiflexion
PS	Plantar flexed (20°)	Plantar flexors (concentric)	Dorsiflexion
S	Plantar flexed, neutral	Pretibials (concentric, isometric)	
Knee joint and tibia			
IC	Extended, externally rotated	Quadriceps (eccentric)	Extension
LR	Flexed (15°), internally rotating	Quadriceps (eccentric)	Flexion
MS	Neutral	Early: quadriceps (concentric); popliteus, gastrocnemius	Neutral
TS	Extended (0°), externally rotating	Hamstrings (eccentric), gastrocnemius (concentric) for knee flexion	Extension
PS	Flexed (45°), externally rotated	Popliteus (concentric), rectus femoris (eccentric)	Flexion
S	Flexed, late: extension (65°)	Early: hamstrings (concentric); late: hamstrings and quadriceps to stabilize knee in extension	

	Anatomical position	Muscles	External moment
Hip joint			
IC	Flexed (35°), slight adduction and external rotation	Hip extensors (eccentric)	Flexion
LR	Flexed, internally rotating	Gluteus maximus, hamstrings (concentric), gluteus medius (eccentric) in frontal plane	Flexion
MS	Early: neutral, internal rotation → external rotation, abduction	Iliopsoas (eccentric), gluteus medius (isometric)	Neutral
TS	Extended, externally rotated	Tensor fasciae latae (eccentric)	Extension
PS	Neutral	Rectus femoris (concentric), adductors (eccentric)	Neutral
S	Early: neutral, flexed, adduction	Psoas (concentric), adductors (eccentric), hamstrings (eccentric)	

IC = initial contact; LR = loading response; MS = midstance; TS = terminal stance; PS = preswing; S = swing; ND = not described

Pelvis

In the sagittal plane, pelvic motion occurs about a mediolateral axis. Motion at the pelvis is identified as anterior and posterior pelvic tilt. At initial contact, the pelvis is anteriorly tilted approximately 7° (Nordin & Frankel 2001). The pelvis is anteriorly rotated until midstance, where it moves posteriorly 8 to 10°.

Hip

The hip joint has key motion in all three planes during the gait cycle. In the sagittal plane, the hip joint is in flexion (30°) at initial contact and then moves into extension as the body progresses over the fixed foot. The hip reaches maximal extension (10°) during terminal stance. It begins to move back into flexion during preswing, reaching a maximal range around 35° during swing and then slightly less flexion just before initial contact.

Knee

The knee joint remains relatively straight during initial contact (figure 17.2). As the loading phase begins, the knee moves into slight flexion to help with shock absorption. At preswing, the knee passively flexes to 45° as the heel lifts off the ground. During initial swing, the knee maximally flexes to 60° to allow the foot to clear the ground. Knee flexion is one of the determinants of gait and is

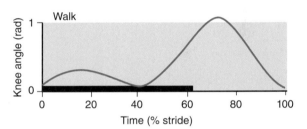

▶ **FIGURE 17.2** Sagittal plane knee joint kinematics. 1 radian = 57.3°.

needed to minimize vertical translation and conserve energy.

Ankle

The ankle (talocrural joint) is relatively neutral at heel strike. The forefoot slowly lowers to the floor as the gait cycle progresses. During loading response and early midstance, the talocrural joint is in relative plantar flexion because the stance leg is in front of the body. As the body progresses over the fixed foot, the talocrural joint moves into relative dorsiflexion (midstance) until the heel leaves the ground, readying it for push-off. Maximal dorsiflexion (10° or greater) is required at terminal stance. If 10° is not available at the talocrural joint, then compensations will take place at the adjacent proximal or distal joints. Compensations may include excessive subtalar joint pronation or hyperextension at the knee. Another compensation may be early heel rise. During swing, the talocrural joint returns to dorsiflexion to clear the ground and to get ready for initial contact.

Frontal Plane

Next we look at frontal plane kinematics of the pelvis, hip, knee, and ankle. Motion in this plane is minimal and when excessive can become injurious to the musculoskeletal system (Willson & Davis 2008).

Pelvis

Pelvic motion in the frontal plane occurs about an anteroposterior axis and is termed lateral pelvic tilt. At initial contact the pelvis is level. During loading response, the pelvis rises on the stance side (4 to 5°) and slightly drops on the swing side. Maximal tilt occurs at preswing and can reach 10°.

Hip

In the frontal plane, the hip moves in relation to the action of the pelvis. At initial contact, the hip is in a neutral position to slight adduction. The hip continues to display slight hip adduction (5°) up until midstance. After midstance, the hip begins to abduct and is in a position of about 10° of hip abduction as the foot leaves the ground. During swing the hip is slightly adducted.

Knee

The knee is stable in the frontal plane because of the ligamentous and bony constraints. Minimal movement into adduction (varus) occurs throughout the majority of stance. During preswing and initial swing, the knee moves into abduction (valgus) but returns to relative neutral during terminal swing.

Ankle

Ankle motion occurs only in the sagittal plane because of the morphological constraints of the talocrural joint.

Subtalar Joint

Most of the motion at the subtalar (ST) joint occurs in the frontal plane and so is discussed in this section. At initial contact in normal gait,

the contact is made on the lateral side of the heel, and the ST joint is minimally supinated. From initial contact through loading response, the ST joint pronates. It reaches maximal pronation at midstance (4 to 6°). After this instance, the ST joint begins to resupinate to get the foot ready for push-off. During terminal stance and preswing, the great toe moves into maximal extension (70°), creating the windlass effect, a passive lifting of the arch due to the plantar fascia.

▶ **KEY POINT**
Maximal pronation of the subtalar joint occurs at midstance and is equivalent to 4 to 6°.

Transverse Plane

The final plane we investigate is the transverse plane kinematics of the pelvis, hip, and knee. Similar to the frontal plane, motion in this plane is minimal and when excessive can become injurious to the musculoskeletal system.

Pelvis

Pelvic rotation in the transverse plane is about a vertical axis and is termed axial rotation or internal/external rotation. In walking, pelvic rotation is important for lengthening the stride limb. The pelvis is maximally rotated forward (5°) (internal rotation) at initial contact to achieve step length. After initial contact, the pelvis rotates backward (external rotation) as the opposite limb begins to move forward over the stance limb. Much of this motion is passive and is driven by the swing limb. The pelvis remains externally rotated until midswing and then begins to reverse direction as the limb swings forward and approaches the ground. The motion of the trunk is opposite the direction of the pelvis in gait mainly due to the reverse arm swing.

Hip

In the transverse plane, at initial contact, the hip is in slight external rotation. After the foot contacts the ground, the primary motion is internal rotation until late midstance, where it begins to externally rotate. This external rotation continues into initial swing. During the majority of swing, the hip remains relatively neutral.

Knee

During initial contact, the tibia mirrors the action of the femur and is externally rotated. As the ST joint pronates and the knee flexes, the tibia internally rotates during loading response and the beginning of midstance. As the heel leaves the ground, the subtalar joint begins to supinate and the tibia externally rotates.

Muscle Activity During Gait

During gait, three rockers can be described that correspond with the stance phases of the gait cycle. A rocker is a phase in the gait cycle where primary motion takes place (Perry 1992). The first rocker occurs from initial contact to loading response and is the period of deceleration. During the first rocker, the joints are absorbing impact. The second rocker is defined from foot flat to midstance and is characterized by the control of the ground reaction force (force absorption). The third rocker occurs during terminal stance and preswing and is a period of acceleration. Beyond the muscle activity that occurs during the three rockers, muscles are also active during the swing phase, which will be discussed.

▶ **KEY POINT**

Three rockers occur during the stance phase of gait, representing periods of acceleration and deceleration.

First Rocker

During first rocker, the hip extensors initially work eccentrically to control hip flexion. Once the foot is flat, the hip extensors concentrically work to pull the body forward over the fixed foot. In the frontal plane, the stance-side gluteus medius is the primary muscle that controls hip and pelvic motion. The adduc-

tor magnus works on a fixed femur to cause internal rotation. At initial contact and loading response, the quadriceps is active to control knee flexion and help with shock absorption. The pretibial muscles (tibialis anterior, extensor digitorum, extensor hallucis longus) are active during first rocker, working eccentrically to control foot slap. As the ground reaction force passes through the heel, an external plantar flexion moment is created. The primary muscles working at the ST joint during first rocker are the tibialis anterior and tibialis posterior. These muscles work eccentrically to decelerate pronation.

Second Rocker

As the hip is extending in second rocker, the gluteal muscles work concentrically. Late in midstance, hip extension is passive and is checked by the iliofemoral ligament and the tensor fasciae latae. Midstance is characterized by minimal to no muscle activity in the quadriceps or hamstrings, as the ligaments about the knee and the external extensor moment help stabilize the joint. The soleus is working eccentrically to control the anterior displacement of the tibia over the fixed foot, and the popliteus is actively controlling transverse plane tibial motion. At the subtalar joint, the tibialis posterior is controlling pronation until the midportion of midstance; it then begins to function concentrically to raise the medial longitudinal arch.

Third Rocker

Third rocker is a period of acceleration. Deceleration of the first two rockers must be counterbalanced by the third rocker. The hip continues to extend into third rocker, where it reaches maximal extension. The tensor fasciae latae is still working to control hip extension. The femur is moving into external rotation, and the adductor longus now works eccentrically to fine-tune this motion. At the knee, the popliteus is concentrically active to unlock the knee by allowing knee flexion. The external moment changes to flexion during terminal stance, and the rectus femoris becomes active to eccentrically control knee flexion. With the foot relatively flat, the soleus continues to work eccentrically to control tibial translation. As the heel rises, the ankle plantar flexes and the gastroc–soleus complex works concentrically for push-off. The push-off is important for maintaining walking velocity and step length (Nordin & Frankel 2001). Toward the end of the stance phase, the foot invertors work concentrically to lock the foot into supination. The peroneals work eccentrically to fine-tune supination and control forefoot motion.

Swing Phase

As swing begins, the hip drives forward into flexion with contraction of the hip flexors, primarily the iliopsoas muscle. The gluteus medius will concentrically raise the pelvis on the swing side to help with foot clearance during late swing. The adductor magnus and hamstrings are active to control hip flexion and abduction. During initial swing, the hamstrings will contract to assist knee flexion; after that point most of the hamstring muscles are active to control the angular acceleration of knee extension. The quadriceps (vasti) work to control knee flexion. The pretibial muscles work concentrically in swing to hold the foot in dorsiflexion to prevent foot drag.

Gait Kinetics

As defined in chapter 2, kinetics is the branch of mechanics that describes motion in terms of forces. With regard to gait, we will discuss ground reaction forces, center of pressure, and joint moments.

Ground Reaction Forces

Newton's third law states that for every action there is an equal and opposite reaction. In terms of gait, the equal and opposite reaction that occurs when the foot is on the ground is the ground reaction force (GRF). The GRF can be resolved into three vector forces: (1) vertical, (2) anteroposterior, and (3) mediolateral (figure 17.3). The vertical GRF can reach

140% of body weight with normal walking. The anteroposterior GRF is 20% and the mediolateral is 5% BW.

The vertical ground reaction force during walking has a classic double-peak profile as shown in figure 17.4. The first peak corresponds to the period between initial contact and loading response and results as the supporting limb is decelerating the body mass. The second peak occurs during terminal stance and preswing (push-off) as the lower limb is accelerating forward. The valley between the two peaks occurs during single-limb support and is less than 100% of body

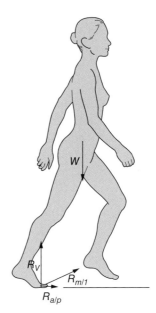

▶ **FIGURE 17.3** Ground reaction forces: one vertical (R_v) and two horizontal ($R_{a/p}$); ($R_{m/l}$).

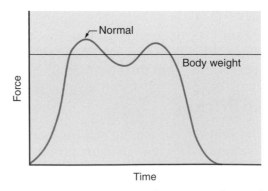

▶ **FIGURE 17.4** Ground reaction force for a normal walking gait.

weight. This is due to elevation of the center of gravity by the opposite swinging leg.

Center of Pressure

The center of pressure (CoP) is the cumulative forces in a given area at an instance in time (figure 17.5). The path of the CoP for walking changes throughout the gait cycle. At initial contact of the heel, the CoP is located just lateral to the midheel. It progresses along the lateral midfoot, which corresponds to the midstance of gait. At terminal stance and heel-off, the CoP is located under the medial forefoot.

Center of pressure can be used to determine the amount of forces distributed over an area, especially when it is excessive. For example, a person with diabetic peripheral neuropathy may develop sores on the feet and be unaware of the incidence. The CoP would help the clinician determine if the insulting forces can be minimized with the use of an orthotic or change in footwear.

Joint Moments

Moments, or torques, are placed on the joints as a result of the GRF. These external moments vary depending on the phase of gait. The external moments are counterbalanced by muscle contraction, which creates an internal moment. Figure 17.6 depicts the external moments at the ankle, knee, and hip at each phase of the stance phase. At initial contact, the external moment at the ankle is posterior, which is in the plantar flexion direction. If unopposed by muscle action, the ankle would plantar flex because of the external moment. However, during this phase of gait, the pretibial muscles are active to control the plantar flexion external moment so that the foot lowers slowly to the ground. At the knee

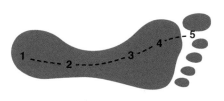

Key

1 = Initial contact
   (heel strike)

2 = Loading response
   (foot flat)

3 = Midstance

4 = Terminal stance
   (heel-off)

5 = Preswing
   (toe-off)

▶ **FIGURE 17.5**   Center of pressure during the gait cycle.

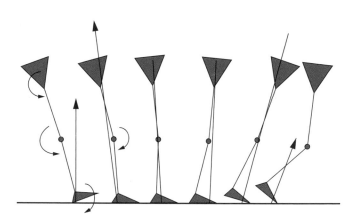

▶ **FIGURE 17.6**   External moments about the hip, knee, and ankle during walking gait.

CLINICAL CORRELATION 17.1

In orthopedics, gait deviations are prevalent after lower extremity injury. A quadriceps avoidance gait pattern has been identified in a cohort of people after anterior cruciate ligament deficiency (Rudolph et al. 2001). These people maintain a significant amount of knee flexion throughout the gait cycle in order to minimize the knee extensor moment that occurs in early stance. This adaptation is hypothesized to reduce quadriceps muscle firing to prevent excessive anterior translation of the tibia. Rehabilitation for this clientele involves improving quadriceps strength and neuromuscular techniques to improve control of the knee (Rudolph et al. 2001).

and hip, the external moment is anterior to the joint axes. At the knee this creates an external extension moment and at the hip an external flexion moment. Again, the muscle activity at these joints will counteract the external moment. The last column in table 17.1 lists the external moment at the hip, knee, and ankle for the stance phase of gait. An imbalance between internal and external moments will disrupt the normal gait cycle. Clinical Correlation 17.1 discusses an example.

Gait Parameters

In quantifying gait, the clinician can measure both distance and temporal measurements. Distance variables deal with length and width of steps or strides, while temporal variables deal with time.

Distance Variables

Step and stride lengths are routinely measured in a gait analysis, and they are relatively easy to record. Right step length is the longitudinal distance from left to right initial contact (figure 17.7). Left step length is the distance from right to left initial contact. The average step length in an adult is 64 to 74 cm (25 to 29 in). Stride length is equivalent to initial contact to initial contact of the same limb. In adults, the average stride length is 1.28 to 1.46 m (4.20 to 4.79 ft); the average equals 1.41 m (4.63 ft), with males on the higher range and females on the lower range (Perry 1992). Stride length should equate to the sum of right and left step lengths. This distance is less in females, aged

persons, and people with musculoskeletal and neurological deficiency (Perry 1992).

Step width is the distance between feet, and it normally is 5 to 10 cm (2 to 4 inches). A wider step width may be indicative of poor balance or peripheral neuropathy. Step angle is the amount of toe-out in the foot. Normally this angle is 4 to 8° (average = 7°).

Temporal Variables

Temporal variables deal with time and how fast a person walks. Gait velocity on average is 1.23 to 1.37 m/s (approximately 3 mph) for adults (Finley & Cody 1970). Step rate, or **cadence**, is the number of steps per time and equates to just less than 2 steps per second, or 111 to 117 steps per minute. On average, the rate for women is usually 6 to 9 steps per minute higher than for men. Gait velocity (speed) = step length × cadence, in units of distance/time. Additionally, the time spent in stance (**stance time**), swing (**swing time**), double support, and single support can be calculated. Reported values for these parameters are as follows:

Stride length (m)	1.33 ± 0.09 to 1.63 ± 0.11
Step length (m)	0.70 ± 0.01 to 0.81 ± 0.05
Step width (cm)	0.61 ± 0.22 to 9.0 ± 3.5
Foot angle (°)	5.1 ± 5.7 to 6.8 ± 5.6
Gait speed (m/s)	0.82 to 1.60 ± 0.16
Cadence (steps/min)	100 to 131
Stance time (s)	0.63 ± 0.07 to 0.67 ± 0.04
Swing time (s)	0.39 ± 0.02 to 0.40 ± 0.04

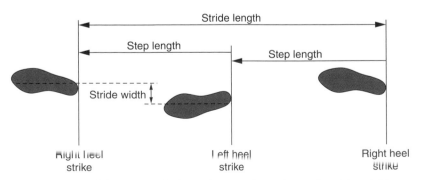

▶ **FIGURE 17.7** Distance variables in stride length, step length, and stride width.

Abnormal Gait

Understanding normal gait is important before examining abnormal gait. In this section, further terminology helps set the stage for describing abnormal gait. The attributes and functional phases of gait are defined.

Attributes of Gait

Five attributes of gait are important for maximizing gait efficiency. These attributes include (1) stability in stance; (2) foot clearance in swing; (3) prepositioning of the foot for initial contact; (4) adequate step length; and (5) energy conservation. Deviations in one or more of the attributes are common in pathological or abnormal gait patterns. An example of a deficiency in stance stability is abnormal foot position such that weight distribution and balance are poor. A child with cerebral palsy who has an equinus (plantar flexed club foot) deformity will be unable to weight-bear on a flat foot, compromising the base of support. This is one reason many people with cerebral palsy use an assistive device. Other deficiencies in the attributes of gait are found in table 17.2.

Functional Phases of Gait

Another way to look at the phases of gait so that gait deviations can be described is to use the functional phases of gait. The three phases are labeled weight acceptance, single-limb support, and swing-limb advance. Weight acceptance includes initial contact and loading response. The single-limb support includes midstance and terminal stance, when only one limb is in contact with the ground. Swing-limb advance includes the last phase of stance, pre-swing, and the entire swing phase. Table 17.3 breaks down the functional phases of gait and the possible gait deviations that can occur at the major joints.

Classic Abnormal Gait Patterns

Many gait abnormalities are identified by the pathological cause of the gait. Listed here are common abnormal gait patterns. However, the clinician should understand that generalizing a gait pattern is not ideal, and it is preferred to use impairment terms such as the ones listed in table 17.3.

Antalgic (Painful) Gait

An antalgic gait occurs as a result of pain in one or both lower extremities. The stance phase of the affected limb will be shorter relative to the opposite side. Because of pain, less time is spent on the involved side. The shortened stance time is a result of a shorter step length. Because of this shortened time, the swing phase of the uninvolved side will also be shorter. In addition, walking velocity and cadence will be decreased.

Arthrogenic Gait

An arthrogenic gait is a result of stiffness in one or more joints of the lower extremity. This gait is characterized by a forward lean,

TABLE 17.2   Attributes of Gait

Attribute	Attribute deficiency
Stability in stance	Abnormal foot position; poor balance
Foot clearance in swing	Loss of knee motion; inadequate dorsiflexion
Prepositioning for initial contact	Inadequate foot position; landing on toes
Adequate step length	Inadequate knee extension; unstable foot; inadequate push-off
Energy conservation	Bouncy gait that increases the center of gravity excursion; uncoordinated movement between the hip, knee, ankle

TABLE 17.3   Gait Deviations During the Functional Phases of the Gait Cycle

Joint	Deviation (impairment) and possible cause
Weight acceptance	
Trunk	Backward lean: to decrease demand on hip extensors (gluteus maximus) Forward lean: due to increased hip flexion (joint contracture or muscle weakness) Lateral lean: right or left weak hip abductors
Pelvis	Contralateral drops: weak hip abductors on reference limb Ipsilateral drops: compensation for shortened limb
Hip	Excessive flexion: hip flexion contracture, excessive knee flexion Limited flexion: weakness of hip flexors, decreased hip flexion
Knee	Excessive flexion: knee pain, weak quadriceps, short leg on opposite side Hyperextension: decreased dorsiflexion, weak quadriceps Extension thrust: intention to increase limb stability
Ankle	Forefoot contact: heel pain, excessive knee flexion, plantar flexion contracture Foot flat contact: dorsiflexion contracture, weak dorsiflexors Foot slap: weak dorsiflexors
Toes	Up: compensation for weak anterior tibialis
Single-limb support	
Trunk	Backward lean: to decrease demand on hip extensors (gluteus maximus) Forward lean: due to increased hip flexion (joint contracture or muscle weakness) Lateral lean: right or left weak hip abductors
Pelvis	Contralateral drops: weak hip abductors on reference limb Ipsilateral drops: compensation for shortened limb Anterior pelvic tilt: hip flexion contracture
Hip	Limited flexion: weakness of hip flexors, decreased hip flexion Internal rotation: weak external rotators, femoral anteversion External rotation: retroversion, limited dorsiflexion Abduction: reference limb longer Adduction: secondary to contralateral pelvic drop
Knee	Excessive flexion: knee pain, weak quadriceps, short leg on opposite side Hyperextension: decreased dorsiflexion, weak quads Extension thrust: intention to increase limb stability Wobbles: impaired proprioception Varus: joint instability, bony deformity Valgus: lateral trunk lean, joint instability, bony deformity
Ankle	Excessive plantar flexion: weak quadriceps, impaired proprioception, ankle pain Early heel-off: tight dorsiflexors, Increased pronation: subtalar joint deformity
Toes	Up: compensation for weak tibialis anterior

Joint	Deviation (impairment) and possible cause
Swing-limb advance	
Trunk	Backward lean: to decrease demand on hip extensors (gluteus maximus)
	Forward lean: due to increased hip flexion (joint contracture or muscle weakness)
	Lateral lean: right or left weak hip abductors
Pelvis	Hikes: to clear swing limb
	Ipsilateral drops: weak hip abductors on contralateral side
Hip	Limited flexion: weakness of hip flexors, decreased hip flexion, hip pain
Knee	Limited flexion: excess hip flexion, knee pain
	Excess flexion: knee contracture, weak quads
Ankle	Excessive plantar flexion: weak quads, impaired proprioception, ankle pain
	Drag: secondary to limited hip flexion, knee flexion, or excess plantar flexion
	Contralateral vaulting: compensation for limited flexion of swing or long swing limb
Toes	Inadequate extension: limited joint motion, forefoot pain, no heel-off
	Clawed or hammered: imbalance of long toe extensors and intrinsics, weak plantar flexion

decreased hip and knee range of motion, and a relative longer limb on the stiff side. Gait deviations could include vaulting or circumduction of the reference limb. The clinician may notice an increase in plantar flexion on the opposite side, to counter the relative long leg.

Ataxic Gait

An ataxic gait is common in people with cerebellar problems. This gait deviation is a result of poor balance. The person will walk with a broad base of support and hands held out to the side to improve stability.

Contracture Gait

In a contracture gait, the person has limited motion at one or more joints of the lower extremity. If the contracture is at the hip in flexion, the person will present with increased lumbar lordosis and knee flexion. Besides these compensations, the person may walk with a flexed trunk because the hip flexors do not allow a full upright posture. If the contracture is with knee flexion, the person will present with increased ankle dorsiflexion on the uninvolved side and early heel rise on the involved side. A plantar flexion contracture at the ankle will create knee hyperextension and forward bending of the trunk.

Gluteus Maximus Gait

This gait is characterized by weakness in the gluteus maximus, which results in a backward lurch position to compensate for the lack of control of hip extension. This gait is sometimes referred to as a myopathic gait.

Gluteus Medius (Trendelenburg) Gait

This gait deviation occurs in the frontal plane and is characterized by hip abductor muscle weakness on the stance side. The primary role of the gluteus medius is to stabilize the pelvis in the frontal plane. When weakness exists in this muscle, excessive hip drop occurs on the opposite side (swing side). In some cases this occurs bilaterally, and the person will walk with a wobbling gait. Possible causes of this gait pattern are hip arthritis, congenital hip dislocation, or coxa vara. Limiting the gluteus medius activation will reduce the compression across the hip joint, thereby lessening the pain.

Hemiplegic Gait

This gait occurs after stroke when one side of the body has been affected by paralysis. This gait pattern is also referred to as a spastic gait. The person will shuffle on the involved side. Many times the upper extremity is held in a tonic position of shoulder adduction, elbow flexion, and wrist flexion, and the fingers are held in a grasping position.

Parkinsonian Gait

As the name implies, this gait is found in people with Parkinson's disease. The gait is characterized by reduced stride length and short, rapid shuffling steps with a slightly increased walking base. In addition, people with this gait pattern will hold their arms out to the side without swinging. They may demonstrate an increased gait velocity once in stride but then have the inability to stop suddenly.

Scissors Gait

The scissors gait is characterized by hip adduction and forceful thrusting of the limb forward with limb advancement. This gait is common in people with spastic paralysis of the lower extremity. It is also referred to as a diplegic gait.

Steppage (Drop Foot) Gait

The steppage gait occurs in people with weak dorsiflexors or peripheral neuropathy. This type of gait is characterized by exaggerated hip and knee flexion in order to lift the foot high to clear the foot from the ground.

Conclusion

This chapter describes the coordinated movement of the pelvis, hip, knee, ankle, and foot during walking. Muscle activities during the three rockers of walking gait are also described. Kinetics of walking gait is presented. Faulty gait biomechanics and common abnormal gaits are listed and described.

REVIEW QUESTIONS

1. List the five phases that make up the stance portion of gait.

2. What is the average gait cycle velocity?

3. What is the second rocker?

4. Name the functional phases of gait.

5. Describe the characteristics of a Trendelenburg gait.

Running Gait

Running is a common mode of aerobic exercise training. Many health benefits are associated with running, including weight reduction, increased bone density, decreased blood pressure, and lower incidence of diabetes. Accompanying this popular activity is a high percentage of injuries. Running injuries occur frequently in athletes who run a high weekly mileage. According to a study by Taunton et al. in 2002, the breakdown of running injuries is as follows: knee (42%), foot and ankle (17%), lower leg (13%), hip and pelvis (11%), Achilles tendon (6%), upper leg (5%), low back (3%), and other (2%). It is important for the clinician to understand the biomechanics of running to help with prevention and treatment of the injuries associated with this activity.

Running Compared With Walking

The running gait cycle is comparable to the walking cycle in that each consists of a stance phase and swing phase. At a certain walking

OBJECTIVES

After reading this chapter, you should be able to do the following:

> Define the various phases of running gait.

> Describe the differences between walking and running gait.

> Describe the three-dimensional kinematics for the pelvis, hip, knee, ankle, and subtalar joint that occur during the different running gait phases.

> Analyze the muscle activity of the major muscles associated with running gait.

> Explain the difference in kinetics between walking and running.

> Describe the most common injuries associated with running.

speed, there is a transition from walking to running gait. This speed varies between people but is commonly between 4.5 and 5.0 mph (7.2 and 8.0 km/h). People run at a variety of speeds. The following list is a common grouping of runners and walkers based on running speed (note: 1 mile = 1.6 km):

- Recreational walker (>15 min/mi)
- Power walker (12 to 15 min/mi)
- Fitness runner (8 to 12 min/mi)
- Competitive runner (6 to 8 min/mi)
- Elite runner (<6 min/mi)

Some contrasts to walking that occur during running are (1) running has a **double float phase** when both feet are off the ground; (2) the running gait cycle is quicker, lasting approximately 0.7 second, whereas the walking cycle takes on the average 1 second; and (3) the magnitude of the ground reaction force can be 2 to 6 times body weight during running compared with 1 to 1.5 times body weight with walking. This greater force is largely due to greater foot impact because of the double float period (Cavanagh 1990). In addition, running requires better balance than walking because there is no double support phase, and the base of support is narrower with running. Running also requires more strength of the lower extremity to decelerate the greater forces that take place with faster locomotion. Further, if you look at the range of motions required for walking and running, you will notice that running requires more motion at the ankle and knee. Knee flexion and ankle dorsiflexion are required to attenuate impact forces. The narrower base of support creates more adduction of the limb.

▶ **KEY POINT**
Differences between walking and running include the following: (1) Running has a double float period when both feet are off the ground. (2) The running cycle is quicker, lasting approximately 0.7 second, whereas the walking cycle takes 1 second. (3) The magnitude of the ground reaction force with running is 2 to 6 times body weight compared with 1 to 1.5 times body weight with walking.

Running Sequence

The distance running sequence includes a wide range of speeds (6 to 12 minutes per mile). The sprint action is distinct from distance running and is not covered in this chapter. The running gait sequence includes the stance and swing phases and is further broken down into foot strike, midsupport, take-off, follow-through, forward swing, and foot descent. Figure 18.1 displays the running cycle. Table 18.1 breaks down the running mechanics into the different phases and presents the joint motions and working muscles for each phase.

▶ **KEY POINT**
Similar to walking, running can be divided into a stance and swing phase. The stance phase consists of foot strike, midsupport, and take-off. The swing phase consists of the follow-through, forward swing, and foot descent.

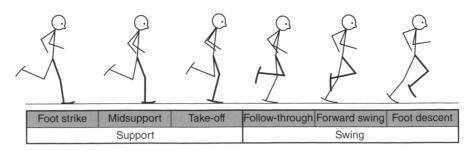

Foot strike	Midsupport	Take-off	Follow-through	Forward swing	Foot descent
Support			Swing		

▶ **FIGURE 18.1** Running gait cycle.

TABLE 18.1 Running Gait

		Hip	Knee	Foot
Foot strike	Joint motion	20-50° flexion	15-40° flexion	10° dorsiflexion
	Muscle activity	Eccentric: gluteus maximus, gluteus medius, tensor fasciae latae, hamstrings Concentric: adductors	Concentric: hamstrings, gastrocnemius, popliteus Quadriceps cocontraction	Eccentric: tibialis anterior, toe extensors
Midsupport	Joint motion	30° flexion	20° flexion	20° dorsiflexion
	Muscle activity	Eccentric: gluteus medius and tensor fasciae latae to control pelvis Eccentric: gluteus maximus, hamstrings to control limb in flexion	Eccentric: quadriceps to control knee flex	Eccentric: gastroc–soleus complex, tibialis posterior
Take-off	Joint motion	10° extension	0° flexion	25° plantar flexion
	Muscle activity	Concentric: hamstrings, gluteus maximus, gluteus medius Eccentric: trunk musculature	Eccentric: quadriceps	Concentric: gastroc–soleus complex, peroneals, toe flexors Eccentric: toe extensors
Follow-through	Joint motion	5° extension	20° flexion	10° plantar flexion
	Muscle activity	Eccentric: adductors to control pelvis, hip flexors to control hip extension, internal rotation of limb	Concentric: medial hamstrings	Concentric: gastrocnemius
Forward swing	Joint motion	10-60° flexion	125° flexion	10° plantar flexion
	Muscle activity	Concentric: iliopsoas, rectus femoris, tensor fasciae latae	Hamstrings and quadriceps cocontraction	Eccentric: pretibial muscles to control ankle
Foot descent	Joint motion	40° flexion	40-20° flexion	10° dorsiflexion
	Muscle activity	Concentric: gluteus maximus and hamstrings decelerate flexing thigh, gluteus medius, tensor fasciae latae	Eccentric: hamstrings	Concentric: pretibial muscles

Stance Phase

The stance phase (also referred to as the support phase) is defined as the point in the gait cycle when the foot is on the ground, and it takes up 38 to 45% of the total running cycle. Three purposes for the stance phase are to (1) establish contact with the supporting surface; (2) provide a stable base for the opposite extremity as it moves through its swing phase; and (3) advance the body forward over a fixed foot. The running cycle stance phase consists of foot strike (initial contact), midsupport (midstance), and take-off.

Foot Strike

Foot strike, or initial contact, is defined as the point when the foot makes the first touch with the ground. At this point the loading forces are controlled eccentrically. Runners may contact the ground with the heel, the midfoot, or the forefoot. For long-distance runners, the preferred contact is with the heel or midfoot. Recent research advocates a midfoot strike for minimizing braking forces and joint forces (Arendse et al. 2004). Forefoot striking at slower speeds has been associated with Achilles tendon stress.

Midsupport

When the foot is flat on the ground, midsupport (MS) begins. From foot strike to midsupport, the body is absorbing the ground reaction forces; this is the braking phase. The ankle and knee are at their maximum flexion angle, and the subtalar joint is pronating as the foot adapts to the ground. The lower limb is primarily working eccentrically to break the forward momentum of the body.

Take-off

Once the runner's body moves anterior to the stance leg, the propulsion phase begins until the foot leaves the ground (take-off). The concentric contraction of the lower limb muscles along with stored potential energy in the tendons help propel the body forward. The arms provide some upward lift, promote efficient movement and balance, and help reduce rotation forces through the body. It is during the stance phase that the greatest risk of injury arises, as forces are acting on the body, muscles are active to control these forces, and joints are being loaded.

Swing Phase

The swing phase (sometimes referred to as the recovery phase) is when the foot is off the ground; it makes up about 55 to 62% of the total running cycle. The purpose of the swing phase is to return the limb to a position that is ready for foot contact. Additionally, during this phase, the swinging limb adds momentum to the body, thereby increasing efficiency.

The swing phase includes follow-through (initial swing), forward swing (midswing), and foot descent (terminal swing). Follow-through is characterized as the end of backward momentum of the leg, where the knee reaches maximal flexion (60°). The limb begins to drive forward as forward swing begins. As the limb prepares for foot contact, foot descent begins. It is during the swing phase that the period of double float, when neither foot is on the ground, happens. Double float occurs at the beginning and end of the swing phase.

Running Kinematics

This section details the lower extremity kinematics that occur in the sagittal, transverse, and frontal plane during running. These motions are somewhat different from those found in walking gait (chapter 17).

Sagittal Plane

The majority of lower extremity motion that occurs with running happens in the sagittal plane. Motion at the pelvis, hip, knee, and ankle is discussed.

Pelvis

In the sagittal plane, pelvic motion occurs about a mediolateral axis. Motion at the pelvis is identified as anterior and posterior pelvic tilt. At foot contact, the pelvis is in slight posterior tilt. As stance continues, the pelvis moves toward anterior tilt, reaching a position of maximal anterior tilt at take-off (Schache et al. 1999). During swing, the pelvis initially moves posteriorly and then tilts anteriorly again during foot descent. During running the pelvic motion is minimal, but as speed increases the pelvis and trunk tilt farther forward. The mean anteroposterior tilt of the pelvis during running is approximately 15°. Ideally, no more motion should take place at the pelvis in order to conserve energy and maintain efficiency (Novacheck 1998).

Hip

At foot contact, the hip is flexed to approximately 45° and immediately begins to move toward extension, reaching around 20° of flexion by midsupport. Hip flexion helps with absorption of impact forces during this initial ground contact. From midsupport to take-off, the hip joint moves from 20° of flexion to 5° of extension. This change in motion helps propel the runner's body forward. As the limb moves into the swing phase, the hip continues to extend to approximately 20°. The hip then changes directions and moves into flexion as it initiates forward swing. The hip reaches maximal flexion (65°) during this portion of the swing. As the foot readies itself for ground contact, the hip moves from 65° of flexion to 40° of flexion. The return of the hip into extension helps reduce the horizontal velocity of the foot before contact and possibly minimizes the ground reaction force (Sinning & Forsyth 1970).

Knee

One difference between walking and running is the knee position at foot contact. With walk-ing the knee is slightly flexed (0 to 5°) at initial heel contact, but with running the knee is flexed to 15 to 20°. The increased flexion that is present in running helps with attenuation of impact forces at foot contact. See Clinical Correlation 18.1 for a comparison between knee flexion during walking and running. As midsupport approaches, the knee continues to flex to approximately 40°. As the body moves over the stance limb, the knee begins to straighten; it reaches 15° at take-off and continues to extend during follow-through. When forward swing begins, the knee moves into a significant amount of flexion, with some runners reaching close to 130° of flexion, depending on running speed. This amount of knee flexion shortens the swing limb, effectively reducing the lever arm.

Ankle

The position of the talocrural joint at foot contact is slight plantar flexion (5°). As the heel drops to the ground, the ankle moves into dorsiflexion, around 10°. During midsupport, the tibia is moving anteriorly over a relatively stable foot, and this requires 20° of ankle dorsiflexion. The ankle will forcefully plantar flex to 25° during take-off. The position of the ankle changes from plantar flexion to slight dorsiflexion during forward swing and then back to slight plantar flexion just before foot contact.

Frontal Plane

There is much less motion in the frontal plane than in the sagittal plane. Excessive frontal plane motion could be a contributing factor to injury in the lower extremity.

CLINICAL CORRELATION 18.1

The greater amount of knee flexion that occurs at foot contact with running as compared with walking has several influences on the musculoskeletal system. The increased knee flexion position will cause an increase in the patellofemoral joint compressive forces. In addition, this added amount of knee flexion requires adequate length of the soleus. Knee flexion places the tibia in internal rotation and subtalar joint pronation, which can influence patellar tracking if the timing of any of these segments is faulty.

Pelvis

Pelvic motion in the frontal plane occurs about an anteroposterior axis and is termed lateral pelvic tilt. At foot contact, the pelvis remains relatively stationary. As stance phase progresses, the pelvis drops on the swing side, and this continues until take-off. The amount of pelvic drop has been documented to range between 5 and 8°. Pelvic tilt is thought to help with shock absorption and in assisting with the descent of the body's center of gravity (James & Brubaker 1973). During swing, the pelvis elevates to obtain foot clearance.

Hip

The hip adducts relative to the pelvis at foot contact. Similar to lateral pelvic tilt, this is a shock-absorbing mechanism. The amount of hip adduction is approximately 6°. From midsupport to take-off, the hip progressively abducts and should be in a position of abduction by the time the foot leaves the ground. During swing, the hip abducts (8°) but returns to adduction just before foot contact. Motion at the hip will mirror the motion of the pelvis in this plane, and this combined motion is thought to help minimize shoulder and head movement (Novacheck 1998).

Knee

Motion at the knee is restricted in the frontal plane by the medial and lateral collateral ligaments. At foot strike, the knee is neutral in the frontal plane and then slightly abducts (valgus) up until midsupport before it begins to adduct. Excessive hip adduction and knee abduction have been associated with patellofemoral pain syndrome in athletes running long distances (Dierks et al. 2008).

Ankle

Frontal plane motion at the ankle is practically nonexistent because of the configuration of the joint.

Subtalar Joint

Although the subtalar (ST) joint acts over the three planes, the primary motion (inversion–eversion) occurs in the frontal plane, so ST joint motion will be discussed here. With a rearfoot contact, the subtalar joint is inverted (supinated) as the foot hits the ground. Almost immediately, the ST joint moves into pronation. Eversion of the calcaneus helps unlock the transverse tarsal joint, increasing foot flexibility and allowing the foot to attenuate ground reaction forces. This pronation occurs until midsupport, at which time the ST joint begins to supinate and remains in a supinated position to provide a rigid lever for push-off. During the stance phase, the ST joint will evert 5 to 15° (Dierks et al. 2010). Maximal pronation occurs at 40% of the stance phase. See figure 18.2 for a graphical representation of subtalar joint motion during running.

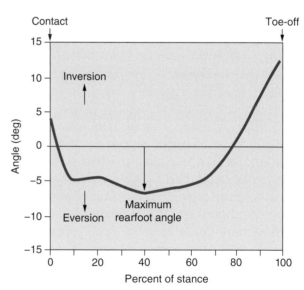

▶ **FIGURE 18.2** Rearfoot motion during the stance phase of running (rearfoot striker).

Transverse Plane

Similar to the frontal plane, motion in the transverse plane should be minimal during the running gait cycle.

Pelvis

Pelvic rotation in the transverse plane is about a vertical axis and is termed axial rotation or internal or external rotation. Internal pelvic rotation is when the reference side of the pelvis is anterior. External pelvic rotation is opposite and occurs when the reference side of the pelvis is posterior.

Transverse plane motion of the pelvis during running is quite different from what is found with walking. During running, the pelvis is externally rotated at foot contact, whereas with walking, the pelvis is internally rotated at initial contact to increase stride length. The runner's pelvis continues to externally rotate after foot contact until it reaches maximal external rotation at midsupport. For the remainder of stance, the pelvis begins to internally rotate so that by take-off the pelvis is relatively neutral in the transverse plane. As swing begins, the pelvis continues to internally rotate until it reaches maximal internal rotation at midswing. The amplitude of transverse plane pelvic movement is reported to be between 16 and 18° (Schache et al. 1999). This patterning is thought to counter the rotation of the shoulders and improve energy efficiency (Novacheck 1998).

▶ **KEY POINT**

During running, the pelvis is externally rotated at foot contact and continues to externally rotate after foot contact until it reaches maximal external rotation at midsupport. For the remainder of stance, the pelvis begins internally rotating so that by take-off the pelvis is relatively neutral in the transverse plane.

Hip

Hip motion in the transverse plane is termed internal and external rotation. Similar to walking gait, the hip is in slight external rotation at foot strike and then moves into internal rotation up until midsupport (Gerringer 1995). The hip then returns to a more neutral position at take-off (Mann 1989). A variation in hip position has been reported in swing (internally rotated versus neutral), although it would seem that a neutral hip would be more advantageous than excessive internal rotation.

Knee

The amount of knee rotation during running needs to be carefully interpreted because three-dimensional analysis of this movement is accompanied by significant measurement error. Internal rotation of the knee is defined as tibial rotation on the femur and is a component motion of knee flexion–extension and ST joint pronation–supination. During stance, the tibia will follow the subtalar joint such that past foot strike as the ST joint is pronating the tibia is internally rotating. Tibial internal rotation continues until midsupport, at which time the tibia begins to externally rotate as the foot is supinating and the knee is extending.

Ankle

Ideally, the ankle position should stay relatively straight with running. Some runners display an abducted foot due to restrictions in dorsiflexion of the talocrural joint. Abduction of the foot promotes pronation at the subtalar joint and can be a source for pronatory faults.

Arm Action During Running

The main function of the upper body and arm action is to provide balance and assist with drive. In long-distance running, however, it is questionable how much the arms assist with forward propulsion (Hinrichs 1990). In the transverse plane, the arms and trunk move to oppose the forward drive of the legs. The arms are held relaxed, with the elbow angle ideally at 90° or less. The hands should remain relaxed. The normal arm action during distance running involves shoulder extension to pull the elbow straight back; then, as the arm comes forward, the hand will move slightly across the body. Excessive crossover will cause too much trunk rotation.

Muscle Activity During Running

Muscle activation is responsible for controlling forces and generating energy with propulsion; therefore, the muscle activity during running for this chapter is broken into two phases: braking phase and propulsion phase. Muscle activation is also important for controlling the limb while it is in swing phase, and this is discussed, as well.

▶ **KEY POINT**

> The stance phase of the running gait can be divided into a braking and propulsion phase. The braking phase begins at foot contact and consists of a period where the body is absorbing ground reaction forces. The propulsion phase is associated with propelling the body forward.

Braking Phase

The braking phase occurs from foot contact to foot flat. It is also referred to as the absorption phase. This phase represents a period of deceleration. The muscles are working eccentrically to control joint motion and gravity. The foot, ankle, and knee motions are coordinated to absorb the vertical landing forces on the body. The primary muscles working are the anterior and posterior calf muscles, quadriceps, hip extensors, hip abductors, and hamstrings.

The muscle activity of the tibialis anterior is similar in walking and running. It becomes active at the instant of foot strike to minimize foot slap in the rearfoot striker. This activity may be absent in the runner who strikes with the midfoot. Also, the amount of activity of the foot is dependent on shoewear (see Clinical Correlation 18.2).

During this phase, the ground reaction force falls medial to the hip and knee. The calf and quadriceps muscles need to work eccentrically to control the knee and subtalar joints, otherwise they would collapse medially, causing limb internal rotation and pronation. In fact

the quadriceps and calf muscles are active before initial contact, and most of their activity occurs between foot contact and midstance to help control the braking forces (Elliott & Blanksby 1979). The quadriceps is the primary shock absorber and absorbs close to 3.5 times as much energy as it produces.

The hip joint extensor, the gluteus maximus, generates power during this phase to pull the body forward by actively extending the hip after the swing. In the frontal plane, the hip abductors are working to control (slow) frontal plane lateral pelvic tilt. The hamstrings are active before foot strike and demonstrate continued EMG activity into the propulsion phase. Their initial activity is responsible for slowing the rapidly extending knee in preparation for contact. See Clinical Correlation 18.3 for further discussion of injuries related to the braking phase of running.

Propulsion Phase

The braking phase is followed by a period of power generation termed the propulsion phase. The propulsion phase occurs from flat foot to toe-off. This phase represents a period of powerful concentric contraction that propels the body forward. During the braking phase, the stretched tendons absorb energy, stores it as potential energy, and then returns 90% of it during the propulsion phase as kinetic energy, which assists with the concentric contraction of the active muscles.

During propulsion the ankle, knee, and hip combine in a triple extension movement to provide propulsion upward and forward. The calf, quadriceps, hamstring, and gluteal activ-

CLINICAL CORRELATION 18.2

Most running shoes are designed to accommodate the rearfoot striker. These shoes are made with a cushioned and somewhat flared heel. The flared heel will decrease the magnitude of impact forces in the rearfoot striker. However, the shape of the heel of a running shoe can affect the point of application of the ground reaction force, and the line of action in relation to the hip, knee, and ankle can also change. A flared heel increases the moment arm of the ground reaction force about the ankle. This will increase the load on the supinators and the dorsiflexors. Recently, there has been a push toward a "minimalist" shoe to be worn by runners, especially those that midfoot or forefoot strike.

CLINICAL CORRELATION 18.3

Sites of chronic injuries may be associated with the forces acting on these tissues during the running cycle. Forces in the patellofemoral joint are estimated at 7 to 11.1 times body weight (BW); 4.7 to 6.9 times BW in the patellar tendon; 6 to 8 times BW in the Achilles tendon; and 1.3 to 2.9 times BW in the plantar fascia. Rehabilitation for these types of chronic injuries requires improved strengthening of the muscle injured, usually in an eccentric fashion. Running gait should also be analyzed for excessive movement that may be occurring at the joint these muscles control (Beck & Osternig 1994).

ity during the propulsion phase is less than during the braking phase because the propulsion energy comes mainly from the recoil of elastic energy stored during the first half of the stance. The gastrocnemius generates the primary propulsion during the propulsive phase (Winter 1983).

▶ **KEY POINT**

The gastrocnemius produces forces between 800 and 1,500 watts during the propulsion phase of running. This is in comparison to forces between 150 and 500 watts that occur during slow to fast walking.

In addition, the hip flexors are dominant in the propulsion phase, and this activity continues through the first half of the swing. This muscle group initially is decelerating hip extension, but it then becomes the primary force generator for forward swing. The gluteus medius contracts concentrically to abduct the hip and provide hip lift (Novacheck 1998). The total amount of power generated increases as speed increases.

Swing Phase

The hip flexors, hamstrings, and ankle dorsiflexors are active both concentrically and eccentrically during the swing phase. The rectus femoris is active in midswing to control the excessive knee flexion. The hamstrings and hip extensors extend the hip in the second half of swing. During foot descent, the hamstrings are active eccentrically at the knee to slow knee extension just before foot contact. The tibialis anterior very actively dorsiflexes during swing to provide foot clearance; it then prepares the foot for ground contact and controls foot slap at the initiation of foot contact.

Throughout the running cycle, the trunk has a slight forward lean and a neutral pelvic tilt. In the frontal plane, the pelvis will tilt laterally from side to side. The pelvis on the opposite side of the stance leg will drop about 5° at midsupport. The gluteus medius muscles (abductors) are of primary importance in providing lateral stability. Their contraction before and during the braking phase prevents the hip from dropping down too far (hip adduction of ipsilateral side) to the swing-leg side. The muscles will be acting eccentrically, or even isometrically, to prevent this movement.

Running Kinetics

The following section discusses the kinetics that occur during running. We present the ground reaction forces and the center of pressure.

Ground Reaction Forces

The ground reaction force (GRF) is the three-dimensional force that acts on the body as the foot contacts the ground. The vertical GRF is the upward push on the runner throughout the stance phase. The magnitude of the GRF increases as running speed increases and ranges between 2 and 6 times body weight. Runners who heel strike on initial contact have a two-peak vertical ground reaction curve compared with a single-peak ground reaction force–time curve that occurs with a midfoot or forefoot strike (figure 18.3). With a heel

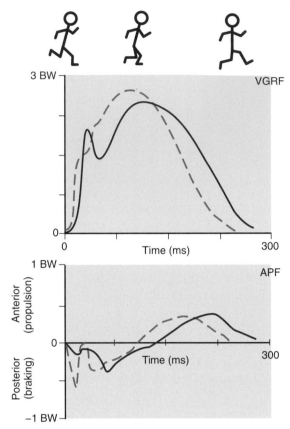

FIGURE 18.3 Ground reaction force–time curves (vertical and horizontal components) for a heel striker (solid line) and forefoot striker (dashed line). BW = body weight; VGRF = vertical ground reaction force; APF = anteroposterior ground reaction force.

Center of Pressure

The point of application of the single vector that represents the GRF applied to the foot is the center of pressure (CoP). The CoP has been mapped for running gait. These plantar pressures are dependent on the strike area of the foot. Rearfoot strikers make initial ground contact with the posterior lateral third of the shoe (figure 18.4). Midfoot strikers make contact on the lateral central third of the shoe. After initial foot contact, the pressures look similar between rearfoot and midfoot strikers—the pressure moves to the medial aspect of the forefoot and continues over the great toe until take-off.

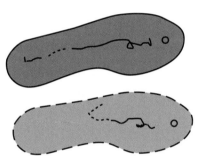

FIGURE 18.4 Center of pressure pattern for the left foot for a heel striker (top footprint) and forefoot striker (bottom footprint).

strike, the initial impact peak occurs at 15 to 25% of stance and is consistent with forces being transmitted directly through the heel and pronation of the foot. The second peak is due to active muscle forces and indicates the beginning of the propulsion phase. With midfoot striking, the single peak occurs later in the stance phase and correlates with attenuation of forces by the eccentric activation of the gastroc–soleus complex and the quadriceps.

The anteroposterior GRF initially pushes backward as the runner makes ground contact. As the runner continues over the planted foot and pushes off the ground, the GRF appears as an anterior force. The mediolateral GRF has more of a lateral push, although this varies between runners.

Running Injuries

Running is a common mode of exercise for millions of people. Unfortunately, accompanying this activity is a high incidence of injury. According to a systematic review by van Gent et al. (2007), the incidence may be as high as 80%. These injuries are commonly due to repetitive tissue loading over many cycles (Nigg 1985). The most common running injuries are patellofemoral pain syndrome, iliotibial band syndrome, plantar fasciitis, tibial stress fracture, and Achilles tendinopathy. A summary of common running injuries and possible contributing factors can be found in table 18.2. Potential causes of running injuries can be classified as intrinsic or extrinsic.

TABLE 18.2 Common Running Injuries

Running injury	Contributing factors	Movement error	Treatment strategy
Anterior knee pain	Laterally tilted patella Weak quadriceps Tight lateral structures Excessive hip IR Rearfoot pronation Weak core	Increased hip adduction and IR Knee valgus Inactive foot and ankle in propulsion	Quadriceps strengthening Hip and core strengthening Running retraining Patellar taping
Iliotibial band syndrome	Adducted gait Ilium anteriorly rotated Weak hip abductors and external rotators Functional leg-length discrepancy Genu varum Limited great toe extension	Excessive femoral rotation Overstriding	Strengthening hip external rotators Soft-tissue massage Superior tibiofibular joint mobilization Cross-training
Exercise-related leg pain	More common in females Higher BMI Leg-length discrepancy Training error	Increased tibial shock Overstriding Increased heel strike	Retraining for softer landing ST joint mobilization and manipulation Calf stretching Hip strengthening Taping Orthotics
Achilles tendinopathy	Facilitated segment L5-S1 Heel-height change in shoes Training or surface errors (hills) Joint mechanics: anterior talus, plantar flexed cuboid	Overstriding Forefoot strike Excessive vertical displacement Abnormal pronation Propulsive whip Poor ankle rocker	Heel lift Slow return to running Core stability Dural stretching Taping Orthotics Strengthening anterior tibialis, soleus, FHL Eccentric heel raises
Plantar fasciitis	Hallux limitus Forefoot varus Subtalar varus Abnormal pronation Tight calf Improper shoewear Tight hamstrings	Strike control Soft strike Active heel-rise retraining Excessive hip IR Dynamic valgus at knee	Arch taping Orthotics Night splint ST joint mobilization and manipulation Calf stretching FHL strengthening
Proximal hamstring strain	Hamstring dominant pattern over gluteal muscle vs. hamstrings) Neural restriction Proximal adhesions Eccentric overload Pelvic malalignment SI hypo- or hypermobility L5 radiculopathy	Overstriding Unilateral strike variance	Eccentric hamstring loading Slump stretching Gluteal strengthening Core stability Hip ROM Soft-tissue massage Kinesiology tape

IR = internal rotation; ST = subtalar; FHL = flexor hallucis longus; BMI = body mass index

Intrinsic Factors

Intrinsic factors are related to body structure, such as anthropometrics, skeletal structure, and fitness level. Some of these (e.g., skeletal structure) are nonmodifiable. Running involves repetitive impact, and each runner has a limit to the amount of stress his tissue can tolerate before breakdown. Even a slight biomechanical abnormality could induce injury (Cook et al. 1985). Several theories implicate structure and function as causative factors for running injuries. Differences in foot structure are associated with common overuse injuries (Williams et al. 2001). A flexible foot allows for more shock absorption than does a rigid foot. However, excessive or prolonged pronation has been attributed to faulty alignment (patellofemoral pain syndrome) and excessive muscle activity (shin splints) (Thijs et al. 2008). When the foot is at maximal pronation, the tibia should be at maximal internal rotation, and the knee should be flexed. This position occurs at the midpoint of midsupport. Past this point the knee begins to extend; however, in some runners the ST joint is still in pronation, and this will cause excessive torsion in the system, especially at the knee (Stergiou et al. 1999; Dierks & Davis 2007). The runner with excessive eversion may also display excessive internal rotation of the tibia, which can also influence knee and hip mechanics (Tiberio 1987).

A runner with a rigid foot is unable to dissipate the ground forces through foot mobility, so this stress is passed on to the tibia and up the leg. It has been reported that these runners may be more prone to stress fractures in the tibia, lateral ankle sprains, and iliotibial band syndrome (Williams et al. 2001).

Recently, attention has been focused on the hip and pelvis as causative factors for a variety of common running injuries. The alignment of the femur and tibia is important for alignment of structures such as the patella. Increased motion of the hip or pelvis in the frontal and transverse planes has been associated with patellofemoral pain syndrome (Dierks et al. 2008; Willson & Davis 2008), iliotibial band syndrome (Ferber et al. 2010), and tibial stress fractures (Milner et al. 2010).

Extrinsic Factors

Extrinsic factors are those characterized outside the body such as the environment, technique, or training and are usually modifiable. Specifically, the research suggests that the biggest predictors of injury are total volume of running and sudden changes in volume or intensity of running (Macera et al. 1989).

Environment

Running terrain can be a factor that contributes to injury. Running surface directly affects the magnitude of the ground reaction forces. Asphalt and concrete are associated with higher forces as compared with a grass surface. An advantage of a harder surface is that better traction occurs between the surface and the runner's shoe or foot. A trade-off with grass, dirt, and sand is less force but less traction, and more joint range of motion that is needed in the lower extremity. A person with an unstable ankle may prefer the harder surface to minimize the chance of injury or reinjury.

Technique

One possible cause of many running injuries is overstriding. This is a common fault in novice runners. Overstriding contributes to excessive braking forces. The correct movement patterns of the hip, knee, and ankle combined with correct activation and strength of the major leg muscles will help control braking forces during running, resulting in a more efficient action using tendon elastic energy and minimizing landing forces. Since running speed is equal to stride length multiplied by stride frequency, one way to improve running rate is to increase stride frequency, or cadence. Recommended cadence for long-distance runners is 160 to 180 steps per minute.

Training

Rapid increases in training pace or volume are common training errors. A common percent used in progression is 10%. The runner should not increase pace or volume more than 10% per week. It is also recommended to increase these factors one at a time. If pace is increased, then mileage should be restricted for a period of time. Also, the long run for a week should not exceed 30% of the total weekly running mileage. In addition, adequate rest between training sessions is important to allow remodeling of the tissues. A rest period of 48 to 72 hours is recommended between successive training sessions. Cross-training can be used to supplement this relative rest period.

▶ **KEY POINT**
Too rapid an increase in training pace or volume is a common training error. Adequate rest (48 to 72 hours) between training sessions is important to allow remodeling of the tissues.

Conclusion

This chapter describes the coordinated movement of the pelvis, hip, knee, ankle, and foot during running. Muscle activities during the braking and propulsion phase are also described. The ground reaction force is presented for various foot strikes. Running injuries and possible intrinsic and extrinsic causative factors are offered.

REVIEW QUESTIONS

1. List differences between walking and running gait.

2. Identify the different phases of the stance and swing phase during running.

3. Describe the sagittal plane pelvic kinematics during running gait.

4. What is the role of the quadriceps during the stance phase of running gait?

5. Differentiate between intrinsic and extrinsic factors related to running injury.

Cutting and Jumping

Jumping and cutting activities are commonplace in almost all competitive sporting activities. Basketball, soccer, volleyball, and football all utilize components of jumping or cutting in almost every play. Additionally, optimal horizontal distances and vertical heights in jumping are seen in track and field during the long jump and triple jump and the high jump, respectively. Proposed mechanisms of injury for knee ligaments, especially the anterior cruciate ligament (ACL), include straight-knee landings and planting and cutting maneuvers. It is thought that loading of the ACL may occur in multiple planes, as anterior tibial translation (Beynnon et al. 1995; Dürselen et al. 1995; Fleming et al. 2001; Hirokawa et al. 1992; Li et al. 1999; Markolf et al. 1990; Renstrom et al. 1986), knee valgus (Bendjaballah et al. 1997), and lower extremity rotation motions (Arms et al. 1984; Berns et al. 1992). Consequently, an enormous amount of scientific literature has been dedicated to optimizing jumping technique and decreasing the risk of knee injury during both jumping and cutting. It does not appear that jumping per se is where the injury risk lies, as it is more commonly the landing that creates injury.

OBJECTIVES

After reading this chapter, you should be able to do the following:

> Describe how jumping, running, and cutting relate to athletic performance.

> Generalize ways in which faulty neuromuscular control will affect lower extremity knee injury when jumping and cutting.

> Understand the difference between muscular and ligament dominance.

> Explain gender differences during jumping and landing during athletic activities.

This chapter highlights what is known about jumping, landing, and cutting as they relate to athletic performance. The large majority of literature related to cutting has been written in response to ACL injury. Consequently, cutting descriptions are infused with literature on evaluating risk factors for ACL injury.

Cutting

Many have heard of the plant-and-cut maneuver, or "faking out" an opponent, during athletic events. The player acts as if she is going to the right but quickly and aggressively plants the right foot and immediately reverses her motion and accelerates to the opposite left side. This occurs in soccer, football, and basketball to get by or away from a defender. Unfortunately, a quick direction change such as this is cited as an injury mechanism for noncontact ACL ruptures (Boden et al. 2000; McNair et al. 1990). Movements such as cutting, rotating, and pivoting occur as often as 70% of the time during the active portion of basketball games (Stacoff et al. 1996). It is during these types of motions that anatomically the leg (knee) can fall into a **valgus collapse**. The closed-chain theory suggests that excessive knee valgus occurs when the leg (thigh) falls into adduction and internal rotation, while the knee (tibia) moves into a position of abduction as the ankle and foot move into eversion during weight-bearing motions.

It is very possible and highly probable that this position is brought about by faulty neuromuscular function, aberrant postural adjustments, or reflex responses (Ford et al. 2005). **Neuromuscular control** of the lower extremity affects not only the knee but also the entire kinematic chain (the foot, ankle, knee, hip, and trunk). This makes females especially vulnerable because they display patterns of ligament dominance. **Ligament dominance theory** is a concept initially developed by Andrews and Axe (1985) to describe their analysis of knee ligament instability. Hewett (Hewett et al. 2002) has expanded its use to describe how during sporting activities, an athlete will allow the knee ligaments, rather than the lower extremity musculature, to absorb a significant portion of the ground reaction forces. **Muscular dominance theory** provides a better option for most athletes because ground reaction forces are absorbed by eccentric control from the lower extremity muscles. The ligament-dominant motor control pattern is seen during cutting or landing when a female athlete allows the ground reaction force to control the direction of motion of the lower extremity. As the ligament accepts an unusually high load of force, the athlete collapses into the position of excessive dynamic valgus, or valgus collapse described earlier.

This collapse has been seen by Malinzak and colleagues (2001) as greater knee valgus angles not only during running but also during sidestep cuts and crossover cuts in female athletes. Females have also been found to have less knee flexion during the stance phase of a sidestep maneuver when compared with males (Malinzak et al. 2001). McLean et al. (1999) found no difference in knee valgus during running but did in regard to increased maximum knee valgus angles during sidestep cuts when compared with male counterparts. Ford and colleagues (2005) assessed unanticipated cutting patterns and found that female athletes had significantly greater knee abduction angles when readying themselves to execute a cutting maneuver when compared with males. This faulty motor pattern could result in injury, as described in Clinical Cor-

CLINICAL CORRELATION 19.1

Both males and females with neuromuscular control deficiencies may land with the lower extremity falling into a valgus collapse. This collapse is believed to put the extremity in a position that does not allow optimal motor recruitment, while additionally placing lower extremity ligaments in a lengthened position and increasing the risk of ligament injury. If a patient exhibits a valgus collapse during an exercise or functional movement pattern during therapy or exercise, the clinician should attempt to have the patient utilize a more optimal motor pattern. This may include education and instruction to make the client aware of this potentially destructive pattern of movement. Further training may require watching video images or performing the pattern while watching in a mirror to allow for immediate feedback to the patient concerning suboptimal patterns.

relation 19.1. Ford reports that these gender differences in knee abduction angle during dynamic cutting movements, and even when adapting the ready position, suggest that women employ an altered muscular control of the lower extremity in contraction patterns (motor control) of the knee and hip abductors and adductors (Ford et al. 2005). In the ready position the trunk, knees, and hips are slightly flexed so that the stance is widened and the base of support is lowered.

Pollard and colleagues (2007) assessed female and male soccer players' hip joint kinematics and kinetics during a sidestep cutting maneuver. Compared with male counterparts, female soccer players demonstrated significantly greater hip internal rotation and a decreased amount of hip flexion. Females also demonstrated significantly greater hip adductor moments as well as decreased hip extensor moments. This suggests that males are better able to engage the hip extensors in order to control the deceleration phase of the cutting maneuver in the sagittal plane (Pollard et al. 2007). Increased use of hip extensors by male athletes may have contributed to the observed gender differences in hip internal rotation. Furthermore, the increased adductor moment seen may be the result of a trunk lean over the stance limb, which would shift the center of mass laterally, thereby increasing the adductor moment (figure 19.1).

It is thought that excessive pronation at the foot may preload the ACL during instances of falling into a valgus collapse because of the coupled effects of foot eversion and internal tibial rotation (Bellchamber & van den Bogert 2000; Loudon et al. 1996; Mundermann et al. 2003; Nyland et al. 1999). Therefore, these altered motor control patterns can occur anywhere in the kinetic chain, from as far proximal as the trunk or hip to the far distal foot and ankle.

Regarding dynamic neuromuscular control, it has been previously shown that females demonstrate greater quadriceps activation and less hamstring activation compared with males during cutting (Malinzak et al. 2001). Hanson and colleagues (2008) demonstrated that female soccer athletes exhibited greater vastus lateralis activation amplitudes and quadriceps–hamstring coactivation ratios than their male counterparts. These findings were seen during the preparatory phase and the loading phase of a running sidestep cutting motion. Hanson also found that females displayed greater gluteus medius activity during the loading phase of sidestep cutting, but not during the preparatory phase.

Colby and colleagues (2000) examined quadriceps and hamstring activity during the eccentric motion of sidestep cutting, crosscutting, stopping, and landing. Their results indicate a high level of quadriceps muscle activation beginning just before foot strike and peaking in mideccentric motion for all conditions. In all these movements, the quadriceps level of muscle activation is greater than that seen in the maximal volitional isometric contraction used to determine activation levels. Hamstring activity was submaximal at and after foot strike. Maximum quadriceps activity

▶ **FIGURE 19.1**   Differences in how (a) females tend to lean the trunk over the stance leg when performing a cutting maneuver as compared with (b) male athletes, who maintain a more upright posture.
Photos courtesy of Michael Reiman.

was at 161% of maximum voluntary contraction, while minimum hamstring muscle activity was at 14%. Increased levels of quadriceps muscle activation coupled with low or diminished levels of hamstring activity during functional activities can produce a significant anterior displacement of the tibia, placing the ACL at risk.

▶ **KEY POINT**

Research has demonstrated that females have a faster quadriceps and slower hamstring recruitment time compared with their male counterparts.

When examining the kinematic differences of male and female adolescent elite soccer players during performance of a crosscutting maneuver, male subjects demonstrated larger knee flexion angles than did females during the stance phase, while women demonstrated greater ankle eversion angles than did males (Landry et al. 2007). Males tended to land with a greater hip extension moment in the

sagittal plane than did females. In regard to frontal plane hip moments, females generated a greater hip adduction moment throughout stance for a crosscut movement (Landry et al. 2007).

Jumping

Many forms of jumping exist and depend on which types of activities the athlete is attempting to perform. Is the athlete jumping for distance such as the long jump or for height as in the high jump? What is the purpose of the jump? A **vertical jump** is performed when an athlete is attempting to jump as high as possible. It is commonly seen in the high jump, where the goal is to elevate the center of mass as high as possible. A maximum vertical jump usually requires performance of a **countermovement jump**, in which the center of mass is quickly lowered to place the hips, knee, and ankles in a more favorable position in terms of the length–tension relationship of the muscles surrounding those joints. The term *countermovement jump* is given because the initial motion is opposite the direction of the actual jump (Reiser et al. 2006). A **horizontal jump** is performed when the athlete wants to cover a maximal horizontal distance. This would include the long jump, broad jump, and single-leg hop. A **squat jump** is performed as a measure of strength and power. It is done from a flexed position. The athlete moves to a static squat position. After a brief moment in this position, the athlete quickly extends the hips, knees, and ankles to jump as high as possible.

Jumping is performed using one of two common patterns of muscle activity. A **simultaneous jumping strategy** is performed when all the muscles of the hip, knee, and foot are synchronously extended (Riewald 2011). The **sequential jumping strategy** is performed when the first muscles to be activated are the hip extensors followed by the knee extensors and then the ankle plantar flexors. This is commonly called the proximal to distal pattern of muscle activation.

Female high school basketball players have demonstrated greater knee valgus angles and motion during the landing of a box-drop vertical jump (Ford et al. 2003). Kernozek and colleagues (2005) found that women experienced knee valgus angles of 24.9° compared with males at 0.7° when landing from a 60 cm (24 in) drop. Houston et al. (2001) showed significantly less knee flexion at initial contact upon landing from 40 cm (16 in) and 60 cm heights in females, but they did not see any differences between males at a 20 cm (8 in) height. Other researchers have not seen these gender differences during landing from a drop jump (Fagenbaum & Darling 2003). This leads some to speculate that the magnitude of

valgus that occurs when landing may depend on experience, with those having more experience incurring less peak valgus (Hughes et al. 2008).

Even more important than the act of jumping is the actual landing from the jump. When landing from a jump, shock attenuation is achieved through eccentric muscle control. It has been reported that between 58 and 63% of noncontact injuries in jumping sports are associated with landings (Gray et al. 1985). Landing from jumping is hard enough, even when the athlete is rested and fresh. Imagine what may occur if the athlete is in a state of fatigue? In fact, most knee injuries occur toward the end of sporting events when participants are tired (Feagin et al. 1987). A theory exists that reduced muscular function through fatigue decreases the shock-absorbing capacity of the body and subsequently can lead to an increased chance of injury (Radin 1986; Verbitsky et al. 1998; Voloshin et al. 1998). Flexing the knees when landing from a jump may have a protective effect as described in Clinical Correlation 19.2.

Assessing lower extremity joint kinematics after a fatigue protocol, Augustsson and colleagues (2006) determined that during take-off for single-leg hops, both hip and knee flexion angles generated power for the knee and ankle joints, and ground reaction forces decreased under fatigued hop conditions. Compared with landing during the nonfatigued condition, hip moments and ground reaction forces were lower for the fatigued hop conditions. Coventry and colleagues (2006) and Madigan and Pidcoe (2003) both found increased range of motion and work of the hip extensors when landing after fatiguing conditions. This may indicate a distal to proximal redistribution of work. This redistribution may allow the larger

CLINICAL CORRELATION 19.2

Landing from a jump with greater degrees of knee flexion may have a protective effect against ACL injury. This increased knee flexion upon landing allows the athlete to dissipate a greater amount of ground reaction forces through the musculoskeletal system of the lower extremities via an eccentric contraction of the hip and knee extensors, which are several of the most powerful muscle groups in the body.

proximal muscles in the lower extremity to contribute more to resisting lower extremity collapse during landing.

During stop-jump tasks, Chappell et al. (2005) found that lower extremity fatigue was associated with increased peak proximal tibial anterior shear force and decreased knee flexion angles. Furthermore, fatigue was associated with an increased knee valgus moment for female subjects, but this result was not seen in male subjects. It is possible that women and men have different lower extremity strategies of motor control during landings, and these differences may be in part responsible for increased risk of noncontact ACL injuries in women.

Benjaminse and colleagues (2008) also found that both males and females demonstrate significantly less maximal knee valgus and decreased knee flexion at initial contact after a fatigue protocol. This decreased knee flexion may be an attempt to increase knee joint stability during landing under fatigue conditions. This "stiffening" of the knee could be what increases anterior shear forces during landing after fatigue. Recently, others have found that fatigue resulted in increased initial and peak knee abduction and internal rotation motions and peak knee internal rotation, adduction, and abduction moments, with the latter being more pronounced in females (McLean et al. 2007). Females compared with males show increased knee flexion velocities and knee joint abduction angles when landing (Gehring et al. 2009). They also show different muscle activation patterns, such as delayed activation of the lateral hamstring and the vastus lateralis muscle during the preparatory phase of the landing, while fatigue led to a reduced preactivation of the medial and lateral hamstrings and gastrocnemius muscle in both males and females.

Conclusion

Jumping and cutting are physical activities used in multiple sports and recreational activities. Optimal jumping, landing, and cutting require proper neuromuscular recruitment patterns. Athletes can easily be injured doing these types of maneuvers because they create tremendous stress and torque about the lower extremities.

REVIEW QUESTIONS

1. In sporting activities such as basketball, what percentage of time are movement patterns involving cutting, rotation, and pivoting occurring?
   a. 50%
   b. 60%
   c. 70%
   d. 90%

2. Which type of athletes have the highest incidence of anterior cruciate ligament tears?
   a. female basketball players
   b. male basketball players
   c. female soccer players
   d. male soccer players

3. During landing in which the lower extremity falls into a valgus collapse, what type of movement occurs at the femur?
   a. internal rotation
   b. external rotation
   c. extension
   d. abduction

4. What is the form of jumping in which the center of gravity is quickly lowered to put extremity musculature in an optimal length–tension relationship?
   a. horizontal jump
   b. countermovement jump
   c. squat jump
   d. ankle jump

APPENDIX
Basic Mathematics

Units and Measurement

There are two systems to describe units of measure, the International System of Units (Systeme Internationale, SI) and the English system. There are variations within each of these systems. For this text, we use the version presented in table A.1. The English system was established first, but the metric system is internationally used. The United States continues to use the English system more than the SI system.

TABLE A.1 Units of Measure

System	Length	Mass	Time
SI	Meter (m)	Kilogram (kg)	Second (s)
English	Foot (ft)	Pound (lb)	Second (s)

There are four fundamental units that are absolute concepts and independent of each other. These fundamental units are length, mass, time, and temperature:

Length (meter = m, ft)

Mass (kilogram = kg, lb): a quantitative property of matter

Time (second = s): temporal measures such as cadence

Temperature (Celsius, Fahrenheit)

Of these four, three are considered base measures (length, mass, and time). From these base measures, all other measurements are defined as denoted in table A.2.

TABLE A.2 Fundamental Units and Derivatives

Quantity	Dimension	SI unit	English
Area	L^2	m^2	ft^2
Volume	L^3	m^3	ft^3
Velocity	L/T	m/s	ft/s
Acceleration	L/T^2	m/s^2	ft/s^2
Force	ML/T^2	$kg\text{-}m/s^2$ (Newton)	ft-lb
Pressure and stress	M/LT^2	N/m^2	lb/in^2
Moment (torque)	ML^2/T^2	N-m	ft-lb
Work and energy	ML^2/T^2	N-m	ft-lb
Power	ML^2/T^3	J/s	ft-lb/s

Conversions of Units

Table A.3 is a conversion table to convert quantities between English and SI.

TABLE A.3 Conversion Table

Quantity	English	SI equivalent
Acceleration	ft/s^2	$0.3048\ m/s^2$
	$in./s^2$	$0.0254\ m/s^2$
Area	ft^2	$0.0929\ m^2$
	$in.^2$	$645.2\ mm^2$
Work and energy	ft-lb	1.356 J
		1 J = N/m
Force	lb-force	4.448 N
	kg-force	9.807 N
Impulse	lb-s	4.448 N × s

> continued

TABLE A.3   > *continued*

Quantity	English	SI equivalent
Length	ft in. mi	0.305 m 25.40 mm (2.54 cm) 1.609 km
Mass	lb mass slug	0.4536 kg 14.59 kg
Moment (torque)	ft-lb	1.356 Nm
Plane angle	degree = 0.0175 rad 1 rev = 360° 1 rev = 6.283 rad	57.3° = 1 rad
Power	lb-ft/s hp	1.356 W 745.7 W 1 watt = J/s
Pressure	lb/ft²	47.88 Pa 1 N/m² = 1 pascal
Temperature	°F = °C × 9/5 + 32	°C = °K – 273.2 °C = 5(°F – 32)/9
Velocity	ft/s in./s mph	0.3048 m/s 0.0254 m/s 0.4470 m/s
Volume	ft³	0.02832 m³

Mathematic Rules

A small introduction to plane geometry and trigonometry is included in this appendix as a refresher for the reader.

Plane Geometry

The following angle rules are commonly used in computing free body diagrams. Actual computation of free body diagrams is covered in chapter 2.

Opposite Angles

Angles (θ) formed by two intersecting lines are of the same value.

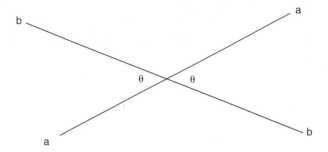

Alternate Angles

Two parallel lines that are intersected by a third line will have alternate angles (θ) that have the same value.

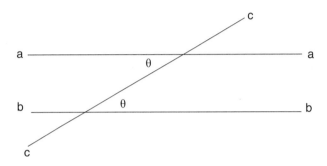

Right Angles

Angles designated as θ are equal. In this case, straight line *cc* is perpendicular to *bb*, and *dd* is perpendicular to *aa*. This geometry is utilized in free body diagrams; *aa* represents the horizontal; *bb* represents the moving limb, which makes an angle with the horizontal; *cc* is perpendicular to the inclined surface; and *dd* is a vertical straight line that represents gravity.

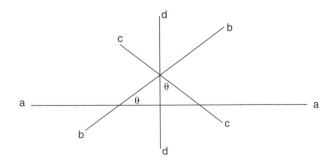

Trigonometry

Trigonometry is the branch of mathematics that is concerned with measurements of sides and angles of triangles and their relationships with each other.

In a right triangle, the side opposite the right angle is called the hypotenuse (h), and it is the longest side of the triangle. With respect to angle α, x is the adjacent side and y is the opposite side.

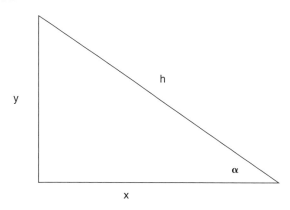

Sine, Cosine, and Tangent

The sine of an acute angle in a right triangle is equal to the ratio of the lengths of the opposite side and the hypotenuse:

Sine: $\alpha = y / h$

The cosine of an acute angle in a right triangle is the ratio of the lengths of the adjacent side and the hypotenuse:

Cosine: $\alpha = x / h$

The tangent of an acute angle in a right triangle is equal to the ratio of the lengths of the opposite side and the adjacent side:

Tangent: $\alpha = y / x$

Law of Sines

For any triangle, such as the one in this figure, the angles and sides of the triangle are related through the law of sines, which states the following:

$b / \sin \beta = a / \sin \alpha = c / \sin \theta$

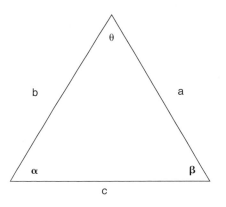

Pythagorean Theorem for a Right Triangle

The Pythagorean theorem states that the square of the length of the hypotenuse of a right triangle is equal to the sum of the squares of the lengths of the other sides of the right triangle:

$c^2 = a^2 + b^2$

Considering the square root of both sides:

$c = \sqrt{a^2 + b^2}$

Law of Cosines

For a non-right-triangle, you can use the law of cosines to solve for angles or sides. This relationship states that the square of the length of any one side of a triangle is equal to the sum of the squares of the other two sides minus twice the product of the lengths of the other two sides and the cosine of the angle opposite the original side. For the previous figure,

$c^2 = a^2 + b^2 - 2ab\cos\theta$

Practice Problems

1. For a right triangle with sides a, b, and c, determine angles α and β, and the length c.

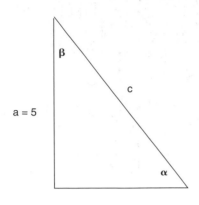

$c^2 = a^2 + b^2$
$c^2 = 25 + 36; c = 7.81$
$\cos \alpha = b / c = 6 / 7.81 = 39.8° = \alpha$
$\cos \beta = a / c = 5 / 7.71 = 50.2° = \beta$

2. For a triangle with sides a, b, and c, determine c.

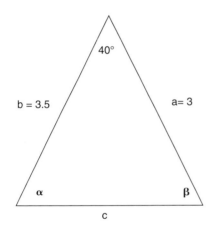

$c^2 = a^2 + b^2 - 2ab(\cos\theta)$
$c^2 = 9 + 12.25 - 21(\cos\theta)$
$c^2 = 9 + 12.25 - 16.1$
$c^2 = 5.15$
$c = 2.27$

3. Convert 65 ft-lb to Nm.

65 ft-lb × (1.3558 Nm / 1 ft-lb) = 88.13 Nm

4. For a right triangle with sides a, b, and c, determine angles α and β and the length c.

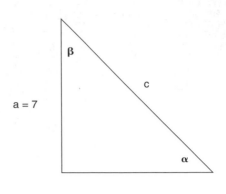

To find c, use the Pythagorean theorem.

c = square root $(7^2 + 4.5^2)$
c = square root (49 + 20.25)
c = square root (69.25)
c = 8.32

To find α, use the equation sin α = a / c

sin α = 7 / 8.32
sin α = 0.841 α = 57.3°

To find β, use the equation sin β = b / c

sin β = 4.5 / 8.32
sin β = 0.541
β = 32.7°

5. For a triangle with sides a, b, and c, determine c.

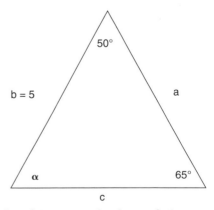

To solve for c, use the law of sines.

c / sin C = b / sin B
c / sin 50 = 5 / sin 65
c / 0.77 = 5 / 0.91
c / 0.77 = 5.49
c = 5.49 × 0.77
c = 4.23

GLOSSARY

acceleration—The time rate of change of velocity (m/s^2, ft/s^2). Average acceleration is equal to change of velocity divided by change of time.

acetabulum—The articulating surface of the hip where the ilium, ischium, and pubis come together.

actin—A protein abundantly present in many cells, especially muscle cells, that contributes to the cell's structure and motility. Actin can very quickly assemble into long polymer rods called microfilaments.

active insufficiency—When a muscle is shortened to the point that its full contraction will not complete a full range of motion.

adipose tissue pad—One of the three types of intra-articular meniscoids in the facet joints of the lumbar spine. They are found at the superior and inferior poles of the joint, consisting of synovial folds.

alar ligaments—Paired ligaments arising on either side of the superior portion of the dens and extending superiorly and laterally to attach to respective sides of the occipital condyles and lateral masses of the first cervical vertebra.

amphiarthrosis—A joint that is held together with cartilaginous tissue.

anatomical neck of the humerus—The space where the shaft of the humerus meets the articular surface of the humeral head.

anatomical snuffbox—The space between the tendons of the extensor pollicis longus on the ulnar side and the extensor pollicis brevis and abductor pollicis longus on the radial side, with the scaphoid forming the floor of the snuffbox.

angle of inclination[1]—The angle formed in the frontal plane between the head of the humerus and the shaft of the humerus.

angle of inclination[2]—The angle between the femoral head and the neck in the frontal plane.

angle of torsion—In the transverse plane, an angle formed between a line bisecting the neck of the femur and a line parallel to the plane of the shaft of the femur.

annulus fibrosus—The fibrous outer ring of the intervertebral disc.

anterior cruciate ligament—Intra-articular ligament that runs anteriorly, distally, and medially from the femur. The primary restraint to anterior tibial translation, especially near extension.

anterior longitudinal ligament—Ligament running along the anterior portion of the vertebral bodies of the entire spine.

anterior sacroiliac ligament—One of the ligaments of the sacroiliac joint. It consists of numerous thin bands that connect the anterior surface of the lateral part of the sacrum to the margin of the auricular surface of the ilium. This ligament is stressed whenever the anterior surfaces of the ilium and sacrum gap.

anteversion—Inclined forward as with the femoral head. An angle of inclination greater than 12° is considered anteverted.

arachnoid mater—One of the three membranes (along with dura and pia mater) enveloping the structures of the central nervous system. It is interposed between the dura and pia mater.

articular eminence—The temporal articulating surface of the TMJ.

atlas—The first cervical vertebra. The atlas is a ringlike structure formed by two lateral masses and sits like a washer between the skull and lower cervical spine.

auriculotemporal nerve—The sensory nerve to the TMJ.

autonomic nervous system—The portion of the peripheral nervous system responsible for innervation of smooth muscle, cardiac muscle, and glands of the body. It primarily functions without a person's conscious control. The two primary components of this system are the sympathetic and parasympathetic divisions.

axis—The second cervical vertebra. The axis serves as a transitional vertebra, as it is the transition from the craniovertebral region to the traditional cervical spine vertebrae.

axon terminal—One of the four functional parts (along with axon, cell body, and dendrites) of the neuron. It is the transmission site for action potentials.

axon—One of the four functional parts (along with dendrites, cell body, and axon terminal) of the neuron. Axons are covered with myelin and conduct information to other nerve cells.

ball-and-socket joint (spheroid)—A joint in which one portion is rounded like a ball and fits into a cuplike socket. This form of joint allows a great deal of motion.

bilaminar zone—A portion of the retrodiscal tissues in the TMJ.

biomechanics—The study of forces and segment movement as they relate to human movement. Biomechanics combines engineering with anatomy and physiology.

body of the rib—The portion of the rib that extends laterally from the neck and curves anteriorly from the rib angle. The bodies of the ribs serve as common muscle attachment sites.

bone—Highly organized tissue composed of organic and inorganic materials and small amounts of water.

bruxism—Grinding of the teeth.

bursa—A fluid-filled sac surrounding a synovial joint that helps reduce friction between tendons, muscles, ligaments, and bones.

cadence—The number of steps per unit time. Also termed stride rate or frequency.

cam—A non-uniform ellipse used to improve the system's mechanical advantage.

cancellous bone—Spongy bone formed around a loose network of tissue that houses red blood marrow.

carpal tunnel—A space within the wrist formed by the carpal bones, specifically by the hook of the hamate medially and the trapezium laterally. The floor is formed by soft tissue: the palmar radiocarpal and palmar ligaments. The roof is covered by the flexor retinaculum, forming the space into a tunnel.

cavus—A rising of a joint, such as the subtalar joint.

cell body—One of the four functional parts (along with axon, dendrites, and axon terminal) of the neuron. The cell body contains the nucleus of the cell and performs integrative functions.

center of mass (gravity)—The point about which a body's mass is equally distributed. In the standing adult, the center of gravity is located anterior to the second sacral vertebra (S2).

center of pressure—The center point at which weight of a body is transferred to the support surface. May or may not be directly under the body's center of gravity. The point of application of force to an object.

central incisors—The front teeth, upper and lower.

central nervous system—One of the two structural divisions (along with the peripheral nervous system) of the nervous system. This division consists of the brain and the spinal cord.

chondrosternal joints—Joints formed by the articulation of the costal cartilages of ribs 1 through 7 anteriorly with the sternum. Chondrosternal joints of ribs 1, 6, and 7 are synchondroses, while chondrosternal joints of ribs 2 through 5 are synovial joints.

closed packed position—The joint position in which the ends of the two surfaces are most congruent. In this position, the ligaments are on maximal tension so that the joint surfaces cannot be separated.

Colles fracture—A fracture of the distal radius, with or without ulnar displacement.

collinear—Two or more points that lie on the same line.

complex joint—A joint composed of three or more bones.

condyloid joint—An oval-shaped condyle that fits into a concavity and allows angular motion but minimal to no rotation.

connective tissue rim—One of the three types of intra-articular meniscoids in the facet joints of the lumbar spine. This structure is a wedge-shaped thickening of the internal surface of the capsule that fills the space left by the curved margin of the articular cartilage.

coplanar—Two or more objects that lie in the same plane.

coracoid process—The superior anterior portion of the scapula that projects anteriorly. It is the attachment site for multiple muscles, tendons, and ligaments.

coronoid—A projecting process on a bone.

coronoid process—The superior projection on the mandible anterior to the ramus for temporalis attachment.

cortical bone—Known also as compact bone, the outer portion of bone that is very dense and forms a protective structure.

costal facets—Facets for rib articulation present on most of the thoracic vertebrae.

costochondral joints—Joints formed by articulations of the 1st to 10th ribs anterolaterally with the costal cartilages. These joints are both synchondroses and synovial joints.

costotransverse joints—Pairs of synovial joints that consist of vertebrae of the 1st through 10th thoracic levels articulating with the rib of the same number.

costovertebral joints—Hyalinated synovial joints formed by the head of the rib, two adjacent thoracic vertebral bodies, and the interposed intervertebral disc.

countermovement jump—The center of mass is lowered quickly to place the hips, knees, and ankles in a better position in terms of their length–tension relationships.

craniovertebral junction—A collective term that refers to the occiput, atlas, axis, and supporting ligaments.

deceleration—Negative acceleration.

demifacets—Costal facets along the posterolateral aspect of the vertebral body for articulation with the adjacent rib.

dendrites—One of the four functional parts (along with axon, cell body, and axon terminal) of the neuron. Dendrites receive information from other nerve cells or the environment.

diarthrosis—A joint that is surrounded by a capsule and filled with synovial fluid.

digastric fossa—The depression on the mandible for digastric attachment.

displacement—The distance and direction of movement from one point to another, measured in meters or feet. Displacement is a vector quantity with magnitude and direction.

distal—Situated away from the middle of the body.

distance—The amount one point travels to a second point, measured in meters or feet. Distance is a scalar quantity that provides magnitude without direction.

double float phase—The phase of running gait during which both lower extremities are off the ground.

double support—The phase of gait during which both lower extremities are in contact with the ground.

dura mater—One of the three membranes (along with arachnoid and pia mater) enveloping the structures of the central nervous system. It forms the dural sac around the spinal cord. It is separated from the bones and ligaments of the vertebral canal by an epidural space.

elasticity—The ability of a tissue to return to its normal resting length after removal of a stretch.

ellipsoidal joint—A joint in which a flattened convex ellipsoid fits into a concave surface.

end plate—Considered to be a component of the disc rather than the vertebral body. The end plate covers the entire nucleus pulposus of the intervertebral disc and is strongly attached to the annulus fibrosis but only weakly attached to the vertebral body.

endurance—The ability to perform low-intensity, repetitive, or sustained activities over a prolonged period of time without fatigue. Endurance can be further broken down into local muscle endurance and general endurance.

energy—The capacity to perform work. Energy is classified as either kinetic or potential energy.

epicondyle—A projection from a long bone near the articulating surface that serves as an attachment site for muscles and ligaments.

extensibility—The ability to be stretched or to increase in length. It might also be described as a muscle's ability to lengthen or stretch beyond resting length.

extrinsic factors—Factors outside the body that may contribute to injury, such as the environment, technique, or training; they are usually modifiable.

false ribs—Ribs 8 through 10, whose anterior attachments are to the costochondral cartilage of the rib superior to them. All the false ribs articulate with the sternum via the 7th rib.

femoral condyles—Two rounded projections off the distal end of the femur. The area of the femur that bears weight during functional weight-bearing activities.

fibroadipose meniscoids—One of the three types of intra-articular meniscoids in the facet joints of the lumbar spine. They project approximately 5 mm into the joint cavity from the inner surface of the superior and inferior capsules.

fibrocartilage—A mixture of connective tissue and cartilage that interposes joints. Provides both stability and support to a joint.

first branchial arch—The embryologic origin of the TMJ.

flat bones—Bones that are flat and have a large surface area for muscular attachments.

floating ribs—Ribs 11 and 12, which do not have an anterior attachment on the rib cage.

foramen magnum—the large opening in the anterior and inferior occipital bone that connects the vertebral canal and the cranial cavity.

force system—Any group of two or more forces acting in relation to an object.

force—The action of one body acting on another.

forefoot—The area of the foot from the metatarsals to the distal phalanx.

fossa—A depression or hollow area in a bone.

friction—The tangential force acting between two bodies in contact that opposes motion or impending motion.

gait cycle—The period from initial contact to initial contact of the same foot.

genu recurvatum—Excessive knee hyperextension range of motion at the knee.

genu valgum—The normal lateral angulation in the frontal plane between the tibia and the femur. Someone with excessive genu valgum is said to have knock-knees.

genu varum—The medial angulation in the frontal plane between the tibia and the femur. Someone with excessive genu varum is said to have bowlegs.

glenoid fossa—The depression on the lateral end of the scapula. The glenoid fossa faces laterally and anteriorly and articulates with the humeral head.

gliding joint (plane)—A joint in which two flat or fairly flat surfaces glide against each other, allowing a gliding, or sliding, motion. Also known as a plane joint.

Golgi tendon organ—A small sensory receptor, located at the junction between a muscle and tendon. The Golgi tendon organ monitors muscle tension.

gomphosis—A joint in which fibrous tissue holds a peg-like portion of the joint in a hole.

ground reaction force—The force created as a result of foot contact with a supporting surface.

Guyon's canal—A narrow space between two carpal bones (the pisiform and the hook of the hamate) through which the ulnar artery and ulnar nerve travel.

hallux abducto valgus—Deviation of the great toe toward the lateral side of the foot

head and neck of the rib—The most posterior, or dorsal, portion of the rib that directly articulates with the vertebral body.

high ankle sprain—Sprain of the anterior tibiofibular ligament and interosseous membrane between the tibia and fibula.

hinge joint (ginglymus)—One of the more basic joint forms in which a convex portion of one bone fits into a concave depression of the other bone. Allows movements in only one plane.

horizontal jump—A jump in which the athlete wants to cover a maximal horizontal distance.

humeral retroversion—The posterior rotation of the humeral head that occurs in the transverse plane.

hyaline cartilage—Very complex, highly ordered structure that is thin and covers the ends of long bones in a synovial joint.

hyoid bone—The horseshoe-shaped bone in the anterior neck that provides a solid juncture between the supra- and infrahyoids.

iliolumbar ligament—On each side of the vertebra, they connect the transverse process of the 5th lumbar vertebra to the ilium. It is generally accepted to form a strong bond between L5 and the ilium in the adult. It functions to restrain flexion, extension, axial rotation, and side bending of the 5th lumbar vertebra on the 1st sacral vertebra.

ilium[1]—The uppermost of the three bones that make up the innominate bone. The ilium forms the superior part of the acetabulum and provides attachment for several muscles, including the obturator internus, the gluteals, the iliacus, and the sartorius. The ilium forms two-fifths of the acetabulum of the hip joint.

ilium[2]—The distal lateral portion of the pelvis that forms the upper portion of the acetabulum.

impulse—The effect of a force acting over a period of time.

inertia—The measure of resistance to change in velocity.

inferior glenohumeral ligament complex—The extensive complex of ligamentous tissue that attaches proximally to the anterior inferior rim of the glenoid and the labrum. This ligament complex helps stabilize the glenohumeral joint.

infrapatellar fat pad—Extensive fat pad structure located between the patellar tendon and the tibia. This pad extends superiorly under the patella and separates it from the tibia.

innominate—The bone of the pelvic girdle consisting of the ilium, ischium, and pubis. The innominate on each side of the body articulates with the sacrum to form the sacroiliac joints bilaterally.

instantaneous center of rotation—The point about which rotation of an object takes place.

interchondral joints—Synovial joints supported by a capsule and interchondral ligament.

intercondylar notch—The location between the two femoral condyles. The attachment sites of the anterior and posterior cruciate ligaments.

interosseous—Located between two bony surfaces.

interosseous (sacroiliac) ligament—The intra-articular ligament within bilateral sacroiliac joints connecting the innominate and the adjacent side of the sacrum.

interspinous ligament—Ligament running the length of the spine between adjacent spinal-level spinous processes. The function of this ligament is controversial, although most authors agree that it resists flexion.

intertransverse ligaments—Paired ligaments connecting adjacent transverse processes on each side of the vertebral body.

intertubercular groove—The groove between the lesser and greater tuberosities. This area contains the long head of the biceps and is covered by the transverse humeral ligament.

intervertebral disc—Provides a separation of the adjacent vertebral bodies. Each IVD is composed of three parts: nucleus pulposus, annulus fibrosus, and end plate. The IVD primarily functions to separate adjacent vertebral bodies, provide compressive and resistive forces for the spine, and assist with motion.

intervertebral foramina—Openings for the spinal nerve to exit laterally off the spinal cord. They are present in all the lower cervical vertebrae, with each one 4 to 5 mm long and 8 to 9 mm high.

intrinsic factors—Factors related to the body structure that may contribute to injury, such as anthropometrics, skeletal structure, and fitness level.

irregular bones—Bones that have an irregular shape and features, such as the vertebrae and facial bones.

irritability—The ability to respond to stimulation from the nervous system. This stimulation is provided by a chemical neurotransmitter.

ischium[1]—One of the three parts of the innominate, which joins the ilium and the pubis to form the acetabulum. The ischium is the dorsal part of the hip bone and is divided into the body of the ischium, which forms the posteroinferior two-fifths of the acetabulum of the hip joint,

and the ramus, which joins the inferior ramus of the pubis.

ischium[2]—One of three bones that compose the pelvis. The ischium makes up the posterior portion of the acetabulum.

isokinetic—A concentric or eccentric muscle contraction in which a constant velocity is maintained throughout the muscle action.

isometric—A static contraction with a variable and accommodating resistance without producing any appreciable change in muscle length.

isotonic—A concentric or eccentric muscle contraction in which movement is performed and therefore change in muscle length occurs.

joints—The areas where the ends of two or more bones meet.

Kienböck's disease—Avascular necrosis of the lunate.

kinematics—The description of motion and the relationship between displacement, velocity, and acceleration.

kinesthesia—The sense that detects body position, weight, or movement of muscles, tendons, and joints.

kinetics—The study of motion and the forces that act to produce the motion.

kyphosis—A posterior curve of the spine as viewed in the sagittal plane, usually seen in the thoracic spine as excessive flexion.

kyphotic—Primary curvature of the spine characteristic of the thoracic and sacral spine, where the concavity of the curvature is anterior and the convexity posterior.

labrum—A fibrous ring that surrounds a joint structure. A ring of fibrocartilage located on one articulating surface of a joint to deepen the socket of the hip and shoulder.

laminae—The region of the vertebral arch that lies posterior to the transverse processes on either side. The laminae project posteriorly on both sides to form the spinous processes.

lateral excursion—Side-to-side translation of the mandible.

lateral ligament—The largest ligament in the TMJ, this ligament suspends the condyle and restricts posteroinferior displacement.

lateral pterygoid plates—The inferior projections of the sphenoid bone on which the lateral and medial pterygoid attach.

lever—A rigid bar pivoted on a fixed point and used to transmit force.

ligament dominance theory—The theory that some athletes rely on ligamentous passive restraint to dissipate ground reaction forces to supply joint stability.

ligaments—Highly organized structures composed of dense connective tissue that connect bones to bones.

ligamentum flavum—A ligament that spans from the anterior surface of one vertebral lamina to the adjacent vertebral lamina to form the smooth posterior surface of the vertebral canal.

ligamentum nuchae—A ligament considered by some to be an extension of the supraspinous and interspinous ligaments, typically from C7 to the occiput.

linea aspera—A line on the side of the femur that is a location for several muscle attachments.

Lister's tubercle—A tubercle on the dorsal aspect of the radius.

long bones—Bones that have a greater length than width.

long dorsal ligament—A ligament of the sacroiliac joint (posteriorly) whose fibers are multidirectional. The ligament is tensed when the sacroiliac joints are counternutated and slackened when nutated.

loose packed position—Also known as the open packed position, the position in which a joint has its greatest capacity or volume. In this position, the joint ends can easily be separated.

lordosis—An anterior curve of the spine as viewed in the sagittal plane.

lordotic—Secondary curvature of the spine characteristic of the cervical and lumbar spine, where the convexity of the curvature is anterior and the concavity posterior.

malleolus—The distal rounded bony prominence of the tibia (medial malleolus) and fibula (lateral malleolus).

mandibular condyle—The inferior bone of the TMJ articulation.

mandibular depression—Lowering of the jaw, which opens the mouth.

mandibular elevation—Raising of the jaw, which closes the mouth.

mandibular protrusion—The movement of the mandible forward so that the lower front teeth move nearer or beyond the upper frontal teeth.

mandibular retrusion—Backward movement of the mandible so that the lower teeth move farther into the oral cavity.

manubriosternal joint—The joint articulation between the most superior portion of the sternum (manubrium) and the largest portion of the sternum (the sternal body).

manubrium—The widest and most proximal portion of the sternum; its proximal border is at approximately the third thoracic vertebra.

mass—The quantity of matter.

mastication—Chewing.

mechanics—The study of forces acting on an object with regard to movement, size, shape, and structure.

mechanoreceptor—Specialized sensory end organ that responds to mechanical stimuli such as tension, pressure, and displacement.

median sacral crest—A midline ridge of bone on the posterior surface of the sacrum.

meniscoid—Synovial intra-articular structures have been observed in these zygapophyseal joints. These structures are vascular and fat filled. They function as space fillers in the cervical spine.

meniscus—Fibrocartilagenous discs between the tibial plateau and the femoral condyles of the femur.

midfoot—The area of the foot formed by the cuboid, the navicular, and the three cuneiforms.

momentum—The quantity of motion.

motor unit—Referred to as the functional unit of movement, a motor unit consists of a motor neuron with an alpha motor neuron and all the terminal braches of the axon, as well as the muscle fiber they innervate.

muscle spindle—A small, complex spindle-shaped sensory receptor located in skeletal muscle that senses how much the muscle is being stretched. It consists of several intrafusal fibers.

muscular dominance theory—The theory that some athletes rely more on the musculoskeletal system to dissipate ground reaction forces to supply joint stability.

mylohyoid line—The line of the mandible for the mylohyoid.

myofilament—Any of the ultramicroscopic filaments made up of actin and myosin that are the structural units of a myofibril.

myosin—A contractile protein that forms the thicker of the two types (actin being the other) of filaments in muscle fibers. These microfilaments interact with actin filaments to alternately unlink and chemically link in a sliding action, producing muscle contraction.

neuromuscular control—The coordinated effort between the nervous and musculoskeletal systems. The ability of the nervous system to recruit muscles to allow appropriate physical function.

nucleus pulposus—The gelatinous mass (except in the cervical spine) found in the center of the intervertebral disc.

odontoid—A bony projection (also known as the dens) that extends superiorly from the body of the axis to just proximal to the atlas, at which point it tapers to a blunt point. The dens functions as a pivot for the upper cervical joints and as the center of rotation for the atlantoaxial joint.

olecranon—The posterior tip of the humerus (elbow joint).

opposition—Movement of the thumb that allows the thumb to contact the distal tip of the little finger. Opposition requires abduction and flexion of the thumb CMC joint.

os coxae—The pelvis, formed by the ilium, ischium, and pubic bones.

osteoblasts—Cells responsible for synthesis, deposition, and mineralization of bone tissue.

osteochondral—Relating to or made of cartilage and bone.

osteoclasts—Cells that bind to bone surfaces and create an acidic environment that causes bone resorption.

osteocytes—Cells that secrete enzymes that remove a thin layer of osteoid covering, allowing osteoclasts to bind to bone and begin resorption.

Pacinian corpuscles—A form of joint receptor that can detect joint compression, pressure, and movement.

parafunctional activities—Repetitive activities that stress the TMJ such as gum chewing, fingernail biting, and pencil chewing.

passive insufficiency—When a muscle is lengthened to the point that full range of motion is not able to be achieved.

patella—The largest sesamoid bone in the body, situated in the anterior portion of the knee. Part of the patellofemoral joint.

patella alta—A high-riding patella; the height of the patella is less than the length of the patellar tendon.

patella baja—A low-riding patella; the height of the patella is more than the length of the patellar tendon.

pedicle—The region of the vertebral arch that lies anterior to the transverse processes on either side. These pedicles transmit tension and force from the posterior elements and the vertebral bodies.

pelvic girdle—Consists of bilateral innominate bones, which articulate anteriorly to form the pubic symphysis and posteriorly with the sacrum on either side to form the sacroiliac joints. Bilateral hip joints are also often considered part of the pelvic girdle.

periosteum—The extreme outer layer of bone that is composed of dense fibrous tissue and permeated by blood vessels and nerve fibers.

peripheral nervous system—One of the two structural divisions (along with the central nervous system) of the nervous system. This division consists of 43 pairs of nerves arising from the central nervous system.

phalanges—The bones of the fingers and thumb.

phalanx—The long bone found in the toe.

pia mater—One of the three membranes (along with arachnoid and dura mater) enveloping the structures of the central nervous system. It is the deepest of the three layers, firmly attached to the outer surface of the spinal cord and nerve roots.

pivot joint (trochoid)—A joint that has a rounded or conical surface on one end and a concave surface or ring on the other. Allows a spinning of the joint surfaces.

plane of the scapula—The angled position of the glenohumeral joint so that the glenoid fossa faces anteriorly about 30 to 45° anterior to the frontal plane.

plantar—The sole of the foot.

planus—A flattening of a joint, such as the subtalar joint.

plicae—Embryonic remnants of synovial chambers formed during fetal development that have not been completely absorbed by the body. Composed of soft-tissue pleats or folds along the inner lining of the synovial joint cavity.

plumb line—A line to which is attached a small weight and that is suspended from overhead to represent a vertical line. The plumb line is used to standardize a starting position for posture assessment.

position—An object's location in space, usually quantified by Cartesian coordinates (x,y) or by polar coordinates (r,θ).

posterior cruciate ligament—Intra-articular ligament that runs posteriorly, laterally, and distally from the femur to the tibia.

posterior glenoid tubercle—The posterior border of the cranial portion of the TMJ.

posterior longitudinal ligament—Ligament running along the posterior portion of the vertebral bodies of the entire spine. This ligament typically has a saw-tooth appearance and runs anterior to the spinal cord.

power—The rate of work, or the amount of work per unit time. Muscular power is the amount of work produced by a muscle per unit time (force × distance / time).

precision handling—A type of grip that requires dexterity of the fingers and thumb in order to hold a particular object.

prehensile—Having the ability to take hold of or grasp objects.

pressure—The force per unit area; pressure = force / area.

primary spinal curves—Curves in the spine present at birth (kyphosis), including the curves in the thoracic and sacral regions.

pronation—Rotation of the segment towards the prone position.

proprioception—The sense of position and movement of the limbs and the sense of muscular tension.

pterygoid fossa—The depression on the anterior aspect of the neck of the condyle for the pterygoid attachment.

pubic symphysis—The articulation between the left and right innominate anteriorly. The joint is classified as a symphysis because it has no synovial tissue or fluid, and it contains a fibrocartilaginous lamina, or disc.

pubis[1]—One of a pair of pubic bones that join at the pubic symphysis and, with the ischium and the ilium, form the pelvis. The pubis forms one-fifth of the acetabulum of the hip joint and is divided into the body, the superior ramus, and the inferior ramus.

pubis[2]—The bone that is the most ventral and anterior of each half of the pelvis.

pulley—A wheel with a grooved rim around which a cord passes. It acts to change the direction of a force or to improve the mechanical advantage of the system.

quadriceps angle (Q angle)—An angle formed between a line connecting the anterior superior iliac spine to the midpoint of the patella and the bisecting line that runs from the tibial tuberosity.

rate coding—The process of modulating the force produced by the associated muscle fibers via the rate of production of sequential action potentials.

rearfoot—The calcaneus and talus. Sometimes referred to as the hindfoot.

recurvatum—Hyperextension of the knees as viewed in the sagittal plane.

retroversion—Inclined backward as with the femoral head. An angle of inclination less than 8° is considered retroverted.

rib cage—Composed of the thoracic spine, adjacent ribs, and articulations anteriorly onto the manubrium and sternum. The primary function of the rib cage is to protect the heart and lungs.

rocker—A phase in the gait cycle where primary motion takes place.

root of the scapula—The medial portion of the scapular spine.

rotator cuff—A collective group of muscles (supraspinatus, infraspinatus, subscapularis, and teres minor) that stabilize the glenohumeral joint.

Ruffini endings—A form of joint receptor that is sensitive to stretch at the extremes of joint motion.

sacral ala—The portion of the sacrum that forms the superolateral portions of the sacral base.

sacral apex—The inferior and posterior portion of the sacrum that articulates with the coccyx.

sacral base—The superior and anterior portion of the sacrum.

sacral cornua—Bilateral downward projections off the sacral hiatus that are connected to the coccyx via the intercornual ligaments.

sacral counternutation—Also referred to as sacral extension, this motion involves movement of the sacral superior base posteriorly and movement of the inferior border anteriorly. This motion is primarily resisted by the long dorsal ligament

sacral nutation—Also referred to as sacral flexion, where the superior base of the sacrum moves anteriorly and the inferior border of the sacrum moves posteriorly. This motion is resisted by various factors: the wedge shape of the sacrum, ridges and depressions on the articular surfaces, the friction coefficient of the joint surface, and the integrity of the soft-tissue structures surrounding the joint.

sacral superior articular processes—The articulating processes on the superior aspect of the sacrum. They are concave and oriented posteromedially, extending upward from the base to articulate with the inferior articular processes of the 5th lumbar vertebra.

sacroiliac (SI) joint—A large diarthrodial joint connecting the spine to the pelvis. This connection consists of joints on each side of the wedge-shaped sacrum and bilateral innominates.

sacrotuberous ligament—The ligament running from the inferior lateral angle of the sacrum to the ipsilateral ischial tuberosity. It is tensed when the sacroiliac joints are nutated and slackened when counternutated.

sacrum—A strong triangular bone located between the two innominates. The sacrum articulates with an innominate on each side, forming the sacroiliac joint bilaterally.

saddle-shaped joint (sellar)—A joint in which each side has both a convex and a concave portion. The reciprocally shaped surfaces are oriented at right angles to each other.

sarcolemma—The cell membrane that covers each fiber in a single muscle.

sarcomere—The functional unit of a myofibril of skeletal muscle.

sarcoplasm—The functional unit of force generation.

scalar—A quantity that has only magnitude. For example, mass, length, and kinetic energy are scalar quantities.

scapular tipping—The movement that occurs as the inferior angle of the scapula moves backward.

scapular winging—A movement in which the medial border of the scapula moves backward.

scapulohumeral rhythm—The motions of the scapula, humerus, and clavicle together to elevate the shoulder in a coordinated fashion.

scoliosis—A lateral curvature of the spine as viewed in the frontal plane.

screw-home mechanism—An obligatory external tibial rotation that occurs in the later ranges of knee extension. Sometimes also called the locking mechanism.

secondary curves—Curves present in the cervical and lumbar spines, typically the result of adaptation to surroundings.

sequential jumping strategy—A jump in which the first muscles to be activated are hip extensors followed by knee extensors and then finally ankle plantar flexors. Sometimes called a proximal to distal pattern of muscle activation.

sesamoid bones—Small pea-shaped bones that are embedded in tendons.

sesamoid—A small accessory bone that improves the lever of a tendon–muscle unit.

short bones—Bones whose width is approximately the same as their length.

shoulder complex—The group of joints (acromioclavicular, glenohumeral, scapulothoracic, and sternoclavicular) that collectively form the shoulder joint.

simple joint—A joint composed of only two bones.

simultaneous jumping strategy—A jump in which all muscles of the hip, knee, and foot are synchronously extended.

single support—The phase of gait during which only one lower extremity is in contact with the ground.

sinus tarsi—A canal formed by a groove in the talus and a groove in the calcaneus.

sinuvertebral nerve[1]—Nerve formed by branches from the anterior or ventral nerve root and the sympathetic plexus. This nerve innervates the outer one-third to one-half of the fibers of the annulus fibrosus.

sinuvertebral nerve[2]—The nerve providing innervations to the outer half of the intervertebral disc and the posterior longitudinal ligament.

somatotype—A way to label body types; includes the classifications of endomorph, ectomorph, or mesomorph.

speed—The time rate of change of distance (m/s, ft/s) without mention of direction. Speed is a scalar quantity.

sphenomandibular ligament—A suspensory ligament that attaches to the medial surface of the mandible and checks forward movement of the ramus.

squat jump—A jump from a flexed position in which the athlete moves to a static squat position for a brief moment before jumping as high as possible.

stance phase—The phase of gait in which one foot contacts the ground and remains in contact with the ground; includes initial contact, loading response, midstance, terminal stance, and preswing.

stance time—The amount of time spent in stance; includes single support and double support.

step length—The distance between initial contact (usually the heel) of one leg and initial contact (usually the heel) of the contralateral leg.

step width—The linear distance between the midpoint of the heel on one foot and the same point on the other foot.

sternal notch—The most superior, or proximal, edge of the manubrium; often referred to as the jugular notch.

sternum—Generally a flat bone consisting of three segments (manubrium, body of sternum, and xiphoid process); it is convex anteriorly and concave posteriorly.

stratum fibrosum—The thick outer layer of a joint capsule.

stratum synovium—The inner layer of the joint capsule, surrounded by a synovial lining.

strength—The ability of a muscle to exert a maximal force or torque at a specified or determined velocity. Strength can be measured in terms of force, torque, or work.

stride length—The distance between two successive events accomplished by the same lower extremity.

styloid process—A bony projection extending from the distal ulna and radius that reinforces the concave surface of the wrist.

stylomandibular ligament—A suspensory ligament for the mandible, it assists in approximating the disc/condyle/eminence complex.

subacromial space—The space between the acromion and the humeral head. A common site of abutment of the rotator cuff muscles surrounding the humeral head.

subtalar—Referring to the joint between the talus and calcaneus bones.

supination—Rotation of the segment towards the supine position.

supraspinous ligament—A cord-like structure that connects the spinous processes from C7 to L3 or L4.

surgical neck of the humerus—The area just slightly below the tuberosities that is frequently the site of fractures.

sustentaculum tali—A structure on the calcaneus that helps support the talus.

swing phase—The phase of gait in which one foot leaves the ground, lasting until that same foot returns to the ground; includes initial swing, midswing, and terminal swing.

swing time—The amount of time spent in swing.

synarthrosis—A type of joint that does not have a synovial cavity. Synarthrosis joints are held together by fibrous or cartilaginous tissue.

synovial fluid—The pale yellow fluid that is secreted from the synovial lining. This fluid contains proteinases, collagenases, hyaluronic acid, and prostaglandins.

tarsals—The seven bones that make up the midfoot.

tendinosis—Degenerative changes in a tendon, usually from overuse.

thoracolumbar fascia—A complex array of dense connective tissue traveling from the spinous process of T12 to the posterosuperior iliac spine and iliac crest. It consists of three layers that envelop the lumbar musculature and separates them into anterior, middle, and posterior layers.

tibial plafond—The articular surface of the distal end of the tibia.

tibial plateau—The uppermost part of the tibia. The medial tibial plateau is concave, while the lateral is flat or slightly convex. A very common location of fractures.

torque—A moment of force or rotational force.

torsion—A twist in a bone or joint, usually seen in the transverse plane.

trabecular—Relating to or consisting of trabeculae, the osseous bars that form the framework of bone.

transverse ligament—Upper cervical ligament that resists posterior translation of the dens into the vertebral foramen and, therefore, the spinal cord.

transverse process—Lateral projection from the junction of the pedicle and lamina on each respective side of the vertebra. These processes have a groove for the spinal nerves exiting the spinal cord. There are two parts of each transverse process: (1) anterior tubercle and (2) posterior tubercle, which is considered the true transverse process.

triangular fibrocartilage complex (TFCC)—A fibrocartilaginous structure at the distal end of the ulna, anchoring the distal ulna and radius. The TFCC cushions weight-bearing forces placed on the hand and transmits axial load from the hand to the forearm.

trigeminocervical nucleus—The sensory nucleus in the C2-C3 region of the spinal cord that has nerve fibers from the cervical and trigeminal nuclei.

trochlea—A smooth articulating bone surface that allows free gliding movement between bones.

trochoid—A rotating, wheel-like joint articulation.

true ribs—Ribs 1 through 7, whose respective cartilages directly attach to the sternum anteriorly.

Type I (slow-twitch) muscle fiber—Slow-contracting and fatigue-resistant muscle fibers of a skeletal muscle. Type I fibers have a contraction time of 100 to 120 milliseconds.

Type II (fast-twitch) muscle fiber—Fast-contracting and more fatigable muscle fibers (compared with Type I fibers) of a skeletal muscle. Type II fibers have a contraction time of 40 to 45 milliseconds. Type II fibers can be divided into Type IIa (fast-contracting and fatigue-resistant) muscle fibers and Type IIb (fast-contracting and fatigable) muscle fibers.

uncovertebral joints—Saddle-shaped diarthrodial joints present from C3 to T1.

upper respiratory obstructions—Any restriction to the normal flow of air in the nasal passages such as inflamed tonsils, adenoids, or a deviated septum.

valgum—Distal segment deviates away from the midline as viewed in the frontal plane.

valgus collapse—The position of the leg when the knee falls into a valgus position upon landing. Usually a position of leg (thigh) adduction and internal rotation, with the tibia moving into abduction.

varum—Distal segment deviates toward the midline as viewed in the frontal plane.

vector—A quantity that has both magnitude and direction. A force is a vector quantity.

velocity—The time rate of change of displacement (m/s, ft/s). Velocity is a vector quantity that includes magnitude and direction. Average velocity is equal to change of position divided by change of time.

ventral—Anatomically, located on the underside or anterior surface of the extremity or core.

vertebral arch—The irregularly shaped posterior arch of the vertebra. The vertebral body is also called the neural arch.

vertebral artery—The artery in the cervical spine that arises from the first part of the subclavian artery and passes upward along the longus colli muscle to enter the transverse foramen of C6. The vertebral artery on each side then travel through the transverse foramen on each side of the cervical vertebrae through C1, pierce the posterior OA membrane, and meet at the foramen magnum.

vertebral body—The cylindrically shaped anterior region of the vertebra. The vertebral body is designed to be the major weight-bearing structure of the spinal column.

vertebral foramen—The area for the spinal cord, bordered anteriorly by the vertebral body and posteriorly by the vertebral arch.

vertical jump—A jump in which the athlete wants to jump as high as possible.

volar plate—The palmar ligament covering the interphalangeal joints, proximally from the neck of each proximal phalanx and distally to the base of its distal phalanx.

work—The force required to move an object over a given distance. Work = force multiplied by distance.

xiphisternal junction joint—The joint articulation between the main body of the sternum and the xiphoid process.

xiphoid—The smallest and most distal portion of the sternum.

zona orbicularis—Fibers of the articular capsule of the hip joint encircling the neck of the femur.

zygapophyseal facet joint—A typical synovial joint on each side of the vertebral body. Each vertebra has two superior and two inferior articular facets.

REFERENCES

Chapter 1

Kaltenborn FM. 1980. *Mobilization of the extremity joints: examination and basic treatment techniques.* Olaf Norlis Bokhandel: Universitetsgate.

Willson JD, Davis IS. 2008. Utility of the frontal plane projection angle in females with patellofemoral pain. *J Orthop Sports Phys Ther* 38:606-615.

Chapter 2

Hamill J, Knutzen K. 2009. *Biomechanical Basis of Human Movement.* 3rd ed. Baltimore: Williams & Wilkins.

Hewett TE, Myer GD, Ford KR, et al. 2005. Biomechanical measures of neuromuscular control and valgus loading of the knee predict anterior cruciate ligament injury risk in female athletes. *Am J Sports Med* 33:492-501.

Mow VC, Ratcliffe A, Poole AR. 1992. Cartilage and diarthrodial joints as paradigms for hierarchical materials and structures. *Biomaterials* 13:67-97.

Nigg BM, Macintosh BR, Master J. 2000. *Biomechanics and Biology of Movement.* Champaign, IL: Human Kinetics.

Tyler TF, McHugh MP, Mirabella MR, et al. 2006. Risk factors for noncontact ankle sprains in high school football players: the role of previous ankle sprains and body mass index. *Am J Sports Med* 34:471-475.

Waterman BR, Belmont PJ Jr., Cameron KL, et al. 2010. Epidemiology of ankle sprain at the United States Military Academy. *Am J Sports Med* 38:797-803.

Chapter 3

Baldwin KM. 1996. Effects of altered loading states on muscle plasticity: what have we learned from rodents? *Med Sci Sports Exerc* 28:S101-S106.

Baldwin KM, Valdez V, Herrick RE, et al. 1982. Biochemical properties of overloaded fast-twitch skeletal muscle. *J Appl Physiol* 52:467-472.

Barker D, Cope M. 1962. The innervation of individual intrafusal muscle fibres. In: Barker D, ed. *Symposium on Muscle Receptors.* Hong Kong: Hong Kong University Press, 263-269.

Borsa PA, Lephart SM, Kocher MS, et al. 1994. Functional assessment and rehabilitation of shoulder proprioception for glenohumeral instability. *J Sport Rehab* 3:84-104.

Carpenter M. 1977. *Human Neuroanatomy.* 7th ed. Philadelphia: Williams & Wilkins.

Chusid JG. 1985. *Correlative Neuroanatomy and Functional Neurology.* 19th ed. Norwalk, CT: Appleton-Century-Crofts.

Cress NM, Conley KE, Balding SL, et al. 1996. Functional training: muscle structure, function and performance in older women. *J Orthop Sports Phys Ther* 24:4-10.

Fawcett DW. 1984. The nervous tissue. In: Fawcett DW, ed. *Bloom and Fawcett: A Textbook of Histology.* New York: Chapman & Hall.

Freeman MAR, Wyke BD. 1967. An experimental study of articular neurology. *J Bone Joint Surg* 49B:185.

Gamble JG. 1988. *The Musculoskeletal System: Physiological Basics.* New York: Raven Press.

Gardner E. 1950. Reflex muscular responses to stimulation of articular nerves in the cat. *Am J Physiol* 161:133-141.

Gregor RJ. 1993. Skeletal muscle mechanics and movement. In: Grabiner MD, ed. *Current Issues in Biomechanics.* Champaign, IL: Human Kinetics.

Henneman E, Somjen G, Carpenter DO. 1965. Functional significance of cell size in spinal motoneurons. *J Neurophysiol* 28:560-580.

Howell JN, Fuglevand AJ, Walsh ML, et al. 1995. Motor unit activity during isometric and concentric-eccentric contractions of the human first dorsal interosseus muscle. *J Neurophysiol* 74:901-904.

Huxley HE, Hanson J. 1954. Changes in the cross striations of muscle during contraction and stretch and their structural interpretation. *Nature* 173:973-1077.

Lieber RL. 1992. *Skeletal Muscle Structure and Function: Implications for Rehabilitation and Sports Medicine.* Baltimore: Williams & Wilkins.

Linnamo V, Moritani T, Nicol C, et al. 2003. Motor unit activation patterns during isometric, concentric and eccentric actions at different force levels. *J Electr Kinesiol* 13:93-101.

McCloskey DI. 1978. Kinesthetic sensibility. *Physiol Rev* 58:763-820.

Milne RJ, Foreman RD, Giesler GJ, et al. 1981. Convergence of cutaneous and pelvic visceral nociceptive inputs onto primate spinothalamic neurons. *Pain* 11:163-183.

Palmer I. 1938. On injuries to ligaments of knee joint: clinical study. *Acta Chir Scand Suppl* 53.

Pratt N. 1996. *Anatomy of the Cervical Spine.* La Crosse, WI: Orthopaedic Section, APTA.

Prochazka A, Gorassini M. 1998. Ensemble firing of muscle afferents recorded during normal locomotion in cats. *J Physiol* 15(507):293-304.

Thomas PK, Olsson Y. 1984. Microscopic anatomy and function of the connective tissue components of peripheral nerve. In: Dyck PJ, Thomas PK, Lambert EH, et al., eds. *Peripheral Neuropathy*. Philadelphia: Saunders.

Vierck CJ, Greenspan JD, Ritz LA. 1990. Long-term changes in purposive and reflexive responses to nociceptive stimulation following anterior-lateral chordotomy. *J Neuro* 10:2077-2095.

Vliet PM, Heneghan NR. 2006. Motor control and the management of musculoskeletal dysfunction. *Man Ther* 11(3):208-213.

Waxman SG. 1996. *Correlative Neuroanatomy*. 24th ed. New York: McGraw-Hill.

Williams PL, Warwick R, Dyson M, et al. 1995. *Gray's Anatomy*. 37th ed. London: Churchill Livingstone.

Wyke BD. 1981. The neurology of joints: a review of general principles. *Clin Rheum Dis* 7:223-239.

Wyke BD. 1985. Articular neurology and manipulative therapy. In: Glasgow EF, Twomey LT, Scull ER, et al., eds. *Aspects of Manipulative Therapy*. 2nd ed. New York: Churchill Livingstone.

Chapter 4

Abbott BC, Bigland B, Ritchie JM. 1952. The physiological cost of negative work. *J Physiol* 117:380-390.

Agre JC, Magness JL, Hull SZ, et al. 1987. Strength testing with a portable dynamometer: reliability for upper and lower extremities. *Arch Phys Med Rehabil* 68:454-458.

Astrand PO, Rodahl K. 1986. *The Muscle and Its Contraction: Textbook of Work Physiology*. New York: McGraw-Hill.

Baechle TR, Earle RW. 2008. *Essentials of Strength Training and Conditioning*. 3rd ed. Champaign, IL: Human Kinetics.

Baratta RV, Solomonow M, Best R, D'Ambrosia R. 1993. Isotonic length/force models of nine different skeletal muscles. *Med Biol Eng Comput* 31:449-458.

Billeter R, Hoppeler H. 1992. Muscle fiber types and morphometric analysis of skeletal muscle in six year old children. *Med Sci Sports* 12:28.

Bohannon RW. 1987a. Hand-held dynamometry: stability of muscle strength over multiple measurements. *Clin Biomech* 2:74-77.

Bohannon RW. 1987b. The clinical measurement of strength. *Clin Rehabil* 1:5-16.

Bohannan RW. 1988. Make test and break test of elbow flexor muscle strength. *Phys Ther* 68:193-194.

Bohannan RW. 1993. Comparability of force measurements obtained with different strain gauge hand-held dynamometers. *J Orthop Sports Phys Ther* 18(4):564-567.

Bohannan RW, Andrews AW. 1989. Accuracy of spring and strain gauge handheld dynamometers. *J Orthop Sports Phys Ther* 10:323-325.

Bonde-Peterson F, Knuttgen HG, Henriksson J. 1972. Muscle metabolism during exercise with concentric and eccentric contractions. *J Appl Physiol* 33:792-795.

Booth F. 1986. Physiologic and biomechanical effects of immobilization on muscle. *Clin Orthoped* 219:15-20.

Brownstein B, Noyes FR, Mangine RE, et al. 1988. Anatomy and biomechanics. In: Mangine RE, ed. *Physical Therapy of the Knee*. New York: Churchill Livingstone.

Byl NN, Richard S, Asturias J. 1988. Intrarater and interrater reliability of the biceps and deltoids using a hand-held dynamometer. *J Orthop Sports Phys Ther* 9:399-405.

Cooper H, Doods WN, Adams ID, et al. 1981. Use and misuse of the tape measure as a means of assessing muscle strength and power. *Rheumatol Rehabil* 20(4):211-218.

Davies G. 1992. *Compendium of Isokinetics in Clinical Usage and Rehabilitation Techniques*. 4th ed. Onalaska, WI: S&S Publishers.

Davies G, Wilk K, Ellenbecker TS. 1997. Assessment of strength. In: Malone TR, McPoil T, Nitz AJ, eds. *Orthopedic and Sports Physical Therapy*. 3rd ed. St. Louis: Mosby.

DeBries HA, Housh TJ. 1994. *Physiology of Exercise for Physical Education, Athletics and Exercise Science*. 5th ed. Dubuque, IA: Brown and Benchmark.

DeLorme T, Wilkins A. 1951. *Progressive Resistance Exercises*. New York: Appleton-Century-Crofts.

Duedsinger RH. 1984. Biomechanics in clinical practice. *Phys Ther* 64:1860-1868.

Eloranta V, Komi PV. 1980. Function of the quadriceps femoris muscle under maximal concentric and eccentric contraction. *EMG Clin Neurophys* 20:159-174.

Escamilla R, Wickham R. 2003. Exercise-based conditioning and rehabilitation. In: Kolt GS, Snyder-Mackler L, eds. *Physical Therapies in Sports and Exercise*. London: Churchill Livingstone.

Fleck SJ, Kraemer WJ. 2004. *Designing Resistance Training Programs*. 3rd ed. Champaign, IL: Human Kinetics.

Garrett WE. 1996. Muscle strain injuries. *Am J Sports Med* 24(6 Suppl):S2-S8.

Garrett WE, Califf JC, Bassett FH. 1984. Histochemical correlates of hamstring injuries. *Am J Sports Med* 12:98-103.

Goldspink DF, Cox VM, Smith SK, et al. 1995. Muscle growth in response to mechanical stimuli. *Am J Physiol* 268:E288-E297.

Hall SJ. 1999. The biomechanics of human skeletal muscle. In: Hall SJ, ed. *Basic Biomechanics*. New York: McGraw-Hill.

Hall C, Thein-Brody L. 1999. Functional approach to therapeutic exercise for physiologic impairments. In: Hall CM, Thein-Brody L, eds. *Therapeutic Exercise: Moving Toward Function*. Philadelphia: Lippincott Williams & Wilkins.

Hettinger T. 1964. *Isometrisches Muskeltraining*. Stuttgart, Germany: Thun.

Heyward VH. 2010. *Advanced Fitness Assessment and Exercise Prescription*. 6th ed. Champaign, IL: Human Kinetics.

Hortobagyi T, Katch FI, Katch VL, et al. 1990. Relationships of body size, segmental dimensions, and ponderal equivalents to muscular strength in high-strength and low-strength subjects. *Int J Sports Med* 11(5):349-356.

Ikai M, Fukunaga T. 1968. Calculation of muscle strength per unit cross-sectional area of human muscle by means of ultrasonic measurement. *Int Z Angew Physiol Arbeitphysiol* 26:26-32.

Janda V. 1994. Muscles and motor control in cervicogenic disorders: assessment and management. In: Grant R, ed. *Physical Therapy of the Cervical and Thoracic Spine*. New York: Churchill Livingstone.

Kendall FP, McCreary EK, Provance PG, et al. 2005. *Muscles: Testing and Function With Posture and Pain*. 5th ed. Philadelphia: Lippincott Williams & Wilkins.

Komi PV. 1992. *Strength and Power in Sport*. London: Blackwell Scientific.

Komi PV, Kaneko M, Aura O. 1987. EMG activity of the leg extensor muscles with special reference to mechanical efficiency in concentric and eccentric exercise. *Int J Sports Med* 8:22-29.

Lederman E. 1997. *Fundamentals of Manual Therapy: Physiology, Neurology and Psychology*. London: Churchill Livingstone.

Leivseth G, Torstensson J, Reikeras O. 1989. The effect of passive muscle stretching in osteoarthritis of the hip. *Clin Sci* 76:113-117.

Levangie PK, Norkin CC. 1978. *Joint Structure and Function: A Comprehensive Analysis*. Baltimore: Davis.

Lexell J, Taylor CC, Sjostrom M. 1988. What is the cause of the ageing atrophy? Total number, size and proportion of different fiber types studied in whole vastus lateralis muscle from 15 to 83-year old men. *J Neurol Sci* 84:275-294.

Lieber RL, Friden J. 2001. Clinical significance of skeletal muscle architecture. *Clin Orthop Rel Res* 383:140-151.

Luttgens K, Hamilton K. 1997. The musculoskeletal system: the musculature. In: Luttgens K, Hamilton K, eds. *Kinesiology: Scientific Basis of Human Motion*. 9th ed. Dubuque, IA: McGraw-Hill.

McCardle WD, Katch FI, Katch VI. 1991. *Exercise Physiology: Energy, Nutrition, and Human Performance*. Philadelphia: Lea and Febiger.

Morris JM, Lucas DB, Bresler B. 1961. Role of the trunk in stability of the spine. *J Bone Joint Surg* 43A:327-351.

Powers SK, Howley ET. 2001. *Exercise Physiology: Theory and Application*. Boston: McGraw-Hill.

Reiman MP. 2006. Training for strength, power, and endurance. In: Manske RC, ed. *Postoperative Orthopedic Sports Medicine: The Knee and Shoulder*. Philadelphia: Elsevier Science.

Sahrmann S. 2002. *Diagnosis and Treatment of Movement Impairment Syndromes*. St. Louis: Mosby.

Sale DG. 1991. Testing strength and power. In: MacDougall JD, Wenger HA, Green HJ, eds. *Physiological Testing of the High Performance Athlete*. 2nd ed. Champaign, IL: Human Kinetics.

Sapega A, Quedenfeld T, Moyer R, Butler R. 1981. Biophysical factors in range-of-motion exercise. *Phys Sportsmed* 9:57-65.

Siff MC, Verkhoshansky YV. 1999. *Supertraining*. 4th ed. Denver: Supertraining International.

Scott SH, Winter DA. 1991. A comparison of three muscle pennation assumptions and their effect on isometric and isotonic force. *J Biomech* 24(2):163-167.

Smith LK, Weiss EL, Lehmkuhl LD. 1996. *Brunnstrom's Clinical Kinesiology*. Philadelphia: Davis.

Tabary JC, Tabary C, Tardieu C, Tardieu G, Goldspink G. 1972. Physiological and structural changes in the cat's soleus muscle due to immobilization at different lengths by plaster cast. *J Physiol* 224:231-244.

Weldon SM, Hill RH. 2003. The efficacy of stretching for prevention of exercise-related injury: a systematic review of the literature. *Man Ther* 8:141-150.

Wessel J, Kaup C, Fan J, et al. 1999. Isometric strength measurements in children with arthritis: reliability and relation to function. *Arthritis Care Res* 12:238-246.

Wessling KC, DeVane DA, Hylton CR. 1987. Effects of static stretch versus static stretch and ultrasound combined on triceps surae muscle extensibility in healthy women. *Phys Ther* 67(5):674-679.

Wilk KE, Arrigo CA, Andrews JR. 1992. A comparison of individuals exhibiting normal grade manual muscle test and isokinetic testing of the knee extension/flexion (abstract). *Phys Ther* 72(6):71.

Williams PE. 1990. Use of intermittent stretch in the prevention of serial sarcomere loss in immobilized muscle. *Ann Rheum Dis* 49:316-317.

Williams PE, Goldspink G. 1978. Changes in sarcomere length and physiological properties in immobilized muscle. *J Anat* 127:459-468.

Williams PE, Goldspink G. 1984. Connective tissue changes in immobilized muscle. *J Anat* 138:343-350.

Wilson GJ, Newton RU, Murphy AJ, et al. 1993. The optimal training load for the development of dynamic athletic performance. *Med Sci Sports Exerc* 25(11):1279-1286.

Witzman FA. 1988. Soleus muscle atrophy induced by cast immobilization: lack of effect by anabolic steroids. *Arch Phys Med Rehabil* 69:81-85.

Chapter 5

Barak T, Rosen ER, Sofer R. 1990. Basic concepts of orthopaedic manual therapy. In: Gould JA, ed. *Orthopaedic and Sports Physical Therapy*. 2nd ed. St Louis: Mosby, 195-211.

Bennell K, Kannus P. 2003. Bone. In: Kolt GS, Snyder-Mackler L, eds. *Physical Therapies in Sports and Exercise*. London: Churchill Livingstone.

Buckwalter JA, Glimcher MJ, Cooper RR, Recher R. 1995. Bone biology. *J Bone Joint Surg* 77A:1256-1277.

Cook CE. 2007. *Orthopedic Manual Therapy: An Evidence-Based Approach*. Upper Saddle River, NJ: Pearson-Prentice Hall.

Frankel VH, Nordin M. 2001. Biomechanics of bone. In: Nordin M, Frankel VH, eds. *Basic Biomechanics of the Musculoskeletal System*. 3rd ed. Philadelphia: Lippincott Williams & Wilkins.

Grieve G. 1998. *Common vertebral joint problems*. 2nd ed. Edinburgh: Churchill Livingstone.

Harris B. 1980. The mechanical behavior of composite materials. In: Vincent JFV, Currey JD, eds. *Mechanical Properties of Biological Materials*. No. 34. Cambridge: Symposia of the Society for Experimental Biology, Cambridge University Press.

Jackson DW, Scheer MJ, Simon TM. 2001. Cartilage substitutes: overview of basic science and treatment options. *J Am Acad Orthop Surg* 9:37-52.

Kaltenborn F. 1980. *Mobilization of the Extremity Joints: Examination and Basic Treatment Techniques*. Oslo: Olaf Noris Bokhandel.

Khan K, McKay H, Kannus P, et al. 2001. *Physical Activity and Bone Health*. Champaign, IL: Human Kinetics.

Neumann DA. 2010. *Kinesiology of the Musculoskeletal System: Foundations for Rehabilitation*. 2nd ed. St. Louis: Mosby.

Noyes FR, Bassett RW, Grood ES, Butler DL. 1980. Arthroscopy in acute traumatic hemarthrosis of the knee: incidence of anterior cruciate tears and other injuries. *J Bone Joint Surg* 62A:687-695,757.

Wolff J. 1982. *Das Gesetz der Tansformation der Knochen*. Berlin: Hirschwald.

Wooden MJ. 1989. Mobilization of the upper extremity. In: Donatelli R, Wooden MJ, eds. *Orthopaedic Physical Therapy*. New York: Churchill Livingstone, 297-332.

Chapter 6

Adams MA, Hutton WC, Stott JRR. 1980. The resistance to flexion of the lumbar intervertebral joint. *Spine* 5:245.

Allia P, Gorniak G. 2006. Human ligamentum nuchae in the elderly: its function in the cervical spine. *J Man Manip Ther* 14:11-21.

Atasoy E. 2004. Thoracic outlet syndrome: anatomy. *Hand Clin* 20(1):7-14.

Bland JH, Boushey DR. 1990. Anatomy and physiology of the cervical spine. *Semin Arthritis Rheum* 20(1):1-20.

Blauvelt CT, Nelson FRT. 1994. *A Manual of Orthopaedic Terminology*. 5th ed. St Louis: Mosby.

Bogduk N. 1997. *Clinical Anatomy of the Lumbar Spine and Sacrum*. 3rd ed. New York: Churchill Livingstone.

Bogduk N, Mercer S. 2000. Biomechanics of the cervical spine. I: normal kinematics. *Clin Biomech* 15:633.

Bogduk N, Windsor M, Inglis A. 1988. The innervations of the cervical intervertebral disks. *Spine* 13:2-8.

Cook C. 2003. Lumbar coupling biomechanics: a literature review. *J Man Manip Ther* 11(3):137-145.

Crisco JJ, Oda T, Panjabi MM, Bueff HU, Dvorak J, Grob D. 1991. Transections of the C1-C2 joint capsular ligaments in the cadaveric spine. *Spine* 16:S474-S479.

Crisco JJ, Panjabi MM, Dvorak J. 1991. A model of the alar ligaments of the upper cervical spine in axial rotation. *J Biomech* 24:607.

Cyriax J. 1982. *Textbook of Orthopaedic Medicine. Volume 1: Diagnosis of Soft Tissue Lesions*. 8th ed. London: Harcourt.

Dvorak J. 1998. Epidemiology, physical examination, and neurodiagnostics. *Spine* 23:2663-2673.

Dvorak J, Panjabi MM. 1987. Functional anatomy of the alar ligaments. *Spine* 12:183.

Dvorak J, Panjabi MM, Gerber M, et al. 1987. CT functional diagnostics of the rotary instability of the upper cervical spine and experimental study in cadavers. *Spine* 12:197-205.

Dvorak J, Schneider E, Saldinger P, et al. 1987. Biomechanics of the craniocervical region: the alar and transverse ligaments. *J Ortho Res* 6:452-461.

Ellis JH, Martel W, Lillie JH, et al. 1991. Magnetic resonance imaging of the normal craniovertebral junction. *Spine* 16:105.

Giles LG, Taylor JR. 1987. Innervation of human lumbar zygapophyseal joint synovial folds. *Acta Orthop Scand* 58:43-46.

Gottleib MS. 1994. Absence of asymmetry in superior articular facets on the first cervical vertebra in humans: implications for diagnosis and treatment. *J Manip Physiol Ther* 17:314-320.

Gray H. 1995. *Gray's Anatomy*. Philadelphia: Lea & Febiger.

Heggeness MH, Doherty BJ. 1993. The trabecular anatomy of the axis. *Spine* 18:1945-1949.

Hertling D, Kessler RM. 1996. *Management of Common Musculoskeletal Disorders*. 3rd ed. Philadelphia: Lippincott.

Hollinshead WH. 1982. *Anatomy for Surgeons: The Back and Limbs*. 3rd ed. Philadelphia: Harper & Row.

Jirout J. 1973. Changes in the atlas–axis relations on lateral flexion of the head and neck. *Neuroradiology* 6:215-218.

Johnson RM, Crelin ES, White AA, et al. 1975. Some new observations on the functional anatomy of the lower cervical spine. *Clin Ortho Rel Res* 111:192-200.

Johnson GH, Zhang M, Jones DG. 2000. The fine connective tissue architecture of the human ligamentum nuchae. *Spine* 25:5-9.

Koebke J, Brade H. 1982. Morphological and functional studies on the lateral joints of the first and second cervical vertebrae in man. *Anat Embryol* (Berl) 164:265-275.

Kokubun S, Sakurai M, Tanaka Y. 1996. Cartilaginous endplate in cervical disc herniation. *Spine* 21:190-195.

Legaspi O, Edmond SL. 2007. Does the evidence support the existence of lumbar spine coupled motion? A critical review of the literature. *J Ortho Sports Phys Ther* 37(4):169-178.

Lundon K, Bolton K. 2004. Structure and function of the bones and joints of the cervical spine. In: Oatis C, ed. *Kinesiology: The Mechanics and Pathomechanics of Human Movement*. Philadelphia: Lippincott Williams & Wilkins.

Lysell E. 1969. Motion in the cervical spine: an experimental study on autopsy specimens. *Acta Orthop Scand* 123S:1.

Magee DJ. 2002. *Orthopedic Physical Assessment*. 4th ed. Philadelphia: Saunders Elsevier.

Maiman DJ, Pintar FA. 1992. Anatomy and clinical biomechanics of the thoracic spine. *Clin Neurosurg* 38:296.

Malanga GA, Landes P, Nadler SF. 2003. Provocative tests in cervical spine examination: historical basis and scientific analyses. *Pain Physician* 6(2):199-205.

McGill S. 2007. *Low Back Disorders: Evidence-Based Prevention and Rehabilitation*. 2nd ed. Champaign, IL: Human Kinetics.

Mercer S. 2004. Kinematics of the spine. In: Boyling JD, Jull GA, eds. *Grieve's Modern Manual Therapy: The Vertebral Column*. Philadelphia: Churchill Livingstone.

Mercer S, Bogduk N. 1993. Intra-articular inclusions of the cervical synovial joints. *Br J Rheumatol* 32:705-710.

Mercer SR, Bogduk N. 1999. The ligaments and annulus fibrosus of human adult cervical intervertebral discs. *Spine* 24:619-626.

Mercer SR, Bogduk N. 2001. The joints of the cervical vertebral column. *J Ortho Sports Phys Ther* 31:174-182.

Myklebust JB, Pintar F, Yoganandan N, et al. 1988. Tensile strength of spinal ligaments. *Spine* 13:526.

Nilsson N, Hartvigsen J, Christensen HW. 1996. Normal ranges of passive cervical motion for women and men 20-60 years old. *J Manip Phys Ther* 19:306.

O'Brien MF, Lenke LG. 1997. Fractures and dislocations of the spine. In: Dee R, Hurst L, Gruber M, et al., eds. *Principles of Orthopaedic Practice*. 2nd ed. New York: McGraw-Hill.

Olszewski AD, Yaszemski MJ, White A. 1996. The anatomy of the human lumbar ligamentum flavum. *Spine* 21:2307.

Orofino C, Sherman MS, Schechter D. 1960. Luschka's joint: a degenerative phenomenon. *J Bone Joint Surg* 5A:853-858.

Pal GP, Routal RV, Saggu SK. 2001. The orientation of the articular facets of the zygapophyseal joints at the cervical and upper thoracic region. *J Anat* 198:431-441.

Pal GP, Sherk HH. 1988. The vertical stability of the cervical spine. *Spine* 13:447.

Panjabi MM, Crisco JJ, Vasavada A, et al. 2001. Mechanical properties of the human cervical spine as shown by three-dimensional load-displacement curves. *Spine* 26:2692-2700.

Panjabi MM, Duranceau J, Goel V, et al. 1991. Cervical human vertebrae: quantitative three-dimensional anatomy of the middle and lower regions. *Spine* 16:861-869.

Panjabi M, Dvorak J, Duranceau J, et al. 1988. Three-dimensional movement of the upper cervical spine. *Spine* 13:727.

Panjabi M, Oda T, Crisco J, Dvorak J, Grob D. 1993. Posture affects motion coupling patterns of the upper cervical spine. *J Ortho Res* 11: 525-536.

Panjabi MM, Oxland TR, Parks H. 1991. Quantitative anatomy of the cervical spine ligaments. *J Spinal Disord* 4:270.

Penning L. 1978. Normal movements of the cervical spine. *J Roentgenol* 130:317-326.

Pick TP, Howden R. 1995. *Gray's Anatomy*. 15th ed. New York: Barnes & Noble Books.

Putz R. 1992. The detailed functional anatomy of the ligaments of the vertebral column. *Anat Anz* 174:40.

Sasso RC. 2001. C2 dens fractures: treatment options. *J Spinal Disord* 14:455-463.

Singh S. 1965. Variations of the superior articular facets of the atlas vertebrae. *J Anat* 99:565-571.

Steindler A. 1955. *Kinesiology of the Human Body*. Springfield, IL: Thomas.

Taylor JR. 1971. Regional variation in the development and position of the notochordal segments of the human nucleus pulposus. *J Anat* 110:131-132.

Van Roy P, Caboor D, de Boelpaep S, et al. 1997. Left–right asymmetries and other common anatomical variations of the first cervical vertebra. Part 1: left–right asymmetries in C1 vertebrae. *Man Ther* 2:24-36.

Walsh R, Nitz AJ. 2001. Cervical spine. In: Wadsworth C, ed. *Current Concepts of Orthopaedic Physical Therapy: Home Study Course*. LaCrosse, WI: Orthopaedic Section, APTA.

Watson D, Trott P. 1993. Cervical headache: an investigation of natural head posture and upper cervical flexor muscle performance. *Cephalgia* 13:272-284.

Werne S. 1958. The possibilities of movement in the craniovertebral joints. *Acta Orthop Scand* 28:165-173.

White AA, Johnson RM, Panjabi MM, et al. 1975. Biomechanical analysis of clinical stability in the cervical spine. *Clin Ortho* 109:85-96.

White AA, Panjabi MM. 1990. *Clinical Biomechanics of the Spine*. 2nd ed. Philadelphia: Lippincott-Raven.

Williams PL. 1995. *Gray's Anatomy*, 38th ed. New York: Churchill Livingstone.

Chapter 7

Alomar X, Medrano J, Cabratosa J, et al. 2007. Anatomy of the temporomandibular joint. *Semin Ultrasound CT* 28(3):170-183.

Buckingham RB, Braun T, Harinstein DA, et al. 1991. TMJ dysfunction syndrome: a close association with systemic joint laxity (the hypermobile joint syndrome). *Oral Surg Oral Med Oral Pathol* 72:514-519.

Dutton M. 2008. *Orthopaedic Examination, Evaluation, and Intervention.* 2nd ed. New York: McGraw-Hill.

Green JH, Silver PHS. 1981. *An Introduction to Human Anatomy.* New York: Oxford University Press.

Grieve GP. 1988. *Common Vertebral Joint Problems.* 2nd ed. New York: Churchill Livingstone.

Heffez LB, Jordan SL. 1992. Superficial vascularity of temporomandibular joint retrodiscal tissue: an element of internal derangement process. *Cranio* 10(3):180-191.

Krause SL. 1994. *Temporomandibular Disorders.* New York: Churchill Livingstone.

La Touche R, Fernández-de-las-Penas C. 2009a. Bilateral mechanical-pain sensitivity over the trigeminal region in patients with chronic mechanical neck pain. *J Pain* (3):256-263.

La Touche R, Fernández-de-las-Penas C. 2009b. The effects of manual therapy and exercise directed at the cervical spine on pain and pressure pain sensitivity in patients with myofascialtemporomandibular disorders. *J Oral Rehabil* 36(9):644-652.

Levangie PK, Norkin CC. 2011. *Joint Structure and Function: A Comprehensive Analysis.* 5th ed. Philadelphia: Davis.

Neumann DA. 2010. Kinesiology of mastication and ventilation. In: Neumann DA, ed. *Kinesiology of the Musculoskeletal System.* 2nd ed. St. Louis: Mosby Elsevier.

Okeson JP. 2003. *Management of Temporomandibular Disorders and Occlusion.* 5th ed. St. Louis: Mosby Elsevier.

Porter MR. 1970. The attachment of the lateral pterygoid muscle to the meniscus. *J Prosthet Dent* 24:555-562.

Rees LA. 1954. The structure and function of the mandibular joint. *Br Dent J* 96:125.

Rocabado M. 2012. *CF 4 – Advanced Craniofacial – State of the Art,* The University of Saint Augustine, Saint Augustine, Florida. May 2012. Class notes and personal communication.

Rocabado M, Iglarsh, ZA. 1991. *Musculoskeletal Approach to Maxillofacial Pain.* Philadelphia: Lippincott.

Travell JG, Simons DG. 1983. *Myofascial Pain and Dysfunction: The Upper Extremities.* Baltimore: Williams & Wilkins.

Ugboko VI, Oginni, FO, Ajike SO, Olasoji HO, Adebayo, ET. 2005. A survey of temporo-mandibular joint dislocation: aetiology, demographics, risk factors and management in 96 Nigerian cases. *Int J Oral Maxillofac Surg* 34(5):499-502.

Chapter 8

American Medical Association. 1988. *Guides to the Evaluation of Permanent Impairment.* 3rd ed. Chicago: AMA.

Arce CA, Dohrmann GJ. 1985. Thoracic disc herniations: improved diagnosis with computed tomographic scanning and a review of the literature. *Surg Neurol* 23:356-361.

Benson M, Burnes D. 1975. The clinical syndromes and surgical treatments of thoracic intervertebral disc prolapse. *J Bone Joint Surg.* 8:457-471.

Bogduk N, Engle R. 1984. The menisci of the lumbar zygapophyseal joints: a review of their anatomy and clinical significance. *Spine* 9:454-460.

Bogduk N, Marsland A. 1988. The cervical zygapophyseal joint as a source of neck pain. *Spine* 13:610.

Boszczyk B, Boszczyk A, Putz R, Buttner A, Benjamin M, Milz S. 2001. An immunohistochemical study of the dorsal capsule of the lumbar and thoracic facet joints. *Spine* 26(15):E338-E343.

Brewin J, Hill M, Ellis H. 2009. The prevalence of cervical ribs in a London population. *Clin Anat* 22(3):331-6.

Buchalter D, Parnianpour M, Viola K, Nordin M, Kahanovitz N. 1988. Three-dimensional spinal motion measurements. Part 1: a technique for examining posture and functional spinal motion. *Spinal Disord* 1(4):279-283.

Butler DS. 1992. *Mobilization of the Nervous System.* New York: Churchill Livingstone.

Cleland J, Selleck B, Stowell T, et al. 2004. Short-term effect of thoracic manipulation on lower trapezius muscle strength. *J Man Manip Ther* 12(2):82-90.

Cook C. 2003. Lumbar coupling biomechanics: a literature review. *J Man Manip Ther* 11(3):137-145.

DeTroyer A, Sampson MG. 1982. Activation of the parasternal intercostals during breathing efforts in human subjects. *J Appl Physiol* 52:524-529.

Dommisse GF. 1974. The blood supply of the spinal cord. *J Bone Joint Surg* 56B:225.

Edmondston SJ, Singer KP. 1997. Thoracic spine: anatomical and biomechanical considerations for manual therapy. *Man Ther* 2:132-143.

Feieretag MA, Horton WE, Norman JT, et al. 1995. The effect of different surgical releases on thoracic spine motion: a cadaveric study. *Spine* 20(14):1604-1611.

Flynn TW. 2001. Thoracic spine and chest wall. In: Wadsworth C, ed. *Current Concepts of Orthopedic Physical Therapy: Home Study Course.* LaCrosse, WI: Orthopaedic Section, APTA.

Galante JO. 1967. Tensile properties of the human lumbar annulus fibrosis. *Acta Orthop Scand* 100(Suppl):1-91.

Gray H. 1995. *Gray's Anatomy.* Philadelphia: Lea & Febiger.

Gregersen GG, Lucas DB. 1967. An in vivo study of the axial rotation of the human thoracolumbar spine. *J Bone Joint Surg Am* 49:247-262.

Horton SJ. 2002. Acute locked thoracic spine: treatment with a modified SNAG. *Man Ther* 7(2):103-107.

Itoi E, Sinaki M. 1994. Effect of back strengthening exercise on posture in healthy women 49 to 65 years of age. *Mayo Clin Proc* 69(11):1054-1059.

Kapandji IA. 1974. *The Physiology of the Joints. Vol. 3. The Trunk and the Vertebral Column.* Edinburgh: Churchill Livingstone.

Kapandji IA. 1978. *The Physiology of the Joints. Vol. 3. The Trunk and the Vertebral Column.* 2nd ed. London: Churchill Livingstone.

Kendall HO, Kendall FP, Boynton DA. 1952. *Posture and Pain.* Baltimore: Williams & Wilkins.

Lee DG. 1988. Biomechanics of the thorax. In: Grant R, ed. *Physical Therapy of the Cervical and Thoracic Spine.* New York: Churchill Livingstone.

Lee DG. 1994. *Manual Therapy for the Thorax: A Biomechanical Approach.* Delta, BC: DOPC.

Lee D. 1996. Rotational instability of the mid-thoracic spine: assessment and management. *Man Ther* 1:234-241.

Legaspi O, Edmond SL. 2007. Does the evidence support the existence of lumbar spine coupled motion? A critical review of the literature *J Orthop Sports Phys Ther* 37(4):169-178.

Loring SH, Woodbridge JA. 1991. Intercostal muscle action inferred from finite-element analysis. *J Appl Physiol* 70:2712-2718.

Lyu RK, Chang HS, Tang LM, et al. 1999. Thoracic disc herniation mimicking acute lumbar disc disease. *Spine* 24:416-418.

Macintosh JE, Bogduk N. 1987. The morphology of the lumbar erector spinae. *Spine* 12:658-668.

Panjabi MM, Brand RA, White AA III. 1976. Mechanical properties of the human thoracic spine as shown by three-dimensional load-displacement curves. *J Bone Joint Surg Am* 58(5):642-652.

Panjabi MM, Hausfeld J, White A. 1981. A biomechanical study of the ligamentous stability of the thoracic spine in man. *Acta Orthop Scand* 52:315-326.

Panjabi MM, Takata K, Goel V, et al. 1991. Thoracic human vertebrae: quantitative three-dimensional anatomy. *Spine* 16(8):888-901.

Panjabi MM, White AA III. 1980. Basic biomechanics of the spine. *Neurosurg* 7:76-93.

Reiman MP, Manske RC, Smith BS. 2008. Immediate effects of soft tissue mobilization and joint manipulation interventions on lower trapezius strength. *J Man Manip Ther* 16(3):166.

Resnick DK, Weller SJ, Benzel EC. 1997. Biomechanics of the thoracolumbar spine. *Neurosurg Clin North Am* 8:455-469.

Romanes GJE. 1981. *Cunningham's Textbook of Anatomy.* Oxford: Oxford University Press.

Saumarez RC. 1986. An analysis of action of intercostal muscles in human upper rib cage. *J Appl Physiol* 60:690-701.

Shacklock M. 2005. *Clinical Neurodynamics: A New System of Musculoskeletal Treatment.* London: Elsevier Health Sciences.

Sinaki M, Itoi E, Rogers JW, Bergstralh EJ, Wahner HW. 1996. Correlation of back extensor strength with thoracic kyphosis and lumbar lordosis in estrogen-deficient women. *Am J Phys Med Rehabil* 75(5):370-374.

Singer KP, Boyle JJW, Fazey P. 2004. Comparative anatomy of the zygapophyseal joints. In: Boyling JD, Jull GA, eds. *Grieve's Modern Manual Therapy: The Vertebral Column.* Philadelphia: Churchill Livingstone.

Singer KP, Day RE, Breidahl PD. 1989. In vivo axial rotation at the thoracolumbar junction: an investigation using low dose CT in healthy male volunteers. *Clin Biomech* 4:137-143.

Singh K, Vaccaro AR, Eichenbaum MD, Fitzhenry LN. 2004. The surgical management of thoracolumbar injuries. *J Spinal Cord Med* 27(2):95-101.

Sirca A, Kosteve V. 1985. The fibre type composition of thoracic and lumbar paravertebral muscles in man. *J Anat* 141:131-137.

Stone J, Lichtor T, Banerjee S. 1994. Intradural thoracic disc herniation. *Spine* 19:1281-1284.

Tawackoli W, Marco R, Liebschner MA. 2004. The effect of compressive axial load preload on the flexibility of the thoracolumbar spine. *Spine* 29(9):988-993.

Taylor A. 1960. The contribution of the intercostal muscles to the effort of respiration in man. *J Physiol* 151:390-402.

Vanichkachorn JS, Vaccaro AR. 2000. Thoracic disk disease: diagnosis and treatment. *J Am Acad Orthop Surg* 8:159-169.

Vicenzino B, Paungmali A, Buratowski S, Wright A. 2001. Specific manipulative therapy treatment for chronic lateral epicondylalgia produces uniquely characteristic hypoalgesia. *Man Ther* 6:205-212.

Wang CH, McClure P, Pratt NE, Nobilini R. 1999. Stretching and strengthening exercises: their effect on three-dimensional scapular kinematics. *Arch Phys Med Rehabil* 80(8):923-929.

White AA. 1969. An analysis of the mechanics of the thoracic spine in man. *Acta Orthop Scand* 127(Suppl):8-92.

White AA III, Panjabi MM. 1990. Kinematics of the spine. In: Cooke DB, ed. *Clinical Biomechanics of the Spine.* Philadelphia: Lippincott.

White AA, Panjabi MM, Thomas CL. 1977. The clinical biomechanics of kyphotic deformities. *Clin Orthop* 128:8-17.

Williams P, Bannister L. 1995. In: Berry M, Collins P, Dyson M, Dussek J, Ferguson M, eds. *Gray's Anatomy.* 38th ed. Edinburgh: Churchill Livingstone.

Chapter 9

Abumi K, Panjabi MM, Kramer KM, et al. 1990. Biomechanical evaluation of lumbar stability after graded facetectomies. *Spine* 15:1142-1147.

Adams MA, Dolan P, Hutton WC. 1988. The lumbar spine in backward bending. *Spine* 13:1019-1026.

Adams MA, Freeman BJ, Morrison HP, et al. 2000. Mechanical initiation of intervertebral disc degeneration. *Spine* 25:1625-1636.

Adams MA, Hutton WC. 1980. The effect of posture on the role of the zygapophyseal joints in resisting intervertebral compressive forces. *J Bone Joint Surg Br* 62:358-362.

Adams MA, Hutton WC. 1981. The effect of posture on the role of the mechanical derangement of the lumbar spine. *Spine* 6:241-248.

Adams MA, Hutton WC, Stott JR. 1980. The resistance to flexion of the lumbar intervertebral joint. *Spine* 5:245-253.

Ahmed AM, Duncan MJ, Burke DL. 1990. The effect of facet geometry on the axial torque-rotation response of lumbar motion segments. *Spine* 15:391-401.

American Medical Association. 1988. *Guides to the Evaluation of Permanent Impairment*. 3rd ed. Chicago: AMA.

Arnoldi CC, Brodsky AE, Cauchoix J, et al. 1976. Lumbar spinal stenosis and nerve root entrapment syndromes. Definition and classification. *Clin Orthop Rel Res* 115:4-5.

Basmajian JV, Deluca CJ. 1985. *Muscles Alive: Their Functions Revealed by Electromyography*. Baltimore: Williams & Wilkins.

Bergmark A. 1989. Stability of the lumbar spine: a study in mechanical engineering. *Acta Orthop Scand* 230:20-24.

Boden SD, Riew KD, Yamaguchi K, et al. 1996. Orientation of the lumbar facet joints: association with degenerative disc disease. *J Bone Joint Surg Am* 78:403-411.

Bogduk N. 1983. The innervations of the lumbar spine. *Spine* 8:286-293.

Bogduk N. 1997. The sacroiliac joint. In: Bogduk N, ed. *Clinical Anatomy of the Lumbar Spine and Sacrum*. 3rd ed. New York: Churchill Livingstone.

Bogduk N, Jull G. 1985. The theoretical pathology of acute locked back: a basis for manipulative therapy. *Man Med* 1:78.

Bogduk N, Macintosh J. 1984. The applied anatomy of the thoracolumbar fascia. *Spine* 9:164-170.

Bogduk N, Twomey LT. 1991. *Clinical Anatomy of the Lumbar Spine*. 2nd ed. New York: Churchill Livingstone.

Bogduk N, Twomey LT. 1997. *Clinical Anatomy of the Lumbar Spine and Sacrum*. 3rd ed. New York: Churchill Livingstone.

Bogduk N, Tynan W, Wilson AS. 1981. The nerve supply to the human lumbar intervertebral discs. *J Anat* 132:39-56.

Bogduk N, Wilson AS, Tynan AS. 1982. The human lumbar dorsal rami. *J Anat* 134:383-397.

Bowen V, Cassidy JD. 1981. Macroscopic and microscopic anatomy of the sacroiliac joint from embryonic life until the eighth decade. *Spine* 6:620-628.

Chow DHK, Luk KDK, Leong JYC, et al. 1989. Torsional stability of the lumbosacral junction: significance of the iliolumbar ligament. *Spine* 14:611-615.

Cook C. 2003. Lumbar coupling biomechanics: a literature review. *J Man Manip Ther* 11(3):137-145.

Cunningham D. 1925. *Textbook of Anatomy*. 5th ed. New York.

Cyriax JH, Cyriax P. 1983. *Illustrated Manual of Orthopaedic Medicine*. London: Butterworths.

Deyo R, Gray D, Kreuter W, Mirza S, Martin BI. 2005. United States trends in lumbar fusion surgery for degenerative conditions. *Spine* 30:1441-1445.

Dreyfuss P, Michaelson M, Pauza K, et al. 1996. The value of medical history and physical examination in diagnosing sacroiliac joint pain. *Spine* 21:2594-2602.

Egund N, Olsson TH, Schmid H, et al. 1978. Movements in the sacroiliac joints demonstrated with roentgen stereophotogrammetry. *Acta Radiol Diagn* 19:833-846.

Erhard R. 1977. The recognition and management of the pelvic component of low back and sciatic pain. *Bull Ortho Sect* 2:4-14.

Fanuele JC, Birkmeyer NJO, Abdu WA, Tosteson TD and Weinstein JN. 2000. The impact of spinal problems on the health status of patients: have we underestimated the effect? *Spine* 25:1509-1514.

Fortin JD, Kissling RO, O'Connor BL, Vilensky JA. 1999. Sacroiliac joint innervations and pain. *Am J Orthop* 28(12):687-690.

Fortin JD, Tolchin RB. 2003. Sacroiliac arthrograms and post-arthrography computerized tomography. *Pain Physician* 6(3):287-290.

Fortin JD, Vilensky JA, Merkel GJ. 2003. Can the sacroiliac joint cause sciatica? *Pain Physician* 6(3):269-271.

Frigerio NA, Stowe RR, Howe JW. 1974. Movement of the sacroiliac joint. *Clin Ortho Rel Res* 100:370-377.

Fritz JM, Cleland JA, Childs JD. 2007. Subgrouping patients with low back pain: evolution of a classification approach to physical therapy. *J Orthop Sports Phys Ther* 37:290-302.

Frymoyer JW, Cats-Baril W. 1987. Predictors of low back pain disability. *Clin Orthop* 221:89-98.

Goode A, Hegedus EJ, Sizer P, et al. 2008. Three-dimensional movements of the sacroiliac joint: a systematic review of the literature and assessment of clinical utility. *J Man Manip Ther* 16(1):25-38.

Gracovetsky S. 1986. The optimum spine. *Spine* 11:543-573.

Gracovetsky S, Farfan H, Helleur C. 1985. The abdominal mechanism. *Spine* 10(4):317-324.

Gray H. 1980. *Gray's Anatomy*. 36th British ed. Philadelphia: Saunders.

Gray H. 1995. *Gray's Anatomy*. Philadelphia: Lea & Febiger.

Grieve GP. 1981. *Common Vertebral Joint Problems*. New York: Churchill Livingstone.

Grob KR, Neuhuber WL, Kissling RO. 1995. Innervation of the sacroiliac join of the human. *Zeitschrift für Rheumatologie* 54:117-122.

Grobler LJ, Robertson PA, Novotny JE, et al. 1993. Etiology of spondylolisthesis: assessment of the role played by lumbar facet joint morphology. *Spine* 1:80-91.

Hansson T, Roos B. 1983. The amount of bone mineral and Schmorl's nodes in lumbar vertebrae. *Spine* 8:266-271.

Harrison DE, Harrison DD, Troyanovich SJ. 1997. The sacroiliac joint: a review of anatomy and biomechanics with clinical implications. *J Manip Physiol Ther* 20(9):607-617.

Hart LG, Deyo RA, Cherkin DC. 1995. Physician office visits for low back pain: frequency, clinical evaluation, and treatment patterns from a US national survey. *Spine* 20:11-19.

Hedtmann A, Steffen R, Methfessel J, et al. 1989. Measurement of human lumbar spine ligaments during loaded and unloaded motion. Spine 14:175-185.

Heisler JC. 1923. *Practical Anatomy*. 3rd ed. Philadelphia.

Heylings DJA. 1978. Supraspinous and interspinous ligaments of the human spine. *J Anat* 125:127-131.

Hukins DWL, Kirby MC, Sikoryn TA, et al. 1990. Comparison of structure, mechanical properties, and function of lumbar spinal ligaments. *Spine* 15:787-795.

Humzah MD, Soames RW. 1988. The human and intervertebral disc. *Anat Rec* 220:337-356.

Kapandji IA. 1991. *The Physiology of Joints, the Trunk and Vertebral Column*. New York: Churchill Livingstone.

Katz JN, Lipson SJ, Brick GW, et al. 1995. Clinical correlates of patient satisfaction after laminectomy for degenerative lumbar spinal stenosis. *Spine* 20:1155-1160.

Katz JN, Stucki G, Lipson SJ, Fossel AH, Grobler LJ, Weinstein JN. 1999. Predictors of surgical outcome in degenerative lumbar spinal stenosis. *Spine* 24:2229-2233.

Kellgren JH. 1938. On the distribution of pain arising from deep somatic structures with charts of segmental pain areas. *Clin Sci* 3:175-190.

Lee DG. 1996. *The Pelvic Girdle: An Approach to the Examination and Treatment of the Lumbo-Pelvic-Hip Region*. 2nd ed. Vancouver: Nascent.

Lee DG. 2004. *The Pelvic Girdle*. London: Churchill Livingstone.

Lee DG, Vleeming A. 2004. The management of pelvic joint pain and dysfunction. In: Boyling JD, Jull GA, eds. *Grieve's Modern Manual Therapy: The Vertebral Column*. Philadelphia: Churchill Livingstone.

Legaspi O, Edmond SL. 2007. Does the evidence support the existence of lumbar spine coupled motion? A critical review of the literature. *J Orthop Sports Phys Ther* 37(4):169-178.

Lewin T. 1964. Osteoarthritis in lumbar synovial joints. *Acta Orthop Scand Suppl* 73:1-112.

Lewin T, Moffett B, Viidik A. 1962. The morphology of the lumbar synovial intervertebral joints. *Acta Morphol Neerl Scand* 4:299-319.

Little JS, Khalsa PS. 2005. Material properties of the human facet joint capsule. *J Biomech Eng* 127(1):15-24.

Long DM, BenDebba M, Torgerson WS, et al. 1996. Persistent back pain and sciatica in the United States: patient characteristics. *J Spinal Dis* 9:40-58.

Lucas D, Bresler B. 1961. *Stability of Ligamentous Spine*. Report no. 40. University of California at Berkley, Biomechanics Laboratory.

Luk KDK, Ho HC, Leong JCY. 1986. The iliolumbar ligament: a study of its anatomy, development and clinical significance. *J Bone Joint Surg* 68B:197-200.

Lundin O, Ekstrom L, Hellstrom M, et al. 1998. Injuries in the adolescent porcine spine exposed to mechanical compression. *Spine* 23:2574-2579.

MacDonald GR, Hunt TE. 1951. Sacro-iliac joint observations on the gross and histological changes in the various age groups. *Canad Med Assoc J* 66:157.

Macintosh J, Bogduk N. 1986. The biomechanics of the lumbar multifidus. *Clin Biomech* 1:205-213.

Meissner A, Fell M, Wilk R, Boenick U, Rahmanzadeh R. 1996. Biomechanics of the pubic symphysis. Which forces lead to mobility of the symphysis in physiological conditions? (abstract). *Unfallchirurg* 99:415-421.

Miller JAA, Schultz AB, Andersson GBJ. 1987. Load displacement behavior of sacro-iliac joints. *J Orthop Res* 5:92-101.

Morris H. 1925. *Human Anatomy*. 8th ed. Philadelphia: Blakiston's.

Nachemson AL, Evans JH. 1968. Some mechanical properties of the third human lumbar interlaminar ligament (ligamentum flavum). *J Biomech.* 1:211-220.

Panjabi MM. 1992a. The stabilizing system of the spine. Part I. Function, dysfunction, adaptation, and enhancement. *J Spinal Disord* 5(4):383-389.

Panjabi MM. 1992b. The stabilizing system of the spine. Part II. Neutral zone and instability hypothesis. *J Spinal Disord* 5(4):390-396.

Panjabi MM, Goel VK, Takata K. 1983. Physiologic strains in the lumbar ligaments: an in vitro biomechanical study. *Spine* 7:192-203.

Pearcy M, Portek I, Shepherd J. 1984. Three dimensional X-ray analysis of normal movement in the lumbar spine. *Spine* 9:294-297.

Penning L, Irwan R, Oudkerk M. 2005. Measurement of angular and linear segmental lumbar spine flexion–extension motion by means of image registration. *Eur Spine J* 14(2):163-170.

Peng B, Wu W, Hou S, Li P, Zhang C, Yang Y. 2005. The pathogenesis of discogenic low back pain. *J Bone Joint Surg Br* 87(1):62-67.

Pitkin HC, Pheasant HC. 1936. Sacrarthrogenic telalgia I: a study of referred pain. *J Bone Joint Surg* 18:111-133.

Pool-Goudzwaard A, Vleeming A, Stoeckart R, Snijders CJ, Mens JM. 1998. Insufficient lumbopelvic stability: a clinical, anatomical and biomechanical approach to "a-specific" low back pain. *Man Ther* 3(1):12-20.

Pope MH, Panjabi M. 1985. Biomechanical definitions of spinal instability. *Spine* 10(3):255-256.

Rhalmi W, Yahia H, Newman N, et al. 1993. Immunohistochemical study of nerve in lumbar spine ligaments. *Spine* 18:264-267.

Richardson CA, Jull GA, Hodges P, et al. 1999. *Therapeutic Exercise for Spinal Segmental Stabilization in Low Back Pain.* London: Churchill Livingstone.

Roberts S, Menage J, Eisenstein SM. 1993. The cartilage end-plate and intervertebral disc in scoliosis: calcification and other sequelae. *J Orthop Res* 11:747-757.

Sashin D. 1930. A critical analysis of the anatomy and pathologic changes of the sacroiliac joints. *J Bone Joint Surg* 12:891-910.

Sato K, Wakamatsu E, Yoshizumi A, et al. 1989. The configuration of the laminas and facet joints in degenerative spondylolisthesis: a clinicoradiologic study. *Spine* 11:1265-1271.

Schonstrom N, Lindahl S, Willen J, et al. 1989. Dynamic changes in the dimensions of the lumbar spinal canal: an experimental study in vitro. *J Orthop Res* 7:115-121.

Schunke GB. 1938. The anatomy and development of the sacro-iliac joint in man. *Anatom Record* 72:313.

Sharma M, Langrana NA, Rodriquez J. 1995. Role of ligaments and facets in lumbar spinal stability. *Spine* 20:887-900.

Smidt GL, McQuade K, Weis S-H, et al. 1995. Sacroiliac kinematics for reciprocal straddle positions. *Spine* 20:1047-1054.

Snijders CJ, Vleeming A, Stoeckart R, et al. 1997. Biomechanics of the interface between spine and pelvis in different postures. In: Vleeming A, Mooney V, Dorman T, et al., eds. *Movement, Stability and Low Back Pain.* Edinburgh: Churchill Livingstone.

Solomonow M, Bing-He Z, Harris M, et al. 1998. The ligamento-muscular stabilizing system of the spine. *Spine* 23:2552-2562.

Sturesson B, Selvik G, Uden A. 1989. Movements of the sacroiliac joints: a roentgen stereophotogrammetric analysis. *Spine* 14(2):162-165.

Sturesson B, Uden A, Vleeming A. 2000. A radiostereometric analysis of the movements of the sacroiliac joints in the reciprical straddle position. *Spine* 25:214-217.

Taylor JR. 1990. The development and adult structure of lumbar intervertebral discs. *J Manual Med* 5:43-47.

Tencer A, Ahmed A, Burke D. 1982. Some static mechanical properties of the lumbar intervertebral joint, intact and injured. *J Biomech Eng* 104:193-201.

Tomkins CC, Battie MC, Hu R. 2007. Construct validity of the Physical Function Scale of the Swiss spinal stenosis questionnaire for the measurement of walking capacity. *Spine* 32:1896-1901.

Truchon M. 2001. Determinants of chronic disability related to low back pain: towards an integrative biopsychosocial model. *Disabil Rehabil* 23:758-767.

Tulsi RS, Hermanis GM. 1993. A study of the angle of inclination and facet curvature of superior lumbar zygapophyseal facets. *Spine* 18:1311-1317.

Twomey LT, Taylor JR. 1987. Age changes in lumbar vertebrae and intervertebral discs. *Clin Orthop* 97-104.

van Tulder MW, Assendelft WJ, Koes BW, et al. 1997. Spinal radiographic findings and nonspecific low back pain. *Spine* 22:427-434.

Van Wingerden JP, Vleeming A, Snijders CJ, et al. 1993. A functional-anatomical approach to the spine–pelvis mechanism: interaction between the biceps femoris muscle and the sacrotuberous ligament. *Eur Spine J* 2:140-142.

Vernon-Roberts B, Pirie CJ. 1973. Healing trabecular microfractures in the bodies of lumbar vertebrae. *Ann Rheum Dis* 32:406-412.

Vleeming A. 1996. The function of the long dorsal sacroiliac ligament: its implication for understanding low back pain. *Spine* 21:556.

Vleeming A, Mooney V, Dorman T, et al. 1997. *Movement, Stability and Low Back Pain.* Edinburgh: Churchill Livingstone.

Vleeming A, Pool-Goudzwaard AL, Stoeckart R, van Wingerden J, Snijders C. 1995. The posterior layer of the thoracolumbar fascia: its function in load transmission from spine to legs. *Spine* 20:753-758.

Vleeming A, Stoeckart R, Volkers ACW, Snijders CJ. 1990. A relation between form and function in the sacroiliac joint. Part I: clinical anatomical features. *Spine* 15:130-132.

Vleeming A, Van Wingerden JP, Snijders CJ, et al. 1989. Load application to the sacrotuberous ligament. *Clin Biomech* 4:204-209.

Walheim GG, Selvik G. 1984. Mobility of the pubic symphysis: in vivo measurements with an electromechanic method and a roentgen stereophotogrammetric method. *Clin Orthop* 191:129-135.

Weinstein JN, Lurie JD, Olson PR, Bronner KK, Fisher ES. 2006. United States' trends and regional variations in lumbar spine surgery: 1992-2003. *Spine* 31(23):2707-14.

Weisl H. 1954. The articular surfaces of the sacroiliac joint and their relation to the movements of the sacrum. *Acta Anatomica* 22:1-14.

Weisl H. 1955. The movements of the sacroiliac joint. *Acta Anatomica* 23:80-91.

Willard FH. 1997. The muscular, ligamentous and neural structure of the low back and its relation to low back pain. In: Vleeming A, Mooney V, Dorman T, et al., eds. *Movement, Stability, and Low Back Pain.* New York: Churchill Livingstone.

Woolf AD, Pfleger B. 2003. Burden of major musculoskeletal conditions. *Bull World Health Organ* 81(9):646-656.

Yahia LH, Garon S, Strykowski H, et al. 1990. Ultrastructure of the human interspinous ligament and ligamentum flavum: a preliminary study. *Spine* 15:262-268.

Yahia LH, Newman N, Richards C, et al. 1988. Neurohistology of lumbar spine ligaments. *Acta Orthop Scand* 59:508-512.

Zheng N, Watson LG, Yong-Hing K. 1997. Biomechanical modeling of the human sacroiliac joint. *Med Biol Eng Comput* 35(2):77-82.

Chapter 10

Abboud JA, Soslowsky LI. 2002. Interplay of the static and dynamic restraints in glenohumeral instability. *Clin Orthop Rel Res* 400:48-57.

American Academy of Orthopaedic Surgeons. 1965. *Joint Motion: Method of Measuring and Recording.* Chicago: AAOS.

An KN. 1991. Three-dimensional kinematics of glenohumeral elevation. *J Orthop Res* 9:143-149.

Atwater AE. 1979. Biomechanics of overarm throwing movements and of throwing injuries. *Exerc Sports Sci Rev* 7:43-85.

Bagg SD, Forrest WJ. 1988. A biomechanical analysis of scapular rotation during arm abduction in the scapular plane. *Am J Phys Med Rehab* 67:238-245.

Bahk M, Keyurapan E, Tasaki A, Sauers EL, McFarland EG. 2007. Laxity testing of the shoulder: a review. *Am J Sports Med* 35(1):131-144.

Basmajian JV, Bazant FJ. 1959. Factors preventing downward dislocation of the adducted shoulder joint. *J Bone Joint Surg* 4A:1182-1186.

Bearn JG. 1967. Direct observation on the function of the capsule of the sternoclavicular joint in clavicular support. *J Anat* 101:159-170.

Bigliani LU, Morrison DS, April EW. 1986. The morphology of the acromion and rotator cuff impingement (abstract). *Orthop Trans* 10:228.

Blakely RL, Palmer ML. 1984. Analysis of rotation accompanying shoulder flexion. *Phys Ther* 64:1214-1216.

Blotter RH, Bruckner JD. 1995. Gouty arthropathy of the acromioclavicular joint. *Am J Orthop* 24(11):859-860.

Boileau P, Walch G. 1997. The three-dimensional geometry of the proximal humerus: implications for surgical technique and prosthetic design. *J Bone Joint Surg Br* 79:857-865.

Bost F, Inman V. 1942. The pathological changes in recurrent dislocation of the shoulder: a report of Bankart's operative procedures. *J Bone Joint Surg* 24:595-613.

Bonsell S, Pearsell AW, Heitman RJ, Helms CA, Major NM, Speer KP. 2000. The relationship of age, gender, and degenerative changes observed in radiographs of the shoulder in asymptomatic individuals. *J Bone Joint Surg Br* 82:1135-1139.

Bosworth BM. 1949. Complete acromioclavicular dislocation. *N Engl J Med* 241:221-225.

Brossmann J, Stabler A, Preidler KW, Trudell D, Resnick D. 1996. Sternoclavicular joint: MR imaging—anatomic correlation. *Radiology* 198:193-198.

Burkhart AC, Debski RE. 2002. Anatomy and function of the glenohumeral ligaments in anterior shoulder instability. *Clin Orthop Relat Res* 400:32-39.

Burkhart SS, Morgan CD, Kibler WB. 2003. The disabled throwing shoulder: spectrum of pathology. Part III: the SICK scapula, scapular dyskinesis, the kinetic chain, and rehabilitation. *Arthroscopy* 19:641-661.

Cave AJ, Brown RW. 1952. On the tendon of the subclavius muscle. *J Bone Joint Surg* 34:466-469.

Cave EF. 1958. Shoulder girdle injuries. In: Cave EF, ed. *Fractures and Other Injuries.* Chicago: Year Book Publishers, 258-259.

Codman EA. 1934. *The Shoulder.* Boston: Thomas Todd.

Conway A. 1961. Movements of the sternoclavicular and acromioclavicular joints. *Phys Ther Rev* 41:421-432.

Cooper DE, Arnoczky SP, O'Brien SJ, Warren RF, DiCarlo E, Allen AA. 1992. Anatomy, histology, and vascularity of the glenoid labrum: an anatomical study. *J Bone Joint Surg Am* 74:46-52.

Cooper KM, Hawyard C, Williams ED. 1993. Calcium pryophospate deposition disease/involvement of the acromioclavicular joint with pseudocyst formation. *Br J Rheum* 32:248-250.

Costic RS, Vangura A Jr., Fenwick JA, Rodosky MW, Debski RE. 2003. Viscoelastic behavior and structural properties of the coracoclavicular ligaments. *Scan J Med Sci Sports* 13:305-310.

Curl LA, Warren RF. 1965. Glenohumeral joint stability: selective cutting studies on the static capsular restraint. *Clin Orthop* 330:54-65.

Dawson PA, Adamson GJ, Pink MM, et al. 2009. Relative contribution of acromioclavicular joint capsule and coracoclavicular ligaments to acromioclavicular stability. *J Shoulder Elb Surg* 18:237-244.

DeLuca C, Forrest W. 1973. Force analysis of individual muscles acting simultaneously on the shoulder joint during isometric abduction. *J Biomech* 6:385-393.

Dempster W. 1965. Mechanics of shoulder movement. *Arch Phys Med Rehab* 45:49-70.

DePalma AF. 1957. *Degenerative Changes in the Sternoclavicular and Acromioclavicular Joints in Various Decades.* Springfield, IL: Thomas.

DePalma AF. 1959. The role of the disks of the sternoclavicular and acromioclavicular joints. *Clin Orthop Rel Res* 13:222-233.

DePalma AF, Callery G, Bennett GA. 1949. Variational anatomy and degenerative lesions of the shoulder joint. *Inst Course Lect* 6:255-281.

Doos SP, Ray GS, Saha AK. 1966. Observation of the tilt of the glenoid cavity of the scapula. *J Anat Soc India* 15:114.

Edelson JG. 1996. Patterns of degenerative change in the acromioclavicular joint. *J Bone Joint Surg* 78B:242-243.

Ferguson SJ, Bryant JT, Ito K. 2001. The material properties of the bovine acetabular labrum. *J Orthop Res* 19:887-896.

Freedman L, Munro R. 1966. Abduction of the arm in the scapular plane: scapular and glenohumeral movements—a roentgenographic study. *J Bone Joint Surg* 48A:1503-1510.

Fung M, Kato S, Barrance PJ, et al. 2001. Scapular and clavicular kinematics during humeral elevation: a study with cadavers. *J Shoulder Elb Surg* 10:278-285.

Gangahar DM, Flogaites T. 1978. Retrosternal dislocation of the clavicle producing thoracic outlet syndrome. *J Trauma* 18:369-372.

Gardner MA, Didstrup BP. 1983. Intrathoracic great vessel injury resulting from blunt chest trauma associated with posterior dislocation of the sternoclavicular joint. *Aust N Z J Surg* 53:427-430.

Graichen H, Stammberger T, Bonel H, Karl-Hans E, Reiser M, Eckstein F. 2000. Glenohumeral translation during active and passive elevation of the shoulder: a 3D open-MRI study. *J Biomech* 33:609-613.

Habermeyer P, Schuller U, Wiedemann E. 1992. The intra-articular pressure of the shoulder: an experimental study on the role of the glenoid labrum in stabilizing the joint. *Arthroscopy* 8:166-172.

Happee R, van der Helm FTC. 1995. The control of shoulder muscles during goal directed movements: an inverse dynamic analysis. *J Biomech* 17:1179-1191.

Harryman DT, Sidles JA, Clark JM, et al. 1990. Translations of the humeral head on the glenoid with passive glenohumeral motion. *J Bone Joint Surg* 72A:1334-1343.

Hertz H, Weinstable R, Grundschober F, Orthner E. 1986. Macroscopic and microscopic anatomy of the shoulder joint and the limbus glenoidalis. *Acta Anat* 125:96-100.

Hollinshead WH. 1982. *Anatomy for Surgeons.* 3rd ed. Philadelphia: Harper & Row.

Howell SM, Galinat BJ. 1989. The glenoid labral socket: a constrained articular surface. *Clin Orthop* 243:122-125.

Howell SM, Imobersteg AM, Seger DH, Marone PJ. 1986. Clarification of the role of the supraspinatus muscle in shoulder function. *J Bone Joint Surg* 68A:398-404.

Huber WB, Putz RV. 1997. Periarticular fiber system of the shoulder joint. *Arthroscopy* 13:680-691.

Iannotti JP, Gabriel JP, Scheck SL, Evans BG, Misra S. 1992. The normal glenohumeral relationships: an anatomical study of one hundred and forty shoulders. *J Bone Joint Surg* 74A:491-500.

Inman B, Saunders J, Abbott L. 1944. Observations of function of the shoulder joint. *J Bone Joint Surg Br* 26:1-32.

Ito N. 1980. Electromyographic study of shoulder joint. *J Jpn Orthop Assoc* 54:53-60.

Johnston T. 1937. The movements of the shoulder joint: a plea for the use of the "plane of the scapula" as the plane of reference in movements occurring in the humero-scapular joint. *Br J Surg* 25:252-260.

Jougon JB, Lepront DJ, Dromer CE. 1996. Posterior dislocation of the sternoclavicular joint leading to mediastinal compression. *Ann Thorac Surg* 61:711-713.

Kapandji IA. 1982. *The Physiology of the Joints. Vol. 1. The Upper Limb.* Edinburgh: Churchill Livingstone.

Kelley MJ. 1995. Biomechanics of the shoulder. In: Kelley MH, Clark WA, eds. *Orthopedic Therapy of the Shoulder.* Philadelphia: Lippincott.

Kelley MH, Clark WA. 1995. *Orthopedic Therapy of the Shoulder.* Philadelphia: Lippincott.

Kronberg M, Nemeth G, Brostom L. 1990. Muscle activity and coordination in the normal shoulder. *Clin Orthop Rel Res* 257:76-85.

Kumar VP, Balasubramaniam P. 1985. The role of atmospheric pressure in stabilizing the shoulder: an experimental study. *J Bone Joint Surg Br* 67:719-721.

Labriola JE, Lee TQ, McMahon PJ. 2005. Stability and instability of the glenohumeral joint: role of shoulder muscles. *J Shoulder Elb Surg* 14(Suppl):32-38.

Lahtinen JT, Leato MU, Kaarela K, Kautiainen HJ, Belt EA, Kauppi MJ. 1999. Radiographic joint space in rheumatoid acromioclavicular joint: a 15 year perspective follow up study in 74 patients. *Rheumatology* 38(11):1104-1107.

Landin D, Myers J, Thompson M, Castle R, Porter J. 2008. The role of the biceps brachii in shoulder elevation. *J Electromyogr Kinesiol* 18(2):270-275.

Levine WN, Flatow EL. 2000. The pathophysiology of shoulder instability. *Am J Sports Med* 28:910-917.

Levy AS, Kelly BT, Lintner SA, Osbahr DC, Speer KP. 2001. Function of the long head of the biceps at the shoulder: electromyographic analysis. *J Shoulder Elb Surg* 10(3):250-255.

Lucas D. 1973. Biomechanics of the shoulder joint. *Arch Surg* 107:425-432.

Ludewig PM, Behrens SA, Meyer SM, Spoden SM, Wilson LA. 2004. Three-dimensional clavicular motion during arm elevation: reliability and descriptive data. *J Orthop Sports Phys Ther* 34:140-149.

Ludewig PM, Cook T. 2000. Alterations in shoulder kinematics and associated muscle activity in people with symptoms of shoulder impingement. *Phys Ther* 80:276-291.

Ludewig PM, Phadke V, Braman JP, et al. 2009. Motion of the shoulder complex during multiplanar humeral elevation. *J Bone Joint Surg Am*. 91:378-389.

Manske RC, Sumler A, Runge J. 2009. Quadrilateral space syndrome. *Athl Ther Today* 14:45-47.

McClure PW, Michener LA, Sennett BJ, Karduna AR. 2001. Direct 3-dimensional measurement of scapular kinematics during dynamic movements in vivo. *J Shoulder Elb Surg* 10:269-277.

McCluskey GM III, Todd J. 1995. Acromioclavicular joint injuries. *J South Orthop Assoc* 4:206-203.

McGregor L. 1937. Rotation of the shoulder: a critical injury. *Br J Surg* 25:425-438.

McQuade KJ, Dawson J, Smidt GL. 1998. Scapulothoracic muscle fatigue associated with alterations in scapulohumeral rhythm kinematics during maximum resistive shoulder elevation. *J Orthop Sports Phys Ther* 28:74-80.

Morrison DS, Frogameni AD, Woodworth P. 1997. Non-operative treatment of subacromial impingement syndrome. *J Bone Joint Surg Am* 79(5):732-737.

Moseley HF. 1958. The clavicle: its anatomy and function. *Clin Orthop* 58:17-27.

Moseley HF, Overgaard B. 1962. The anterior capsular mechanism in recurrent anterior dislocation of the shoulder: morphological and clinical studies with special reference to the glenoid labrum and gleno-humeral ligaments. *J Bone Joint Surg Br* 443:913-927.

Neer CS II. 1972. Anterior acromioplasty for the chronic impingement syndrome in the shoulder: a preliminary report. *J Bone Joint Surg* 54A:41-50.

Neumann DA. 2002. *Kinesiology of the Musculoskeletal System: Foundations for Physical Rehabilitation*. St. Louis: Mosby.

Nishida K, Hashizume H, Toda K, Inoue H. 1996. Histologic and scanning electron microscopic study of the glenoid labrum. *J Shoulder Elb Surg* 5:132-138.

Noda M, Shiraishi H, Misuno K. 1997. Chronic posterior sternoclavicular dislocation causing compression of the subclavian artery. *J Shoulder Elb Surg* 6:564-569.

O'Brien SJ, Neves MC, Arnvoczky SP, et al. 1990. The anatomy and histology of the inferior glenohumeral ligament complex of the shoulder. *Am J Sports Med* 18:449-456.

O'Brien SJ, Schwartz RS, Warren RF, Torzilli PA. 1995. Capsular restraints to anterior-posterior motion of the abducted shoulder: a biomechanical study. *J Shoulder Elb Surg* 4:298-308.

Ono K, Inagawa H, Kiyota K, Teradda T, Suzuki S, Maekawa K. 1998. Posterior dislocation of the sternoclavicular joint with obstruction of the innominate vein: a case report. *J Trauma* 44:381-383.

Oppenheimer A. 1943. Arthritis of the acromioclavicular joint. *J Bone Joint Surg* 25A:867-870.

Pagnani MJ, Deng XH, Warren RF, Tozilli PA, O'Brien SJ. 1996. Role of the long head of the biceps brachii in glenohumeral stability: a biomechanical study in cadaver. *J Shoulder Elb Surg* 5(4):255-262.

Palmer ML, Blakely RL. 1986. Documentation of medial rotation accompanying shoulder flexion. *Phys Ther* 66:55-58

Perry J. 1978. Normal upper extremity kinesiology. *Phys Ther* 58:265-278.

Perry J. 1988. Biomechanics of the shoulder. In: Rowe C, ed. *The Shoulder*. New York: Churchill Livingstone.

Pitchford KR, Cahill BR. 1997. Osteolysis of the distal clavicle in overhead athletes. *Oper Tech Sports Med* 5(2):72-77.

Poppen NK, Walker PS. 1976. Normal and abnormal motion of the shoulder. *J Bone Joint Surg* 58A:195-201.

Poppen NK, Walker PS. 1978. Forces at the glenohumeral joint in abduction. *Clin Orthop* 58:165-170.

Pratt NE. 1994. Anatomy and biomechanics of the shoulder. *J Hand Ther* 7:65-76.

Prodromos CC, Ferry JA, Schiller AL, Zarins B. 1990. Histological studies of the glenoid labrum from fetal life to old age. *J Bone Joint Surg* 72A:1344-1348.

Rayan GM. 1994. Compression brachial plexopathy caused by chronic posterior dislocation of the sternoclavicular joint. *J Okla State Med Assoc* 87:7-9.

Reinold MM, Macrina LC, Wilk KE, et al. 2007. Electromyographic analysis of the supraspinatus and deltoid muscles during 3 common rehabilitation exercises. *J Athl Train* 42(4):464-469.

Rievtveld AB, Daanen HA, Rozing PM, et al. 1988. The lever arm in glenohumeral abduction after hemiarthroplasty. *J Bone Joint Surg Br* 70:561-565.

Rockwood CA Jr., Wirth MA. 1998. Disorders of the sternoclavicular joint. In: Rockwood CA Jr., Matsen FA, eds. *The Shoulder*. 2nd ed, vol. 1. Philadelphia: Saunders, 555-609.

Rodriques D, Pessan MA, Kawano MM, Stabile GR, Cardoso JR. 2008. Electromyographic analysis of deltoid muscle fatigue during abduction on scapular and frontal planes. *Electromyogr Clin Neurophysiol* 48(6-7):293-300.

Romanes GJE. 1981. *Cunningham's Textbook of Anatomy*. Oxford: Oxford University Press.

Rundquist P, Anderson DD, Guanche CA, Ludewig PM. 2003. Shoulder kinematics in subjects with frozen shoulder. *Arch Phys Med Rehab* 84:1473-1479.

Saha AK. 1961. *Theory of Shoulder Mechanism: Descriptive and Applied*. Springfield, IL: Thomas.

Saha AK. 1971. Dynamic stability of the glenohumeral joint. *Acta Orthop Scand* 42:491-505.

Saha AK. 1973. Mechanics of elevation of glenohumeral joint: its application in rehabilitation of flail shoulder in upper brachial plexus injuries and poliomyelitis and in replacement of the upper humerus by prosthesis. *Acta Orthop Scand* 44:668-678.

Saha AK. 1983. Mechanisms of shoulder movements and a plea for the recognition of "zero position" of the glenohumeral joint. *Clin Orthop* 173:3-10.

Sahara W, Sugamoto K, Murai M, Yashikawa H. 2007. Three-dimensional clavicular and acromioclavicular rotations during arm abduction using vertically open MRI. *J Orthop Res*. 25:1243-1249.

Sakurai G, Ozaki J, Tomita Y, Nishimoto K, Tamai S. 1998. Electromyographic analysis of shoulder joint function of the biceps brachii muscle during isometric contraction. *Clin Orthop Rel Res* 354:123-131.

Schwartz E, Warren RF, O'Brien SJ, Fronek J. 1987. Posterior shoulder instability. *Orthop Clin N Am* 18:409-419.

Soslowsky LJ, Flatow EL, Bigliani L, Pawluk RJ, Ateshian GA, Mow VC. 1992. Quantification of an in situ contact area at the glenohumeral joint: a biomechanical study. *J Orthop Res* 10:524-534.

Spencer EE, Kuhn JE, Carpenter JE, Hughes RE. 2002. Ligamentous restraints to anterior and posterior translation of the sternoclavicular joint. *J Shoulder Elb Surg* 11:43-47.

Steindler A. 1955. *Kinesiology of the Human Body Under Normal and Pathological Conditions*. Springfield, IL: Thomas.

Stokdijk M, Eileers PHC, Nagels J, Rozing PM. 2003. External rotation in the glenohumeral joint during elevation of the arm. *Clin Biomech* 18:296-302.

Tham A, Purchase R, Kelly JD IV. 2009. The relation of the coracoids process to the glenoid: an anatomic study. *Arthroscopy* 25(8):846-848.

Teece RM, Lunden JB, Lloyd AS, Kaiser AP, Cieminski CJ, Ludewig PM. 2008. Three-dimensional acromioclavicular joint motions during elevation of the arm. *J Orthop Sports Phys Ther* 38:181-190

Terry GC, Hammon D, France P, Norwood LA. 1991. The stabilizing function of passive shoulder restraints. *Am J Sports Med* 19:26-34.

Ticker JB, Bigliani LU, Soslowsky LJ, Pawluk RJ, Flatow EL, Mow VC. 1996. Inferior glenohumeral ligament: geometric and strain-rate dependent properties. *J Shoulder Elb Surg* 5:269-279

Turkel SJ, Panio MW, Marshall JL, et al. 1981. Stabilizing mechanisms preventing anterior dislocation of the glenohumeral joint. *J Bone Joint Surg* 63A:1208-1217.

van der Helm FCT, Pronk G. 1995. Three-dimensional recording and description of motions of the shoulder mechanism. *J Biomech Eng* 117:27-40.

van der Helm FCT, Veeger HEJ, Pronk GM. 1992. Geometry parameters for musculoskeletal modeling of the shoulder mechanism. *J Biomech* 25:129-144.

Warner JJP, Deng X, Warren RF, Torzilli PA. 1992. Static capsuloligamentous restraints to superior-inferior translation of the glenohumeral joint. *Am J Sports Med* 20:675-685.

Warner JJP, Paletta G, Warren RF. 1991. Biplanar roentgenographic evaluation of glenohumeral instability and rotator cuff tears. Presented at the Annual Meeting of the American Academy of Orthopaedic Surgeons, Anaheim, CA.

Warrwick R, Williams P, eds. 1973. *Gray's Anatomy*. 35th ed. London: Longman.

Williams P, Bannister L, Berry M, et al. 1995. *Gray's Anatomy: The Anatomical Basis of Medicine and Surgery*. Br. ed. London: Churchill Livingstone.

Wuelker N, Korell M, Thren K. 1998. Dynamic glenohumeral joint stability. *J Shoulder Elb Surg* 7:43-52.

Yamaguchi K, Riew KD, Galatz SM, Syme JA, Neviaser RJ. 1997. Biceps activity during shoulder motion: an electromyographic analysis. *Clin Orthop Relat Res* 336:122-129.

Yoshizaki K, Hamada J, Tamai K, Sahara R, Fujiwara T, Fujimoto T. 2009. Analysis of the scapulohumeral rhythm and electromyography of the shoulder muscles during elevation and lowering: comparison of dominant and nondominant shoulders. *J Shoulder Elb Surg* 18:756-763.

Chapter 11

American Academy of Orthopaedic Surgeons. 1965. *Joint Motion: Methods of Measuring and Recording*. Chicago: AAOS.

An KN, Hui FC, Morrey BF, et al. 1981. Muscles across the elbow joint: a biomechanical analysis. *J Biomech* 14:659-69.

Conway JE, Jobe FW, Glousman RE, Pink M. 1992. Medial instability of the elbow in throwing athletes. *J Bone Joint Surg Am* 74(1):67-84.

Ellenbecker TS, Mattalino AJ. 1997. *The Elbow in Sport*. Champaign, IL: Human Kinetics.

Indelicato PA, Jobe FW, Kerlan RK, Carter VS, Shields CL, Lombardo SJ. 1979. Correctable elbow lesions in professional baseball players: a review of 25 cases. *Am J Sports Med* 7:72-75.

Ishizuki M. 1979. Functional anatomy of the elbow joint and three dimensional quantitative motion analysis of the elbow joint. *J Jpn Orthop Assn* 53:989-996.

Joyce ME, Jelsma RD, Andrews JR. 1995. Throwing injuries to the elbow. *Sports Med Arthrosc* 3:224-236.

London JT. 1981. Kinematics of the elbow. *J Bone Joint Surg Am* 63:529-535.

Loudon JK. 2000. Lateral epicondylitis. *Phys Ther Case Rep* 3:163-170.

Morrey BF. 1993. *The Elbow and Its Disorders*. 2nd ed. Philadelphia: Saunders.

Morrey BF, An KN. 1983. Articular and ligamentous contributions to stability of the elbow joint. *Am J Sports Med* 11:315-319.

O'Driscoll SW, Morrey BF, An KN. 1990. The pathoanatomy and kinematics of the posterolateral instability of the elbow. *Orthop Trans* 14:306.

Palmer AK, Glisson RR, Werner FW. 1982. Ulnar variance determination. *J Hand Surg* 7:376.

Chapter 12

American Academy of Orthopaedic Surgeons. 1965. *Joint Motion: Method of Measuring and Recording*. Chicago: AAOS.

Berger RA. 2001. The anatomy of the scaphoid. *Hand Clin* 17:525-532.

Dutton, M. 2008. *Orthopaedic Examination, Evaluation and Intervention*. 2nd ed. New York: McGraw-Hill.

Ginanneschi F, Filippou G, Milani P, Biasella A, Rossi A. 2009. Ulnar nerve compression neuropathy at Guyon's canal caused by crutch walking: case report with ultrasonographic nerve imaging archives of physical medicine and rehabilitation. *Arch Phys Med Rehab* 90(3):522-524.

Hoppenfeld S. 1976. *Physical Examination of the Spine and Extremities*. New York: Appleton-Century-Crofts.

Kapandji IA. 1970. *The Physiology of the Joints*. Edinburgh: Livingstone.

Levangie PK, Norkin CC. 2011. *Joint Structure and Function: A Comprehensive Analysis*. 5th ed. Philadelphia: Davis.

Loudon J, Swift M, Bell S. 2008. *The Clinical Orthopedic Assessment Guide*. 2nd ed. Champaign, IL: Human Kinetics.

Magee, DJ. 1997. *Orthopedic Physical Assessment*. 3rd ed. Philadelphia: Saunders.

Neumann DA. 2010. Ankle and foot. In: Neumann DA, ed. *Kinesiology of the Musculoskeletal System*. 2nd ed. St. Louis: Mosby Elsevier.

Nordin M, Frankel VH. 2001. *Basic Biomechanics of the Musculoskeletal System*. 3rd ed. Baltimore: Lippincott Williams & Wilkins.

Norkin C, White D. 2009. *Measurement of Joint Motion: A Guide to Goniometry*. 4th ed. Philadelphia: Davis.

Unglaub F, Thomas SB, Wolf MB, et al. 2010. Cartilage cell proliferation in degenerative TFCC wrist lesions. *Arch Orthop Trauma Surg* 130(8):953-956. Epub 2009 May 5.

Windischi G, Grechenig W, Peicha G, Tesch NP, Seibert FJ. 2001. Capsular attachment to the distal radius for extracapsular placement of pins. *Surg Radiol Anat* 23(5):313-316.

Chapter 13

Beall DP, Sweet CF, Martin HD, et al. 2005. Imaging findings of femoroacetabular impingement syndrome. *Skeletal Radiol* 34:691-701.

Beck M, Kalhor M, Leunig M, Ganz R. 2005. Hip morphology influences the pattern of damage to the acetabular cartilage: femoroacetabular impingement as a cause of early osteoarthritis of the hip. *J Bone Joint Surg Br* 87:1012-1019.

Bergmann G, Graichen R, Rohlmann A. 1993. Hip joint loading during walking and running measured in two patients. *J Biomech* 26:969-90.

Clark JM, Haynor DR. 1987. Anatomy of the abductor muscles of the hip as studied by computed tomography. *J Bone Joint Surg Am* 69:1021-1031.

Cyriax J. 1982. *Textbook of Orthopaedic Medicine. Volume 1: Diagnosis of Soft Tissue Lesions*. 8th ed. Boston: Harcourt.

Delp SL, Hess WE, Hungerford DS, Hones LC. 1999. Variation of rotation moment arms with hip flexion. *J Biomech* 32:493-501.

DiStefano LJ, Blackburn JT, Marshall SW, Padua DA. 2009. Gluteal muscle activation during common therapeutic exercises. *J Orthop Sports Phys Ther* 39:532-540.

Ferber R, Noehren B, Hamill J, Davis I. 2010. Competitive female runners with a history of iliotibial band syndrome demonstrate atypical hip and knee kinematics. *J Orthop Sports Phys Ther* 40:52-58.

Gottschalk F, Kourosh S, Leveau B. 1989. The functional anatomy of tensor fasciae latae and gluteus medius and minimus. *J Anatomy* 166:179-189.

Ito K, Minka MA II, Leunig M, Werlen S, Ganz R. 2001. Femoroacetabular impingement and the cam-effect: a MRI-based quantitative anatomical study of the femoral head-neck offset. *J Bone Joint Surg Br* 83:171-176.

Johnston RC, Smidt GL. 1970. Hip motion measurements for selected activities of daily living. *Clin Orthop* 72:205.

Kendall FP, McCreary EK, Provance PG. 1993. *Muscles testing and function*. 4th ed. Baltimore: Williams & Wilkins.

Kim YT, Azuma H. 1995. The nerve endings of the acetabular labrum. *Clin Orthop* 320:176-181.

Lyons K, Perry J, Gronlcy JK, Barnes L, Antonelli D. 1983. Timing and relative intensity of hip extensor and abductor muscle action during level and stair ambulation. *Phys Ther* 63:1597-1605.

Neumann DA. 2010. Kinesiology of the hip: a focus on muscular actions. *J Orthop Sports Phys Ther* 40:82-94.

Nordin M, Frankel VH. 2001. *Basic Biomechanics of the Musculoskeletal System*. 3rd ed. Baltimore: Lippincott Williams & Wilkins.

Philippon MJ. 2001. The role of arthroscopic thermal capsulorrhaphy in the hip. *Clin Sports Med* 20:817-829.

Powers CM. 2003. The influence of altered lower-extremity kinematics on patellofemoral joint dysfunction: a theoretical perspective. *J Orthop Sports Phys Ther* 33:639-646.

Siffert RS, Levy RN. 1981. Trabecular patterns and the internal architecture of bone. *Mt Sinai J Med* 48:221-229.

Ward SR, Winters TM, Blemker SS. 2010. The architectural design of the gluteus muscle group: implications for movement and rehabilitation. *J Orthop Sports Phys Ther* 40:95-102.

Winter D. 2005. *Biomechanics and Motor Control of Human Movement*. Hoboken, NJ: Wiley.

Chapter 14

Aglietti P, Insall JN, Ceruilli G. 1983. Patellar pain and incongruence I. Measurements of incongruence. *Clin Orthop Rel Res* 176:217-224.

Almquist PO, Arnbjornsson A, Zatterstrom R, Ryd L, Ekdahl C, Friden T. 2002. Evaluation of an external device measuring knee joint rotation: an in vivo study with simultaneous roentgen stereometric analysis. *J Orthop Res* 20:427-432.

American Academy of Orthopaedic Surgeons. 1965. *Joint Motion: Method of Measuring and Recording*. Chicago: AAOS.

Amis AA, Bull AM, Gupte CM, Hijazi I, Race A, Robinson JR. 2003. Biomechanics of the PCL and related structures: posterolateral, posteromedial and meniscofemoral ligaments. *Knee Surg Sports Traumatol Arthosc* 11:271-281.

Amis AA, Dawkins GP. 1991. Functional anatomy of the anterior cruciate ligament: fiber bundle actions related to ligament replacements and injuries. *J Bone Joint Surg* 73B(2):260-267.

Ateshian GA, Soslowsky LJ, Mow VC. 1991. Quantification of articular surface topography and cartilage thickness in knee joints using stereophotogrammetry. *J Biomech* 24:761-776.

Bach JM, Hull ML, Patterson HA. 1997. Direct measurement of strain in the posterolateral bundle of the anterior cruciate ligament. *J Biomech* 30:281-283.

Besier TF, Draper CE, Gold GE, Beaupre GS, Delp SL. 2005. Patellofemoral joint contact area increases with knee flexion and weight-bearing. *J Orthop Res* 23;345-350.

Beynnon BD, Fleming BC, Johnson RJ, et al. 1995. Anterior cruciate ligament strain behavior during rehabilitation exercises in vivo. *Am J Sports Med* 23.24-34.

Beynnon BD, Yu J, Huston D, et al. 1996. A sagittal plane model of the knee and cruciate ligaments with application of a sensitivity analysis. *J Biomech Eng* 118:227-239.

Boden BP, Pearsall AW, Garrett WE Jr., Feagin JA Jr. 1997. Patellofemoral instability: evaluation and management. *J Am Acad Orthop Surg* 5:47-57.

Brantigan OC, Voshell AF. 1941. The mechanics of the ligaments and menisci of the knee joint. *J Bone Joint Surg* 23:44-66.

Brattstrom H. 1964. Shape of the intercondylar groove normally and in recurrent dislocation of the patella. *Acta Orthop Scand* 68(Suppl):S1-S44.

Butler DL, Noyes FR, Grood ES. 1980. Ligamentous restraint to anterior-posterior drawer in the human knee. *J Bone Joint Surg* 62A:259-270.

Caylor D, Fites R, Worrell TW. 1993. The relationship between quadriceps angle and anterior knee pain syndrome. *J Orthop Sports Phys Ther* 17:11-16.

Chatain F, Adeleine P, Chambat P, Neyret P. 2003. A comparative study of medial versus lateral arthroscopic partial meniscectomy on stable knees: 10-year minimum follow-up. *Arthroscopy* 19(8):842-849.

Covey DC, Sapea AA, Riffenburgh RH. 2008. The effects of sequential sectioning of defined posterior cruciate ligament fiber regions on translational knee motion. *Am J Sports Med* 36:480-486.

Cowan DN, Jones BH, Frykman PN, et al. 1996. Lower limb morphology and risk of overuse injury among military infantry trainees. *Med Sci Sports Exerc* 28:945-952.

DeFrate LE, Sun H, Gill TJ, Rubash HE, Guoan L. 2004. In vivo tibiofemoral contact analysis using 3D MRI-based knee models. *J Biomech* 37:1499-1504.

Dye SF, Vaupel GL, Dye CC. 1998. Conscious neurosensory mapping of the internal structures of the human knee without intra-articular anesthesia. *Am J Sports Med* 26:773-777.

Edwards A, Bull AM, Amis AA. 2007. The attachments of the anteromedial and posterolateral fibre bundles of the anterior cruciate ligament. Part 1: tibial attachment. *Knee Surg Sports Traumatol Arthrosc* 15:1414-1421.

Edwards A, Bull AM, Amis AA. 2008. The attachments of the anteromedial and posterolateral fibre bundles of the anterior cruciate ligament. Part 2: femoral attachment. *Knee Surg Sports Traumatol Arthrosc* 16:29-36.

Eichenblat M, Nathan H. 1983. The proximal tibio fibular joint: an anatomical study with clinical and pathological considerations. *Int Orthop* 7:31-39.

Fairbank JCT, Pynsent PB, Poortvliet JA, Phillips H. 1984. Mechanical factors in the incidence of knee pain in adolescents and young adults. *J Bone Joint Surg* 66B:685-693.

Feller JA, Amis AA, Andrish JT, Arendt EA, Erasmus PJ, Powers CM. 2007. Surgical biomechanics of the patellofemoral joint *Arthroscopy* 23:542-553.

Freeman MAR, Pinskerova V. 2005. The movement of the normal tibio-femoral joint. *J Biomech* 38:197-208.

Fukubayashi T, Torzilli PA, Sherman MF, Warren RF. 1982. An in vitro biomechanical evaluation of anterior-posterior motion of the knee: tibial displacement, rotation, and torque. *J Bone Joint Surg* 64:258-264.

Fulkerson JP. 1997. *Disorders of the Patellofemoral Joint.* 3rd ed. Baltimore: Williams & Wilkins.

Garrett WE Jr., Swiontkowski MF, Weinstein JN, et al. 2006. American Board of Orthopaedic Surgery practice of the orthopaedic surgeon. Part 2: certification examination case mix. *J Bone Joint Surg* 88A:660-667.

Gigris FG, Marshall JL, Monajem ARS. 1975. The cruciate ligaments of the knee joint: anatomical, functional, and experimental analysis. *Clin Orthop* 106:216-231.

Gold GE, Besier TF, Draper CE, Asakawa DS, Delp SL, Beaupre GS. 2004. Weight-bearing MRI of patellofemoral joint cartilage contact area. *J Magn Reson Imaging* 20:526-530.

Goodfellow J, Hungerford DS, Zindel M. 1976. Patella-femoral joint mechanics and pathology: 1. functional anatomy of the patella-femoral joint. *J Bone Joint Surg* 58B:287-290.

Grelsamer RP, Klein JR. 1998. The biomechanics of the patellofemoral joint. *J Orthop Sport Phys Ther* 28:286-298.

Grelsamer RP, Weinstein CH. 2001. Applied biomechanics of the patella. *Clin Orthop Rel Res* 389:9-14.

Grood ES, Noyes FR, Butler DL, et al. 1981. Ligamentous and capsular restraints preventing straight medial and lateral laxity in intact human cadaver knees. *J Bone Joint Surg* 63A:1257-1269.

Guerra JP, Arnold MJ, Gajdokik RL. 1994. Q-angle: effects of isometric quadriceps contraction and body position. *J Orthop Sports Phys Ther* 19:200-204.

Haim A, Yaniv M, Dekel S, et al. 2006. Patellofemoral pain syndrome: validity of clinical and radiological features. *Clin Orthop Relat Res* 451:223-228.

Heegaard J, Layvraz PF, Curnier A, Rakotomanana L, Huiskes R. 1995. The biomechanics of the human patella during passive knee flexion. *J Biomech* 28(11):1265-1279.

Hemmerich A, Brown H, Smith S, Marthandam SS, Wyss UP. 2006. Hip, knee, and ankle kinematics of high range of motion activities of daily living. *J Orthop Res* 24:770-781.

Hill PF, Vedi V, Williams A, Iwaki H, Pinskerova V, Freeman MA. 2000. Tibiofemoral movement 2: the loaded and unloaded living knee studied by MRI. *J Bone Joint Surg* 82B:1196-1198.

Horton MG, Hall TL. 1989. Quadriceps femoris muscle angle: normal values and relationships with gender and selected skeletal measures. *Phys Ther* 69:897-901.

Hsu RW, Himeno S, Coventry MB, et al. 1990. Normal axial alignment of the lower extremity and load bearing distribution at the knee. *Clin Orthop Relat Res* 255:215-227.

Huberti HH, Hayes WC. 1984. Patellofemoral contact pressures. The influence of Q-angle and tendofemoral contact. *J Bone Joint Surg* 66A:715-724.

Insall J, Falvo KA, Wise DW. 1976. Chondromalacia patellae: a prospective study. *J Bone Joint Surg* 58A:1-8.

Johnson DL, Swenson TM, Livesay GA, Aizawa H, Fu F, Harner CD. 1995. Insertion-site anatomy of the human menisci: gross arthroscopic and topographical anatomy as a basis for meniscal transplantation. *Arthroscopy* 11:386-394.

Jordan SS, DeFrate LE, Nha KW, Papannagari R, Gill TJ, Li G. 2007. The in vivo kinematics of the anteromedial and posterolateral bundles of the anterior cruciate ligament during weightbearing knee flexion. *Am J Sports Med* 35:547-554.

Kaltenborn FM. 2011. *Manual Mobilization of the Joints: The Kaltenborn Method of Joint Examination and Treatment. Vol. I: The Extremities.* 7th ed. Oslo, Norway: OPTP.

Kernozek TW, Greer NL. 1993. Quadriceps angle and rear-foot motion: relationships in walking. *Arch Phys Med Rehabil* 74:407-410.

Kettelkamp DB, Jacobs AW. 1972. Tibiofemoral contact area: determination and implications. *J Bone Joint Surg* 54A:349-356.

Komistek RD, Dennis DA, Mabe JA, Walker SA. 2000. An in vivo determination of patellofemoral contact positions. *Clin Biomech* 15:29-36.

Krevolin JL, Pandy MG, Pearce JC. 2004. Moment arm of the patella tendon in the human knee. *J Biomech* 37:785-788.

Levangie PK, Norkin CC. 2005. *Joint Structure and Function: A Comprehensive Analysis.* 4th ed. Philadelphia: Davis.

Liu F, Kozanek M, Hosseini A, et al. 2010. In vivo tibio-femoral cartilage deformation during the stance phase of gait. *J Biomech* 43(4):658-665.

Livingstone LA. 1998. The quadriceps angle: a review of the literature. *J Orthop Sports Phys Ther* 28:105-109.

Livingstone LA, Mandigo JL. 1997. Bilateral within subject Q-angle asymmetry in young adult females and males. *Biomec Sci Instrum* 33:112-117.

Lorbach O, Pape D, Maas S, et al. 2010. Influence of the anteromedial and posterolateral bundles of the anterior cruciate ligament on external and internal tibiofemoral rotation. *Am J Sports Med* 38(4):721-727.

Luyckx T, Didden K, Vandernneucker H, Labey L, Innocenti B, Bellemans J. 2009. Is there a biomechanical explanation for anterior knee pain in patients with patella alta? Influence of patellar height on patellofemoral contact force, contact area and contact pressure. *J Bone Joint Surg* 91B:344-350.

Magee DJ. 2008. *Orthopedic Physical Assessment.* 5th ed. St. Louis: Saunders.

Markolf KL, Feeley BT, Jackson SR, McAllister DR. 2006. Biomechanical studies of double-bundle posterior cruciate ligament reconstructions. *J Bone Joint Surg* 88A:1788-1794.

Markolf KL, Mensch JS, Amstutz HC. 1976. Stiffness and laxity of the knee: the contributions of the supporting structures—a quantitative in vitro study. *J Bone Joint Surg* 58A: 583-594.

McMahon PJ, Skinner HB. 2003. Sports medicine. In: Skinner HB, ed. *Current Diagnosis and Treatment in Orthopedics*. 3rd ed. New York: McGraw-Hill.

McMurray T. 1942. The semilunar cartilages. *Br J Surg* 29:407-414.

Mendes E, Vieira da Silva M. 2006. Anatomy of the lateral collateral ligament: a cadaver and histological study. *Knee Surg Sports Traumatol Arthroscopy* 14(3):221-228.

Messier SP, Davis SE, Curl WW, Lowery RB, Pack RJ. 1991. Etiologic factors associated with patellofemoral pain in runners. *Med Sci Sports Exerc* 23:1008-1015.

Messner K, Gao J. 1998. The menisci of the knee joint: anatomical and functional characteristics, and rationale for clinical treatment. *J Anat* 193:161-178.

Miyasaka KC, Daniel DM, Stone ML, Hirshman P. 1991. The incidence of knee ligament injuries in the general population. *Am J Knee Surg* 4(1):3-8.

Mizuno Y, Kumagai M, Mattessich SM, et al. 2001. Q-angle influences tibiofemoral and patellofemoral kinematics. *J Orthop Res* 19:834-840.

Mossber KA, Smith LK. 1983. Axial rotation of the knee in women. *J Orthop Sports Phys Ther* 4:236-240.

Nakagawa S, Kadoya Y, Kobayashi A, Tatsumi I, Nishida N, Yamano Y. 2003. Kinematics of the patella in deep flexion: analysis with magnetic resonance imaging. *J Bone Joint Surg* 85A:1238-1242.

Neumann DA. 2010. *Kinesiology of the Musculoskeletal System: Foundations for Rehabilitation*. 2nd ed. St. Louis: Mosby.

Nguyen AD, Shultz SJ. 2007. Sex differences in clinical measures of lower extremity alignment. *J Orthop Sports Phys Ther* 37:389-398.

Ogden JA. 1974. The anatomy and function of the proximal tibiofibular joint. *Clin Orthop* 101:186-191.

Patel VV, Hall K, Ries M, et al. 2004. A three-dimensional MRI analysis of knee kinematics. *J Orthop Res* 22:283-292.

Petersen W, Tillmann B. 2002. Anatomy and function of the anterior cruciate ligament. *Orthopade* 31:710-718.

Powers CM. 2003. The influence of altered lower-extremity kinematics on patellofemoral joint dysfunction: a theoretical perspective. *J Orthop Sports Phys Ther* 33:639-646.

Radakovich M, Malone T. 1982. The superior tibiofibular joint: the forgotten joint. *J Orthop Sports Phys Ther* 3(3):129-132.

Recondo JA, Salvador E, Villanua JA, Barrera MC, Gervas C, Alustiza JM. 2000. Lateral stabilizing structure of the knee: functional anatomy and injuries assessed with MR imaging. *RadioGraphics* 20:S91-S102.

Reider B, Marshall JL, Warren RF. 1981. Clinical characteristics of patellar disorders in young athletes. *Am J Sports Med* 9:270-274.

Reilly DT, Martens M. 1972. Experimental analysis of the quadriceps muscle force and patella-femoral joint reaction force for various activities. *Acta Orthop Scand* 43:126-137.

Romeyn RL, Davies GJ, Jennings J. 2006. The multiple ligament–injured knee: evaluation, treatment and rehabilitation. In: Manske RC, ed. *Postsurgical Orthopedic Sports Rehabilitation: Knee and Shoulder*. St. Louis: Mosby.

Roy S, Irvin R. 1983. *Sports Medicine: Prevention, Evaluation, Management, and Rehabilitation*. Englewood Cliffs, NJ: Prentice Hall.

Salsich GB, Perman WH. 2007. Patellofemoral joint contact area is influenced by tibiofemoral rotation alignment in individuals who have patellofemoral pain. *J Orthop Sports Phys Ther* 37:521-528.

Salsich GB, Ward SR, Terk MR, Powers CM. 2003. In vivo assessment of patellofemoral joint contact area in individuals who are pain free. *Clin Orthop Rel Res* 417:277-284.

Sanchez AR, Sugalski MT, LaPrade RF. 2006. Anatomy and biomechanics of the lateral side of the knee. *Sports Med Arthrosc* 14:2-11.

Schmidt GL. 1973. Biomechanical analysis of knee flexion and extension. *J Biomech* 6:79-92.

Schulthies SS, Francis RS, Fisher AG, Van de Graaff DM. 1995. Does the Q angle reflect the force on the patella in the frontal plane? *Phys Ther* 75:24-30.

Seebacher JR, Inglis AE, Marshall JL, Warren RF. 1982. The structure of the posterolateral aspect of the knee. *J Bone Joint Surg* 64A:536-541.

Seedhom BB, Dawson D, Wright V. 1974. Proceedings: functions of the menisci—a preliminary study. *Ann Rheum Dis* 33:11.

Stranding S. 2009. *Gray's Anatomy: The Anatomical Basis of Clinical Practice*. 40th ed. St. Louis: Elsevier.

Sutton JB. 1897. *Ligaments: Their Nature and Morphology*. London: MK Lewis.

Takai S, Woo SL, Livesay GA, Adams DJ, Fu FH. 1993. Determination of the in-situ loads on the human anterior cruciate ligament. *J Orthop Res* 11(5):686-695.

Thomee R, Renstrom P, Karlsson J, Grimby G. 1995. Patellofemoral pain syndrome in young women: a clinical analysis of alignment, pain parameters, common symptoms and functional activity level. *Scan J Sports Med* 5:237-244.

Warren RF, Marshall JL, Girgis F. 1974. The prime static stabilizer of the medial side of the knee. *J Bone Joint Surg* 56A:655-670.

Winby CR, Lloyd DG, Besier TF, Kirk TB. 2009. Muscle and external load contribution to knee joint contact loads during normal gait. *J Biomech* 42(14):2294-2300.

Witvrouw E, Lysens R, Bellemans J, Lysens R, Danneels L, Cambier D. 2000. Intrinsic risk factors of the development of anterior knee pain in an athletic population: a two-year prospective study. *Am J Sports Med* 28:480-489.

Woodland LH, Francis RS. 1992. Parameters and comparisons of the quadriceps angle of college-aged men and women in the supine and standing positions. *Am J Sports Med* 20:208-211.

Zantop T, Peterson W, Sekiya JK, Musahl V, Fu F. 2006. Anterior cruciate ligament anatomy and function relating to anatomical reconstruction. *Knee Surg Sports Traumatol Arthrosc* 14(10):982-992.

Chapter 15

Alfredson H, Pietila T, Jonsson P, Lorentzon R. 1998. Heavy-load eccentric calf muscle training for treatment of chronic Achilles tendinosis. *Am J Sports Med* 26(3):360-366.

American Academy of Orthopaedic Surgeons. 1965. *Joint Motion: Methods of Measuring and Recording.* Chicago: AAOS.

Bouche RT, Johnson CH. 2007. Medial tibial stress syndrome (tibial fasciitis): a proposed pathomechanical model involving fascial traction. *J Am Podiatr Med Assoc* 97:31-36.

Bulucu C, Thomas KA, Halvorson TL, et al. 1991. Biomechanical evaluation of the anterior drawer test: the contribution of the lateral ankle ligaments. *Foot Ankle* 11:389.

Burdett RG. 1982. Forces predicted at the ankle during running. *Med Sci Sports Exerc* 14:308-316.

Cyriax J. 1982. *Textbook of Orthopaedic Medicine. Volume 1: Diagnosis of Soft Tissue Lesions.* 8th ed. Boston: Harcourt.

Donatelli R. 1996. *The Biomechanics of the Foot and Ankle.* Philadelphia: Davis.

Eickhoff CA, Hossain SA, Slawski DP. 2000. Effects of prescribed foot orthoses on medial tibial stress syndrome in collegiate cross-country runners. *Clin Kines* 54(4):76-80.

Hicks JH. 1954. The mechanics of the foot II: the plantar aponeurosis and the arch. *J Anat* 88:25.

Inman VT. 1976. *Joints of the Ankle.* Baltimore: Williams & Wilkins.

Jahss MH, Michelson JD, Desai P, et al. 1992. Investigations into the fat pads of the sole of the foot: anatomy and histology. *Foot Ankle* 13:227.

Klaue K, Hansen ST, Masquelet AC. 1994. Clinical quantitative assessment of first tarsometatarsal mobility on the sagittal plane and its relation to hallux valgus deformity. *Foot Ankle Int* 15:9-13.

Mahan KT, Carter SR, 1992. Multiple ruptures of the tendo Achillis. *J Foot Surg* 31:548-559.

Manter JT. 1941. Movements of the subtalar and transverse tarsal joints. *Anat Rec* 80:397.

Quinn K, Parker P, de Bie R, Rowe B, Handoll H. 2000. Interventions for preventing ankle ligament injuries. *Cochrane DB Syst Rev* 2:CD000018.

Sarrafian, SK. 1983. *Anatomy of the Foot and Ankle.* Philadelphia: Lippincott.

Sangeorzan BJ, 1991. Arthrodesis of the ankle with modified distraction-compression and bone-grafting. *JBJS* 73:790-798.

Scott SH, Winter DH, 1990. Internal forces of chronic running injury sites. *Med Sci Sports Exerc* 22:357-369.

Tweed JL, Avil SJ, Campbell JA, Barners MR. 2008. Etiologic factors in the development of medial tibial stress syndrome. *J Am Podiatric Med Assoc* 98:107-111.

Chapter 16

Adams MA, Hutton WC. 1985. The effect of posture on the lumbar spine. *J Bone Joint Surg Br* 67:625-29.

Bradford DS. 1995. Kyphosis in the elderly. In: Lonstein JE, Bradford DS, Winter RB, et al., eds. *Moe's Textbook of Scoliosis and Other Spinal Deformities.* Philadelphia: Saunders.

Janda V. 1987. Muscles and motor control in low back pain: assessment and management. In: Twomey LT, ed. *Physical Therapy of the Low Back.* New York: Churchill Livingstone.

Janda, V. 1988. Muscles and cervicogenic pain syndromes. In: R. Grand, ed. *Physical Therapy of the Cervical and Thoracic Spine.* New York: Churchill Livingstone.

Janda V, VaVrova. 1996. Sensory motor stimulation. In: Liebenson C, ed. *Rehabilitation of the Spine.* Baltimore: Williams & Wilkins.

Kendall F. 2005. *Muscles: Testing and Function With Posture and Pain.* 5th ed. Baltimore: Lippincott Williams & Wilkins.

Magee DJ. 1997. *Orthopedic Physical Assessment.* 3rd ed. Philadelphia: Saunders.

Neumann DA. 2010. Kinesiology of the hip: a focus on muscular actions. *J Orthop Sports Phys Ther* 40:82-94.

Nordin M, Frankel VH. 2001. *Basic Biomechanics of the Musculoskeletal System.* 3rd ed. Baltimore: Lippincott Williams & Wilkins.

Page P, Frank C, Lardner R. 2010. *Assessment and Treatment of Muscle Imbalance: The Janda Approach.* Champaign, IL: Human Kinetics.

Sahrmann S. 2002. *Diagnosis and Treatment of Movement Impairment Syndromes.* St. Louis: Mosby.

Sahrmann S. 2011. *Movement System Impairment Syndromes of the Extremities, Cervical and Thoracic Spines.* St. Louis: Mosby, Elsevier.

Wilke HJ, Neef P, Caimi M, Hoogland T, Claes LE. 1999. New in vivo measurements of pressures in the intervertebral disc in daily life. *Spine* 24:755-762.

Chapter 17

Finley FR, Cody KA.1970. Locomotion characteristics of urban pedestrians. *Arch Phys Med Rehab* 51:423-426.

Nordin M, Frankel VH. 2001. *Basic Biomechanics of the Musculoskeletal System*. 3rd ed. Baltimore: Lippincott Williams & Wilkins.

Perry J. 1992. *Gait Analysis: Normal and Pathological Function*. Thorofare, NJ: Slack.

Rudolph KS, Axe MJ, Buchanan TS, Scholz JP, Snyder-Mackler L. 2001. Dynamic stability in the anterior cruciate ligament deficient knee. *Knee Surg Sports Traumatol Arthrosc* 9:62-71.

Willson JD, Davis IS. 2008. Lower extremity mechanics of females with and without patellofemoral pain across activities with progressively greater task demands. *Clin Biomech* (Bristol, Avon) 23:203-211.

Chapter 18

Arendse RE, Noakes TD, Azvedo LB, Romanov N, Schwellnus MP, Fletcher G. 2004. Reduced eccentric loading of the knee with the pose running method. *Med Sci Sports Exerc* 36:272-277.

Beck B, Osternig L. 1994. Medial tibial stress syndrome: the location of muscles in the leg and relation to symptoms. *J Bone Joint Surg Am* 76:1057-1061.

Cavanagh PR, ed. 1990. *Biomechanics of Distance Running*. Champaign, IL: Human Kinetics.

Cook SD, Kester MA, Brunet ME, Haddad RJ Jr. 1985. Biomechanics of running shoe performance. *Clin Sports Med* 4:619-626.

Dierks TA, Davis I. 2007. Discrete and continuous joint coupling relationships in uninjured recreational runners. *Clin Biomech* 22:581-591.

Dierks TA, Davis IS, Hamill J. 2010. The effects of running in an exerted state on lower extremity kinematics and joint timing. *J Biomech* 43:2993-2998.

Dierks TA, Manal KT, Hamill J, Davis IS. 2008. Proximal and distal influences on hip and knee kinematics in runners with patellofemoral pain during a prolonged run. *J Orthop Sports Phys Ther* 38:448-456.

Elliott BC, Blanksby BA, 1979. The synchronization of muscle activity and body segment movements during a running cycle. *Med Sci Sports Exerc* 11:322-327.

Ferber R, Noehren B, Hamill J, Davis I. 2010. Competitive female runners with a history of iliotibial band syndrome demonstrate atypical hip and knee kinematics. *J Orthop Sports Phys Ther* 40:52-58.

Gerringer SR. 1995. The biomechanics of running. *J Back Musculoskelet Rehab* 5:273-279.

Hinrichs RN. 1990. Upper extremity function in distance running. In: Cavanagh PR, ed. *Biomechanics of Distance Running*. Champaign, IL: Human Kinetics.

James SL, Brubaker CE. 1973. Biomechanics of running. *Orthopaedic Clin North Am* 4:605-615.

Macera CA, Pate RR, Powell KE, Jackson KL, Kendrick JS, Craven TE. 1989. Predicting lower-extremity injuries among habitual runners. *Arch Intern Med* 249:2565-2568.

Mann RA. 1989. *Biomechanics of running*. In: D'Ambrosia RD, Drez D, eds. *Prevention and Treatment of Running Injuries*. 2nd ed. Thorofare, NJ: Slack, 1-20.

Milner CE, Hamill J, Davis IS. 2010. Distinct hip and rearfoot kinematics in female runners with a history of tibial stress fracture. *J Orthop Sports Phys Ther* 40:59-66.

Nigg BM. 1985. Biomechanics, load analysis and sports injuries in the lower extremity. *Sports Med* 2:367-379.

Novacheck T. 1998. The biomechanics of running. *Gait Posture* 7:77-95.

Schache AG, Bennell KL, Blanch PD, Wrigley TV. 1999. The coordinated movement of lumbo-pelvic–hip complex during running: a literature review. *Gait Posture* 10:30-47.

Sinning WE, Forsyth HL. 1970. Lower-limb actions while running at different velocities. *Med Sci Sports* 2:28-34.

Stergiou N, Bates BT, James SL. 1999. Asynchrony between subtalar and knee joint function during running. *Med Sci Sports Exerc* 31:1645-1655.

Taunton et al. 2002. A retrospective case-control analysis of 2002 running injuries. *Br J Sports Med* 36:95-101.

Thijs Y. DeClercq D, Roosen P, Witvrouw E. 2008. Gait-related intrinsic factors for patellofemoral pain in novice recreational runners. *Br J Sports Med* 42:466-471.

Tiberio D. 1987. The effect of excessive subtalar joint pronation and patellofemoral mechanics: a theoretical model. *J Orthop Sports Phys Ther* 9:160-165.

van Gent RN, Siem D, van Middelkoop M, van Os AG, Bierma-Zeinstra SMA, Koes BW. 2007. Incidence and determinants of lower extremity running injuries in long distance runners: a systematic review. *Br J Sports Med* 41:469–480.

Williams DS III, McClay IS, Hamill J. 2001. Arch structure and injury patterns in runners. *Clin Biomech* 16:341-347.

Willson JD, Davis IS. 2008. Lower extremity mechanics of females with and without patellofemoral pain across activities with progressively greater task demands. *Clin Biomech* 23:203-211.

Winter D. 1983. Moments of force and mechanical power in jogging. *J Biomech* 16:91-97.

Chapter 19

Agel J, Arendt EA, Bershadsky B. 2005. Anterior cruciate ligament injury in National Collegiate Athletic Association basketball and soccer: a 13-year review. *Am J Sports Med* 33(4):24-330.

Andrews JR, Axe MJ. 1985. The classification of knee ligament instability. *Orthop Clin North Am* 16:69-82.

Arms SW, Pope MH, Johnson RJ, Fischer RA, Arvidsson I, Eriksson E. 1984. The biomechanics of anterior cruciate ligament rehabilitation and reconstruction. *Am J Sports Med* 12(1):8-18.

Augustsson J, Thomee R, Linden C, Folkesson M, Tranberg R, Karlsson J. 2006. Single-leg hop testing following fatiguing exercise: reliability and biomechanical analysis. *Scand J Med Sci Sports* 16:111-120.

Bellchamber TL, van den Bogert AJ. 2000. Contributions of proximal and distal moments to axial tibial rotation during walking and running. *J Biomech* 33:1397-1403.

Bendjaballah MZ, Shirazi-Adl A, Zukor DJ. 1997. Finite element analysis of human knee joint in varus–valgus. *Clin Biomech* (Bristol, Avon) 12(3):139-148.

Benjaminse A, Habu A, Sell TC, et al. 2008. Fatigue alters lower extremity kinematics during a single-leg stop-jump task. *Knee Surg Sports Traumatol Arthrosc* 16:400-407.

Berns GS, Hull ML, Patterson HA. 1992. Strain in the anteromedial bundle of the anterior cruciate ligament under combination loading. *J Orthop Res* 10(2):167-176.

Beynnon BD, Fleming BC, Johnson RJ, Nichols CE, Renstrom PA, Pope MH. 1995. Anterior cruciate ligament strain behavior during rehabilitation exercises in vivo. *Am J Sports Med* 23(1):24-34.

Boden BP, Dean GS, Feagin JA, Garrett WE. 2000. Mechanisms of anterior cruciate ligament injury. *Orthopedics* 23:573-578.

Chappell JD, Herman DC, Knight BS, Kirkendall DT, Garrett WE, Yu B. 2005. Effect of fatigue on knee kinetics and kinematics in stop-jump tasks. *Am J Sports Med* 33(7):1022-1029.

Colby S, Francisco A, Kirkendall D, Finch M, Garrett W. 2000. Electromyographic and kinematic analysis of cutting maneuvers: implications for anterior cruciate ligament injury. *Am J Sports Med* 28(2):234-240.

Coventry E, O'Connor KM, Hart BA, Earl JE, Ebersole KT. 2006. The effect of lower extremity fatigue on shock attenuation during single-leg landing. *Clin Biomech* 21:1090-1097.

Dürselen L, Claes L, Kiefer H. 1995. The influence of muscle forces and external loads on cruciate ligament strain. *Am J Sports Med* 23(1):129-136.

Fagenbaum R, Darling WG. 2003. Jump landing strategies in male and female college athletes and the implications of such strategies for anterior cruciate ligament injury. *Am J Sports Med* 31:233-240.

Feagin JA Jr., Lambert KL, Cunningham RR, et al. 1987. Consideration of the anterior cruciate ligament injury in skiing. *Clin Orthop Rel Res* 216:13-18.

Fleming BC, Renstrom PA, Ohlen G, et al. 2001. The gastrocnemius is an antagonist of the anterior cruciate ligament. *J Orthop Res* 19(6):1178-1184.

Ford KR, Myer GD, Hewett TE. 2003. Valgus knee motion during landing in high school female and male basketball players. *Med Sci Sports Exerc* 35:1745-1750.

Ford KR, Myer GD, Toms HE, Hewett TE. 2005. Gender differences in the kinematics of unanticipated cutting in young athletes. *Med Sci Sports Exerc* 37(1):124-129.

Gehring D, Melnyk M, Gollhofer A. 2009. Gender and fatigue have influence on knee joint control strategies during landing. *Clin Biomech* 24:82-87.

Gray J, Taunton J, Taunton DC, et al. 1985. A survey of injuries to the anterior cruciate ligament of the knee in female basketball players. *Int J Sports Med* 6:314-316.

Hanson AM, Padua DA, Blackburn JT, Prentice WE, Hirth CJ. 2008. Muscle activation patterns during side-step cutting maneuvers in male and female soccer athletes. *J Athlet Train* 43(2):133-143.

Hewett TE, Paterno MV, Myer GD. 2002. Strategies for enhancing proprioception and neuromuscular control of the knee. *Clin Orthop* 402:76-94.

Hirokawa S, Lolomonow M, Lu T, Lou ZP, D'Ambrosia R. 1992. Anterior-posterior and rotational displacement of the tibia elicited by quadriceps contraction. *Am J Sports Med* 20(3):299-306.

Houston LJ, Vibert B, Ashton-Miller JA, Wojtys EM. 2001. Gender differences in knee angle when landing from a drop-jump. *Am J Knee Surg* 14:215-219.

Hughes G, Watkins J, Owen N. 2008. Gender difference in lower limb frontal plane kinematics during landing. *Sports Biomech* 23:289-299.

Kernozek TW, Torry MR, Wyland DJ, et al. 2005. Gender differences in frontal and sagittal plane biomechanics during drop landings. *Med Sci Sports Exerc* 37(6):1003-1012.

Landry SC, McKean KA, Hubley-Kozey CL, Stanish WD, Deluzio KJ. 2007. Neuromuscular and lower limb biomechanical differences exist between male and female elite adolescent soccer players during an unanticipated run and crosscut maneuver. *Am J Sports Med* 35(11):1901-1911.

Li G, Rudy TW, Sakane R, Kanamori A, Ma CB, Woo SL. 1999. The importance of quadriceps and hamstring muscle loading on knee kinematics and in-situ forces in the ACL. *J Biomech* 32(4):395-400.

Loudon JK, Jenkins W, Loudon KL. 1996. The relationship between static posture and ACL injury in female athletes. *J Orthop Sports Phys Ther* 24:91-97.

Madigan ML, Pidcoe PE. 2003. Changes in landing biomechanics during a fatiguing landing activity. *J Electromyog Kines* 13:491-498.

Malinzak RA, Colby SM, Kirkendall DT, Yu B, Garrett WE. 2001. A comparison of knee joint motion patterns between men and women in selected athletic tasks. *Clin Biomech* 16:438-445.

Markolf KL, Gorek JF, Kabo JM, Shapiro MS. 1990. Direct measurement of resultant forces in the anterior cruciate ligament: an in vitro study performed with a new experimental technique. *J Bone Joint Surg* 72A(4):557-567.

McLean SG, Felin RE, Suedekum N, Calabrese G, Passerallo A, Joy S. 2007. Impact of fatigue on gender-based high-risk landing strategies. *Med Sci Sports Exerc* 39:502-514.

McLean SG, Neal RJ, Myers PT, Walters MR. 1999. Knee joint kinematics during the sidestep cutting maneuver: potential for injury in women. *Med Sci Sports Exerc* 31:959-968.

McNair PJ, Marshall RN, Matheson JA. 1990. Important features associated with acute anterior cruciate ligament injury. *N Z Med J* 103:537-539.

Mundermann A, Nigg BM, Humble RN, Stefanyshyn DJ. 2003. Foot orthotics affect lower extremity kinematics and kinetics during running. *Clin Biomech* 18:254-262.

Nyland J, Caborn DN, Shapiro R, Johnson DL, Fang H. 1999. Hamstring extensibility and transverse plane knee control relationship in athletic women. *Knee Surg Sports Traumatol Arthrosc* 7:257-261.

Pollard CD, Sigward SM, Powers CM. 2007. Gender differences in hip joint kinematics and kinetics during side-step cutting maneuver. *Clin J Sports Med* 17(1):38-42.

Radin EL. 1986. Role of muscles in protecting athletes from injury. *Acta Med Scand* 711(Suppl):143-147.

Reiser R, Rocheford E, Armstrong CJ. 2006. Building a better understanding of basic mechanical principles through analysis of the vertical jump. *Strength Cond J* 28(4):70-80.

Renstrom P, Arms SW, Stanwyck TS, Johnson RJ, Pope MH. 1986. Strain within the anterior cruciate ligament during hamstring and quadriceps activity. *Am J Sports Med* 14(1):83-87.

Riewald SA. 2011. Applied biomechanics of jumping. In: Magee DJ, Manske RC, Zachazewski J, et al., eds. *Athletic and Sports Issues in Musculoskeltal Rehabiitation*. St. Louis, MO: Mosby Elsevier.

Stacoff A, Steger J, Stussi E, Reinshmidt C. 1996. Lateral stability in sideward cutting movements. *Med Sci Sports Exerc* 28:350-358.

Verbitsky O, Mizrani J, Voloshin A, Treiger J, Isakov E. 1998. Shock transmission and fatigue in human running. *J Appl Biomech* 14:300-311.

Voloshin A, Mizrani J, Verbitsky O, Isakov E. 1998. Dynamic loading on the human musculoskeletal system-effect of fatigue. *Clin Biomech* 13:513-520.

INDEX

Note: The italicized *f* and *t* following page numbers refer to figures and tables, respectively.

427

Janice K. Loudon, PT, PhD, ATC, is an associate professor in the Doctor of Physical Therapy Division in the Department of Community and Family Medicine at Duke University in Durham, North Carolina.

Loudon has more than 25 years of sports medicine clinical work and has worked as a physical therapy instructor for over 20 years. Previously, she was an associate professor and director of the Doctorate of Physical Therapy (DPT) program at the University of Kansas Medical Center in Kansas City. In 2007, Loudon was named an *Outstanding Physical Therapy Faculty* by the department of physical therapy education at the University of Kansas Medical Center.

She has published more than 25 articles within refereed journals, 4 book chapters, and co-authored two editions of *The Clinical Orthopedic Assessment Guide* (Human Kinetics, 1998, 2008). She is a frequent invited presenter at national, state, and local conferences.

Loudon is a member of the American Physical Therapy Association. She resides in Durham, North Carolina. In her spare time she enjoys tennis, cycling, and gardening. Photo courtesy of Janice Loudon.

Robert C. Manske, PT, DPT, MEd, SCS, ATC, CSCS, is an associate professor of physical therapy at Wichita State University. He earned a doctoral degree in physical therapy in 2006 from the MGH Institute of Health Professions. Manske was also a sport physical therapy fellow, training under the guidance of George J. Davies in one of the first sport physical therapy residency programs. As a practicing physical therapist, Manske has over 18 years of clinical experience in orthopedic rehabilitation and actively researches knee and shoulder rehabilitation and sport performance enhancement.

Manske has published over 30 articles in peer-reviewed journals, edited 3 home study courses, co-authored 12 home study course chapters, co-authored 30 book chapters, and authored or co-authored 5 books, all related to orthopedic or sports rehabilitation. He is a board-certified sport physical therapist, certified athletic trainer, and certified strength and conditioning specialist. He is also a member of the American Physical Therapy Association and the National Athletic Trainers' Association. He serves as Vice President of the Sports Section of the APTA. Manske presents multiple weekend courses on various shoulder and knee topics throughout the year and still remains active in clinical practice.

He also serves as an associate editor for the *International Journal of Sports Physical Therapy*. He is a reviewer for several journals including *American Journal of Sports Medicine, Journal of Sport Rehabilitation, Journal of Orthopaedic and Sports Physical Therapy, Physical Therapy in Sports, Sports Health, Athletic Training and Sports Health Care,* and *Physiotherapy Theory and Practice.*

In 2007, Manske received the Sports Section Excellence in Education Award. He has also received the Kansas Physical Therapy Educator Award from the Kansas Physical Therapy Association (2003) and the Rodenberg Teaching Award from the College of Health Professions at Wichita State University (2004).

Manske and his wife, Julie, live in Wichita. He enjoys spending time with his family, exercising, and watching college and professional sports. Photo courtesy of Robert Manske.

Michael P. Reiman, PT, DPT, MEd, OCS, SCS, ATC, FAAOMPT, CSCS, is an associate professor in the Doctor of Physical Therapy Division in the Department of Community and Family Medicine, at Duke University Medical Center in Durham, North Carolina. He also serves as an adjunct assistant professor in the Department of Physical Therapy at Wichita State University. He also serves as clinical faculty in the Duke University Medical Center Manual Therapy Fellowship Program.

Reiman has published over 30 articles within peer-reviewed journals as well as 9 book chapters, 3 home study courses, and co-authored one text, *Functional Testing in Human Performance* (Human Kinetics, April 2009), with Robert C. Manske. He has given numerous presentations at national, regional, and local conferences.

Reiman is a member of the American Physical Therapy Association, American Academy of Orthopaedic and Manual Physical Therapists, Kansas Physical Therapy Association, National Athletic Trainers' Association, National Strength and Conditioning Association, and Alpha Eta Society. He also serves as an associate editor for the *Journal of Physical Therapy*, and is a member of the editorial boards for the *International Journal of Sports Physical Therapy* and the *Journal of Sport Rehabilitation*. He is a reviewer for several journals including *British Journal of Sports Medicine, Journal of Sport Science and Medicine, Physiotherapy Theory and Practice, Journal of Sport Rehabilitation, Journal of Manual and Manipulative Therapy, Journal of Orthopaedic and Sports Physical Therapy, Clinical Anatomy,* and the *Journal of Athletic Training.*

Reiman is a Level 1 Track and Field Coach and a Level 1 Olympic Weightlifting Club Coach. He works with athletes of various skill level for both rehabilitation and performance enhancement.

Reiman resides in Hillsborough, North Carolina, where he enjoys spending time with his family, hiking in the surrounding hills, and wakeboarding with his children. Photo courtesy of Michael Reiman.

Sue Klein, PT, MTC, is a graduate of the University of Kansas with a bachelor's degree in physical therapy. She has a certification in manual therapy from the University of Saint Augustine and has studied craniomandibular and craniovertebral dysfunction with Mariano Rocabado. Sue is currently an adjunct faculty member at Rockhurst University in Kansas City, Missouri, and is a clinician with an emphasis on craniomandibular, craniovertebral, and spinal dysfunction at Performance Rehab in Overland Park, Kansas.

Neena Sharma, PT, PhD, is an assistant professor in the Department of Physical Therapy and Rehabilitation Science at the University of Kansas Medical Center. She has completed core courses and a mentorship program in orthopedic manual therapy and has received manual therapy certification from the North American Institute of Orthopaedic Manual Therapy (NAIOMT). She teaches courses in musculoskeletal physical therapy. Her research interests relate to understanding the mechanisms of chronic pain syndromes and conservative treatments, including exercise and manual therapy using brain imaging methods. She serves as a reviewer for several journals and is a member of the editorial board for the *Journal of Novel Physiotherapies*. She is a member of the American Physical Therapy Association and the Society for Neuroscience.